PIPELINE TRANSPORTATION OF CARBON DIOXIDE CONTAINING IMPURITIES

Mo Mohitpour
Patricia Seevam
Kamal K. Botros
Brian Rothwell
Claire Ennis

With contribution by:
Professor Martin Downie
Dr. Julia Race

ASME PRESS

Library of Congress Cataloging-in-Publication Data

Pipeline transportation of carbon dioxide containing impurities / Mo Mohitpour ... [et al.].
p. cm.
Includes bibliographical references and index.
ISBN 978-0-7918-5983-4
1. Gas pipelines. 2. Carbon dioxide—Transportation. I. Mohitpour, Mo.
TA660.P55P59 2011
665.7'44—dc23

2011042723

DEDICATION

This book is dedicated to the memory of a great pipeline research scientist

Dr Wojciech Studzinski, P.Eng.

whose pioneering scientific research has led to many improvements in pipeline transportation technology

TABLE OF CONTENTS

PREFACE

This book is a sequel to and augments the series of ASME initiated pipeline books first published in 2000. These books are:

- *Pipeline Design and Construction — A Practical Approach*; Mohitpour, Golshan, and Murray, 3rd Edition, 2007
- *Pipeline Operation and Maintenance — A Practical Approach;* Mohitpour, van Hardeveld, Peterson and Szabo, 2nd Edition, 2010
- *Energy Supply and Pipeline Transportation — Challenges and Opportunities*; Mohitpour, 2008
- *Pipeline Pumping and Compression Systems — A Practical Approach*; Mohitpour, Botros, and Van Hardeveld, 2008
- *Pipeline Integrity Assurance — A Practical Approach*; Mohitpour, Murray, McManus, and Colquhoun, 2010

The mitigation of anthropogenic carbon dioxide (CO_2), a combustion by-product of fossil fuel, is an important step toward a cleaner environment.

Carbon dioxide capture and storage (CCS) is a process for reducing such greenhouse gas (GHG) emissions by extracting CO_2 from flue stacks and transporting it for enhanced oil recovery (EOR) or for sequestration.

Pipeline systems play the safest means of transporting such captured CO_2. However, the phase diagram for a CO_2 stream containing impurities is very sensitive to the level of these impurities which in turn affect pipeline design and the boundaries where CO_2 pipelines can be operated without affecting the facilities design as well as delivery conditions.

The largest network of CO_2 pipelines is in North America (mostly in the USA), the oldest one constructed in 1963 (Denbury's 82 km, Cranfield pipeline from Mississippi to Louisiana). Such pipelines transport CO_2 (after processing and dehydration) predominantly from naturally occurring underground reservoirs to oil fields for use in CO_2 EOR and sequestration. An exception is the 324 km of Weyburn pipeline (on-stream in year 2000) that transports CO_2 captured from flue stacks of gasification plant (in North Dakota, USA) to oil fields in Weyburn, Saskatchewan.

Design and operating experience for pipeline transportation of CO_2 with various impurities is thus limited and sparsely spread. These books bring together the entire spectrum of design and operating needs for a pipeline and network of facilities that would transport CO_2 containing impurities safely without adverse impact on people or the environment.

Chapters of the book have been written with the intent that each chapter could stand alone on the subject matter presented without referencing other chapters. In this book, both imperial and metric units have been used. This was justified because the industry continues to use the unit systems interchangeably. A conversion table is provided within this section.

The authors have exercised care to ensure correctness of the content, acknowledgement of other publications, copyright permissions, and referencing documents and names. It is not intended that specific techniques, examples, or applications be applied or copied for turnkey use. Readers are very much encouraged to check and assess all details before use and application. The authors and ASME welcome notification of corrections, omissions, and attributions. These will be attended to in the next edition and contributors acknowledged as such.

Mo Mohitpour, White Rock British Columbia, Canada
Patricia Seevam, London SW, United Kingdom
Kamal K. Botros, Calgary, Alberta, Canada
Brian Rothwell, Calgary, Alberta, Canada
Claire Ennis, Calgary, Alberta, Canada
2011

SPECIAL ACKNOWLEDGMENTS

Sincere acknowledgements are due to ***Professor Martin Downie*** and ***Dr Julia Race*** of the **Newcastle University**, United Kingdom for their important research into aspects of carbon capture and storage (CCS) and appertaining transportation systems. This has included research contributing to Chapters 2, 3 and 4 of this book. Their contribution to the industry and support in our efforts is immeasurable. THANK YOU.

Mo Mohitpour
Patricia Seevam
Kamal K. Botros
Brian Rothwell
Claire Ennis

ACKNOWLEDGMENTS

The book was conceptualized in discussion between ***Dr Andrew Cosham, Dr Patricia Seevam, Dr Mo Mohitpour and John Tiratsoo*** at the time of the 2nd International Seminar on Carbon Capture and Geological Storage (**CCGS**), Salvador, Brazil (Sept 9-11, 2008) organized by Petrobras University, Brazil.

The outcome of the discussion was to bring about in one document all concepts of design and operational aspects of pipeline transportation of carbon dioxide containing impurities, a subject even to date is under considerable research and evaluations and design alternatives debates.

Thank you Dr Cosham for your insight and John Tiratsoo (of *Scientific Surveys Ltd, UK*)for solid support over the years in the industry leading to the publication of many articles on pipeline technology issues including the subject matter in the Journal of Pipeline Engineering with associate publisher BJ Lowe, Clarion Technical Publishers, Houston, TX, USA.

Thanks are due to our colleagues in the industry who, in many ways, encouraged and contributed to the review of preparatory manuscripts, updating, corrections, additions, and the supply of materials for this book. The authors wish to express sincere thanks to and acknowledge the valuable contribution of all of these colleagues including:

Dr. Arne Dugstad, Institute for Energy Technology (IFE), Norway
Dr. Beryl Edmunds, INFOCHEM, UK
Charles E Fox, PE. Kinder Morgan CO_2 Company (KMOC$_2$), LP.
Dr. Michel Berghmans, NOVA Chemicals Corporation
Chris Foy, P.Eng, NOVA Chemicals Corporation
Dr. Alan Glover, P.Eng
Dr. Hossein Golshan, TransCanada PipeLines Limited
John Kazakoff, P.Eng, Silver Fox Engineering Consultants
Dr. Robert McBrien, P.Eng, Nova Chemicals Corporation
Stewart Midwinter, P.Eng, Energy Solutions International (ESI)
Tony Moorwood, INFOCHEM, UK
Hal Oliver, PE
Dr. Tony Oosterkamp, Polytec, Norway
Penspen Limited, UK
Dr. Julia Race, Newcastle University, United Kingdom.
Connell Rader, Denbury Inc, USA
Marco Roncal, Kinder Morgan CO_2 Company (KMCO$_2$), LP.
J.H. (Jim)Russell, P.Eng, formerly of Enbridge
Schlumberger, UK
SPT Group, Norway
Charlie B Wallace

We also acknowledge the leadership, management, and unyielding support of the ASME Press and staff who have made this book possible. Acknowledgments are due to Mr. Philip DiVietro, Ms. Mary Grace Stefanchik, and Ms. Tara Collins Smith for their tremendous encouragement, dedication, commitments, and above all their patience toward the completion of this book. And where would we be without ASME's marketing efforts? Thank you Annette Missouri and John Yelavich, and the staff for your huge support over the years and for the future sale of this book. THANK YOU.

Mo Mohitpour, Ph.D., PE, P.Eng, F. I. Mech. Eng, FASME, FEIC
mo.mohitpour@shaw.ca
Patricia Seevam, Ph. D., C.Eng.
Patricia.Seevam@uk.bp.com
Kamal K. Botros, Ph D., P.Eng.
botrosk@novachem.com
Brian Rothwell, P.Eng
brian.rothwell@shaw.ca
Claire Ennis, E.I.T.
enniscj@novachem.com

PERMISSION ACKNOWLEDGMENTS

The authors wish to thank all the organizations and individuals that granted their permission to use and reproduce figures, data, and other material in this book. Some have been obtained from their Web sites and papers and thus references have been given accordingly. We apologize for any reference or attributions that we may have overlooked.

The authors gratefully acknowledge the following organizations for use of material and permission granted to reproduce various items that appear in the text:

British Geological Survey (BGS)
ChemicaLogic Corporation, www.chemicaLogic.com
DRESSER RAND
The DYNAMIS and ECOFYS project
Energy Solutions International, Stewart Midwinter P.Eng
European Communities
GASSCO, Norway
GE Oil and Gas
FLOWSERVE CORPORATION
Intergovernmental Panel on Climate Change (IPCC), Working Group III
International Energy Agency Green House Gas (IEAGHG), UK
JRC, Institute for Energy, Petten, The Netherlands
MANN Turbo
National Grid, UK
Society of Petroleum Engineers (SPE)
UKCCSC (United Kingdom Carbon Capture and Storage Consortium)

FOREWORDS

Foreword From TransCanada Pipelines

A significant engineering challenge of the 21st century is to reduce emissions which are predominantly made up of carbon dioxide (CO_2), a by-product of burning fossil fuels.

One method to accomplish this is to extract CO_2 from the flue stacks, transport the processed product for storage underground and in enhanced oil recovery (EOR).

Pipeline systems play an important role in transporting captured CO_2 to distant storage fields. Unfortunately, hydraulic and mechanical designs are very sensitive to the level of impurities contained in the CO_2 product stream. This will affect the limits in which the CO_2 pipeline can be operated safely and without affecting the delivery of the end product.

This book brings together the entire spectrum of design and operating parameters for a pipeline and network of facilities that would transport CO_2 containing impurities safely without impacting people or the environment.

The authors, well known in the pipeline industry have previously published a series of five comprehensive pipeline books covering Design, Construction, Operation, Maintenance, Energy Supply, Pumping and Compression, and Pipeline Integrity Assurance all of which have had a tremendous impact on our industry.

We are very pleased to continue our support for this series of ASME Press books and the efforts made by the authors in researching and bringing together the knowledge so important as a reference guide for the safe design and operation of pipeline and facilities when transporting CO_2.

Andrew Jenkins, P.Eng
Vice President, TransCanada Pipelines Limited

TransCanada
In business to deliver

Foreword From Newcastle University

The subject of pipeline transportation of carbon dioxide (CO_2) is becoming increasingly important as Carbon Capture and Storage (CCS) schemes worldwide are moving from the pilot to the demonstration phases. Although it is expected that the development of pipeline transportation networks for CCS will build on the large amount of expertise available in the pipeline transportation of natural gas and also on the existing experience of transporting high-pressure CO_2 for enhanced oil recovery (EOR) by pipeline, there will also be some new challenges that must be overcome.

This publication focuses on the pipeline transportation of CO_2 containing impurities or non-CO_2 components. Among the challenges to be faced by the pipeline community is the minimization of the constraints imposed by the CO_2 stream composition on the development of transport systems. Ideally, the pipeline should be able to accept as wide a specification as possible, within the limits imposed by health, safety, and environmental considerations throughout the whole CCS chain. In order to meet this challenge, confidence is required in the models and procedures currently used in pipeline design, maintenance, and operation to ensure that these approaches can be safely transferred to a CO_2 pipeline infrastructure. Many of these considerations are explored in the chapters of this book.

Perhaps, the greatest challenge of all, however, is that of the scale of global infrastructure required in order that CCS can make an effective contribution to climate change mitigation. Less attention and research effort has been directed toward defining the problems to be encountered in the design and operation of large-scale CO_2 pipeline infrastructures in which multiple sources incorporating different capture technologies, and therefore stream compositions, are connected to onshore and offshore storage sites. In addition to the technical requirements, the development of such infrastructures will require rational, strategic planning at both national and international governmental levels to allow safe and economic networks to be developed with minimal environmental impact. Ultimately, this will require the establishment of sound regulatory frameworks and financial incentive schemes.

It is to be hoped that, as an industry, we can rise to these challenges and therefore, the publication of a book dedicated to the subject of the transportation of CO_2 by pipeline is to be welcomed by the pipeline community. Newcastle University is pleased to have been able to contribute to the preparation and review of this timely and significant work.

Prof Martin Downie and Dr Julia Race
School of Marine Science and Technology
Newcastle University, United Kingdom.

Newcastle University

Foreword From Kinder Morgan

The authors have performed a service to the industry in writing this book on pipeline transportation of CO_2, and I'm pleased to say I provided some small assistance at the beginning of their effort. Patricia contacted me for help in 2006 because I co-authored the ***Society of Petroleum Engineers*** monograph, "**Practical Aspects of CO_2 Flooding**," and manage the world's largest CO_2 pipeline system. I put them in touch with my company's experts and wished them good luck. Five years later, after much effort on their part, "Pipeline Transportation of Carbon Dioxide Containing Impurities" is ready for publication.

This book will assume a prominent place on the shelf of engineers involved in handling CO_2 for enhanced oil recovery and carbon storage. These technologies involve capturing or producing the CO_2, transporting it and finally injecting it into the ground and managing its movement in the subsurface. Until now, the transportation aspect has been unjustly neglected. The authors from various different backgrounds and organizations related to pipeline engineering have assembled the state of the art and science.

Like all companies, my employer constantly faces staff turnover, and we struggle to pass along the knowledge of CO_2 transportation to the newcomers. People who have vast experience with transporting natural gas, crude oil, or refined products may be unfamiliar with handling CO_2 and need help to get started. This reference will help assure that experts always operate our pipeline system. I encourage others who plan, design or operate CO_2 pipelines to obtain this book and use it.

Charles E. Fox
Vice President, Operations and Engineering
Kinder Morgan CO_2 Company, LP

METRIC CONVERSION OF SOME COMMON UNITS

To Convert From Customary Unit	To Define Unit As	Symbol	Multiply By
barrel per hour	liters per second	L/s	0.044 163
barrel per day	cubic meters per day	m^3/d	0.158 987
MMBOD	cubic meters per day	m^3/d	$0.158\ 987 \times 10^6$
Btu/second	kilowatt	kW	1.055 056
Btu/hour	watt	W	0.293 071
Btu/lb_m	kilojoule per kilogram	kJ/kg	2.326
Btu/lb_m-°F	kilojoule per kilogram-Kelvin	kJ (kg K)	4.1868
Btu/lb_m-mole-°R	joule per mole-Kelvin	J/(mol K)	4.1868
Btu/°R	kilojoule per Kelvin	kJ/K	1.8991
Btu/ft^2-hr.	joule per sq. meter-second	J / (m^2 s)	3.154 591
Btu/ft-hr-°F	joule per meter-second-kelvin	J / (m/s K)	1.730 735
Btu/ft-hr^2-°F	joule per square meter-second Kelvin	J / (m^2 s K)	5.678 263
foot-pound force (ft. lb_f)	joule	J	1.355 818
$foot^2$	square meter	m^2	0.092 903
$foot^3$	cubic meter	m^3	0.028 316 85
$foot^3$/minute	liter per second	L/s	0.471 947
$foot^3$/hour	cubic meter per day	m^3/d	0.679 604
MMSCFD	cubic meter per second	m^3/s	0.327 774
gallon/minute (GPM)	liter per second	L/s	0.063 090
$inch^2$	square centimeter	cm^2	6.451 600
$inch^3$	cubic centimeter	cm^3	16.387 064
kilowatt-hour (kWh)	megajoule	MJ	3.6
mile per hour	kilometer per hour	km/h	1.609 344

pound	kilogram	Kg	0.453 592 37
pound mass/foot^3 (lb_m/ft^3)	kilogram per cubic meter	kg/m^3	16.018 463
pound mass/gallon	kilogram per liter	kg/L	0.119 826
pound mass/hour	kilogram per hour	kg/h	0.453 592
psi	kilopascal	kPa	6.894 757
psi/foot	kilopascal per meter	kPa/m	22.620 59
psi/mile	Pascal per meter	Pa/m	4.284 203
Watt-hour	kilojoule	kJ	3.6
yard^2	square meter	m^2	0.836 127
yard^3	cubic meter	m^3	0.764 555
acre	square meter	m^2	4046.856
atmosphere (std)	kilopascal	kPa	101.325
barrel (42 gal)	cubic meter	m^3	0.158 987
Btu (International Table)	kilojoule	kJ	1.055 056
calorie (Thermochemical)	joule	J	4.184
degree F	degree Celsius	°C	5/9 (°F-32)
degree R	degree Kelvin	K	5/9
foot	meter	m	0.3048
gallon (US liquid)	liter	L	3.785 412
horsepower (US)	kilowatt	kW	0.7457
inch (US)	millimeter	mm	25.4
inch of mercury (60°F)	kilopascal	kPa	3.376 85
inch of water (60°F)	kilopascal	kPa	0.248 843
mil	micrometer	μm	25.4
mile (US Statute)	kilometer	km	1.609 344
ounce (US fluid)	milliliter	mL	29.573 53
poise	Pascal-second	Pa.s	0.1
stokes	square centimeter per second	cm^2/s	1
ton, long (2240 lb_m)	ton	t	1.016 047
ton, short (2000 lb_m)	ton	t	0.907 184 74
ton of refrigeration	kilowatt	kW	3.516 853
yard (US)	meter	m	0.9144

Chapter 1

OVERVIEW AND CHALLENGES OF CO_2 TRANSPORTATION

1.1 INTRODUCTION

Carbon dioxide (CO_2) is a combustion by-product of fossil fuels (oil, natural gas, coal) that are used for electricity production, transportation, heating, and industrial applications. It is also released when solid waste, wood, and wood particles/products are burned.

CO_2 is one of six anthropogenic greenhouse gases (GHGs) that can contribute to global warming. The other five associated gases of concern are as follows:

- methane (CH_4),
- nitrous oxide (N_2O),
- hydrofluorocarbons (HFCs),
- perfluorocarbons (PFCs), and
- sulfur hexafluoride (SF6).

Carbon dioxide capture and storage (CCS) is a process for reducing such greenhouse gas (GHG) emissions by extracting CO_2 from flue stacks, and transportation for storage purposes and sequestration operation including enhanced oil recovery (EOR).

1.1.1 Background to CO_2 Capture, Transportation, and Sequestration

A significant engineering challenge of the 21st century is to mitigate climate change, caused by greenhouse gas (GHG) emission [1]. Underground storage of gaseous emissions from fossil fuel-fired power plant (coal, natural gas, and other power plants), industrial furnaces, cement plants, engine exhausts, and lime kiln exhausts, etc. is a method of limiting greenhouse gas emissions. The method can have the potential for immediate, worldwide deployment, provided there is compatibility between the impure CO_2 available from the source and CO_2 sequestration acceptability. In any case, such CO_2 gases must be transported through pipeline from the source of supply to the storage sites. This may require that CO_2 gathered/captured be processed to acceptable compositional level for pipeline transmission and indeed for any subsequent sequestration.

In the past four decades, the CO_2 storage has been focused on the injection of a stream of pure CO_2 into depleted gas and oil storage and fields [2]. Such ventures have been successful mostly because pure CO_2 has been transported by pipelines in a liquid form and then injected in the subsurface and generally does not contain incompatible impurities with that of the reservoir condition. An injection stream contaminated by effluent gases such as CO, O_2, H_2, SO_x, and NO_x may, on the other hand, not be suitable for the reservoir and also may behave quite differently than pure CO_2 once injected.

A transportation of CO_2 from naturally occurring sources generally comprise of wet and dry CO_2 gathering and processing facilities and subsequent pipeline transmission systems [2]. Two-phase CO_2 is initially pipelined from CO_2 producing well clusters in the CO_2

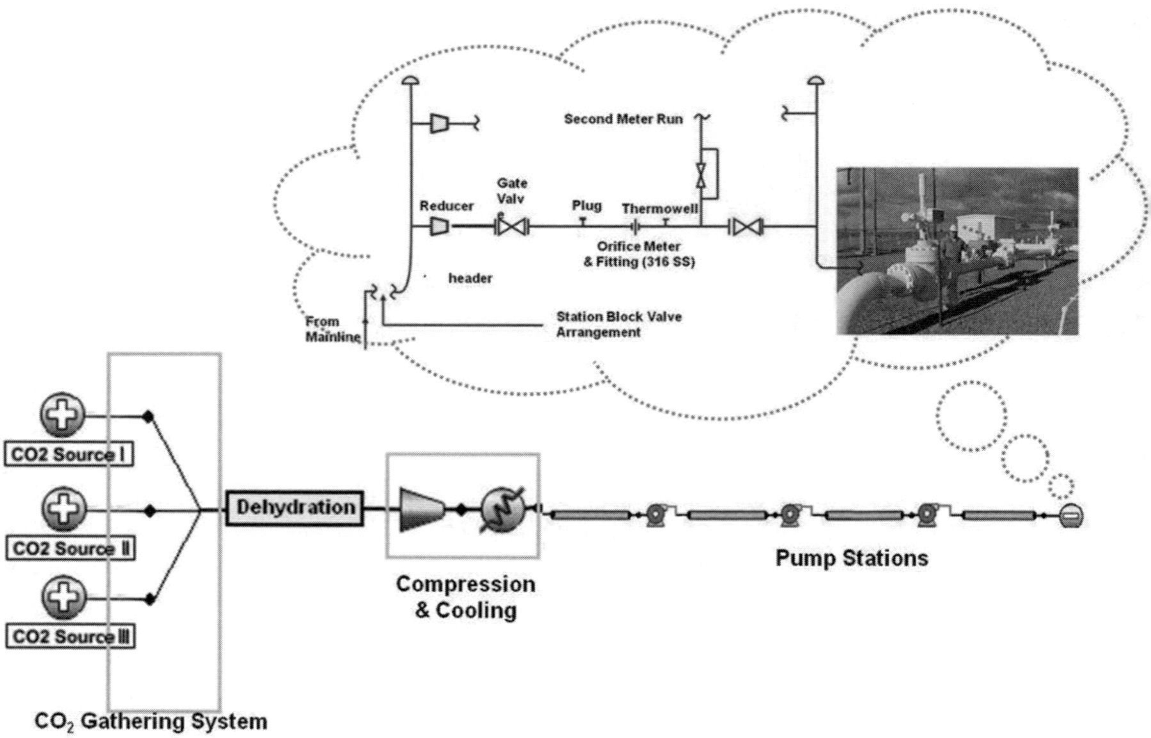

Figure 1-1. A typical naturally occurring CO_2 gathering and transportation system

fields to a CO_2 central processing facility. At such a facility, the CO_2 is processed, dried to reduce water content to acceptable levels, and then compressed. The dry CO_2 is transported from a central processing facility to metering and dispatch stations. It is then transported through a pipeline in liquid form, consisting of various pumping and valving arrangements to delivery locations, usually for injection purposes, as shown in Figure 1-1.

A stream of CO_2 generally available from flue stacks is contaminated by effluent gases such as CO, O_2, SO_x, and NO_x. The gathering from industrial processes and location, and thus transportation by pipeline (for subsequent use) will require a different treatment than CO_2 from naturally occurring sources. The impure CO_2 may behave quite differently than pure CO_2 once injected. In particular, impurities may cause the injection stream to be a low-density gas or a two-phase mixture of gas and liquid in the subsurface. The impurities may also change significantly the character of reservoir (e.g., from sweet to sour). As an example, formation of multiple phases near the injection well has been observed in pilot projects and limits the rate at which CO_2 can be injected [1]. Additionally, impurities in CO_2 stream affect pipeline transportation system design as well as operational conditions of sequestration process.

A typical system for gathering and transportation of CO_2 containing impurities is generally similar to that for gathering, processing, and transportation of naturally occurring CO_2. However, the capture processes are different and depend on the level of impurities and suitability for pipeline transportation and, the most important of all, compatibility with storage reservoir use. Streams of CO_2 with varied impurities cannot be comingled for transportation or sequestration as the resulting CO_2 properties (and phase diagram) may be significantly altered to be acceptable or accommodated for either transportation or sequestration.

1.1.2 Historical Timelines/CO_2 Chronology

Greenhouse gas (GHG) emission describes the atmospheric gases that contribute to the greenhouse effect by absorbing infrared radiation produced by solar warming of the Earth's surface This includes between 9% and 26% carbon dioxide (CO_2), 4-9% methane (CH_4), and other components, such as carbon monoxide (CO),nitrous oxide (NO_2), sulphur dioxide (SO_2), sulfur hexafluoride (SF_6), and water vapor. Water vapor creates about 36-72% of the greenhouse effect on Earth.

Greenhouse gases occur naturally in the atmosphere. However, the elevated levels especially of carbon dioxide and methane that have been observed in recent decades are directly related, at least in part, to human activities, such as the burning of fossil fuels and the deforestation of tropical forests. It is for the latter reason that capture, transportation, and sequestration of CO_2 gases (making up the largest components of greenhouses gases and are mostly emanated flue stacks from burning of fossil fuels) are prominent in reducing the GHG.

Establishment of greenhouse gas emission goes as far back as 1896 when Svante Arrhenius (1859-1927), a Swedish scientist, postulated that fossil fuel combustion may eventually result in enhanced global warming. He proposed a relation between atmospheric carbon dioxide concentrations and temperature that is the forefather of present day.

The following list of historical events is the important precursor to the Kyoto Protocol, a major world event for climate control [3].

- **Pre-history** Carbon discovery.
- **300 BC** Documenting human activity affecting climate change by Theophrastus, a student of Greek philosopher Aristotle.
- **1580-1644** Discovery of "gas sylvestre" (carbon dioxide), given off by burning charcoal By Jan van Helmont.
- **1722** Transformation of iron into steel through the absorption of some substance, now known to be carbon, (René Antoine Ferchault de Réaumur (28 February 1683 La Rochelle-17 October 1757).
- **1750** Start of industrial revolution, Manchester, England. Coal powers the mills.
- **1772** First demonstration that diamonds are from carbon by Antoine Lavoisier.
- **1786** First demonstration graphite from carbon by French scientists Claude Louis Berthollet, Gaspard Monge and C. A. Vandermonde.
- **1789** Recognition of carbon as an element by Antoine Lavoisier.
- **1824** First description of "greenhouse effects" by French physicist, Joseph Fourier.
- **1825** High-level CO_2, maxima > 400 ppm [4].
- **1834** First observation of dry ice (solid CO_2) by Charles Thilorier.
- **1837** First theory of climate changes in Earth's history by Louis Agassiz.
- **1859** First discovery of that some gases block infrared radiation by John Tyndall.
- **1878** First study of CO_2 toxicity by Friedlander and Herter.
- **1896** Thomas Chamberlain (1841-1896) and Svante Arrhenius, establishing Earth warming by CO_2 additions to atmosphere.
- **1905** First investigation of atmospheric carbon dioxide *Brown and Escombe* [4].
- **1910-1915** Origin of phase diagram.
- **1911** Relating Ice ages to Earth orbit change by Serbian mathematician, Milutin Milankovitch, proven in 1966.
- **1930s** 1st reporting of Global warming trend since late 19th century.
- **1937** The term "greenhouse effect" is coined by Glenn Thomas Trewartha.

- **1940** Development of infrared spectroscopy, measuring long wave radiation.
- **1942** High level CO_2 recorded, maxima > 400 ppm [4].
- **1950s** Establishing carbon dioxide has an atmospheric lifetime of approximately10 years.
- **1952** The first patent for CO_2 EOR technology was granted to Whorton, Brownscombe, and Dyes of the Atlantic Refining Company [5].
- **1955** Gilbert Plass (1921-2004) concluded adding CO_2 to atmosphere would intercept infrared radiation.
- **1957** Launch of Soviet Sputnik satellite.
- **1958** Bert Bolin and Erik Eriksson describe ocean sequestration of atmospheric CO_2.
- **1960** Charles Keeling (1928-2005) postulated accurate measurement of CO_2.
- **1970** First Earth Day, Creation of US National Oceanic and Atmospheric Administration.
- **1979** World's first climate conference declaring world's governments "to foresee and prevent potential man-made changes in climate that might be adverse to the well-being of humanity." *Atmospheric carbon dioxide concentration recorde*d: 337 ppm.
- **1979** U.S. National Academy of Sciences landmark report on the greenhouse effect to global warming.
- **1980** First appearance of global mean annual temperature curve.
- **1985** Discovery of fullerene, an allotrope of carbon.
- **1987** Verification that human activities cause global warming by adding CO_2.
- **1987** Warmest year since records began.
- **1988** Creation of Intergovernmental Panel on Climate Change (IPCC) by United Nations.
- **1988** First acknowledgement of warmer global temperature from 1880.
- **1990** First IPCC Climate assessment report.
- **1990** CO_2 Australia Carbon Sequestration Program.
- **1990** Warmest year on record [6].
- **1992** Creation of the UN Framework Convention Climate Change (UNFCCC).

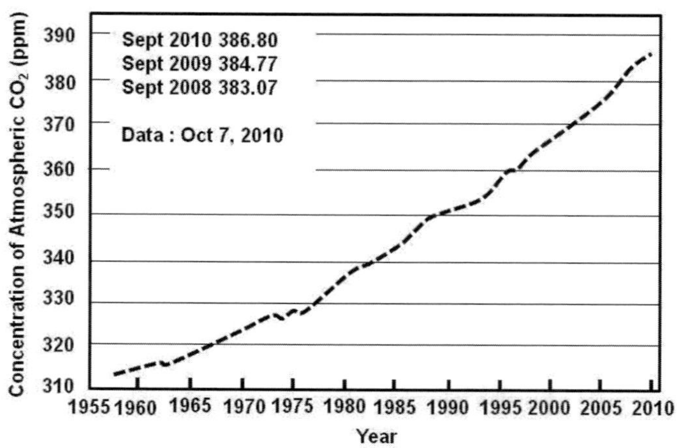

Figure 1-2. Atmospheric CO_2, August 1958 to August 1975 (Note: Data from Scripps 1958-1974 and NOAA-ERSL since 1974) [7]

- **1993** The Keeling Curve measuring the progressive buildup of carbon dioxide, a greenhouse gas, in the atmosphere concentration curves for atmospheric CO_2.
- **1995** Hottest year recorded to date.
- **1995** The first Conference of Parties (COP), was held in Bonn, Germany.
- **1998** 1st "Super El Niño" causes weather disasters.
- **1998** Kyoto Protocol, for reduction of CO_2, CH_4, N_2O, HFCs, PFCs, and SF_6 by 5% below 1990 level (between 2008 and 2012).
- **2000** Atmospheric carbon dioxide concentration is 369 ppm.
- **2005** Hurricane Katrina devastates New Orleans.
- **2005** Kyoyo treaty in effect.
- **2008** Measured level of CO_2 in atmosphere 384 ppm, August.
- **2009** Measured level of CO_2 in atmosphere 385 ppm, Mean Global Temperature 14.5 °C, August.
- **2009** Second warmest year on record.
- **2010** Measured level of CO_2 in atmosphere 388.15 ppm, August, 386.8 ppm, September (Figure 1-2).

From Figure 1-2, it may be noted that atmospheric concentration of CO_2 has increased by about 23.5% since 1960.

1.2 CO_2 SCIENCE

As previously alluded to, globally, there are concerns over global warming and CO_2 and greenhouse gas (GHG) emissions. Consequently, various carbon capture and sequestration (CCS) initiatives have been considered as viable options for managing man-made industrial greenhouse gases, forecasted to rise to about 44,000 million metric tons by 2030 [8]. However, carbon (C), carbon cycle, carbon dioxide (CO_2), climate change, greenhouse gases, and other similar phrases are often referred to without defining the terms or used interchangeably which can mask their connotation for reference in engineering applications.

As indicated above, CO_2 is either extracted from natural sources in underground geological formations or is typically a by-product of combustion processes from natural gas processing, power generation, cement manufacture, and other similar processes. Naturally occurring CO_2 from underground sources is processed to remove undesirable components (such as water) or particulates that affect design and operational issues while transported through pipelines. Such a CO_2 stream is usually over 99% pure and is transported in liquid form (Mohitpour, [8]; see Section 1.4 for details).

The impurities can significantly affect the CO_2 condition, influence pipeline capacity, pipe fracture control properties, and, most importantly, pipeline operation and emergency response related to release and dispersion of CO_2 and toxic impurities, such as H_2S. Such impurities are prevalent in CO_2 stream from flue stacks.

1.2.1 Definitions

In order to dispel misconception and understand the significance of influencing factors that affect pipeline design and operational conditions, it is important to separate issues through the definition of terms given below:

1.2.1.1 Carbon The name "carbon" comes from Latin language *carbo*, for coal and charcoal [9]. The theoretical phase diagram for carbon is shown in Figure 1-3, which shows the state of matter for varying temperatures and pressures and some typical types of

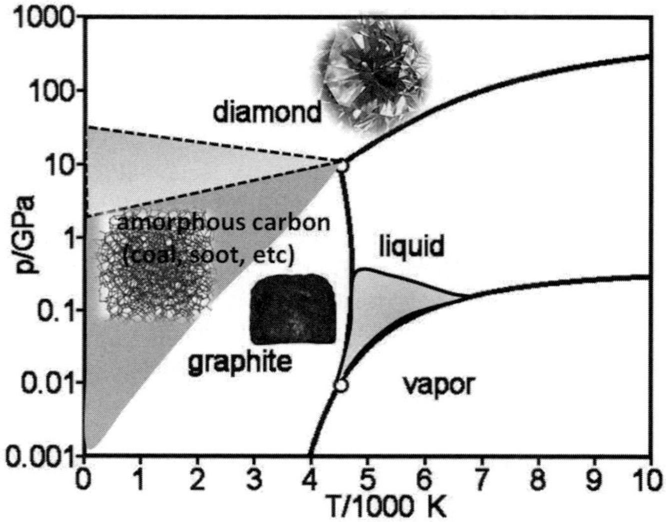

Figure 1-3. Carbon phase diagram and typical formation

carbonaceous materials formation. The darker regions indicate conditions under which one phase is metastable, so that two phases can coexist. There are several allotropes of carbon. The different forms or allotropes of carbon include the hardest naturally occurring substance, diamond, and also one of the softest known substances, graphite.

The physical properties of carbon vary widely with the allotropic form. Carbon in pure form does not cause a serious threat to environment as although it forms an extraordinary variety of compounds, most forms of carbon are comparatively unreactive under standard conditions. It is, however, the compounds of the carbon such as carbon monoxide (CO) that can displace oxygen and can, therefore, be hazardous. In the latter context, it may be noted that all practical forms of hydrogenated carbon—including cigarette smoke, wood fire smoke, chimney soot, mined coal, such as bitumen and anthracite—contain large amounts of polycyclic aromatic hydrocarbon tars, and are therefore carcinogenic.

1.2.1.2 Carbon Cycle The conversion of one element to another is very rare under terrestrial conditions. The processes that use carbon would obtain it somewhere and dispose of it somewhere else. For example, plants draw carbon dioxide out of their environment and use it to build biomass. This causes the amount of carbon on Earth to remain constant.

The paths that carbon follows in the environment make up the carbon cycle (Figure 1-4). As can be seen from Figure 1-4, carbon cycle is the combined processes, including photosynthesis, decomposition, and respiration, by which carbon as a component of various compound cycles between its major reservoirs, the atmosphere, oceans, and living organisms. Some of this biomass is eaten by animals, whereas some carbon is exhaled by animals as carbon dioxide (CO_2). The carbon cycle has a long loop; for example, some carbon dioxide is dissolved in the oceans; dead plant or animal matter may become petroleum or coal, which can burn with the release of carbon, should bacteria not consume it.

From Figure 1-4, it may be noted that the values given are the mass of the pure *carbon.*

To compare with carbon dioxide values, the values may be multiplied by 3.67. This is because CO_2 has one carbon atom with atomic weight 12 and two oxygen atoms with atomic weight 16, and $(2 * 16 + 12) / 12 = 3.67$.

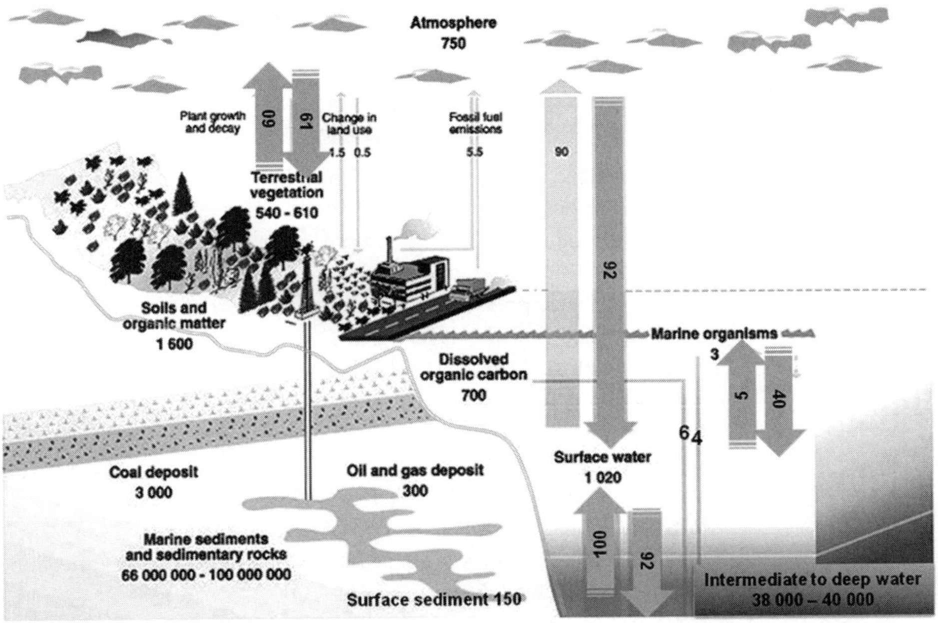

Figure 1-4. Carbon cycle process (volumes and exchanges in billion tonnes of carbon—average for period 1980 to 1989) [10]

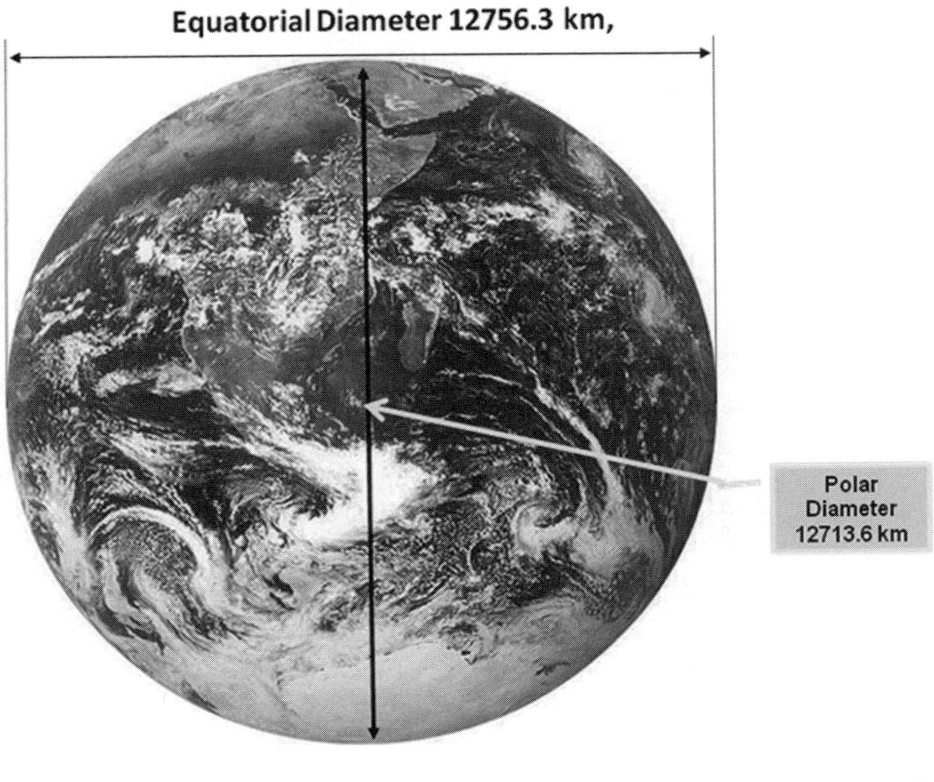

Figure 1-5. Earth Info: mean radius 6371.0 km, Equatorial radius 6378.1 km, polar radius 6356.8 km

Calculating the tonnage of CO_2 in atmosphere would render its significance in contributing to the global greenhouse gas (GHG) effects and hence the carbon cycle process: This can be attained by considering the average Earth radius (R) of 6371 km and the surface area = $4 \times \pi \times R^2$ as follows:

- The surface area of the earth (Figure 1-5): $4 \times 3.14 \times (6371)^2$ km² = 5.098×10^{14} m²
- The concentration of CO_2 by weight = 0.000582 (current value). The concentration per volume is 0.038% (=380 ppm). It may be noted that CO_2 is slightly heavier than air.
- Atmospheric pressure is 101.325 kPa = 101.325×10^3 N/m² = 101.325 / (9.809×10^4) kg/cm² = 10331.905 kg/m².
- The Earth surface area multiplied with the weight of the atmosphere:
- = 5.098×10^{14} m² × 1.0331905×10^4 kg/m² = 5.2672×10^{18} kg.

Thus the mass of the carbon dioxide = 5.2672×10^{18} kg × 0.000582 (i.e., the fraction of CO_2 in the atmosphere) = 3.0655×10^{15} kg, or about 3000 gigatons.

It may be noted that the excess of CO_2 since 1850 is (1− (280/380)) × 3000 gigatons = 789 gigatons. This assumes a linear increase in atmospheric emission of CO_2 per annum, refer to Figure 1-2 and 1-6 for values of CO_2.

1.2.1.3 Carbon Dioxide Carbon dioxide (CO_2) is the most important greenhouse gas because its impacts are large mostly because of human activities. Each year, human activities generate more than 6 billion tons of CO_2 [8, 11, 12]. This is emitted from the burning of fossil fuels for electricity, to power cars, and from burning wood products, etc. One molecule can warm the atmosphere for up to 120 years. Historical/recorded and projected CO_2 concentrations in air (in ppm) are shown in Figure 1-6. The figure includes the actual recoded CO_2 concentration in the atmosphere, "The Keeling Curve" taken at Mauna Loa Station in Hawaii. It is the longest continuous record of atmospheric carbon dioxide in the world and is considered a reliable indicator of global trend in the mid-level troposphere.

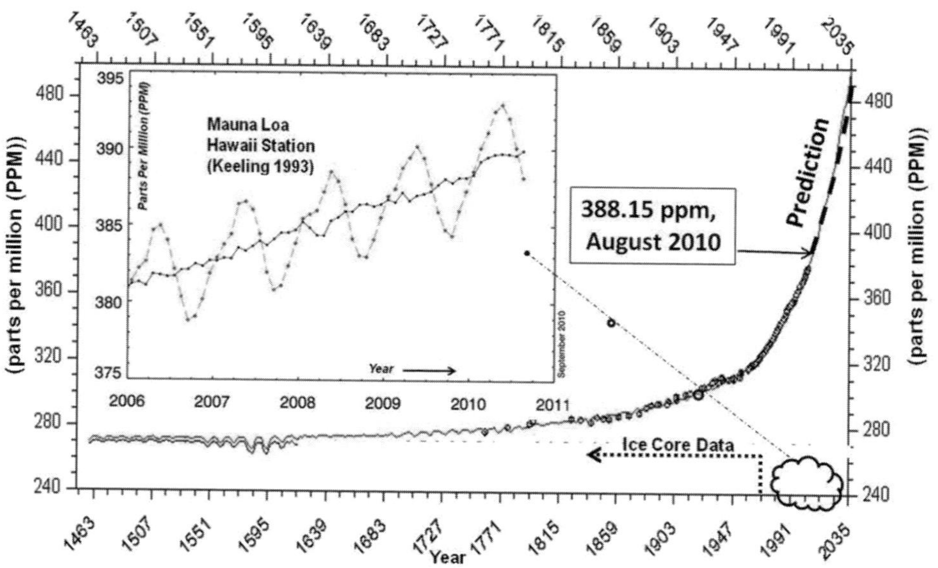

Figure 1-6. Carbon dioxide (CO_2) concentration record, 1751–2004 and projection (Source: Oak Ridge National Laboratory, refs [12, 13]; Inset Indicates recent data, [11])

1.3 CO_2 AND CLIMATE CHANGE

1.3.1 Global Warming

Global warming refers to the increase in the average temperature of the Earth's near-surface air and oceans in recent decades and its projected continuation. It is caused by emission of various gases when fossil fuels are used in plants (coal power and other fossil fuel-fired plants). In such cases, one of the most significant emitted gases is carbon dioxide (CO_2), a gas that traps heat in the Earth's atmosphere.

Global average surface temperature has risen about 1°C to 1.2°C since reliable records began in the late 1850s (Figure 1-7). The global average air temperature near the Earth's surface has risen about 0.74 ± 0.18°C (1.33 ± 0.32°F) during the past 100 years. It may be noted that, "most of the observed increase in globally averaged temperatures since the mid-20th century is very likely due to the observed increase in anthropogenic greenhouse gas concentrations such as CO_2" via the greenhouse effect, [6]. CO_2 is an anthropogenic gas.

IPCC [6] indicate that average global surface temperature will likely rise a further 1.1 to 6.4°C (2.0 to 11.5°F) during the 21st century.

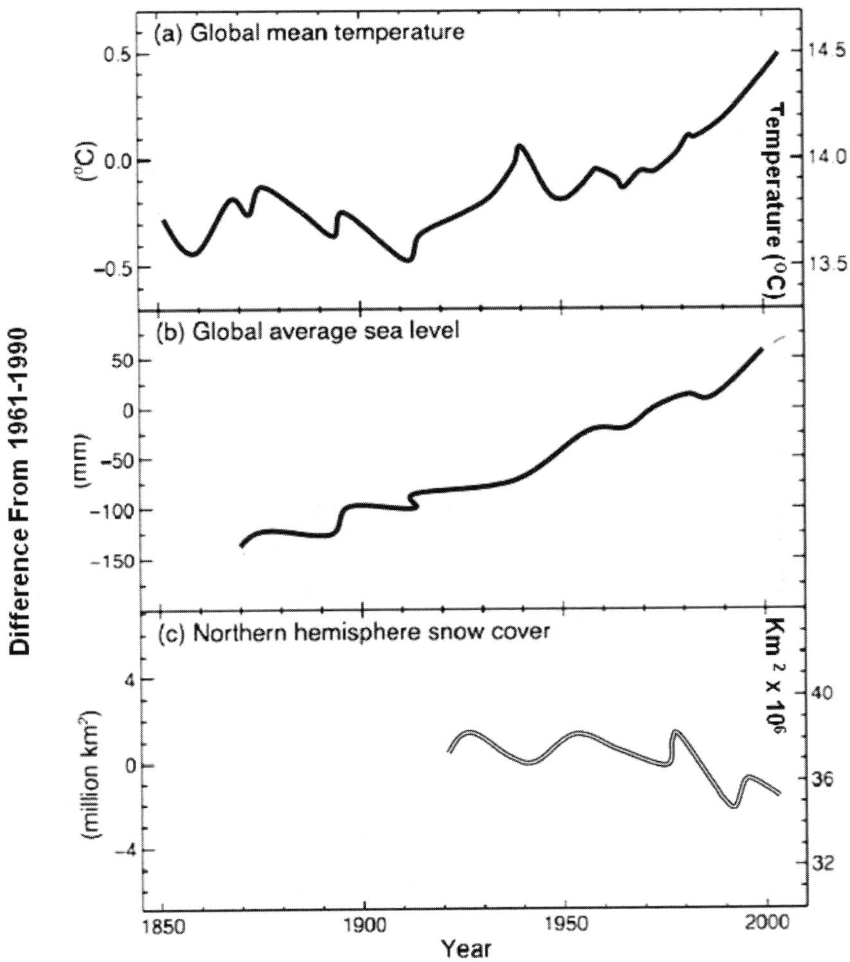

Figure 1-7. Observed Average Global Environmental Changes1961-1990 (adapted from ref [6])

Increasing global temperatures will cause sea level to rise and is expected to increase the intensity of extreme weather events and to change the amount and pattern of precipitation. Other effects of global warming include changes in agricultural yields, glacier retreat, species extinctions, and increases in the ranges of disease vectors.

1.3.2 Greenhouse Effects

The warming of the atmosphere is by the trapping of long-wave infrared radiation being radiated to space. The gases most responsible for this effect are water vapor and carbon dioxide (CO_2) (Figure 1-8).

Some of the human activities that cause and increase greenhouse gas effects are as follows:

- Combustions (through burning of fossil fuel),
- Deforestation,
- Industrialization,
- Motor vehicles,
- Biomass burning and use change,
- Stationary sources,
- Power plants,
- City gridlocks.

Some of the natural factors include the following:

- Volcanic eruption,
- Tundra,
- Permafrost.

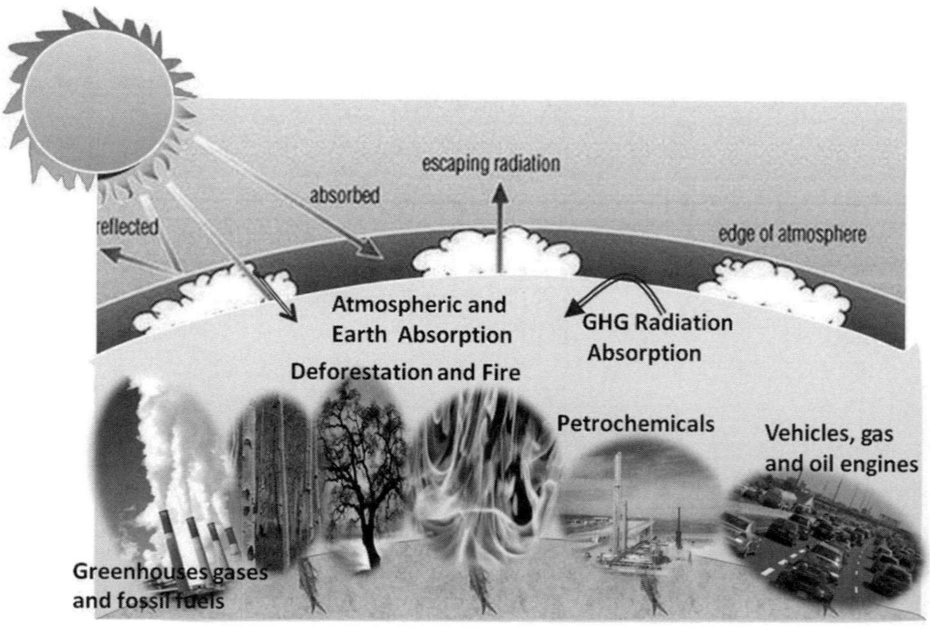

Figure 1-8. Greenhouse gas effects (generation, emission, radiation, and absorption)

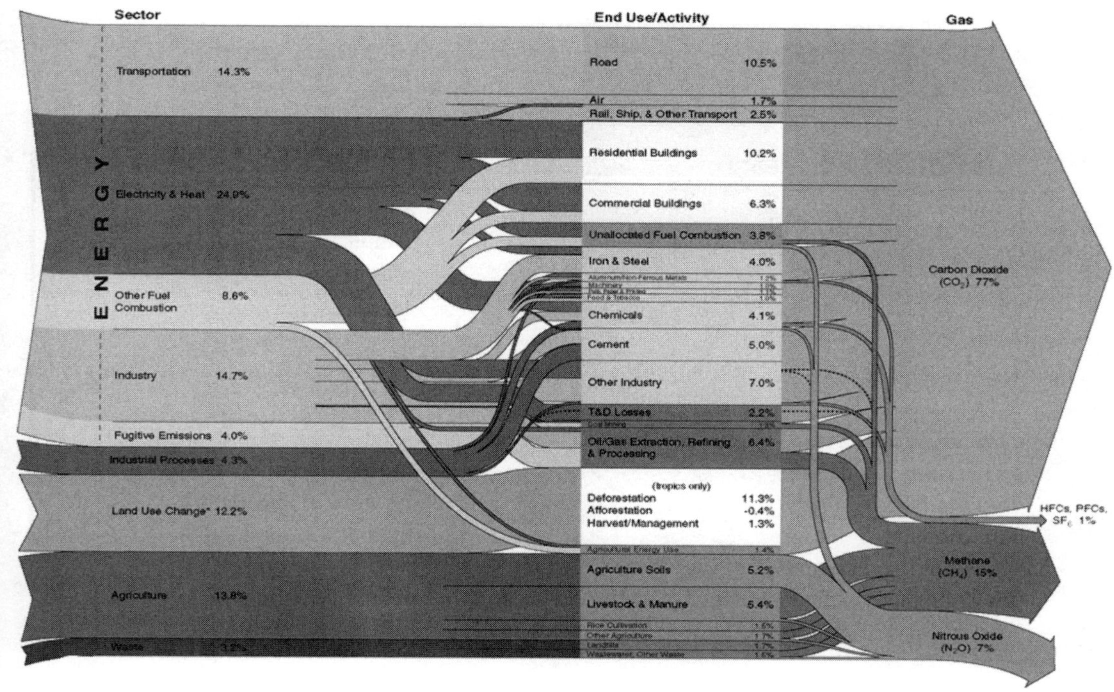

Figure 1-9. World greenhouse gas emission contribution—2005 ([12] Courtesy of WRI)

One of the greatest challenges relating to global warming is that greenhouse gases result—directly or indirectly—from almost every major human industry and activity. World greenhouse gas (GHG) emission chart is shown in Figure 1-9, which indicates percentage contribution of various sectors to CO_2, methane, nitrous oxide, and other gases. This chart is a comprehensive view of global, anthropogenic greenhouse gas (GHG) emissions which indicates that CO_2 makes up over 61% of GHG emissions caused by various forms of energy use. With land use changes, this contribution increased to 77%. Total world GHG emission in 2005 amounted to *44,153 MtCO$_2$ equivalent* [14].

Herzog [14] indicates that total global emissions grew 12.7% between 2000 and 2005, an average of 2.4% a year. However, individual sectors grew at rates between 40% and near zero, and there are substantial differences in sectoral growth rates between developed and developing countries.

1.3.3 Fossil Fuel Usage and CO_2 Emission

Fossil fuels are fuels containing carbon—coal, oil and gas—that were formed over millions of years through the decay, burial, and compaction of rotting vegetation on land and of marine organisms on the sea floor. Burning fossil fuels is the major way in which humans add to the greenhouse gases in the atmosphere.

Over a period of 50 to 60 years, while consumption of fossil fuels grew substantially, the world undertook a transition in its usage of fossil fuels, from solids (coal), to liquids (oil) to gases (natural gas) (Figure 1-10). While coal accounted for 52% of all fossil fuel consumption in 1950, this share dropped to 22% in 2000, equivalent to the share of natural gas. The share of oil substantially increased between 1950 and 2000, where it peaked at 40% of fossil fuel use in 1970, then declined further. Figure 1-10 also looks at the world energy over a long span over 200 years. Usage of one form of energy will need adoption span of at least 50-60 years.

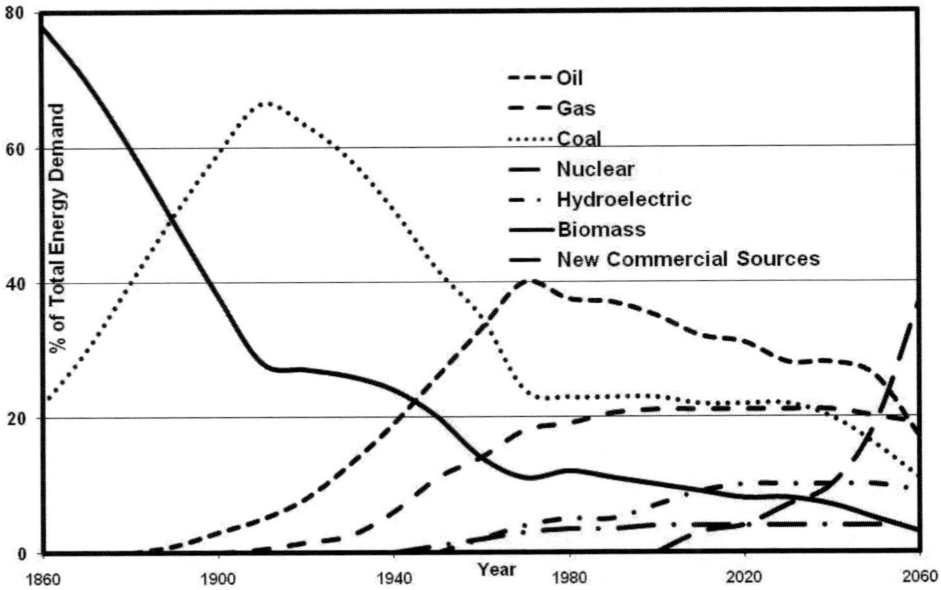

Figure 1-10. Fuel/energy usage market share [8, 15]

World energy consumption per capita in terms of tonnes of oil equivalent is indicated in Figure 1-11.

It may be noted that carbon dioxide (CO_2) emission and world energy consumption are very much tied together. Increasing energy consumption also increases CO_2 emission, as evidenced in comparing Figure 1-9 with that of CO_2 emissions per capita in Figure 1-12.

It may be noted that since 1750 to 2003, the Earth's CO_2 levels have increased from 280 to 380 ppm (see previous Figure 1-6 for trend). However, human contribution to the CO_2 in the atmosphere is anything but small.

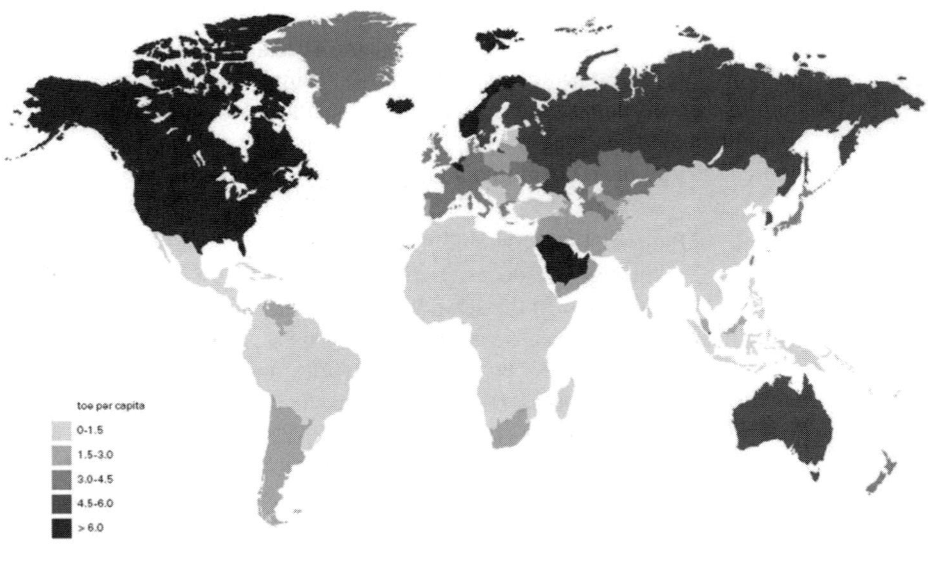

Figure 1-11. World energy consumption in tons of oil equivalent (TOE)

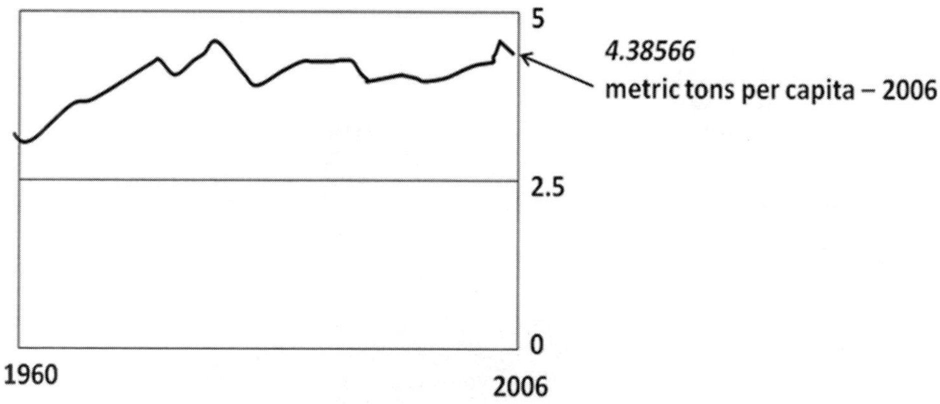

Figure 1-12. Carbon dioxide emissions per capita million tons of oil equivalent (MTOE) [16]

1.4 CO_2 SOURCES, USES, AND SAFETY ISSUES

1.4.1 Sources

Carbon dioxide is one of the earliest "natural" products. High concentrations of carbon dioxide are and have been produced by the volcanic activities. Indeed, most "natural" carbon dioxide sources which exist on the Earth today are produced from magma deep underneath the Earth's surface.

Natural-occurring carbon dioxide can be found in underground wells at concentrations of 90% to almost 100% depending on the location of the well. Large carbon dioxide wells exist in various locations in the United States (e.g., in Colorado, Mississippi, New Mexico, Utah, and Wyoming) and in Europe (e.g., at Répcelak and Oelboe in Hungary and at Bad Driburg-Herste and Rottenburg in Germany). Geothermal carbon dioxide is found in numerous locations across Spain and Italy (e.g., at Torre Alfina).

The largest source of impure CO_2 is from flue stacks due to the combustion of fossil fuels such as coal, oil, and gas in power plants, industrial facilities, and other sources. Figure 1-13 indicates the percentage of CO_2 emission caused by various forms of activities. This chart is further detailed in Figure 1-14 which indicates the CO_2 emitted across the entire U.S. economy [17]. In Figure 1-14, each square represents about 1 million tons of CO_2 emissions. Total U.S. CO_2 emission in 2007 was 726 million tons.

As indicated in Figure 1-13 and 1-14, maximum amount of CO_2 (> 30%) is produced from the burning of coal in power plants and other forms of combustion/burning of coals followed by burning of gasoline/diesel/petroleum products in various forms of transportations (including cars, trucks, airplanes, water crafts, etc.). While diverse activities, such as burning trash (in landfill) and feeding cattle or agricultural activities generate greenhouse gases, the primary source of CO_2 emission is, by far, the burning of fuel for heat and power. The trend is consistent with that of world greenhouse gas emission chart indicated in Figure 1-9.

Other industrial sources include chemical processes, such as the processing of phosphate rocks which can release carbon dioxide. The impurity profile of this source of carbon dioxide can be unusual since it may contain traces of phosphines (PH_3) or radon gas (due to uranium decays).

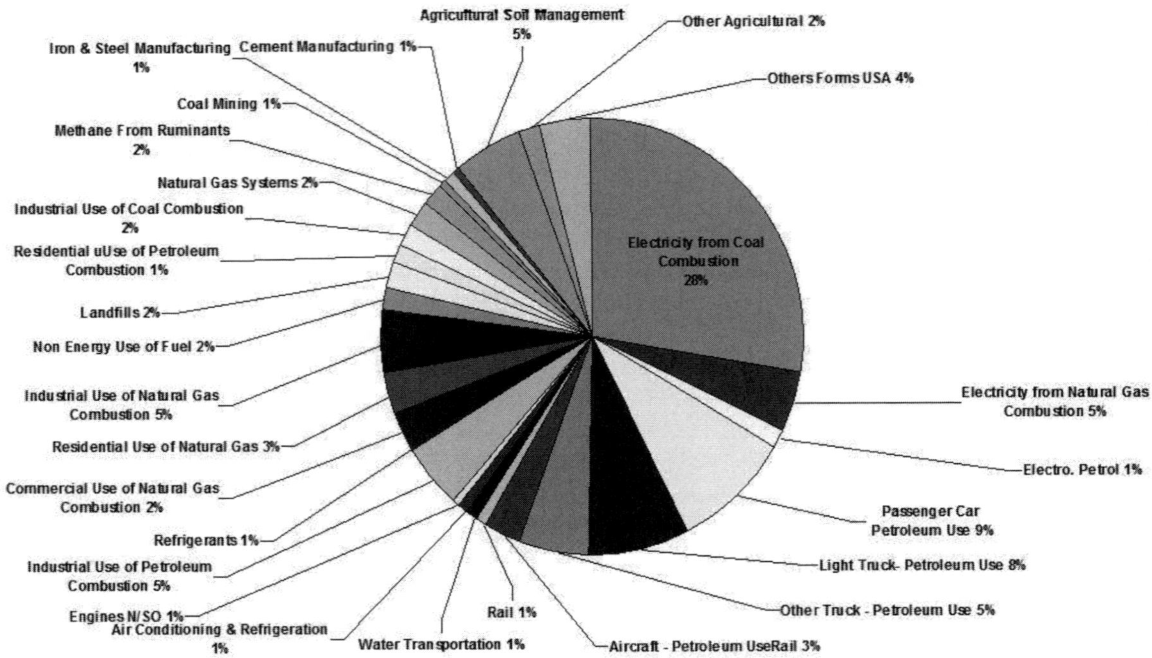

Figure 1-13. Causes of CO_2 emission by fossil fuel—2007

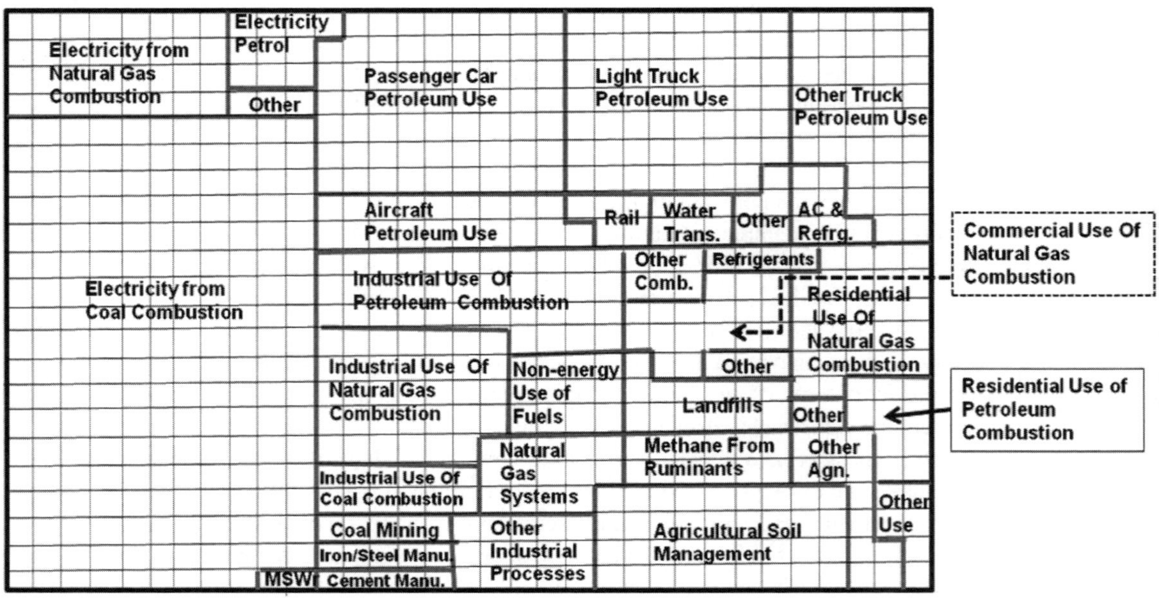

Figure 1-14. US CO_2 emission by activities, [18] (Courtesy of ASME Mechanical Engineering)

1.4.2 Uses of CO_2

Gaseous, liquid, and solid carbon dioxide in pure form is generally used for the following [19]:

- carbonated beverages,
- aerosol propellants,
- recharging/conservation of natural mineral water, wine, and beer
- enrichment of air in greenhouses,
- chemical industry,
  - preparation of sodium carbonate,
  - manufacturing foam rubber,
  - dehydration,
  - tanning,
  - manufacturing of paint and vanishes,
  - Use of CO_2 for manufacturing ultrathin polymers/plastics.
- shielding gas for welding,
- Mineral carbonation,
- fire extinguishers,
- gas analysis as carrier gas,
- dry ice for refrigeration, carbon dioxide snow,
- water treatment,
- fracturing and acidizing oil wells,
- tertiary oil recovery, miscible flooding.

In the oil recovery process, CO_2 (transported from naturally occurring sources) is injected into the formation/reservoir where it dissolves in the oil to form a miscible fluid. The injection causes the oil to swell, which results in reduced viscosity and increased density. Carbon dioxide presence exerts an acidic effect on the reservoir rock, and in some cases, vaporizes some of the oil. As a rough rule of thumb, approximately 6-10 MCF (170-280 m^3) of CO_2 is required for recovery of one barrel of (0.16 m^3) oil. Carbon dioxide miscible flooding generally recovers about 10% to 15% of the oil remaining in oil reservoirs [20]. Other uses include the following:

- Biological conversions of CO_2 to fuel or other useful chemicals;
- Chemical conversions of CO_2 to fuel or other useful chemicals;
- Use of CO_2 as a heat exchange fluid or working fluid;
- Cushion (or base) gas (e.g., for natural gas storage or Compressed Air Energy Storage (CAES));
- Use of CO_2 as a solvent;
- Use of CO_2 as a fumigant, propellant, or inert gas;
- Concrete processing.

An example of direct usage is use of impure CO_2 for natural gas cushioning (Oldenburg, [21]). When demand for natural gas exceeds supply (e.g., during cold weather months when heating requirement peaks or during hot weather months when electricity usage peaks due to air conditioning usages, etc.), natural gas is withdrawn from underground gas storage reservoirs to supplement the pipeline supply and meet the peak demand. On the other hand, when demand for natural gas is low, gas is taken from the pipeline and injected back into the gas storage reservoir. Impure CO_2 (with refined properties) can thus be utilized as a cushioning gas during the natural gas withdrawal and injection process from storage gas reservoirs.

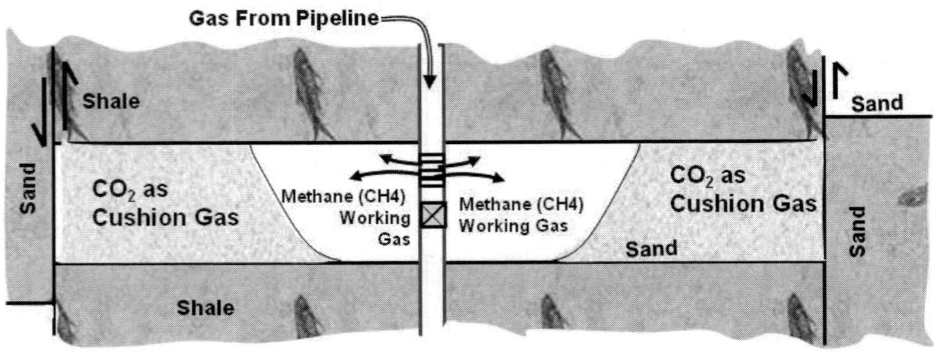

Figure 1-15. Schematic of a single-well natural gas storage with working gas and CO_2 cushion gas (Oldenburg, [21])

A cross-section schematic of an idealized gas storage in a depleted gas reservoir is shown in Figure 1-15 (Oldenburg, [21]), where the working gas is methane (CH_4) from a pipeline. Critical to the operation of gas storage reservoirs is the use of a cushion gas, i.e., a gas that compresses and expands as the working gas is injected or withdrawn but which is itself not produced.

1.4.3 Safety Data

CO_2 is essential for life, being a critical component in photosynthesis. Greenhouses purposely elevate CO_2 levels in order to "fertilize" the plant.

At low concentrations (1% by volume), CO_2 causes no ill effects on humans, fauna, or flora. At concentrations of about 6% by volume, CO_2 can cause nausea, vomiting, diarrhea, and irritation to mucous membranes, skin lesions, and sweating.

At about 10% by volume, it causes asphyxiation, refer to Table 13-2 Chapter 13.

1.5 IMPURE CO_2 CAPTURE TECHNOLOGIES

CO_2 capture is generally divided into four different groups [22, 23, 24], these are the following:

- Post-combustion capture,
- Pre-combustion capture,
- Oxy-fuel combustion capture,
- Other industrial separation processes.

The capture methods are shown in Figure 1-16 [24, 25] and the detailed description of Post, Pre, and Oxy are provided in Chapter 2.

In general, it must be noted that the practical recovery of CO_2 from flue gas is sufficiently different from other gas treating applications, such as natural gas plant and refinery gas sweetening and purification of ammonia and hydrogen plant synthesis gas that requires its own specialized processes. Candidate capture processes must be active at low CO_2 partial pressure and tolerate oxygen and NO_X. Flue gases from coal-fired boilers also contain SO_X, soot, and fly ash, which must be dealt with [26].

Different end-uses have differing requirements from a product quality perspective. There are examples of EOR operations utilizing a relatively clean CO_2 transportation

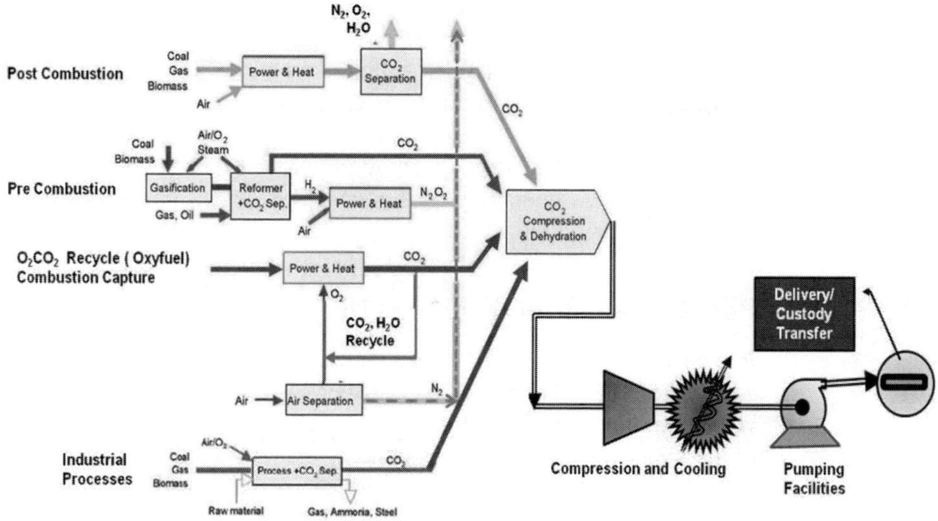

Figure 1-16. Carbon dioxide capture technologies (from refs. [24, 26, 27]

system, such as in the Permian Basin. There are point to point EOR operations utilizing CO_2 with a high content of H_2S (1-2%), but there are currently no examples of a CO_2 pipeline transporting a product which contain concentrations of H_2, CO, NO_x, SO_x, or inorganic trace elements. The CO_2 capture concepts thus are devised to meet the end-use requirement. The three major capture technologies are described below.

1.5.1 Pre-Combustion (Figure 1-17)

The first of the three major capture technologies—pre-combustion—allows operators of power plants to capture CO_2 and maximise power output. With this method, an air separation unit produces a stream of almost pure oxygen. The oxygen flows into the gasifier and

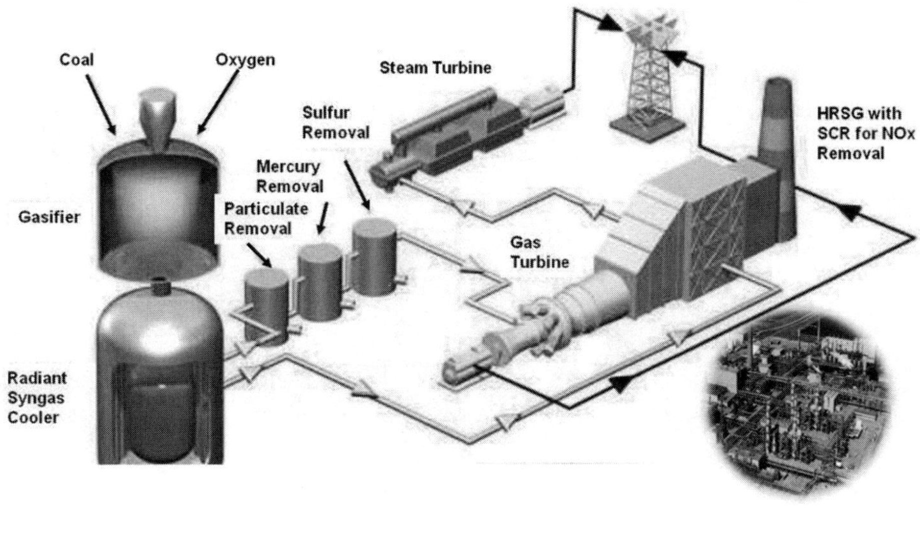

Figure 1-17. Typical pre-combustion CO_2 capture technology [24], Inset: Coal Gasification Plant, USA

reacts with pulverized coal to form synthetic gas, or "Syngas." Steam is then added to the Syngas in a shift reactor converting the carbon monoxide to hydrogen and CO_2. Using a physical wash, the CO_2 is then captured from the gas stream and, after dehydration and compression, is ready for transport and storage. Today, the hydrogen is burnt in power turbines and makes electricity; tomorrow, it could also be used as fuel for transport.

The flue gas that results from the hydrogen-powered turbines passes through a Heat Recovery Steam Generator (HRSG), which powers steam turbines, re-using this energy and optimizing energy output.

1.5.2 Post-Combustion (Figure 1-18)

In a post-combustion technology, a mixture of coal (or gas) and air is blasted into a boiler and ignited. Usually, the coal is "washed" before it is fed into the boiler. The process involves passing coal through a series of liquids of varying densities which removes many of the impurities.

The heat from the combustion of the coal and air mixture generates steam, driving a steam turbine. The flue gas, a by-product of burning coal, is removed from the boiler and filtered. The steam turbines then generate electricity which is then transmitted to the distribution grid. Once the steam has passed through the turbine, it arrives at a condenser. This unit uses cool water to condense the steam back into water, allowing it to be piped back into the boiler and be re-heated.

This is the first of several "cleaning" processes that the flue gas will pass through. At this point, small particles called "fly ash" are removed from the gas. Sulphur is also removed from the flue gas before it enters the CO_2 absorber where it needs to be cooled. This stage, using water, lowers the temperature of the gas. Here, the gas stream is typically passed though a liquid sorbent, which reacts with the CO_2, chemically binds with it, and removes it from the flue gas.

Once the CO_2 is captured, the sorbent is moved to a desorber to be "regenerated," which usually involves heating the sorbent to release the captured CO_2.

Figure below indicate a post-combustion process for CO_2 removal from flue stacks (using an amine absorption process) which is similar to the technology described above in coal-fired power stations.

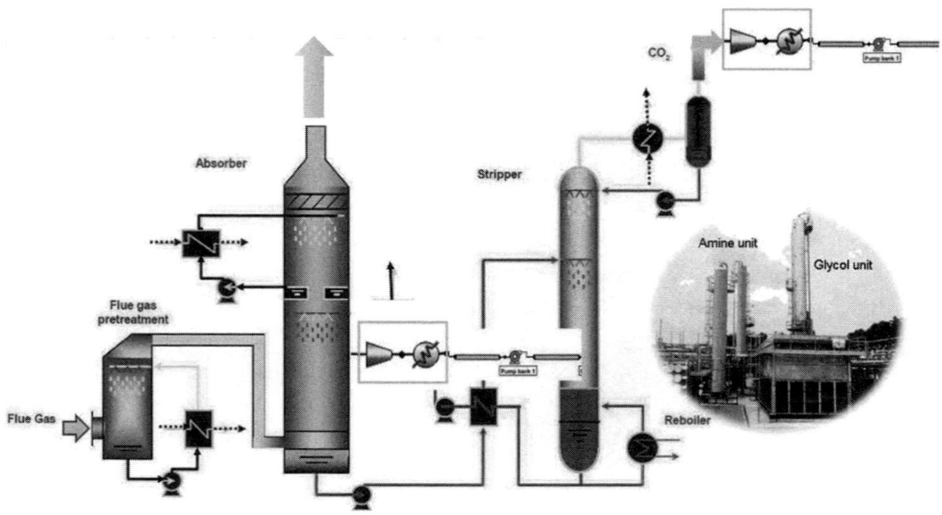

Figure 1-18. Typical post-combustion amine (CO_2) capture technology, [24] (inset amine glycol unit, [28])

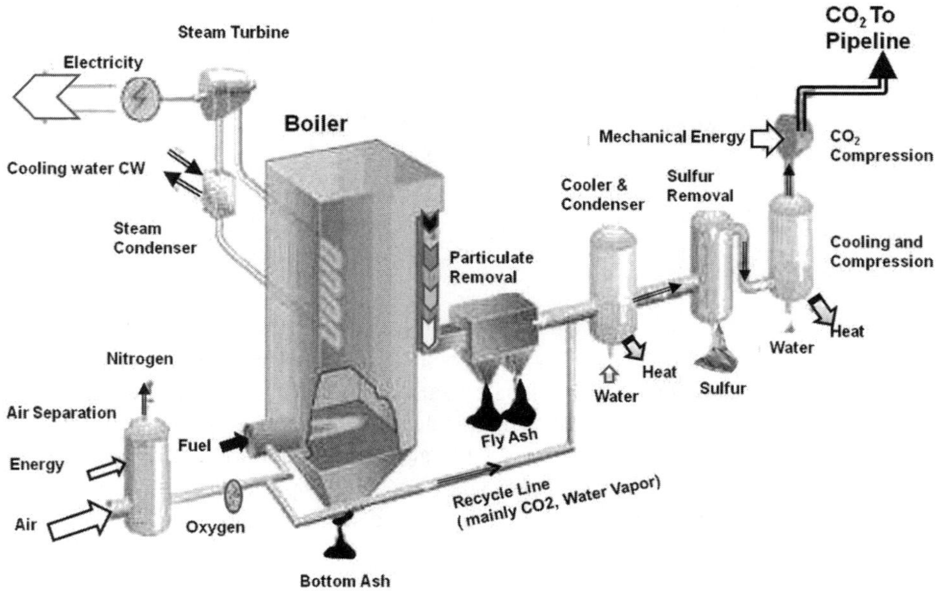

Figure 1-19. Oxyfuel CO_2 capture technology and process [24]

1.5.3 Oxyfuel (Figure 1-19)

The Oxyfuel technology consists of burning fuel in a mixture of pure oxygen and re-circulated flue gas instead of air. This process increases the CO_2 concentration in the flue gas, thereby making it more efficient to remove before processing for transportation and sequestration.

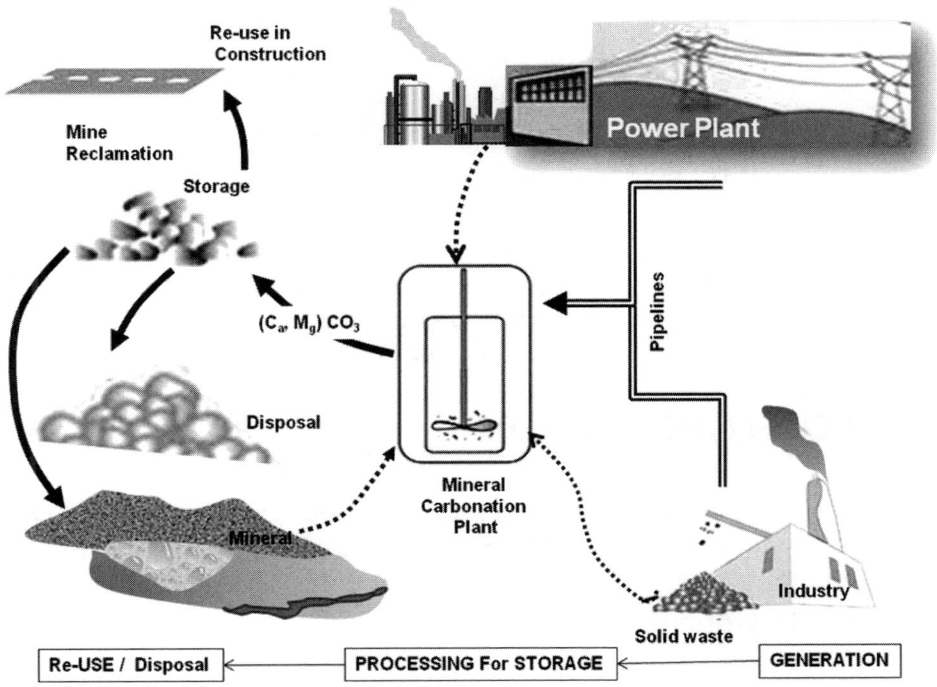

Figure 1-20. Typical industrial carbonation process (reproduced from Mazzotti [29])

TABLE 1-1. Typical component concentration from CO_2 capture technologies [30]

	Component Concentrations from CO_2											
Capture Method	**CO_2**	**CH_4**	**N_2**	**H_2S**	**C_2+**	**CO**	**O_2**	**NO_x**	**SO_x**	**H_2**	**Ar**	**H_2O**
Pre-Combustion	>95.6 vol%	350 ppm	<0.6 vol%	<3.4 vol%	<0.01 vol%	<0.4 vol%	trace	–	–	<3 vol%	<0.05 vol%	N/A
Post, Amine	>99 vol%	<100 ppm	<0.17 vol%	trace	<100 ppm	<10 ppm	<0.01 vol%	<50 ppm	<10 ppm	trace	trace	N/A
Oxyfuel	>90 vol%	–	<7 (0.6–7)	trace	–	trace	<3 vol%	<0.25 vol%	< 2.5 vol%	trace	<5 vol%	N/A

This technology deploys an air separation unit that removes nitrogen from the air, producing oxygen. This oxygen alongside the fuel is then injected into a boiler where combustion takes place. Steam is generated and used to power turbines and make electricity. Meanwhile, the flue gas, CO_2, and water vapor are re-circulated to control boiler temperature and gradually cool. This leaves the captured CO_2 to be compressed and dehydrated, ready for pipeline transportation and sequestration.

1.5.4 Industrial Processes

CO_2 can be used either directly or as a feedstock [31] in chemical processes that produce valuable carbon containing products ([29]; Figure 1-20). The utilization of CO_2 thereby can establish an inventory of stored CO_2, primarily in the form of carbon-containing fuels, chemicals, and other products. The production and use of these products involve a variety of different "life cycles" (i.e., the chain of processes required to manufacture a product from raw materials to use the product for its intended purpose and ultimately to dispose of it or to reuse it in some fashion).

Depending on the product life cycle, CO_2 may be stored for varying periods of time and in varying amounts. As long as the recycled carbon remains in use, this carbon pool successfully stores carbon. Withdrawal from this pool, by decay or by disposal, typically re-injects this carbon into the atmospheric pool.

1.5.5 Typical Composition from Capture Technologies (Pre, Post, and Oxyfuel)

Table 1-1 illustrates the range of component concentrations from CO_2 capture technologies at various combustion gas capture stages. Further details are provided in Chapter 3.

It may be noted that the composition of the capture stream for CCS varies in different literature and that the same capture technology may produce different combinations of impurities depending on the process "clean up" regime.

1.6 CO_2 SEQUESTRATION

Sequestration options are summarized below, however details are provided in Chapter 2.4 as "Carbon Dioxide Sinks".

1.6.1 Options

Carbon dioxide capture and storage (CCS) is a technology with the potential to reduce the GHG problem, while allowing the continued use of fossil fuels. With CCS, the CO_2 arising from fossil fuel use is captured and then stored in secure geological formations. The

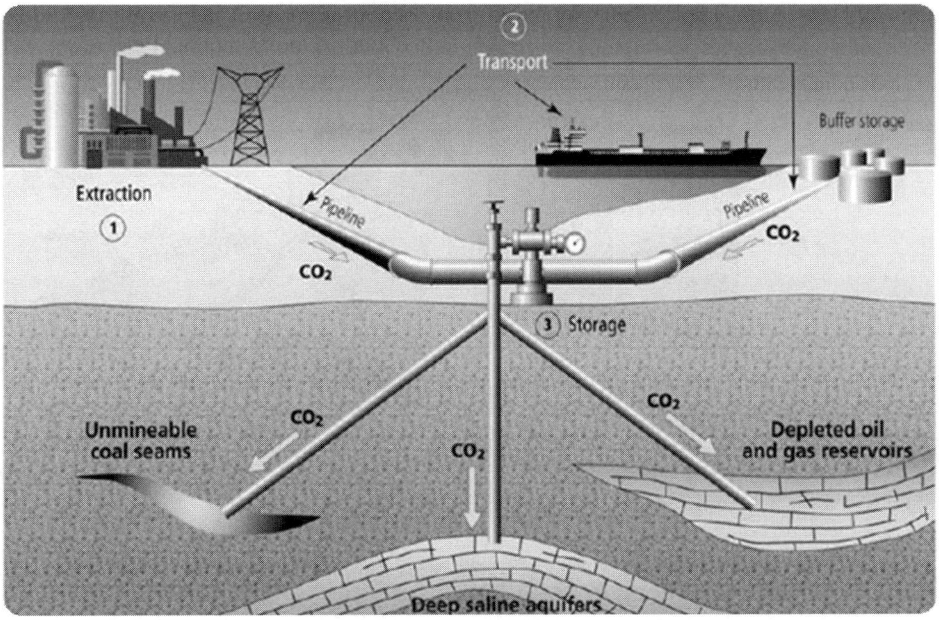

Figure 1-21. Carbon dioxide capture, transportation and storage options (from Leopardi [32])

three main options for CCS to avoid CO_2 release into the atmosphere are illustrated in Figure 1-21.

The geological sequestration of CO_2 can occur by either physical or chemical trapping mechanisms. Four types of trapping mechanisms control the storage density and leakage potential of the CO_2 injected into geologic sinks such as coal seams [33]. These four trapping mechanisms include the following (Figure 1-22):

- Physical/adsorption trapping,
- Hydrodynamic trapping,
- Solubility trapping,
- Mineral trapping.

These trapping mechanisms are briefly described below:

- **Physical/Adsorption Trapping:** In physical adsorption, CO_2 molecules are immobilized or trapped at near liquid-like densities on micropore wall surfaces of coal organic matter, kerogen, or minerals. The hydrostatic pressure in the formation controls the gas adsorption process. Coal seams and shales are types of geologic sinks where physical adsorption trapping occurs.
- **Hydrodynamic Trapping:** In hydrodynamic trapping, CO_2 occupies the pore space of the rock comprising the geologic sink. The geologic sink for hydrodynamic trapping is a porous rock layer capped by an essentially impermeable rock layer.
- **Solubility Trapping:** In solubility trapping, the CO_2 simply dissolves in the formation water or reacts with the water to form carbonic acid and other aqueous carbonate species.
- **Mineral Trapping:** In mineral trapping, dissolved CO_2 undergoes chemical reactions with silicate minerals rich in Ca, Mg, and Fe, resulting in the formation of a solid carbonate mineral phase. Mineral trapping results in the most stable, permanent form of geologic CO_2 sequestration.

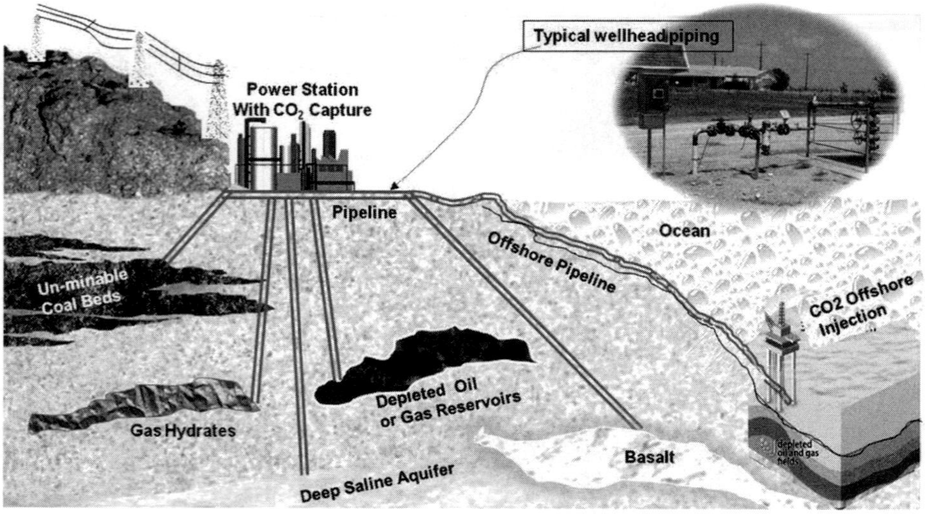

Figure 1-22. Geological sequestration/trapping options for CO_2 (from McGrail [34])

It also may be noted that geologies have differing primary and secondary trapping mechanisms.

However, potential reservoirs (for CO_2 storage) are not a constraint in most locations, but careful assessment must be made of their suitability for injection of CO_2 containing varied impurities. For example, an indication of the extent of potential reservoirs for se-

- Deep saline aquifers **400 - 10 000 GtCO$_2$**
- Un-minable coal seams **40 GtCO$_2$**
- Depleted gas fields **690 GtCO$_2$**
- Depleted oil fields/CO2- EOR **120 GtCO$_2$**

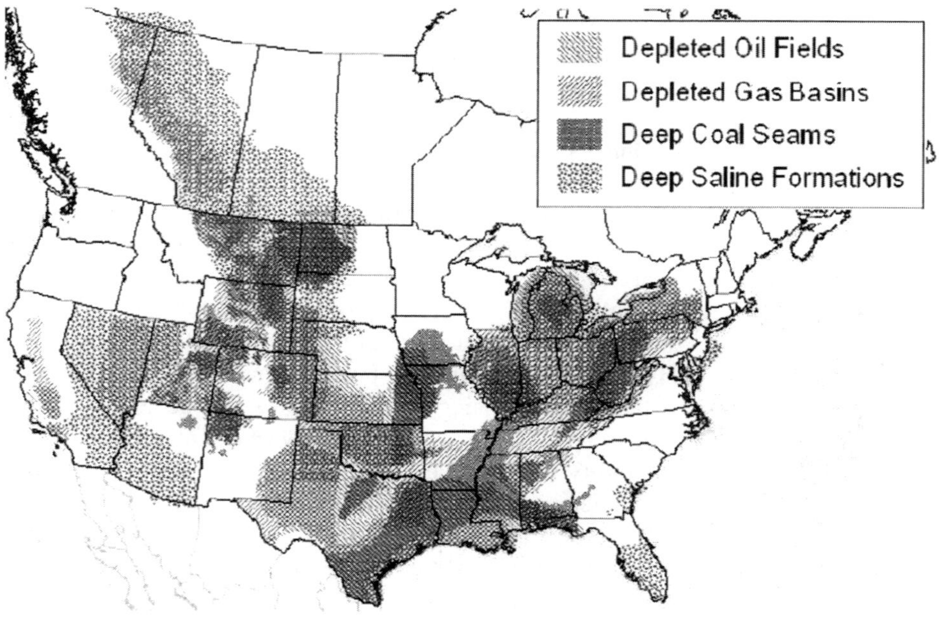

Figure 1-23. Potential storage reservoirs in North America for CO_2 sequestration [35])

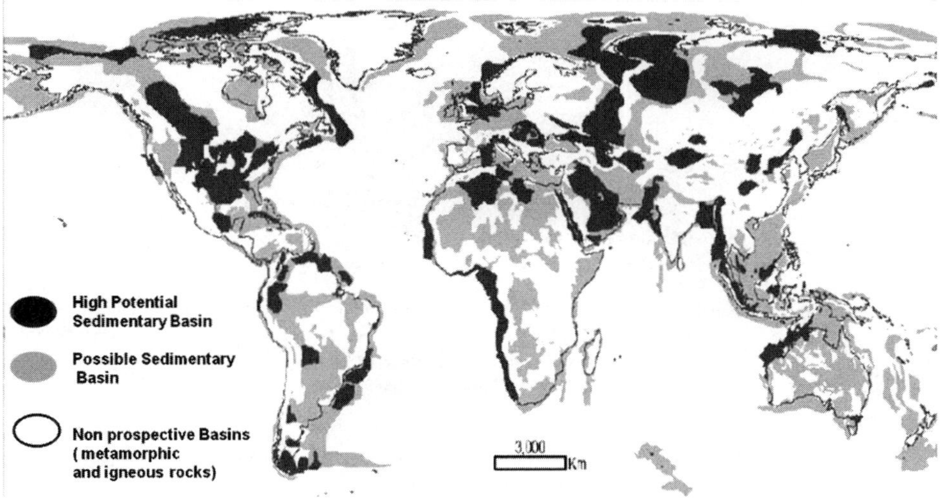

Figure 1-24. Potential global CO_2 storage sites (Beck, [27], Source: IPCC [6])

questration of CO_2 in North America is indicated in Figure 1-23 [35]. There are more than 330 potential onshore geological CO_2-storage reservoirs in North America. These candidate geological CO_2-storage reservoirs have more than 3,800 $GtCO_2$ of potential storage capacity comprising:

- Deep saline formations (DSF), **3,700 $GtCO_2$**
- Deep un-mineable coal seams, **65 $GtCO_2$**
- Depleted gas fields, **40 $GtCO_2$**
- Depleted oil fields with potential for enhanced oil recovery (EOR), **13 $GtCO_2$**.

In comparison, Beck [4] indicates that global potential storage capacity for CO_2 is as follows (Figure 1-24):

- Deep saline aquifers, **400 to 10,000 $GtCO_2$**
- Un-mineable coal seams, **40 $GtCO_2$**
- Depleted gas fields, **690 $GtCO_2$**
- Depleted oil fields/CO_2-EOR, **120 $GtCO_2$**

Dooley et al [35] indicate that CO_2 point sources are large in the same region. However, it may be noted from the above data that only 3% of reservoirs are coal, gas, and oil fields where the deep injections of CO_2 stream with impurities may not pose a significant concern. Injectivity of CO_2 stream with impurities affect storage capacity (depending on the level of impurities) and potentially other geological parameters that impact CO_2 storage. Such sequestration is highly site-specific and injectivity is likely to range over an order of magnitude even within the same reservoir. The core of any CO_2 sequestrations must be the matching of the identified point sources with candidate storage reservoirs.

As indicated previously (Figure 1-21), from physical sequestration points of view, CO_2 can be trapped through two fundamental methods. This can be achieved through physical means by geological, hydrodynamic, or cavern trapping. It can also be achieved through chemical means by solubility trapping, ionic trapping, or adsorption trapping. Figure 1-25 further shows a schematic illustration of these processes [36].

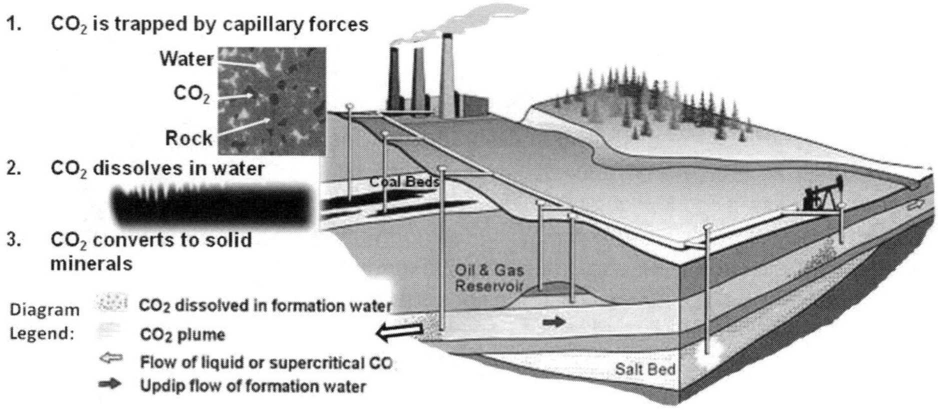

Figure 1-25. CO_2 sequestration (Source: Alberta Geological Survey [36], modified from Bruant et al [37], Beck, [27])

1.6.2 Potential Reservoir Siting and Issues

CO_2 can be safely sequestered in geologic formations; that CO_2 sequestration is as safe as activities that have been ongoing for decades, such as EOR, natural gas storage, and deep underground disposal of acid gas [38].

The successful sequestration of CO_2 will require proper site selection using accurate subsurface information, an effective monitoring and verification program, as well as a reasonable legal and regulatory framework. Generally, consideration must also be given for protection of groundwater deposits and specific characteristics of CO_2 injection and storage sites. Careful project siting will ensure that no geologic formation with an unsafe or uncharacterized geologic seal will be used for CO_2 sequestration. Storage safety can be further enhanced by means of appropriate mitigation measures to stop or control CO_2 releases should they arise.

Depleted oil and gas reservoirs, in particular, have a number of attractive features as CO_2 storage reservoirs including the following:

- Proven traps, known to have held liquids and gasses for years;
- Known geology;
- Least exploration cost;
- Numerous existing sites.

Geological storage of CO_2 is analogous to natural CO_2 fields that occur in sedimentary basins and thus regulatory process includes similar consideration, such as:

- A rigorous characteristics of the storage site and surrounding area.
- The construction of geological model of the storage site and surrounding area.
- Simulation of CO_2 injection into the storage reservoir.
- Risk assessment and management process.
- Monitoring program for the stored CO_2.

US experience has demonstrated that geological storage of CO_2 is as safe as natural gas storage and enhanced oil recovery. There is already a strong base of industry experience in enhanced oil recovery, where water and CO_2 are pumped into depleted oil wells to repressurize wells and increase oil production.

1.7 CO_2 TRANSPORTATION

1.7.1 Optimal Transportation Conditions

CO_2 can be transported as liquid, solid, dense phase, or in supercritical gas phase. However, in order to transport large quantities of CO_2 efficiently, the CO_2 that comes from flue stacks must be transformed into a form with higher density [39].

Figure 1-26 shows the various methods for CO_2 transport and how they relate to CO_2 densities as a function of temperature and pressure.

1.7.1.1 Ship Transportation Generally, CO_2 transportation by ship (Figure 1-27) is very cost effective due to large loading capacity. However, experience of large-scale ship transportation of CO_2 is limited [40]. So far, ships up to 1500 m^3 have been used to transport CO_2. Generally, they are more economical for smaller quantities of CO_2 at lower pressures as high pressure vessels will have very high costs.

There are three types of CO_2 transportation by ships [41]:

1. Pressure type/fully pressurized operating at high pressure to prevent CO_2 gas from boiling under ambient temperature conditions;
2. Low-temperature type/fully refrigerated ship operating at a sufficiently low temperature to keep CO_2 as a liquid (or) under atmospheric pressure a solid;
3. Semi-refrigerated type/semi-pressurized combined conditions of temperature and pressure necessary for gas to be kept as a liquid.

Typically, the offshore transportation of CO_2 involves taking the gas from the carbon capture plant/system (as previously shown in Figure 1-16), dehydrating, compressing, and liquefying it for storage.

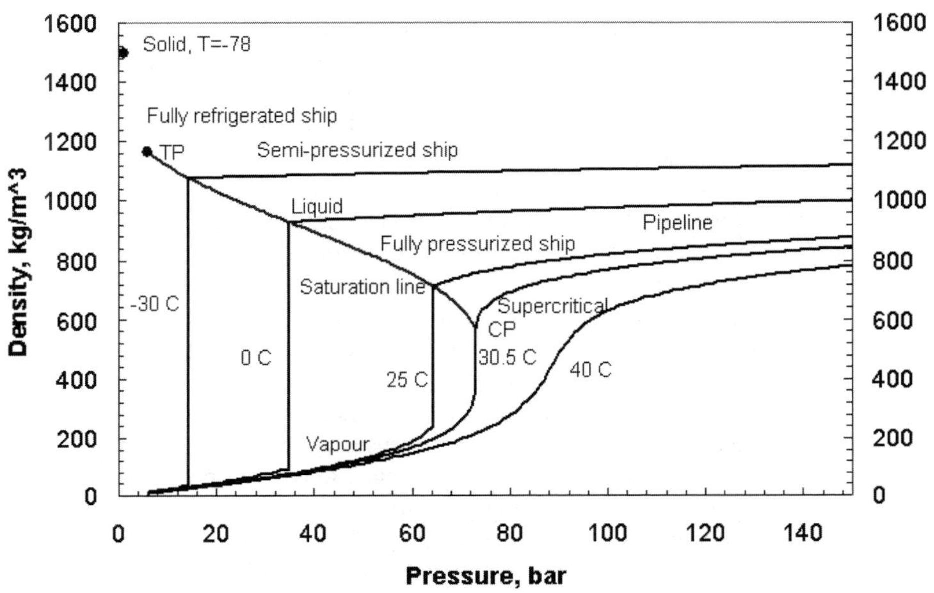

Figure 1-26. Optimal pure CO_2 transportation pressure for shipping in relation to phase diagram (adapted from Aspelund et al [39])

Figure 1-27. Typical CO_2 transportation vessel (courtesy of Anthony Veder http://www.co2-shipping.com/)

- **Liquefaction plant (with Liquefaction train):** Liquefying CO_2 involves using a sub-critical liquefaction process, utilizing an external refrigerant operating on an economized condensing vapor compression cycle, with multiple evaporating temperatures. Since usually the carbon capture plant is designed to turn down with the daily demand cycle of the power plant, two or multi-train approach is usually considered for this.
- **Refrigeration cycle:** the refrigeration cycle can be a conventional, economized vapor compression cycle condensing against cooling water. Two or more levels of refrigeration are generally utilized.
- **Storage:** the CO_2 is stored in cryogenic storage bullets. Total storage capacity depends on the logistical solution the shipping concept has to offer.
- **Loading:** in most cases, a single liquid loading arm is used due to the relatively low loading rate which is necessary for a smaller ship. The ship's vapor is returned to the liquefaction plant via the piggyback arrangement on the vapor return arm.
- **Offloading:** CO_2 is usually discharged into a pipeline system or offshore, into a depleted gas field.

Ships have the advantage of introducing flexibility in the CO_2 value chain, allowing collection of concentrated CO_2 from various sources at volumes below the critical size for pipeline transportation. A disadvantage would be the requirement for liquefaction plants and intermediate storage facilities, thus adding cost. If an onshore CO_2 hub were to be introduced, intermediate storage facilities matching the incoming amounts of liquefied CO_2 would also be needed there.

1.7.1.2 Truck and Railway Liquefied CO_2 can be transported by road tank trucks (provided with trailers), or railway trucks stored in cryogenic vessels (Figure 1-28). These forms of transportation provide much lower transport capacities than pipelines or ships. Technically, the CO_2 transportation is a feasible option where CO_2 is transported as liquid, typically at a temperature of −20 to −30°C and at 1.7 to 20 MPa pressure.

The tanks have an inner vessels or "liquid container" which is surrounded and supported by an outer vessel or "vacuum jacket." The space between the two is filled with a natural material that provides insulation. The delivery system includes piping which carries CO_2 gas from the vessel through the vacuum jacket to the outside, controlled by gauges and

Figure 1-28. Typical CO_2 tank truck trailer and railcar for oil field applications [42, 43]

valves mounted outside the tank. These vessels are available in various sizes ranging from 2 to 30 tons to suit customer's requirements.

In context of long distance and large amounts of CO_2 transportation for CCS, these options are unattractive options and uneconomical in comparison to other options like ship or pipeline.

Only on a very small scale or where flexibility plays an important role these options may be an alternative [44].

1.7.1.3 Pipelines Pipelines for the transportation of pure CO_2 have been used since the early 1970s in the enhanced oil recovery (EOR) industry. The first CO_2 pipeline (354 km) was completed and put into operation in 1972 by the Canyon Reef Carriers (CRC) for transportation of naturally occurring CO_2 to the SACROC oilfield in Texas, USA. The largest existing CO_2 pipeline is the 808 km (API 5LX-65 carbon steel, NPS 30 (762 mm OD)) Cortez pipeline transporting naturally occurring CO_2 from Cortez, Colorado to Denver City, Texas. It delivers 20 Mt of CO_2 per year. Other examples are the 648 km Sheep Mountain pipeline and 38 km Bravo pipeline in the USA [2]. *See Chapter 2, Section 2.5 for the list and the extent of Major CO_2 pipelines.*

However, the pipeline that transports CO_2 containing impurities is the 330 km (NPS 14 (14")) Weyburn pipeline which transports CO_2 captured from flue stacks of the Great Plains Synfuels Plant (GPSP) near Beulah, North Dakota, USA to EnCana's oil field in Weyburn, Saskatchewan, Canada (Figure 1-29). The pressure of the CO_2 leaving the plant is 17 MPa (2500 psig) and the delivery pressure at Weyburn is 14.8 MPa (2175 psig).

The Weyburn CO_2 Project is an integrated CO_2 monitoring and EOR project. Injection of CO_2 is into a carbonate oil reservoir in southeastern Saskatchewan, which began on September 22, 2000.

Emission from various sources (i.e., flue stacks of power generation stations, brick kilns, cement plants, etc.) will have different range of impurities. Additionally, range of impurities from the four groups of capture and processing technologies (previously shown in Figure 1-16 and Table 1-1) can generate CO_2 stream with different levels of impurities. Source(s) of CO_2 and market(s) for the CO_2 need to be known so that compatibility can be established

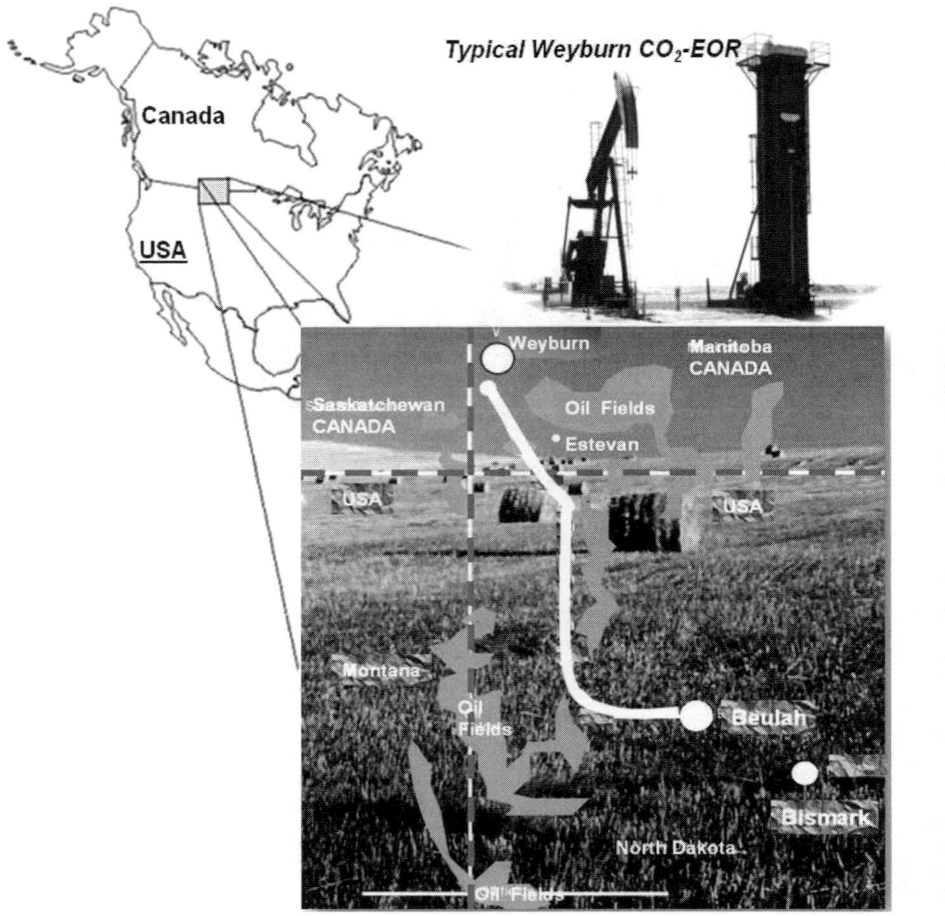

Figure 1-29. Weyburn pipeline

(considering the capture economy for each capture technology) and product specification developed for an appropriate pipeline transportation specification and system. Thus, recommendations on CO_2 quality for transportation purposes must be considered not only from emission source and economics of capture technology but also in close connection with the limits set from an end-use perspective, EOR versus pure storage (sequestration).

Also, from pipeline transportation points of view, limits on impurities relate to efficiency of transportation as high level of lighter component impurities, such as H_2, N_2, Ar, etc. reduces available transport capacity.

Typical pipeline systems covering CO_2 capture from a plant through to separation, processing, including dehydration, compression, metering, pipeline transportation, and injection system are conceptually shown in Figure 1-30.

Issues that generally relate to the pipeline transportation of CO_2 with impurities include the following [45]:

- Effect of impurities on phase diagram of CO_2 and equations of state impacting pipeline design and pipe material fracture control properties;
- Use of existing pipeline infrastructure—pressures, integrity, availability;
- Limited experience for operation of impure CO_2 pipelines;
- No experience in transporting CO_2 offshore long distance;
- Flow assurance—risk of hydrate formation and two-phase flow;

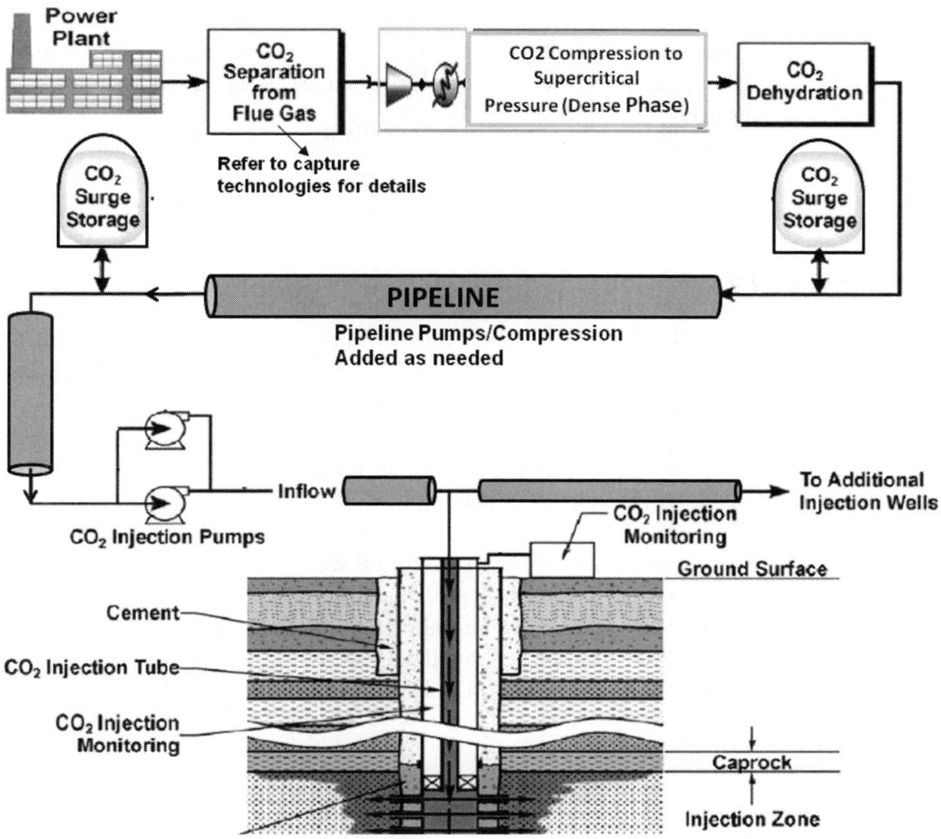

Figure 1-30. Typical pipeline and sequestration systems for transportation and injection of CO_2 containing impurities (adapted from Process Energy Solutions [46])

- Risk of routing through high population areas;
- Risk of long-running brittle fractures (due to cooling effects around leaks) and long-running ductile fractures (due to phase changes during depressurisation).

1.8 PIPELINE TECHNOLOGY FOR TRANSPORTATION OF CO_2 WITH IMPURITIES

CO_2 pipelines are needed for both [47]:

- Geological sequestering of CO_2 and for compliance and/or marketable credits/offsets of greenhouse gas,
- Carbon capture storage (CCS) can readily optimize values from both markets.

While the largest network of gas and hazardous liquid pipelines are located in North American continent, so is the initiation of first CO_2 pipelines for enhaced oil recovery as well as carbon capture of anthropogenic CO_2 for sequestration purposes.

Figure 1-31 is an illustration of the extent network of gas and hazardous liquid pipelines in USA.

For oil reservoir maintenance and oil recoveries from depleted fields, the practice of gas as well as air, nitrogen injection and water flooding preceeds that of CO_2 injection. The natural gas re-injection also helps to reduce flaring quite common in the early gas and oil production and processing. However, it is the depletion of the oil reservoirs that has intiated the injection processes and the CO_2 technology development. For example, air injection is a technique to enhance oil recovery that surged in popularity in the second half of the 1980s.

1.8.1 CO_2 Pipeline Transportation Chronology

The following is the chronology of major events that has lead to CO_2 sequestration and pipeline transportation development. Major events are the 1967 clean air act and the creation of Environmental Protection Agency (EPA).

- **347 CE** Earliest known oil wells drilled in China (use bamboo poles);
- **1594** Oil wells hand dug at Baku, Persia, up to 115 ft [35 m deep];
- **1859** Colonel Drake Strikes Oil;
- **1862** Use of iron pipe for oil wells;
- **1865** The first oil pipeline in the USA;
- **1879** The First Trunkline in USA, Tidewater;
- **1905** Crude Oil Pipelines;
- **1917** Introduction of Enhanced Oil Recovery [48];

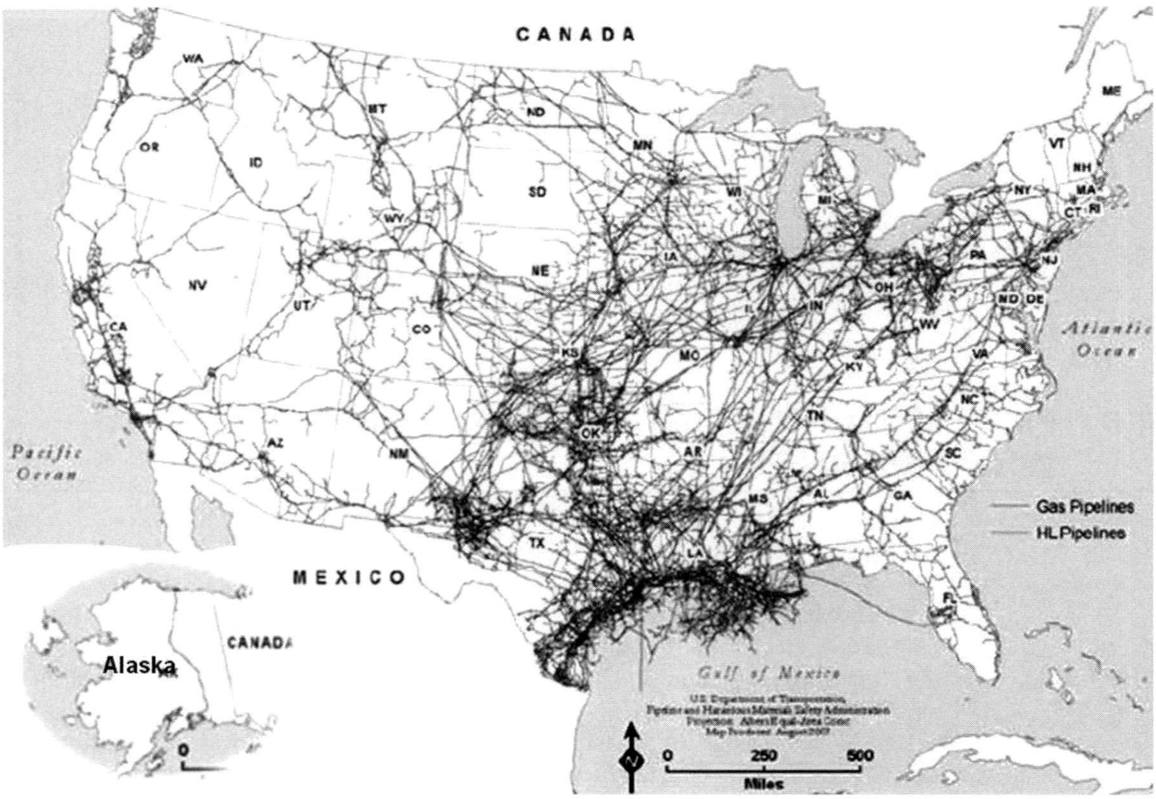

Figure 1-31. Network of gas and hazardous liquid pipelines, USA (Aug 2008)

- **1926** Discovery of the Yates field USA Oct;
- **1947** Drilling of first offshore oil well;
- **1952** 1st EOR Patent [5];
- **1964** Field test at the Mead Strawn Field (injection of a large slug of CO_2 (25% of the hydrocarbon pore volume or HCPV);
- **1967** The U.S. Congress passes the Clean Air Act;
- **1970** Creation of the U.S. Environmental Protection Agency;
- **1970-1977** The Trans Alaska Pipeline System (TAPS);
- **1970s-1990s** The Advent of Specialty Pipes (High grade steel);
- **1970s** Low-Volume Geologic Storage of CO_2 (Onshore) for EOR;
- **1970s** Removal of CO_2 from Flue gas of Power Plants;
- **1972** 1st commercial EOR Injection at SACROC (Scurry Area Canyon Reef Operators Committee) Unit of the Kelly-Snyder Field in Scurry County, West Texas;
- **1972** 1st CO_2 Pipeline, Canyon Reef Carriers (225 km, 5.2 $MtCO_2$/year Capacity);
- **1972** The North Cross (Devonian) Unit CO_2 flood by Shell (US First major CO_2 flood, (Scurry County, Texas);
- **1973** Worldwide energy crisis;
- **1976** 1st CO_2 injection early June;
- **1977** Creation of US DOT;
- **1979** 1st Detection of Corrosion Due to Use of Carbon Dioxide for Enhanced Oil Recovery;
- **Mid1980-90** Pioneering application of air injection in oil fields the United States;
- **1982-1983** Cortez CO_2 Pipeline Excavation [49];
- **1983** Bati Raman CO_2 Pipeline, Turkey (90 km, 1.15 $MtCO_2$/year Capacity);
- **1984** Bravo CO_2 Pipeline (350 km 7.3 $MtCO_2$/year Capacity);
- **1985** 1st Reservoir re-pressurization with water injection Sept;
- **1986** Cortez CO_2 Pipeline Operation (808 km,19.3 $MtCO_2$/year Capacity);
- **1988** Sheep Mountain CO_2 Pipeline (660 km,9.5 $MtCO_2$/year Capacity);
- **1998** Val Verde CO_2 Pipeline (130 km, 2.5 5 $MtCO_2$/year Capacity);
- **1989** Acid Gas Injection;
- **1996** First Offshore Saline Aquifer Injection (Statoil);
- **2000** Weyburn CO_2 Pipeline (USA/Canada), 328 km (5.05 $MtCO_2$/year Capacity);
- **2009** Delhi CO_2–EOR Project (Louisiana) with 128 km, CO_2 Pipeline;

1.8.2 Summary Status of CO_2 Pipeline Technology

- CO_2 pipelining is mature and has been in existence since the early 1970s, mostly in the USA. However, over 95% of the pipelines carry CO_2 from naturally occurring sources (Table 1-2).
- Despite the existence of over 7000 km of CO_2 pipelines, only the 340 km of the Weyburn–Mildale pipeline system (extending between North Dakota and southern Saskatchewan) carries CO_2 with significant impurities, refer to Section 1.9, Table 1-5. CO_2 is captured from a coal gasification flue stack and transported for EOR.
- The review of current CO_2 pipeline technology indicates that there are no real insurmountable obstacles for pipeline transportation of CO_2.
- The safety risks of transporting CO_2 stem mainly from its highly concentrated state during transport, and the potential for leakage. Compressed CO_2 is heavier than air, which causes it to sink near the ground in low lying areas, displacing surrounding oxygen. In a highly concentrated state, CO_2 can cause respiratory problems or suffocation, and harm to the surrounding environment.

- Further specific technical aspects result from the fact that CO_2, when transporting in supercritical or liquid state, is cooling down to about −78°C (when depressurized) and may result in either brittle or ductile long running fractures which must be prevented by suitable material/wall thickness specification or fracture arrestors.
- Thus, the arrest of propagating ductile fractures is an important criterion that needs to be considered when designing CO_2 pipelines. Steel toughness values may be much higher than for pipes used for natural gas service due to much higher vapor pressure. This is one of the major obstacles for the conversion of existing natural gas pipelines to CO_2.
- Design of equipment such as pumps and compressors for CO_2 service requires specific attention. Suction pressure and compression/pumping intervals need to be strategized to avoid two-phase flow regions in CO_2 pipeline transmission.
- From a safety point of view, CO_2 and impurities concentrations determine the safe exposure limits for the CO_2 stream and thus pipeline design. Design must be such that excessive concentrations of CO_2 gas and any toxic impurities, such as H_2S, can be avoided during any unforeseen rupture or scheduled event such as blow-down. Potential issues during blow-down are the following:
    - Cooling of the line pipe material below the design temperature;
    - Dry ice and hydrate formation; and
    - Safety: dispersion of CO_2 cloud and pipeline integrity.
- Incidents related to CO_2 operations are rare. No injuries or fatalities associated with operations have been reported for CO_2 pipelines. The only incident of extremely significant importance relates to CO_2 release from a non-pipeline source.

1.8.3 The Effect of Impurities on Pipeline Transportation

The industry has been facing increasing technological challenges to access and develop new gas fields resources globally. This is because major natural gas sources have generally the following challenges:

- High CO_2 content (> 30% Asia),
- High H_2S content (Middle-East and Russia).

The above in turn has implied a growing part in atmospheric emissions and thus in the specific management of H_2S and CO_2.

TABLE 1-2. Naturally occurring CO_2 stream composition from major dome fields in the USA — after processing [30]

Field	CO_2	CH_4	N_2	H_2S	C_2+	CO	O_2	NO_x	SO_x	H_2	Ar	H_2O_2
Canyon Reef Carriers	85–98%	2–15% $+C_6H_{14}$	<0.5%	<200 ppm	–	–	–	–	–	–	–	50 ppm wt
Central Basin Pipeline	98.5%	0.2%	1.3%	<20 ppm spec	–	–	<10 ppm (spec)	–	–	–	–	240 ppm wt
Sheep Mountain	96.8–97.4%	1.7%	0.6–0.9%	–	0.3–0.6%	–	–	–	–	–	–	120 ppm wt
Bravo Dome Source	99.7%	–	0.3%	–	–	–	–	–	–	–	–	–
Cortez Pipeline	95%	1–5	4%	0.002%	trace	–	–	–	–	–	–	240 ppm wt
Jackson Dome	98.7–99.4	trace	trace	trace	–	–	–	–	–	–	–	–

Carbon dioxide stream purity is an important element in developing the infrastructure for large-scale CO_2 sequestration from capture, processing, transportation, and storage. The properties of CO_2, which in turn affect CO_2 mobility, buoyancy, and reservoir sequestration capacity, are strongly dependent on the purity and quality of the CO_2 stream (i.e., the amount and type of other gases present in the CO_2 stream).

Impurities in the CO_2 product stream affect many aspects of CO_2 pipeline transportation. This is because the presence of impurities affects the following:

- The physical and transport properties of CO_2;
- Pipeline route;
- Pipeline hydraulics and thus capacity (number of compressors, compressor power and temperature transients, etc);
- Pipe fracture propagation;
- Corrosion and pipeline protection;
- Operational risk assessment.

The impurities change the solubility characteristics of water in CO_2 stream; non-metallic component deterioration; the formation of hydrates and clathrates and, also the pipeline capacity itself [2].

CO_2 (pure or impure) can collect in topographic low points. The CO_2 stream containing beyond 20% is life threatening. Hydrogen sulphide (H_2S) and/or sulphur dioxide (SO_2) may also be impurities in higher quantities in the CO_2 stream captured from power generation and can be life threatening in certain concentration as well. In case of free water CO_2 and these impurities are highly corrosive.

Impurities, such as SO_x, NO_x, and argon in the CO_2, have not been transported before in a CO_2 pipeline, and thus their effects are not fully evaluated. As well, there is little published work on transportation of CO_2 with these impurities. The impurities change the design condition and the operating regime of the pipeline, which typically have to be operated at a higher pressure than what would be required for pure CO_2 in order to maintain it as single-phase supercritical or dense phase.

1.8.4 CO_2 Pipeline Codes and Regulations

1.8.4.1 Codes With the planning and implementation of CO_2 capture and storage, cross-country pipelines are required to connect CO_2 "production" and the storage location, often several hundred kilometres of length.

The worldwide (but specifically USA, Canada, and Australia) broad experience with CO_2 transportation has provided inputs into codes and standards for CO_2 pipeline design, construction, and operation.

Internationally applied norms and regulations for pipeline design, construction, and operation are the following:

- **ISO 13623.** Petroleum and natural gas industries—pipeline transportation systems (this norm correlates with DIN EN 14161).
- **ASME B31.4.** Pipeline transportation systems for liquid hydrocarbons and other liquids.
- **ASME B31.8.** Gas transmission and distribution piping systems.

Other codes are the following:

- **UK Health and Safety Executive** directives [50];
- **CSA Z662**, Canadian Code: Oil and Gas Pipeline Systems [51];

- **AS 2885-2:2008,** Australian, Pipelines—gas and liquid petroleum;
- **DNV-RP-J202, 2010** [52] Design and Operation of CO_2 Pipelines;
- **Institute of Petroleum** (UK), Pipeline Code IP6;
- **BS EN 14161, 2003**: Petroleum and natural gas industries—Pipeline transportation systems);
- **BS PD 8010 2004**: Code of practice for pipelines. Subsea pipelines;
- **DNV OS-F101:** Submarine Pipeline Systems.

The above-mentioned codes are all applicable to pipelines including those transporting CO_2, some, however, categorizing CO_2 as a non-flammable, non-toxic fluid which is gaseous at ambient temperature and pressure. IP6 also treats it as a gas. Most standards refer to associated standards (Figure 1-32), however, none of these associated standards are specific to CO_2 pipeline transportation.

Existing CO_2 pipeline facilities have been designed to meet the current above-mentioned gas and/or oil pipeline systems codes and standards. However, most of the items mentioned in the above standards are applicable also to CO_2 pipelines, while some of standards provide specifics of CO_2 requiring a particular consideration. For example, AS 2885 applies to high-pressure steel pipelines that are used to transport single and multiphase hydrocarbon fluids. The standard is a risk-based standard, principles of which apply to CO_2.

Almost all codes require additional safety and design information. These also require design aspects to be covered by a supplementary regulative guidance [53]. This includes the following:

- Route selection (particular topography and populated areas);
- Pipe/material strength/toughness requirements (fracture control);

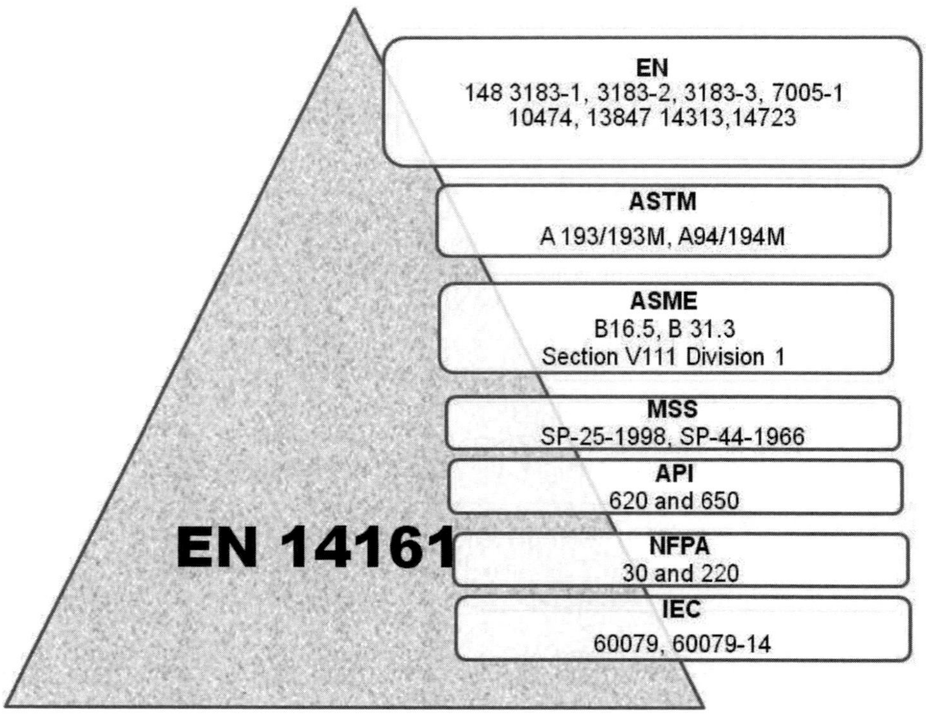

Figure 1-32. Example of European Standard EN 14161 and Associated Standards (from CO_2 Europipe [54])

- Section of isolation valves (spacing of valve stations);
- Emergency shutdown systems (de-pressurization, venting);
- Corrosion management (drying procedures, absolute avoidance of free water);
- Risk assessment (among others dispersion models to be developed).

The standards are extensive for pipeline in general and no single CO_2 pipeline standards exist. However, the recent recommended practice (RP) by DNV [52] addresses some of the major gaps and currently is striving towards a multitude of experimentation to verify fracture control properties of high-strength carbon steel pipe in high-pressure CO_2 transportation environment [55]. As indicated previously, while CO_2 is a substance with well-known characteristics, the same cannot be said for pure CO_2-containing impurities.

DNV-RP-J202 [52] has however been developed to address the need for guidance for how to manage risks and uncertainties specifically related to transportation of CO_2 in pipelines.

1.8.4.2 Regulation

A. North American Regulatory Oversight

In North America, existing CO_2 pipeline facilities have been designed to meet current gas and/or oil pipeline systems codes and standards. No specialty fluid transmission code is applicable or available. CO_2 Pipelines were unregulated until 1986. Generally, ASME/ANSI B31.4 and B31.8 are applicable as appropriate. In Canada, CSA Z 662 covers CO_2 pipelines. However depending on H_2S content, the design and safety issues are addressed appropriately.

In the US, CO_2 pipelines are regulated by US Department of Transportation (DOT)—Code of Federal Regulation (CFR) Section 195 Liquids Pipelines (for the transportation of CO_2 in Liquid Form) [56]. Under US DOT CFR 195, CO_2 is regulated as a "hazardous material and carbon dioxide."

In Canada, NATURAL RESOURCES CODE, Chapter 117 "Hazardous Liquids or Carbon Dioxide Pipeline Transportation Industry," 2005 as well CSA Z662, 2011 applies.

CO_2 pipelines are considered High Volatile/Low Hazard and Low Risk facility. However, US Department of Transportation (DOT) consistently has adopted the language of "hazardous materials and carbon dioxide." This means that a higher level of inspection is required for CO_2 pipelines than for crude oil pipelines.

Regulations specifically call for 26 inspections (generally monitoring of pipeline right of way—ROW) per year. CO_2 is classified as a Class "L" material—highly volatile, non-flammable/non-toxic. No judgment is made as to the CO_2 safety risk.

There is a specific call to mitigate fracture propagation with a fracture arrestor. Generally, fracture arrestors are specified where materials do not have sufficient toughness to arrest running ductile fracture. Valve material compatibility in CO_2 service is also a requirement.

Because CO_2 is being treated as hazardous, but not declared hazardous, all the review requirements for high-risk hazardous pipelines apply if the pipeline is greater than 457 mm (NPS18) or passes through a populated area. The "populated area" means a density greater than 1000/square mile (2630/km^2).

However, from a Canadian regulatory perspective the following are applicable:

- All CO_2 pipeline applications are subject to a mandatory review by regulatory bodies to ensure technical completeness and operational safety
- Since CO_2 is non-toxic—no specific requirements for setbacks, emergency response planning (ERP), or leak detection are required
- Because it is heavier than air and tends to accumulate in low areas (ventilation is poor), Pipeline Operators are required to be trained in safe working procedures

in oxygen (O_2)-deficient atmospheres as well as in the handing of fluids that undergo phase change under pressure (asphyxiation occurs in concentrations greater than 5%).

- Internal Inspections Requirements (every 5 years):
  - Smart Pig (Difficult in CO_2 P/L),
  - Direct Assessment.

From a governmental policy point of views, specific legislative proposals are being reviewed by legislative assemblies that reflect the current perception that CO_2 capture probably represents the largest technological hurdle to implementing widespread CCS, and that CO_2 transportation by pipelines does not present as significant a barrier. While these perceptions may be accurate, industry and regulatory bodies are identifying important policy issues related specifically to CO_2 pipelines which may require government attentions [57].

Pipelines that transport carbon dioxide are generally required to conform to two sets of U.S. regulations as listed below (US DOTCFR, 2010):

- Transportation of Hazardous Liquids by Pipeline:
  - U.S. Department of Transportation (DOT) Code 49 CFR195
- Transportation of Natural and Other Gas by Pipeline:
  - Minimum Federal Safety Standards U.S. DOT Code 49 CFR 192

However, in the USA, pipeline safety regulations as applicable by each State of the federation also come into play.

In the US, CO_2 pipelines are regulated by US Department of Transportation (DOT)—Code of Federal Regulation (CFR) Section 195 Liquids Pipelines (for the transportation of CO_2 in Liquid Form). Under US DOT CFR 195 CO_2 is regulated as a "hazardous material and carbon dioxide."

Although CO_2 pipelines are classified as hazardous, CO_2 is not defined as a hazardous substance. It is a Class L, highly volatile, non-flammable/non-toxic material (CFRg, CFRe, Appendix B, Table 1.4).

CO_2 pipelines are treated as hazardous and are reviewed as high-risk hazardous pipelines when they have a diameter greater than 457 mm (NPS 18, 18 in) or when they pass through High-Consequence Areas.

States certified to regulate intrastate pipelines are as follows:

- Alabama,
- Arizona,
- California,
- Louisiana,
- Maryland,
- Minnesota,
- Mississippi,
- New York,
- Oklahoma,
- New Mexico,
- Texas,
- Virginia,
- Washington, and
- West Virginia

Some states have their own regulations usually emulating CFR 49 part 195. Example includes: Texas Natural Resources Code CHAPTER 117. HAZARDOUS LIQUID OR CARBON DIOXIDE PIPELINE TRANSPORTATION INDUSTRY—2005.

49 CFR § 195.2 defines low-stress pipeline as a hazardous liquid pipeline that is operated in its entirety at a stress level of 20% or less of the specified minimum-yield strength of the pipeline (CFRf). 49 CFR § 195.2 defines rural area as an area outside the limits of any incorporated or unincorporated city, town, village, or any other designated residential or commercial area, such as a subdivision, a business or shopping center, or community development. The rural areas are considered to be the non-environmentally sensitive areas.

An easement is a limited perpetual interest in land that allows the pipeline owner to construct, operate, and maintain a pipeline across the land. An easement does not grant an unlimited entitlement to use the right of way. The rights of the easement owner are set out in the easement agreement.

Eminent domain is the power of government to take private land for public use. Under current law, there is no federal eminent domain power granted for the construction of CO_2 pipelines. A number of states, however, do allow the use of eminent domain for CO_2 pipeline construction under certain conditions.

B. European and Other Regulations

Generally, regulations around pipelines are well established as are the design, construction, and operation codes. However, none of the regulations address carbon dioxide as a specific named substance, but it is considered as a "dangerous fluid" or "dangerous substance for the purpose of public safety and environmental considerations."

As indicated in Figure 1-32, the European standard EN 14161 "Petroleum and Natural Gas Industries — Pipeline Transportation Systems" and associated standards are accepted throughout Europe. The associated standards are used to complement the core code EN 14161 as applicable.

1.9 CO_2 PROPERTIES AND PHASE DIAGRAM

1.9.1 Pure and Impure Carbon Dioxide (CO_2)

Pure CO_2 is a chemical compound composed of two oxygen atoms covalently bonded to a single carbon atom. Pure carbon dioxide (with purity of 90% to almost 100%) is from underground reservoirs throughout the world. However, impure CO_2 is generally from combustion processes that are used in power plants to produce energy by the reaction of a hydrocarbon with oxygen. This reaction results in the production of a by-product flue gas stream consisting of nitrogen and carbon dioxide. Since the hydrocarbon fuel is not commonly purified prior to combustion, the carbon dioxide product of this process can contain an unusual range of impurities (e.g., mercury, organochlorides, and HCN).

While there are quite a large experience on the design and operation of pure CO_2, i.e., CO_2 from natural underground sources which makes up some 95% to 96% world CO_2 pipelines (some 7400KM), there is limited experience on pipeline transportation of CO_2 with impurities. Besides, such an experience is further limited to a small range of impurities, and only single pipelines are involved. This is further explained below.

"In the US, naturally occurring CO_2 is routinely transported for considerable distances overland, although mostly through sparsely-populated regions, for the purpose of enhanced oil recovery (EOR). There is also some limited transport of captured, or 'anthropogenic' now the subject of much attention for capture, transportation, and sequestration collectively known as CCS."

1.9.2 Physical Conditions and Properties of Pure CO_2

To understand the issues relating to the transport of CO_2, it is necessary to be aware of its physical conditions. In brief, pure CO_2 is a colorless, odorless, and non-flammable substance which can be transported as a solid, liquid, gas, or dense-phase liquid. The most efficient state of CO_2 for pipeline transport is as a dense-phase liquid or supercritical gas. In the latter condition, the fluid has the density of a liquid, but the viscosity of a gas. In this "supercritical" mode, captured CO_2 has to be compressed to a pressure above the critical pressure prior to transport (Figure 1-33).

Properties of pure CO_2 are indicated in Tables 1-3 and 1-4 below.

Carbon dioxide has no liquid state at pressures below 5.11 atmospheres. At 1 atmosphere (near mean sea level pressure) and at temperatures below −78°C (−108°F; 195.1 K), CO_2 gas deposits directly to a solid form. This solid sublimes directly to a gas above −78°C. In its solid state, carbon dioxide is commonly called dry ice.

Figure 1-34 shows the density and viscosity of CO_2 as the pressures and temperatures vary [58]. These properties are indicated for typical pipeline operation.

Above the critical pressure of 7.38 MPa and at temperatures lower than 20°C, CO_2 would have a density between 800 and 1200 kg/m^3 (compare this to the density of water which is 1000 kg/m^3). A higher density is favorable when transporting liquid CO_2, as it is easier to move a dense liquid than a gas. Therefore, it is typical to compress CO_2 to above 7.38 MPa for efficient transport. A detailed Pressure-Enthalpy diagram for pure CO_2 is shown in Figure 1-35.

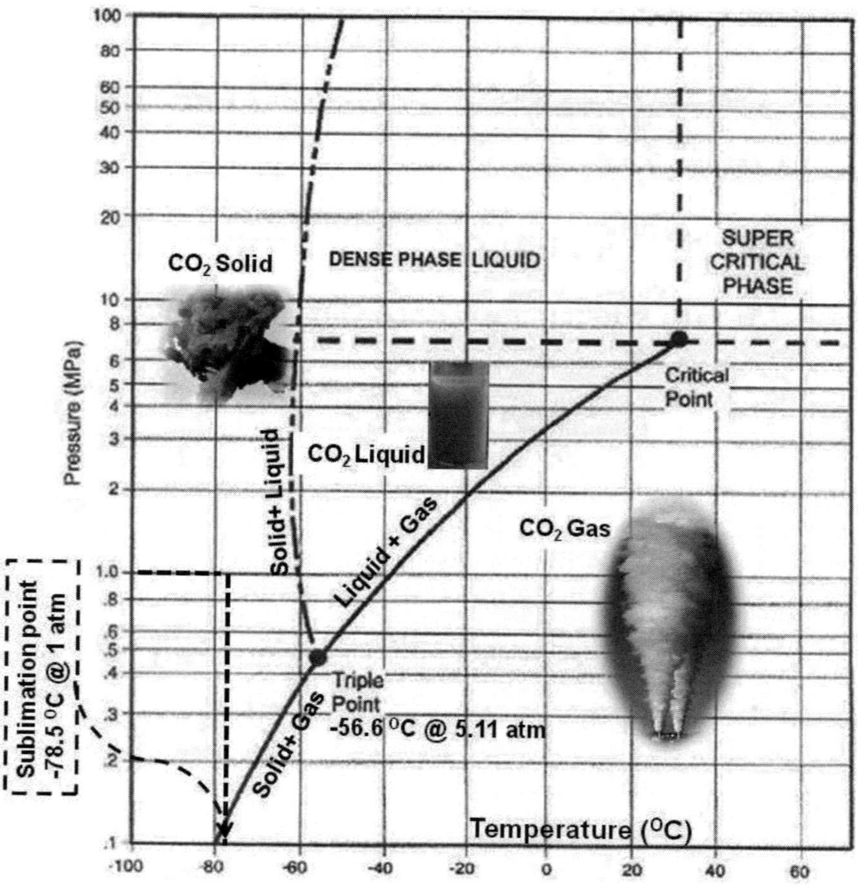

Figure 1-33. CO_2 Phase Conditions

TABLE 1-3. Pure CO_2 Properties at 0°C and 1.013 bar (101.325 kPa), Source: modified from [44]

	Molecular weight	44.01
	Critical temperature	31.1°C
	Critical pressure	73.9 bar
	Critical density	467 kg/m$_3$
	Triple point temperature	–56.5°C
	Triple point pressure	5.18 bar
	Boiling (sublimation) point (1.013 bar)	–78.5°C
Solid Phase		
	Density of carbon dioxide snow at freezing point	1562 kg/m^{-3}
	Latent heat of vaporization (1.013 bar at sublimation point)	571.1 kJ kg^{-1}
Liquid Phase		
	Vapour pressure at 20°C	56.5 bar
	Liquid density at –20°C and 19.7 bar	1032 kg/m^{-3}
	Viscosity at STP	99 μN·s/m^2
Gas Phase		
	Gas density	2.814 kg/m^3
	Gas density at STP*	1.976 kg/m^3
	Specific volume	0.506 m^3/kg
	Heat capacity Cv (constant volume) at STP	0.0364 kJ (mol^{-1} K^{-1})
	Heat capacity Cp (constant pressure) at STP	0.0278 kJ (mol^{-1} K^{-1})
	Viscosity at STP	13.72 μN·s/m^2
	Thermal conductivity at STP	14.65 mW (m K^{-1})
	Solubility in water at STP	1.716 vol vol^{-1}
	Enthalpy at STP	21.34 kJ mol^{-1}
	Entropy at STP	117.2 J mol K^{-1}
	Entropy of formation	213.8 J mol K^{-1}

1.9.3 Anthropogenic/Impure CO_2

While the following provide a summary of compositional anthropogenic CO_2, Chapter 3 will provide complete details and compositional scenarios.

1.9.3.1 Compositional Make Up CO_2 that is captured from power plant and other anthropogenic sources is not pure, and the amount and type of impurities are dependent on the nature of the capture process and the capture technology used. Depending on the source of the flue gas, the CO_2 stream recovered from it may contain the following components:

- Free Water (H_2O),
- Hydrogen Sulphide (H_2S),
- Carbon Monoxide (CO),
- Oxygen (O_2),
- Methane (CH_4),

TABLE 1-4. Vapor pressures (boiling points) for CO_2 [59]

Temperature, °C	Vapor pressure atm	Temperature, °C	Vapor pressure atm
–56.6 (triple pt)	5.11	–10	26.13
–55	5.48	–5	30.05
–50	6.74	0	34.38
–45	8.22	5	39.16
–40	9.93	10	44.41
–35	11.88	15	50.17
–30	14.10	20	56.50
–25	16.61	25	63.45
–20	19.44	30	71.14
–15	22.60	31.1 (crit pt)	72.95

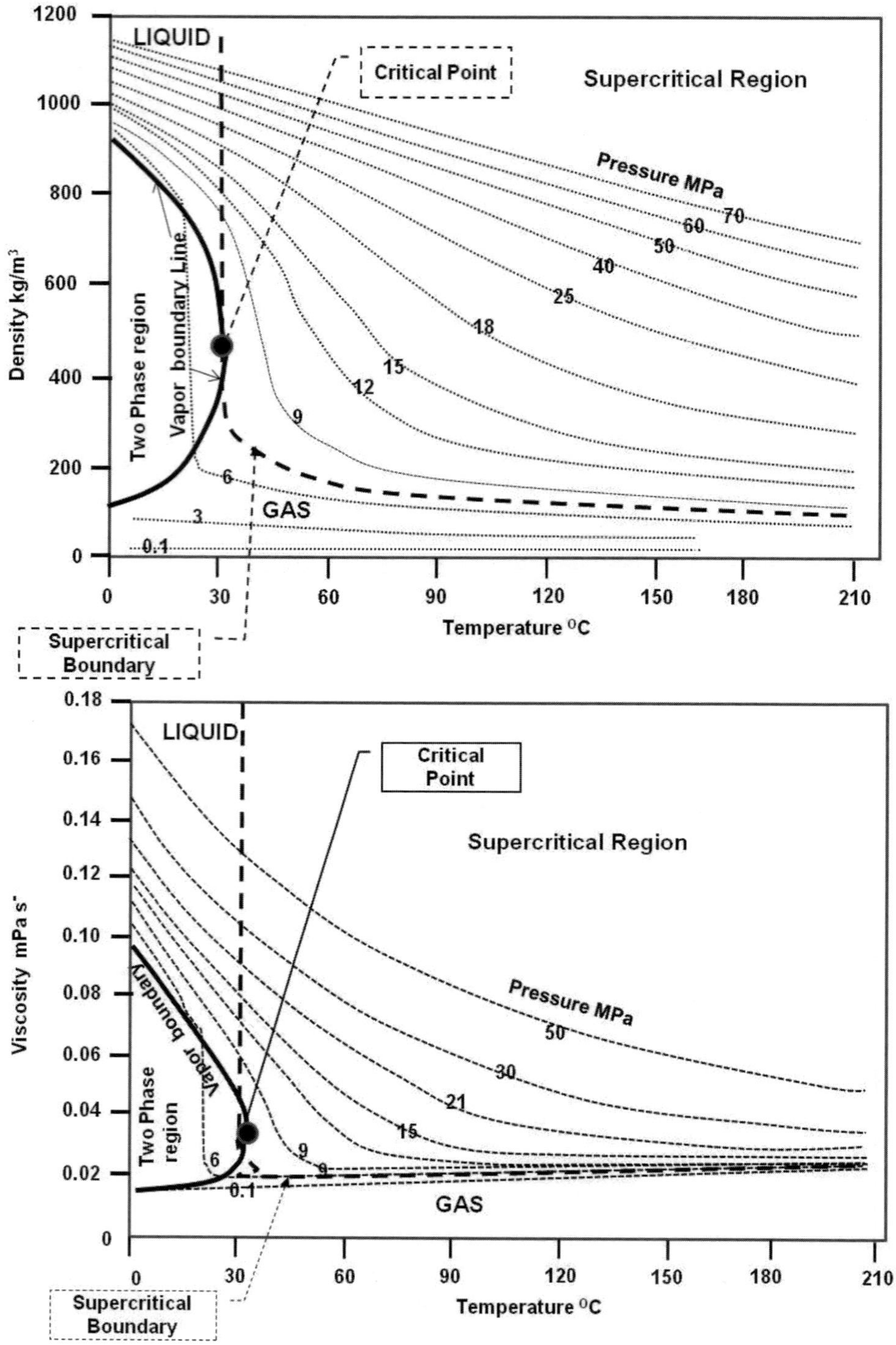

Figure 1-34. Density and viscosity of pure CO_2 [58]

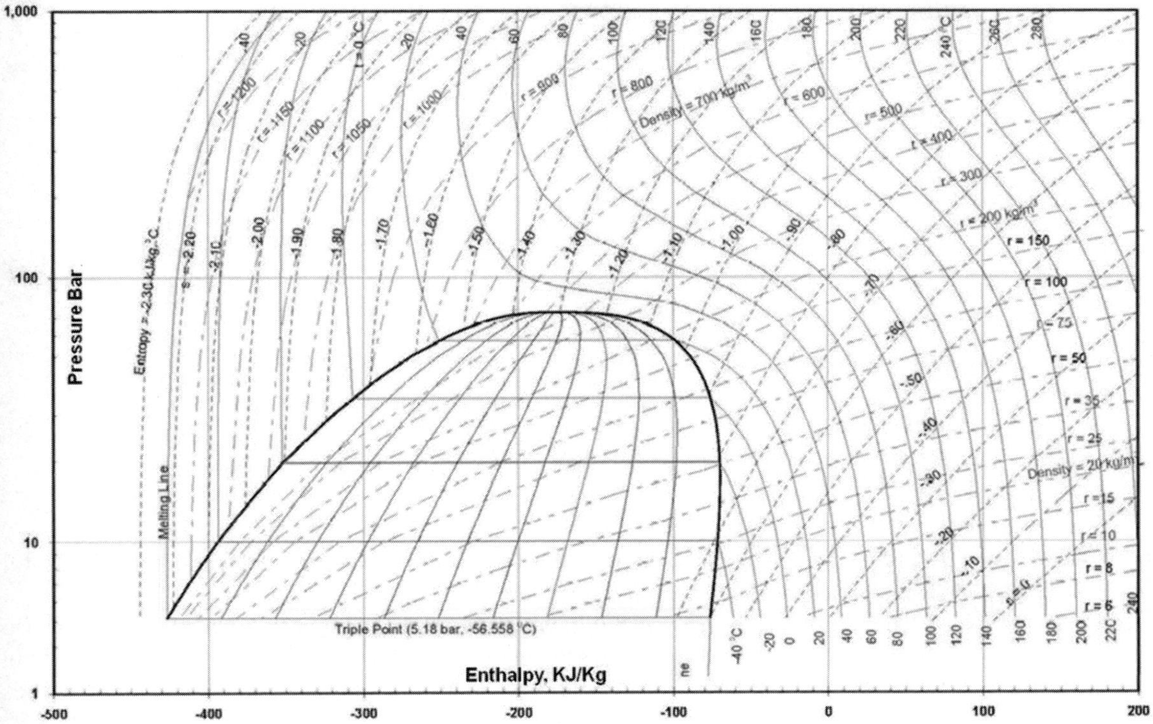

Figure 1-35. Pressure–enthalpy diagram for pure CO_2 (Courtesy of chemicaLogic Corporation www.chemicaLogic.com)

- Nitrogen (N_2),
- Ammonia (NH_3),
- Argon (Ar),
- Hydrogen (H_2),
- SO_x,
- NO_x,
- CO_2.

These impurities might have an impact on the physical state of the rich CO_2 stream and hence the operation of the facilities including compressors, pipelines, and storage tanks. Additionally, some impurities in the CO_2-rich gas composition, such as hydrogen or toxic hydrogen sulphide (H_2S), present a broad degree of technical, environmental, and safety challenges associated for pipeline transportation. Typical composition of anthropogenic captured CO_2 is depicted in Table 1-5.

CO_2 stream components level must thus indicate an acceptable concentration range from both safety and pipeline transportation as well as injection quality points of view.

As an example, limits for H_2S are set by safety considerations, rather than by technical limits. The carbon monoxide (CO) level is set to assure safe transportation of CO_2 by pipeline. Oxygen (O_2) is not generally expected to be present in the CCS stream but a limit is set based on the underground effects of O_2 on EOR.

Levels of both SO_2 and NO_2 in CO_2 mixture are also limited from a health and safety perspective.

The total volume of non-condensable gases (N_2, H_2, CH_4, O_2, Ar) is set to least affect the EOR process. However, the amount of H_2 is limited as much as possible because of its high energy content and market value.

TABLE 1-5. Anthropogenic CO_2 properties from US sources [22]

Pipeline Name and Location	Dakota Gas (ND/SK)	Exxon (WY)
Components		
Carbon Dioxide	95% min	95% min
Water	< 100 ppm	< 30 lbs/MMSCF
Methane	< 0.5%	----------
Ethane	< 1%	---------
Propane	< 0.5%	Not available
Hydrocarbons	Not available	Not available
Hydrogen sulfide	< 2.0%	< 20 ppm
Oxygen	< 0.5%	< 10 ppm
Nitrogen	< 1%	< 4%
Mercaptans	< 250 ppm	--------
Total sulfur	-----------	< 35 ppm
Glycol	-----------	< 0.3 gal/MMscf
Temperature	-----------	< 120°F

While the effect of CH_4 on the solubility of water in CO_2 is significant, it is not harmful for transportation of CO_2 at some concentrations levels.

Thus, different impurities will have different impacts on the critical properties of a CO_2 mixture. Table 1-6 summarizes the individual effect that common impurities found in CO_2 produced from power plants and other industrial sites have on the critical properties of the mixture.

These recommendations on CO_2 quality for transportation need to be considered in close connection with the limits set from an end-use perspective, EOR versus pure sequestration.

The durability of the transport infrastructure is ensured by avoiding free water formation, hydrate formation, and corrosion. Limits for impurities relate also to efficient transport, as a high level of impurities reduces available transport capacity.

1.9.3.2 Impure CO_2 Phase Diagram While pure CO_2 may be transported in the liquid phase, CO_2 gas stream containing impurities resulting from industrial processes (gasification, combustion, petroleum processing, etc.) contains chemical impurities that make it impractical for it to be transported in liquid form. Depending on their level, the impurities can open up a two-phase zone in the phase diagram and change the critical point (Figure 1-36; [45]). Generally, the critical pressure increases and the critical temperature decreases

TABLE 1-6. Degree on impurities affecting critical properties in CO_2 stream

Component	Impact	Comments
CH_4	Moderate	Methane slightly increases compressibility of mixture. However a quantity of methane in the CO_2 stream can help to flare the mixture when blow-down is required.
C_2+	Minimal	Ethane and heavier hydrocarbons will not usually be an issue, as they are removed due to their value. They slightly decrease the critical temperature and pressure but have no significant impact on compressibility.
H_2S	Very Minimal	The quantity of hydrogen sulphide in the CO_2 mixture should be minimized because of its toxicity; however, the impact to critical properties is minimal. It has no significant impact on compressibility.
N_2	Moderate	Nitrogen has a moderate impact on the critical properties of CO_2 mixtures. Nitrogen will increase the critical pressure and decrease the critical temperature.
H_2	Major	Hydrogen increases the compressibility of a CO_2 mixture. As well, it significantly decreases the critical temperature and significantly increases the critical pressure. This can affect pipeline capacity.

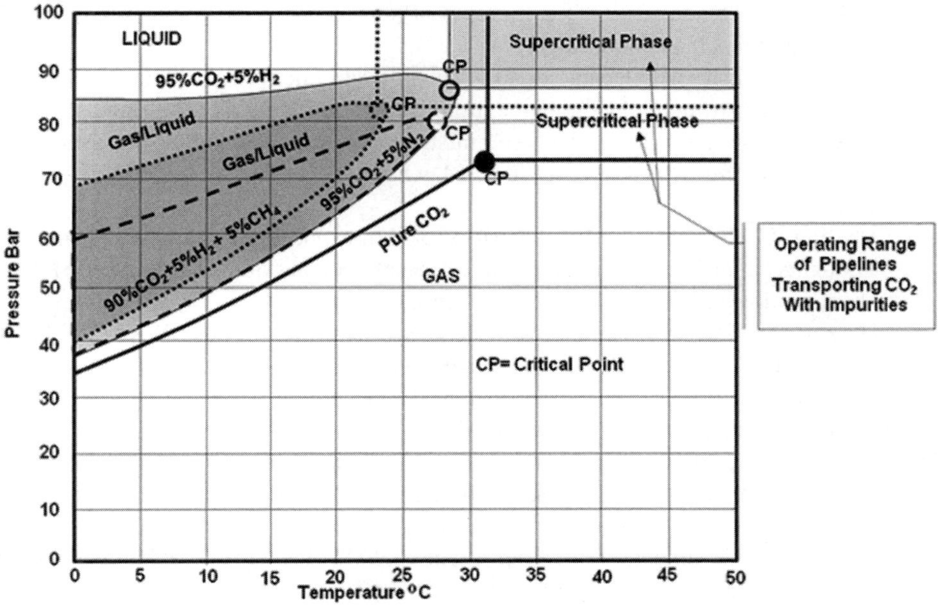

Figure 1-36. Influences of impurities on CO_2 phase diagram and pipeline operating range (from Race [45])

as impurity levels increase. This, in turn, affects the operating range of a pipeline and the way CO_2 is transported (i.e., liquid versus supercritical dense phase).

Vapor pressures for natural gas and CO_2 gas with impurities are widely different. Therefore, significant differences in fracture control requirements exist between the two systems. It is for these reasons that natural gas pipelines cannot be utilized on a turnkey basis for the transportation of CO_2.

REFERENCES

[1] LaForce, T., 2008 "Reducing Greenhouse Gas Emissions: Geological Storage of CO_2" *Nature Proceedings*, February, http://precedings.nature.com/users/bd76958c3217b5462541d0d5d37bacef.

[2] Mohitpour, M., Golshan, H. and Murray, A. 2007 "Pipeline Design and Construction—A Practical Approach," *3rd Edition*, ASME Press, New York.

[3] Maslin M., 2004 "Global Warming: A Very Short Introduction," Oxford University Press, paperback 2008, http://www.amazon.ca/Global-Warming-Very-Short-Introduction/dp/0199548242/ref=dp_ob_title_bk.

[4] Beck, E. G., 2007 "180 Years of atmospheric CO_2 gas analysis by chemical methods" Energy & Environment, Vol. 18 No. 2 2007.

[5] Whorton, L. P., Brownscombe, E. R., and Dyes, A. B., 1952, "Method for Producing Oil by Means of Carbon Dioxide," U.S. Patent 2,623,596.

[6] IPCC (The Intergovernmental Panel on Climate Change). 2007 "Report on the state of Earth's climate," http://www.wunderground.com/education/ipcc2007.asp.

[7] CO2Now.Org, 2010 "Atmospheric CO2 for September 2010," http://co2now.org/.

[8] Mohitpour, M., 2008 "Energy Supply and Pipeline Transportation – Challenges and Opportunities," ASME Press, New York.

[9] Senese, F., 2000 "Who discovered carbon?". Frostburg State University. http://antoine.frostburg.edu/chem/senese/101/inorganic/faq/discovery-of-carbon.shtml.

[10] UNEP Grid Arendal, 2010 "The present Carbon Cycle", Vital graphics, http://www.grida.no/publications/vg/climate/page/3066.aspx.

[11] NOAA (US National Oceanic & Atmospheric Administration), 2010 "Mauna Loa CO_2 annual mean data" US Department of Commerce, ftp://ftp.cmdl.noaa.gov/ccg/co2/trends/co2_mm_mlo.txt.

[12] EIA (US Energy Information Administration) 2008 "Greenhouse Gases, Climate Change, and Energy" Brochure #: DOE/EIA-X012, May http://www.eia.doe.gov/bookshelf/brochures/greenhouse/Chapter1.htm.

[13] Keeling, C. D., 1993 "A Brief History of Atmospheric CO_2 Measurements and Their Impact on Thoughts about Environmental Change; Speech: Winner of the Second Blue Planet Prize http://www.af-info.or.jp/eng/honor/bppcl_e/e1993keeling.txt.

[14] Herzog, T., 2009 "World Greenhouse Gas Emissions in 2005" WRI (World Resources Institute), report. http://pdf.wri.org/working_papers/world_greenhouse_gas_emissions_2005.pdf.

[15] McKnight, J., 1998 "Transportation in the 21st Century" I. Mech. Eng. Presidential Address, Latimer Trend & Company Ltd, Plymouth, England.

[16] World Bank, 2010 "World Development Indicators," http://www.google.com/publicdata?ds=wb-wdi&met=en_atm_co2e_pc&idim=country:CAN&dl=en&hl=en&q=%22carbon+dioxide+emissions+per+capita+%22#met=en_atm_co2e_pc&tdim=true.

[17] US EPA (Environmental Protection Agency), 2007 "Inventory of US Greenhouse Gas Emissions and Sinks: 1990-2005" EPA Report # EPA 430-R-07-002. http://www.epa.gov/climatechange/emissions/downloads06/07CR.pdf.

[18] Winters J., 2007 "Carbon Unloaded" ASME Mechanical Engineering, Vol129 #4.

[19] L'Air Liquide, 1976 "Gas Encyclopedia, Elsevier/North Holland Inc, NY, USA 10017," pp. 333–368.

[20] Recht D. L., 1986 "CO_2 Line Design Needs," Pipeline and Gas Journal, Jan. Vol 213, pp. 41–44.

[21] Oldenburg, C. M., 2003 "Carbon Dioxide as Cushion Gas for Natural Gas Storage" American Chemical Society, Jr. of Energy & Fuels, Vol. 17, No 1, pp. 240–246, http://www.lbl.gov/tt/publications/2570pub.pdf.

[22] Apt. J., Gresham, L., Morgan, M. G., and Newcomer A., 2007 "Incentives for Near-Term Carbon Dioxide Geological Sequestration," Carnegie Mellon Electricity Industry Centre, Tepper School of Business, Oct 9. Pen, USA, www.cmu.edu/electricity.

[23] Dillon, J., 2008 "Status of CO_2 Capture and Sequestration in Canada", Hatch Energy May 19.

[24] Phillips, I., 2010 "Where does Carbon Capture and Storage fit in?," Licensing Executives Society(LES) Presentation, May 13, Galsgo http://www.les-bi.org/scotland/documents/IanPhillipsPresentation.pdf.

[25] Jordal, K., Anheden M., Yan, J and Strömberg., L., 2004 "*Oxy-fuel Combustion* Technology for *Coal-fired Power Generation*-Opportunities and Challenges" http://uregina.ca/ghgt7/PDF/papers/peer/054.pdf.

[26] Chapel D. G., Ernest, J., and Mariz C. L., 1999 "Recovery of CO2 from Flue Gases: Commercial Trends" Presentation at the Canadian Society of Chemical Engineers annual meeting October 4–6, Saskatoon, Saskatchewan, Canada.

[27] Beck, B., 2007 "CCS Technology: Capture, transport and storage of CO2", *IEA GHG, CCS-CDM Africa, September,* http://www.ccs-africa.org/fileadmin/ccs-africa/user/docs/Dakar_6_9/Dakar_Gaborone_Beck_06sept07.pdf.

[28] Robertson, E. P., 2007 "Analysis of CO_2 Separation from Flue Gas, Pipeline Transportation, and Sequestration in Coal" Idaho National Laboratory(INL) Idaho Falls, Idaho 83415 Report# INL/EXT-08-13816.

[29] Mazzotti, M., 2005 "Mineral carbonation and industrial uses of carbon dioxide" *IPCC Special Report on Carbon dioxide Capture and Storage*, http://www.ipcc.ch/pdf/special-reports/srccs/srccs_chapter7.pdf.

[30] Oosterkamp, A., 2008 "CO_2 Pipeline Transmission-A Broad State Of The Art," Polytech, November 07. http://www.docstoc.com/docs/43753551/Offshore-vindkraft-og-Polytec.

[31] Aresta, M. and Dibenedetto, A., 2007 "Utilization of CO_2 as a chemical feedstock: opportunities and challenges" *Dalton Transactions,* www.rsc.org/dalton, DOI: 10.1039/b700658f.

[32] Leopardi, A., 2010 "The cost of integration of zero emission power plants—A case study for the island of Cyprus," Salento University, http://oldsite.ing.unisalento.it/seo-start/page/home.bacheca_allegato/news_id/f9241ebc30d25c4434d2095564625a9b/no_header/seo-stop/index.php?format=.pdf.

[33] White, C. M., Straziser, B. R., Granite, E. J., Hoffman, J. S., and Pennline, H. W., 2003, "Separation and capture of CO_2 from large stationary sources and sequestration in geological formations—coalbeds and deep saline aquifers" Journal of the Air & Waste Management Association, v. 53, p. 645–715.

[34] McGrail, P. B., 2005 "Overview CO_2 Sequestration Science & Technology in the U.S" NAURC Summer meeting, July 25, Austin TX, USA, http://www.narucmeetings.org/Presentations/ERE_McGrail_s05.pdf.

[35] Dooley, J. J., Dahowski, R. T., Davidson, C. L., Bachu S., Gupta N., Gale, J., 2004 "A CO_2 Storage supply curve for north america and its implications for the deployment of carbon dioxide capture and storage systems" Greenhouse Gas Control Technologies (GHGT-7 Conference, Sept. Vancouver, Canada and Elsevier, Amsterdam, Netherlands. http://www.globalchange.umd.edu/staff/jdooley/publications/.

[36] Alberta Geological Survey, 2008 "Mechanisms, Means and Duration of CO_2 Sequestration in Geological Media" August.

[37] Bruant, R. G., Jr., Guswa, A. J., Celia, M. A., and Peters, C. A., 2002 "Safe storage of CO_2 in deep saline aquifers" Environmental Science and Technology, v. 36, no. 11, p. 240A–245A.

[38] US CSC (United stated Carbon Sequestration Council), 2009 "Is CO2 Sequestration Safe?," April, http://www.southwestcarbonpartnership.org/_Resources/PDF/3%5B1%5D.%20Is%20CCS%20Safe.pdf.

[39] Aspelund, A., Sandvik, T. E., Krogstad, H., and De Koeijer, G., 2006 " Liquifaction of Captured CO_2 For Ship-Based Transport," EnergyConsulting, Norway, http://www.energyconsulting.no/subpages/Pdf/Konferanser/GHGT-7-Liquefaction%20of%20CO2.pdf.

[40] Barrio, M. Aspelund, A., Weydahl, T., Sandvik, T. E., Wongraven, L. R., Krogstad, H. and Henningsen R., Mølnvik, M., and Eide, S.I., 2004 "Ship base Transport of CO_2," energy consulting. no http://energy consulting.no/subpages/Pdf/Konferanser/GHGT-7-Ship%20based%20transport%20of%20CO2.pdf.

[41] Agustsson, H. and Grinestaff, G. H., 2004 A study of IOR by CO_2 injection in the Gullfaks Field, offshore Norway. Presented at the *2004 SPE/DOE Fourteenth Symposium on Improved Oil Recovery.* Tulsa, Oklahoma, U.S.A., 17–21 April 004.

[42] Wong, S., 2005 "CO_2 Compression and Transportation to Storage Reservoir" Report prepared for *Asia-Pacific Economic Cooperation by Alberta Research Council (*APEC Reference #205-RE-01.3), March; http://science.uwaterloo.ca/~mauriced/`691-duss/CO2_Materials_From_ARC_APEC_Beijing_2006/CarSeq_Module4.pdf.

[43] Illinois State Geological Survey, 2004 "Carbon Dioxide Capture and Transportation Options in the Illinois Basin" Technical Report, (*U.S. DOE Contract: DE-FC26-03NT41994) http://sequestration.org/publish/MGSC_year1report.pdf.*

[44] IPCC (The Intergovernmental Panel on Climate Change) 2005 "Special Report on Carbon Dioxide Capture and Storage" Technical Summary., http://www.ipcc.ch/pdf/special-reports/srccs/srccs_technicalsummary.pdf

[45] Race, J., M., 2006 "Designing a Safe Carbon Dioxide Transport Infrastructure for the UK" Presentation to CCSA – 24th July Newcastle University, School of Marine Science & Technology UK.

[46] Process Energy Solutions 2008 "World CO_2 History" http://cpe.kmutt.ac.th/wiki/index.php?title=....World_CO2_History...&redirect=no&printable=yes.

[47] Moore, M., 2010 "Legislative and Regulatory Activities for CO_2 Pipelines in the US", ASME International Pipeline Conference, Calgary, AB Canada Sept 27, CO_2 Panel discussion.

[48] Squires, F., 1917 "Method of Recovering Oil & gas" U.S. patent 1,238,355, aug 28 Turner, R., Hardy, N & Hooper., B, 2006 "Risks associated with a CO2 pipeline—Methodology and case study" GHGT – 8,Trondheim 19th -22nd June, https://extra.co2crc.com.au/modules/pts2/download.php?file_id=951&rec_id=369.

[49] Marshall, M. P., 1985 "The Excavation of the Cortez CO2 Pipeline" Report for the Office of Contract Archeology, University of New Mexico, Albuquerque, New Mexico Project 185-161A, C,D,M. http://www.azwater.gov/Adjudications/documents/HopiContestedCaseDisclosures/Navajo%20Supplemental%20Disclosure/CDV007%5CBNDR002%5C004%5CNN028884.pdf.

[50] HSE (UK Health and Safety Executive), 2008 "Pipeline design codes and standards for use in UK CO2 Storage and Sequestration projects" http://www.hse.gov.uk/pipelines/resources/designcodes.htm.

[51] CSA (Canadian Standard Association), 2007 "CSA z662: Oil and Gas Pipeline Systems" www.csa.ca.

[52] DNV (Det Norsk Veritas), 2010 "Design and Opeartion of CO_2 Pipelines" Recommended Practice (RP) DNV-RP-J202, April, http://exchange.dnv.com:6389/dynaweb/offshore/rp-j202/@Generic__BookTextView;hf=0;cs=default;ts=default.

[53] Heinz, C., 2011 "Applicable Norms and Required Improvements for CO2 Pipelines 6th Pipeline Technology Conference, April, HANNOVER MESSE, Germany http://www.pipeline-conference.com/abstract-database/item/159-applicable-nor.

[54] CO_2 Europipe 2009 "Towards a transport infrastructure for large-scale CCS in Europe" WP2.1 Report-existing Infrastructure Rev 1http://www.co2europipe.eu/Publications/D2.1.1%20-%20CO2 Europipe%20Report%20-%20Existing%20infrastructure.pdf.

[55] Johnsen, K., Helle, K., Torbergsen, L. E., and Eldevik, F, 2010 "DNV Recommended Practice: Design and Operation of CO2 Pipelines," Proceeding of ASME 8th International Pipeline Conference, IPC, Track 3 Design and Construction, paper # IPC2010-31241, Sept 27-Oct 1,Calgary, AB Canada.

[56] US DOT OPS (US Department of Transportation, Office of Pipeline Safety), 2010 "Code of Federal Regulation (CFR) Title 49-Transportation of Hazardous Liquid By Pipeline" http://setonresource center.com/49CFR/Docs/wcd00007/wcd00773.asp.

[57] Parfomak, P. W. & Foldger, P., 2007 "Carbon Dioxide (CO_2) Pipelines for Carbon Sequestration: Emerging Policy Issues," CRS Report for the US Congress, RL33971http://www.iepa.com/ETAAC/ETAAC%20Handouts%208-8-07/CRS%20-%20Report%20CO2%20Pipelines%20for%20CCS%20k%20davis.pdf.

[58] Metz, B., O., Davidson, H. C. de Coninck, M. Loos, and Meyer, L. A. (eds.), 2005 "Carbon Dioxide Capture and Storage" IPCC Special Report, Prepared by Working Group III of the Intergovernmental Panel on Climate Change, Cambridge University Press, Camb. UK and, NY, USA.

[59] Brubacher, L., 2006 "Obtaining Reliable Data Onine—Phase Diagram For carbon Dioxide" *Chem 13 News*, Chemistry Department, Waterloo Univ. ON. November.

[60] Reed. J., 2010 "Beneficial Use of Carbon Dioxide" California Carbon Capture and Storage Review Panel Technical Advisory Committee Report, *October* http://www.climatechange.ca.gov/carbon_capture_review_panel/meetings/2010-10-21/white_papers/Beneficial_Use_of_Carbon_Dioxide.pdf.

[61] Rekacewicz, P., 2000 "Images for *Philippe Rekacewicz*" http://www.google.ca/images?hl=en&q=Philippe+Rekacewicz&rlz=1R2GGLL_enCA358&um=1&ie=UTF-8&source=univ&ei=6CKJTKXWGY7CsAOX2aCYCg&sa=X&oi=image_result_group&ct=title&resnum=4&ved=0CDAQsAQwAw&biw=1263&bih=539.

[62] US DOT OPS, 2010 "CFR49 part 192-Transportation of Natural and Other Gas by Pipeline: http://setonresourcecenter.com/49CFR/Docs/wcd00006/wcd00645.asp.

[63] Wikipedia, 2010, "Carbon" http://en.wikipedia.org/wiki/Carbon, retrieved 4th September.

Chapter 2

CARBON CAPTURE, TRANSPORT AND STORAGE — A BACKGROUND

2.1 CARBON CAPTURE AND STORAGE

As indicated in Chapter 1, carbon capture and storage (CCS) involves capturing CO_2 mainly from fossil fuel power plants (or other large stationary industrial sources), compressing and then transporting it via pipelines or ships and finally storing it underground in a geological site, such as saline aquifers or depleted oil and gas fields or alternately using it for enhanced oil recovery (EOR), enhanced gas recovery (EGR) or enhanced coal bed methane (ECBM). The CCS techniques were pictorially depicted and descriptively summarized in Chapter 1. The chapter also depicted the role of pipelines in the CCS; however, an overview of the entire CCS process is shown schematically in Figure 2-1. This chapter in a consolidated form provides detail descriptions of all aspects of carbon capture technologies and carbon dioxide sinks: storage and sequestration, namely, enhance oil recovery (EOR).

The CCS method provides a potential way of reducing carbon emissions and achieving a stabilized level of CO_2 in the atmosphere. It has been predicted that the total emissions avoided would be approximately 1 GT of carbon per year minimum by 2054 [1]. It can be said that CCS will provide a bridge to the fully renewable technologies while meeting the demand for energy worldwide.

The CO_2 storage site has to be selected so that it is geologically secured for significant timescales. Storage sites for CO_2 are available in depleted oil and gas reservoirs and saline aquifers. The other potential opportunity is to use the captured CO_2 for enhanced oil recovery (EOR), where it has been practiced onshore for over the past 4 decades in the United States. The advantage of CO_2 injection for EOR is that it mobilizes oil that conventional recovery methods are unable to remove. Therefore, using CO_2 for EOR has the dual benefit of increasing oil production, particularly from fields nearing the end of their life, while also reducing finding a permanent storage space (i.e., sink) for CO_2 instead of emitting it to the atmosphere.

Clearly, the connection between the sources and sinks for carbon capture and storage would involve the use of transport systems by which the captured anthropogenic CO_2 will be transported via pipelines or ships to the storage or EOR sites. It is therefore important to understand both the physical and hydraulic properties of anthropogenic CO_2 in the context of its transport.

2.2 BASIC PROPERTIES OF CO_2

The phase diagram for pure CO_2 is shown in Figure 2-2. It contains two distinct features which are a "triple point" at 5.2 bar, –56°C; and a "critical point" at 74 bar, 31°C. At the triple point CO_2 can exist as one of three phases, solid, liquid, or gas. The curve connecting

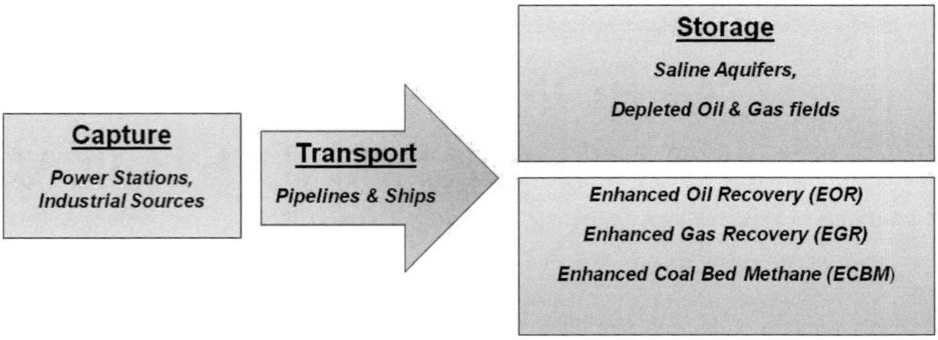

Figure 2-1. An overview of the carbon capture and storage process

the triple point and the critical point in Figure 2-2 is the vapor–liquid line separating the gaseous and liquid phases. The "critical pressure" is the pressure above which liquid and gas phases cannot be distinguished. The "critical temperature" is the vapor pressure at the critical pressure. The region above the critical temperature and critical pressure is labeled "supercritical region" while the region above the critical pressure, but below the critical temperature, is called the "dense phase region" [2].

In the supercritical region, CO_2 no longer exists in distinct gaseous and liquid phases. In a supercritical phase, it has the density of a liquid but the viscosity of a gas. Increases in pressure no longer produce liquids at temperatures exceeding the critical temperature. In the dense phase region, CO_2 density increases with decreasing temperature. The most efficient way in terms of cost and throughput of transporting CO_2 by pipeline is in its supercritical or dense phase where it has the density of a liquid with the viscosity of a gas [3, 4, 5].

Anthropogenic CO_2 sources refers to sources of CO_2 emissions such as from fossil fuel burning power plants, ammonia plants, chemical facilities, refineries, and other

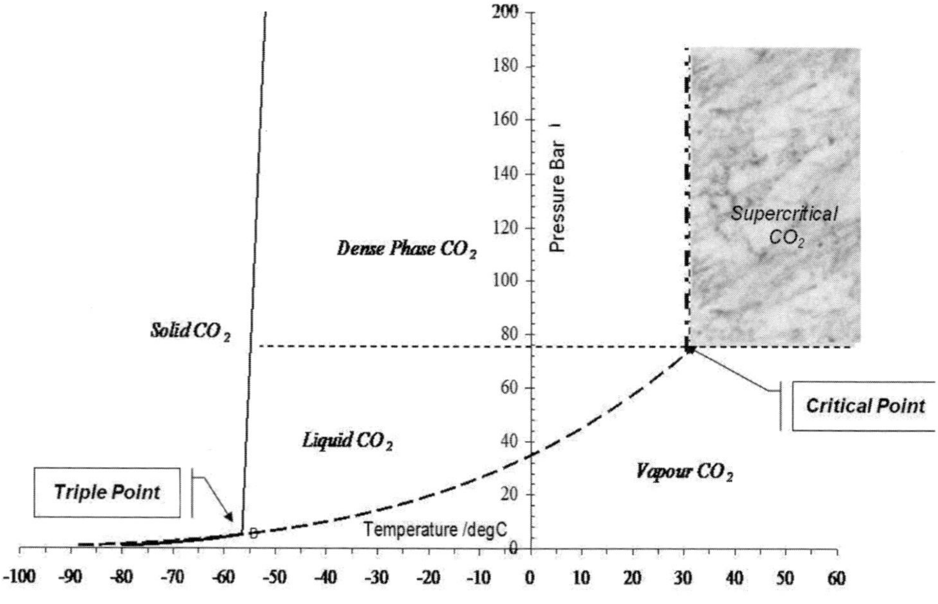

Figure 2-2. Phase diagram for pure CO_2

stationary industrial sources. The three different capture technologies are post-combustion, pre-combustion, and oxyfuel. Each capture technology will have varying purity of CO_2 due to the different stages and techniques involved in the capture process. The captured anthropogenic CO_2 will then be transported to specific sink(s) such as depleted oil and gas reservoirs, saline aquifers, or oil and gas reservoirs for the purpose of Enhanced Oil Recovery (EOR) as mentioned earlier.

2.2.1 Supercritical CO_2 Experience and Comparison

CO_2 has been handled within industry, especially in ammonia and urea production, at high pressures (155 bar) and temperatures (180°C) which are well within the supercritical region of CO_2. Since 1948 [6], supercritical CO_2 has also been used extensively as a working fluid for applications in advanced versions of the gas turbine cycle, in advanced nuclear reactors [7], and also supercritical water reactors with minimum temperatures of 550°C and pressures of 200 bar.

Major differences exist between the conditions for CO_2 transport and other industrial uses of CO_2. The processes mentioned above all operate or utilize CO_2 well into the supercritical region without approaching the region around the critical point. As mentioned earlier, this region is characterized by rapid changes in the CO_2 properties. However, CO_2 pipeline operations can extend into the critical region and are in close proximity to the critical point. In contrast, fluids such as methane, nitrogen, water, and oil are transported at pressures and temperatures away from their respective critical points. The critical conditions and typical pressures for some of these fluids are shown in Table 2-1 [5], and it can be observed that the pipeline temperatures for nitrogen and methane are above their critical point and therefore, no phase change occurs. In addition, the operating pressure of oil and water are much lower than their respective critical points and therefore produce no change in phase being liquid in this case.

In contrast, current CO_2 pipelines (especially in the United States) have a compressor discharge temperature of 38°C and flowing temperatures ranging from 24°C to 4°C [5], which are all in the vicinity of the critical temperature. This region of the phase diagram has the greatest discontinuities in physical properties associated with it and is unavoidable without temperature control.

Gas phase transport is undesirable for long distance transportation of CO_2 by pipelines because of the low density and consequent requirement for a large diameter pipeline [8,9,10]. Consequently, current operating practices for CO_2 pipelines are to maintain the pressure well above the critical pressure, i.e., in the supercritical or dense phase regions. However, the presence of impurities in anthropogenic CO_2 alters the physical and transport properties of CO_2 due to the changes in the phase behavior, critical temperatures, and pressures of the transported fluid. The components and concentrations of these impurities depend on the capture technology and the degree of purification in the process as will be discussed in the next section.

TABLE 2-1. Critical temperatures and pressures for various fluids transported by pipelines [5]

Fluid	Critical Temperature (°C)	Critical Pressure (bar)	Typical Pipeline Operating Pressure (bar)
Pure CO_2	31	73.8	86.2
Methane	83	40.7	48.3
Nitrogen	−147	34	48.3
Oil (MW −353)	539	9.9	17.3
Water	374	221.2	17.3

*Reaction between H_2S and SO_2 to produce elemental sulphur and water.

2.3 CO_2 CAPTURE TECHNOLOGY

It has been suggested that the early implementation of CCS will involve large stationary anthropogenic sources (>0.1 $MtCO_2$/year), and will involve the power and industrial sector where 60% of the CO_2 emissions originate [11]. The new CO_2 sources will have to be built with capture facilities whereas existing sources will require a retrofit of these facilities. Consequently, coal and natural gas power plants have been projected to be the dominant fuel for power generation in 2020 [11].

The three different capture technologies (post-combustion, oxyfuel, and pre-combustion) require the removal of CO_2 from a gas stream, and this is achieved by using five main technologies in which the choice depends on process and plant parameters, such as the volume, concentration, and pressure of the CO_2 that is to be captured. The five main technologies [11,12] are the following:

- Chemical or physical solvent scrubbing,
- Adsorption,
- Absorption,
- Membrane separation,
- Cryogenic separation.

The selection of a method of capturing CO_2 and then separating it from the gas stream has implications in terms of cost, technological development, plant efficiency, flexibility, energy requirements, type of facilities required, and final quality of the CO_2 stream that is to be transported for use in a specific application. The main priorities in terms of capture technology development are the minimization of energy requirements and the improvement of plant efficiency. This would reduce the overall capture cost and at the same time minimize environmental impacts, as a result of increased plant efficiency [11].

The three main capture processes are discussed in the following sub-sections along with the basic processes involved for each capture technology and the respective composition of the captured stream that has to be transported. The information contained below is based on a literature survey of reports and publications from the International Energy Agency [13, 14], the Department of Climate change in the UK, and the International Panel on Climate Change [11].

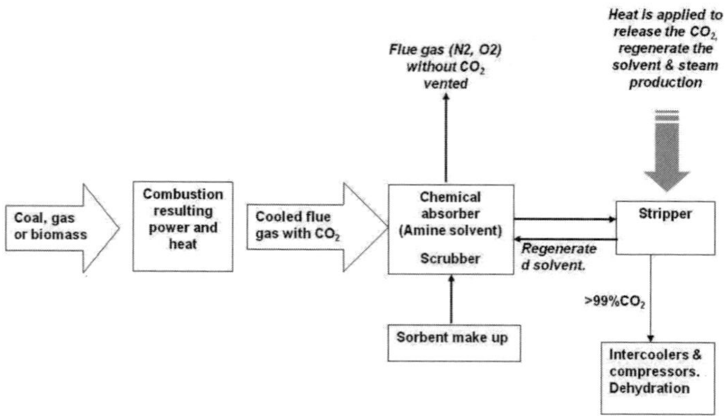

Figure 2-3. Process diagram of post-combustion CO_2 capture from flue gas by chemical absorption

2.3.1 Post-Combustion Capture

Post-combustion capture involves the combustion of fossil fuels and biomass using air. Figure 2-3 shows a basic schematic diagram of the post-combustion capture process using amine absorption. Comparative studies [14, 15, 16] have shown that absorption based on chemical solvents is one of the most favored options for post-combustion CO_2 capture. The flue gas is not discharged to the atmosphere but is passed through a chemical absorber usually an amine [11] with an acid or sour gas. The chemical solvent binds to the CO_2, separating it from the flue gas. The remaining flue gas is then released into the atmosphere. The resulting CO_2-rich solvent is then pumped to the top of a stripper column via a heat exchanger, whereby the CO_2 is removed using steam, and the chemical solvent is regenerated. This process occurs at elevated temperatures of 100°C to 140°C and pressures slightly higher than atmospheric pressure. The CO_2 saturated with water vapor then leaves the stripping column at about 80°C [17]. The mixture is subsequently cooled, possibly in several stages, and the water is removed. Heat is constantly supplied to the reboiler to ensure optimum regeneration conditions are maintained. The CO_2 is then cooled, dried, and compressed for transport. Heat is also recovered through intercooling stages in the CO_2 compression train and reflux condensers. Currently, chemical solvents are the preferred option due to high capture efficiency and selectivity with the lowest energy consumption and reduced cost relative to other capture technologies [11].

It should be noted that additional equipment may be required to ensure the quality of the solvent due to the formation of degradation and corrosion products along with the presence of particles. Equipments, such as filters, carbon beds, and thermally operated reclaimers, are used to ensure the quality of the solution which has been a key aspect in the development of the absorption process [11, 13, 18].

The composition of the CO_2 stream that will be transported depends on a number of factors. The main factors are the type and grade of fuel used, which determines the composition of the flue gas, the solvent, and stripping element used as this will in turn determine the waste by-product and the cleanup process itself. The outlet stream from the capture process contains other elements besides CO_2, i.e., *impurities*. This has its origins in EOR for which there are a number of reasons supporting the use of pure CO_2 that do not apply to simple storage. The label of "impurity" has cultural negative connotations that should not be transferred to the CO_2 stream to be transported for storage. Some of its components may, or may not, be deleterious to the transportation process or the environmental interaction with the storage medium, but it is not the case that impurities per se should be minimized in the CO_2 stream. The important point is that the outlet stream should contain only components that can be safely transported and stored, and that have an acceptable environmental impact, but that being acknowledged, there may well be an economic case for including as many as possible.

It may be noted that in this chapter, consistent with the literature, all components in the captured outlet flow that are not CO_2 will be referred to as impurities.

Table 2-2 shows the impurities and their potential source from within the post-combustion process that could exist in the captured CO_2 that is to be transported. The impurities indicated as N/A in Table 2-2 could however be present in trace amounts for various reasons including the type of solvent technology used, the composition of the flue gas, and leakage. Some of the impurity levels will need to be reduced to acceptable levels as they may react preferentially with the solvent. The type of fuel is a major determinant of the type, combination, and levels of impurities present. Examples of such impurities are sulphur, SO_2, NO_2, fly ash, and soot. It should be noted that the quantity of the impurities will once again depend on the cleanup (purification) scheme implemented within the process which would in turn depend on a full cost benefit analysis of the entire CCS value chain [19].

TABLE 2-2. Potential impurities and their possible source in the post-combustion CO_2 transport stream [11, 13, 20]

	Potential Source of Impurity from the Post-Combustion Process	
Imp.	**Gas**	**Coal**
CH_4	Trace from feed	N/A
N_2	Air used in combustion, solvent form physical absorption of scrubbing	Air used in combustion, solvent from physical absorption of scrubbing
H_2S	Fuel quality — Trace as Claus reaction* will take place ($H_2S + SO_x$)	Fuel quality — Trace as Claus reaction* will take place ($H_2S + SO_x$)
C2+	Natural Gas Content	N/A
CO	Flue gas	Flue gas
O_2	Flue gas or air	Coal combustion/flue gas or air
NO_X	Air/capture method — separation membranes	Coal combustion/flue gas/capture method - separation membranes
SO_X	Fuel quality/flue gas/capture method used-separation membranes	Coal combustion/flue gas/capture method used-separation membranes
H_2	N/A	N/A
Ar	N/A	N/A
S	Source of gas/reaction of H_2S and SOx will be removed as product	Type of coal, reaction of H_2S and SO_x will be removed as product
H_2O	Steam-stripping gas/coal combustion/leakage	Steam-stripping gas/coal combustion/leakage
Trace	N/A	HCl, HCF, Mercury

*Reaction between H_2S and SO_2 to produce elemental sulphur and water.

2.3.2 Oxyfuel Capture

Oxyfuel combustion involves the combustion of fossil fuel in pure O_2 or in a mixture containing CO_2 and O_2. It is also a key aspect of the pre-combustion technology. This is because the combustion temperature of fossil fuels with pure O_2 is very high, typically about 3500°C which exceeds current power plant material limitations. The presence of CO_2 or recycled flue gas and gaseous or liquid water limits the combustion temperature to reasonable ranges that are similar to combustion with air, for example 1300 to 1400°C in a typical gas turbine cycle and 1900°C in an oxyfuel coal fired boiler. Since air is not used directly in the combustion process nitrogen, its oxides are therefore greatly reduced from the flue gas. The oxyfuel process can be integrated with boilers and gas turbines that have a specific design which enables it to work with high concentration of CO_2. An air separation unit is required to produce pure O_2 for combustion and this increases the amount of energy needed and consequently the associated cost.

Figure 2-4 illustrates the basic working principle of capturing CO_2 using an oxyfuel process. Fossil fuels are burnt in the combustion chamber which could be a boiler for pulverised coal with a steam cycle or in the combustion section of a gas turbine combined cycle or a steam turbine cycle. The O_2 is normally supplied through an air separation unit. The path of the combustion product depends on the power cycle used. In a pulverized coal plant using a supercritical steam cycle, the combustion product will have three paths. The first involves recycling the CO_2-rich flue gas back to the combustor, in this case, a boiler to limit the combustion temperature, the second path involves the recycling of the CO_2 stream as a drying gas in a coal mill, and the third path leads to capture and transport. These options are shown by the dashed line in Figure 2-4.

However, in a gas turbine combined cycle, or a steam turbine power generation cycle, after combustion in an O_2-enriched environment, the heat from combustion is used in the steam, gas, or combined cycle to produce electricity. The combustion product is then freed of any particulates and passed through a heat exchanger where the heat removed from the product is re-used in the power cycles (dotted line). Water is also condensed here. The

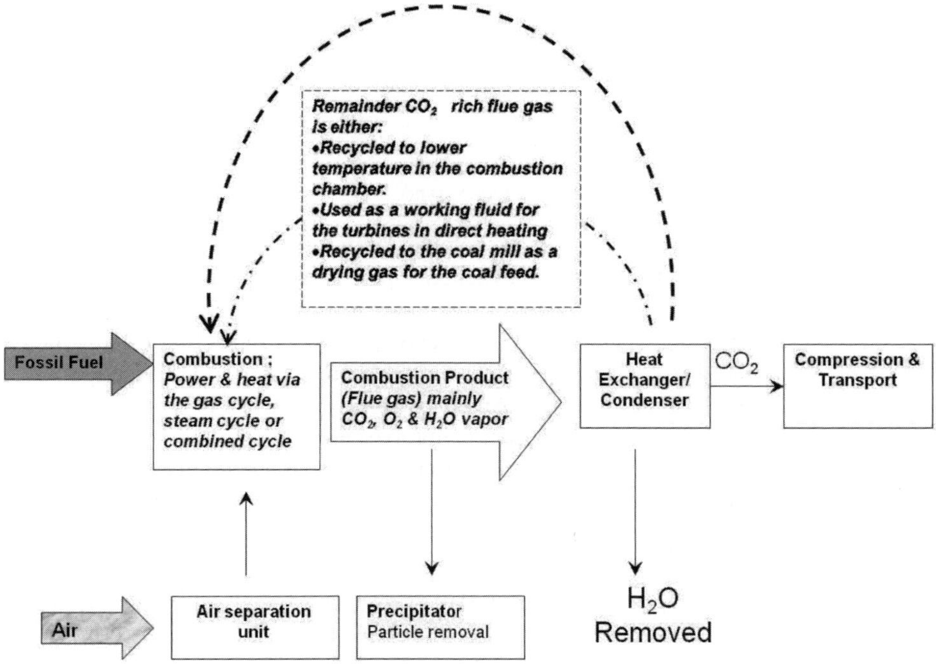

Figure 2-4. Schematic diagram of the oxyfuel capture process

CO_2 is then compressed for pipeline transport. It should be noted that a lot of variation exists within the oxyfuel process itself in terms of cleanup schemes, flue gas recycling, working fluid used by the turbines, and energy penalties due to the requirement for pure O_2 production.

Table 2-3, below, shows the potential impurities and their sources that can be generated during the oxyfuel process. For example, impurities, such as SO_X, NO_X, HCl, and Hg, are dependent on the fuel used, whereas N_2, Ar and O_2 could potentially be derived from the O_2 stream or air leakage into the system. It should be noted that other sources and types of impurities that are not listed in Table 2-3 may be present for a given capture method or other power generation processes that are currently being developed. The type, quantity,

TABLE 2-3. Potential impurities and their possible source in the oxyfuel CO_2 transport stream [11, 13, 20]

	Potential Source of Impurity from the Oxyfuel Capture Process	
Impurity	**Gas**	**Coal**
CH_4	Gas	Fuel
N_2	Oxygen feed or air leakage into system	Oxygen feed or air leakage into system
H_2S	Fuel	Fuel
C2+	Cryogenic O_2 production method using air-trace	Cryogenic O_2 production method using air-trace
CO	N/A	N/A
O_2	Oxygen feed or air leakage into system	Oxygen feed or air leakage into system.
NO_X	Fuel used and possible air leakage (N_2)	Fuel used and possible air leakage (N_2)
SO_X	Fuel used	Fuel used
H_2	Syngas from direct heating gas turbine cycle	Syngas from direct heating gas turbine cycle
Ar	Oxygen feed or air leakage into system	Oxygen feed or air leakage into system
S	Presence of SO_x and H_2S, Claus Reaction	Presence of SO_x and H_2S, coal type
H_2O	Flue gas, steam/boiler	Flue gas, steam/boiler
Trace	Fuel source- Hg, HCl	Fuel source - Hg, HCl

and combination of impurities depend mainly on the fuel used, the combustion process itself, purity of the O_2 feed, the CO_2 capture method employed, and the cleanup train implemented, such as flue gas desulphurisation (FGD) and De-NO_X.

2.3.3 Pre-Combustion Capture

Pre-combustion capture involves capturing CO_2 prior to combustion to produce a stream of H_2 which is then used to generate power and heat. Pre-combustion can only be used with the Integrated Gasification Combined Cycle (IGCC) process. The first step involves steam reforming whereby steam is added to the fuels and is followed by the second reaction, which is called "gasification" if solid fuels are used, or "partial oxidation" if liquid and gaseous fuels are used. Both approaches involve the same principles.

The first stage reaction involves the production of syngas (CO + H2) from the fuel via steam reforming in which steam is added as shown in Eq. (2-1), or partial oxidation/gasification in which O2 is added shown in Eq. (2-2). Both reactions produce syngas.

$$C_xH_y + xH_2O \Leftrightarrow xCO + (x+y/2)\ H_2 \text{ (Steam Reforming)} \qquad (2-1)$$

$$C_xH_y + x/2O_2 \Leftrightarrow xCO + (y/2)\ H_2 \text{(Gasification/partial oxidation)} \qquad (2-2)$$

The second stage of the process involves a "shift" reaction in which the CO is converted to CO_2 using steam as shown in Eq. (2-3).

$$CO + H_2O \Leftrightarrow CO_2 + H_2 \qquad (2-3)$$

The CO_2 is then separated and captured, and H_2 is combusted in a gas turbine to provide heat and power. Nitrogen or steam from the air separation unit is sometimes introduced along with the hydrogen stream to reduce the flame temperature in the gas turbine combustor. Figure 2-5 illustrates this basic pre-combustion process. The amount of CO_2 captured from the gasifier/reformer output ranges from 15% to 60% and has pressures typically in the ranges of 20 to 70 bar. However, variations do exist in the way in which the power and heat is produced from H_2 and the method used to capture CO_2.

Sulphur is first removed from the feed as it is known to poison the nickel-based catalyst, and this is followed by the addition of steam. The reformed syngas is then cooled in a waste heat boiler to produce steam for the CO shift system in which CO is converted to CO_2 and H_2 is produced. However, some CO may still be present in the syngas stream or in the H_2 stream. The CO_2 is then removed using a chemical absorption process such as an amine scrubber or using a pressure swing absorber (PSA) where O_2 instead of steam is used to produce syngas. The absence of N_2 (from air) in the syngas reduces cost as separation is not required. On the other hand, if the H_2 is to be used for de-carbonization applications, such as in a gas turbine, N_2 from the air separation unit is mixed with the H_2 in order to lower the flame temperature in a gas turbine combustor. This option is also illustrated in Figure 2-5 by the dotted line.

The gasification of coal, petroleum residues, biomass, and industrial waste, all of which are low-value feedstock is a process used to make high-value products such as electricity, clean synthetic fuels, and chemicals. The products of coal gasification include synthetic natural gas, diesel, gasoline, ether, and methanol. One of the early implementations of coal gasification was the conversion of coal to fuel gas for domestic heating [12]. Gasification mainly involves the combustion of solid fuels, and the reaction is given in Eq. (2-2). The basic

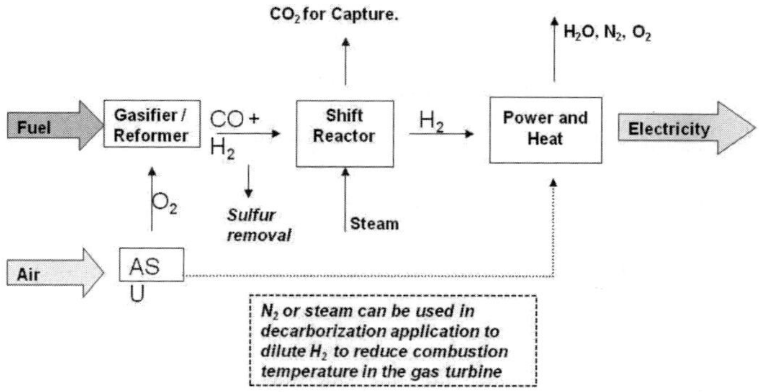

Figure 2-5. The pre-combustion capture process

process for gasification as with the other H_2 production methods is shown in Figure 2-5. The output of the gasifier contains CO, H_2, CO_2, H_2O, and impurities, such as N_2, COS (carbonyl sulphide), H_2S, NH_3, and volatile mineral traces. These impurities need to be produced at levels acceptable by safety regulations and, perhaps in the future, to levels also acceptable for pipeline transport. Examples of gasification plants are the numerous ammonia plants in China and the North Dakota Coal gasification plant which produces synthetic natural gas from coal and also produces large amounts of pure CO_2. An estimated amount of 3.3 Mt CO_2/year is captured using the methanol-based physical solvent scrubbing process in which about 1.5 Mt CO_2/year is transported to the Weyburn field in Canada for EOR and the remaining CO_2 vented. H_2S is also captured along with the CO_2 from the sulphur-tolerant shift reactor (sour shift) by physical absorption. In the case for the North Dakota plant, the CO_2 is transported with a certain amount of H_2S for EOR as part of an acid gas management strategy.

TABLE 2-4. Potential impurities and their possible source in the pre-combustion CO_2 transport stream [11, 13, 20]

	Potential Source of Impurity from the Oxyfuel Capture Process	
Impurity	**Gas**	**Coal**
CH_4	From the shift reaction	Shift reaction
N_2	From the shift reaction	Gasification of coal, petroleum residue, or biomass
H_2S	High sulphur fuel, syngas, Physical Solvent scrubbing process — main impurity	High sulphur fuel, syngas, physical solvent scrubbing process, lignite, and bituminous coal — main impurity
C2+	Steam reforming of gas and light hydrocarbons using PSA	Trace — fuel used
CO	From the shift reaction catalyst Physical Solvent scrubbing process	From the shift reaction catalyst. Gasification of coal, petroleum residue, or biomass. Physical solvent scrubbing process
O_2	Leakage-trace amounts	Leakage-trace amounts
NO_X	Hydrogen combustion/removed	Hydrogen combustion/removed
SO_X	IGCC plant	Fuel
H_2	Steam reforming of gas and light hydrocarbons using PSA. Physical solvent scrubbing process	Gasification of coal, petroleum residue or biomass. Physical solvent scrubbing process
Ar	–	Type of coal used
S	From reducing of H_2S	If oil or coal is used especially bituminous coal. Coal gasification reduction of H_2S to S.
H_2O	Shift reaction	Gasification of coal, petroleum residue or biomass
Trace elements		Gasification of coal, petroleum residue or biomass
Others		HCN, NH_3,COS, Hg

The pre-combustion technologies described above are all essentially similar in their conversion routes with their main differences existing in the method for the production of syngas, the use of H_2, the end product formed, the removal of impurities, and the method employed to capture the CO_2. As demonstrated, there are various methods of producing H_2 and also capturing the CO_2, therefore, the type, amount, and combination of impurities present will depend mainly on these factors and the cleanup scheme deployed. Table 2-4 shows the various potential sources of impurities that could possibly exist in the oxyfuel process flow, with some having the potential to be introduced inadvertently into the CO_2 transport stream.

2.3.4 Capture Technology and CO_2 Quality

The composition of the captured output stream plays a major part in the transport and storage of CO_2 in terms of health, safety, physical properties, hydraulics, design issues, regulations, and cost. Impurities in the capture stream are highly dependent on the capture technology used, type of fuel, and also cleanup method implemented at the capture plant. The CO_2 from the solvent scrubbing process in the post-combustion technology contains low level of impurities when compared to pre-combustion and oxyfuel [13]. The 2003 IEA GHG report describing the gasification combined cycle power generation method states that the capture of CO_2 from the pre-combustion physical solvent scrubbing process could potentially contain 1% to 2% H_2 and CO along with H_2S which depends on the amount of the sulphur compound present in the fuel. The CO_2 stream from the oxyfuel process would contain O_2, N_2, Ar, S, NO_X, and other trace impurities [11]. N_2 up to 2% could be present in the oxyfuel stream [21].

It should be noted that the type, amount, and combination of these impurities will vary for the various reasons mentioned above, and a similar variation exists in the published literatures. It is difficult to determine the exact levels of impurities that may be present in the captured CO_2 as this will be affected not only by the process, as described above, but also by economics (i.e., the increased capture cost associated with the removal of impurities to low levels) and also legislative and regulatory requirements, EOR specifications, and safety considerations. A number of studies and assessments such as the IPCC [11], ENCAP [22], DYNAMIS [23], [24], Ecofys [25], and the IEA [20] attempting to determine the type, combination and ranges of impurities that could potentially exist in the CO_2 stream have been reviewed. However, uncertainties still exist in this area. In addition, the specifications for CO_2 streams will need to be determined according to the requirements of transport optimization and storage.

2.4 CARBON DIOXIDE SINKS

The captured and transported CO_2 can be mainly managed or stored in saline aquifers, depleted oil, and gas fields used for enhanced oil (EOR) or gas (EGR) recovery and also for enhanced coal bed methane (ECBM) purposes. Other storage options such as mineral carbonization and ocean storage have also been proposed [11, 26].

2.4.1 Storage

The following provides a discussion on storage options in salineaquifers, depleted oil and gas fields, and EOR. This is because these options are relatively well understood and are the likely options to be implemented in the earliest ages of CCS [11].

Table 2-5 shows the global capacity estimated by the International Energy Agency and the percentage of CO_2 emissions that can be stored. It should be noted that the estimate for deep saline aquifers were made in 1990, and it is predicted that currently in North West Europe alone, about 800 Gt CO_2 can be stored in saline aquifers [27]. The large capacity

TABLE 2-5. Natural reservoirs suitable for CO_2 Storage — 1990 estimate [27]

Storage option	Global Capacity	
	Gt CO_2	% of emissions to 2050
Depleted oil and gas fields	920	45
Deep saline reservoirs	400–10,000	20–500
Unminable coal measures	>15	>1

of depleted oil and gas fields and saline aquifers makes these two options even more likely for storage in the future.

Geological CO_2 reservoirs exist naturally in underground cavities as a result of volcanic activity, biological activity in the past (geological timescales), and chemical reactions between fluids in the earth's crust and carbonate rocks [11]. The first injection of CO_2 into a geological structure for storage was conducted in the 1970s in Texas, USA. This was part of an enhanced oil recovery project, and the practice has since become more widespread, especially in the USA where CO_2 has been used for 73 different oil recovery projects. The CO_2 supplied for these projects has been predominantly from naturally occurring CO_2 in underground domes and a few anthropological sources of CO_2 from coal gasification plants, such as the Dakota Gasification plant and also gas plants such as the Val Verde and La Barge gas plants [28]. It should be noted that all of the CO_2 captured in the USA has been used for enhanced oil recovery.

Aquifers are made of porous and permeable reservoir rocks that store and transmit water [29]. Adjacent pore spaces in aquifers are linked by narrow pathways that are called pore throats. This allows the fluid in reservoir rocks (aquifers) to pass through it and also to be stored within it. Figure 2-6 illustrates the structure and storage mechanism of a typical sandstone aquifer. The presence of faults can further enhance the reservoir rocks permeability and to a much lesser extent the permeability.

Figure 2-6. Typical sandstone reservoir (Sherwood Sandstone) rock ([29], British Geological Survey with permission)

TABLE 2-6. Current and potential future CO_2 storage projects [31]

Year	Location	CO_2 Injection Rate (Million Tonnes/Year)	Project Type	Operator
1996	Sleipner, Norway	1	Storage	Statoil
2004	In Salah, Algeria	1.3	Storage	BP
2004	K12B	0.2	Storage	Gaz de France
2006	Snøhvit	0.7	Storage	Statoil
2009	Gorgon, Australia	3.3	Storage	Chevron

Natural reservoirs containing pure CO_2 can be found worldwide particularly in Permian Basins, volcanic and faulted areas. A number of these fields exist in the USA, such as the Bravo Dome, the McElmo Dome and the Jackson dome, which are currently being used to supply CO_2 for EOR operations. The Sherwood sandstone shown in Figure 2-6 is analogous to reservoir rock from the natural CO_2 reservoirs in the Permian Basin such as the Bravo Dome [30]. Therefore, some experience can be drawn from the early studies of these reservoirs in terms of CO_2 leakage, seal integrity, and storage mechanisms [11].

The largest CO_2 saline aquifer storage project is Statoil's Sleipner project in the North Sea, which separates CO_2 from the natural gas and stores it underground in the Utsira formation. The In-Salah project in Algeria involves storing CO_2 in a depleted gas reservoir. The CO_2 is separated from the high CO_2 content natural gas field. Table 2-6 summarizes some of these storage projects.

The cost and technical feasibility of all the proposed methods of storing CO_2 vary from one location to another. Factors such as reservoir or aquifer depth, number of injection wells, whether the project is onshore or offshore while taking into account the risks to population in case of a leak and also monitoring techniques need to be considered. Also, it is important to address the current knowledge gap in terms of accurate estimates on regional storage capacity, leakage rates, regulations governing storage, and better understanding of CO_2 mobility in geological formations.

Depleted or abandoned oil and gas fields are recognized as prime candidates for CO_2 storage [11]. There are a number of reasons for this. Firstly, oil and gas reservoirs have already been characterized and therefore there are large amounts of data and computer models available with regards to its geology, movement of hydrocarbons, and also chemistry. Second, hydrocarbons have been trapped within these reservoirs for millions of years which demonstrates the integrity and security of these reservoirs. Third, most of the injection equipment and facilities have already been put in place and could potentially be used for CO_2 storage. The only drawback of this storage option is the plugging of these fields which have already been done at the time of abandonment. These cement plugs were designed as plugs for mud laden fluid filled into depleted wells and not designed to cap buoyant and reactive fluids such as CO_2. However, the integrity of all wells needs to be assessed on a case by case basis, and the reservoir pressure needs to be maintained below the fracture pressure. Although, storage in reservoirs of depths less than 800 m is deemed possible, the low storage capacity of these reservoirs could pose a problem [11, 29, 32]. Also, at depths above 800 m, the storage space is optimally and efficiently utilized by the CO_2 [33]. It should be noted that aquifers tend to overlie oil and gas reservoirs [34]. When oil production decreases so does the pressure in the reservoir. This in turn reduces the pressure difference between the reservoir and the well used to drive the fluids toward the wells. Given that some aquifers tend to overlie reservoirs, water from the aquifer sometimes invades the reservoir. This is referred to natural water drive or aquifer influx. In such type of reservoirs, the injected CO_2 must force out the invaded water which in turn reduces the injectivity, where injectivity is defined as a quantitative measure of the ease with which the fluid can be placed in a reservoir formation. Injectivity is directly proportional to permeability and inversely proportional to viscosity [35]. In contrast, a closed type reservoir in which the water invasion is at a very minimum with an ineffective water drive would not require the CO_2 to displace the water and therefore increase injectivity [36]. In terms of an abandoned or depleted reservoir, it may be more feasible to reopen the field for production using the CO_2 for EOR instead of just for storage purposes given the increased injectivity of CO_2.

2.4.2 Enhanced Oil Recovery (EOR)

Existing oil and gas sites may also be viable for EOR. Any storage or enhanced oil recovery CO_2 storage capacity estimates of possible oil and gas fields that can be used as CO_2 sinks provided in the literature may have to be revisited in the future due to changing oil and gas market trends.

Enhanced oil or gas recovery is actually a potential form of storage. Approximately 60% of the CO_2 injected is stored in the reservoir [37]. As mentioned earlier, most EOR projects using CO_2 are predominantly in North America with majority of the CO_2 originating from natural sources and EOR technology using CO_2 has been well established [11]. The two main methods in recovering oil are miscible and immiscible recoveries. The method selection is dependent on the properties of the fluid injected and the behavior properties of the fluid at reservoir conditions. EOR using CO_2 is typical of tertiary recovery in which the reservoir has already undergone primary (pumps and reservoir drive) and secondary recovery such as water flooding or natural gas injection. At this point, the reservoir is already approaching its end of field life. Primary recovery is able to recover about 5% to 40% of the original oil in place (OOIP) [38, 39]. Secondary recovery by water flooding is capable of a further 10% to 20% recovery of OOIP [40]. Tertiary recovery, using various fluids which include CO_2 produce about 7% to 23% recovery OOIP [41, 42]. The amount of OOIP that is recovered using CO_2 varies in literature. For example, Tzimas et al [39] reports that 5% to 15% of OOIP are recovered where Gozalpour et al [37] reports over 8% recovery. Fields in North America report an incremental recovery of 4% to 12% OOIP [43]. Obviously, the amount of oil able to be recovered using any of the recovery methods are strongly dependant mainly on the type of reservoir and the properties of the oil [37, 39]. Therefore, the recovery factor may vary throughout the literature.

Enhanced oil recovery using CO_2 can be performed via two processes, miscible and immiscible displacements. The miscible process is usually used for recovery of medium or light oils where the immiscible process is usually used for heavy oils [44]. The CO_2 flooding improves recovery by expanding the volume of the oil which reduces the viscosity, reducing the interfacial tension between the oil and reservoir rock, developing multiple complex phase changes (e.g., vaporization of oil) that improve its fluidity, and also increasing the mobility of the oil which improve volume sweep (replacement) efficiency [39, 44–46]. CO_2 injection techniques such as water alternating gas (WAG), simultaneous injection of CO_2 and water, CO_2 injection with other gases, continuous CO_2 injection and gravity stable gas injection (GSGI) have been suggested and used for oil recovery using CO_2 [39, 46, 47]. The selection of one technique over the other strongly depends on the reservoir characteristics, such as pressure and temperature, composition, and properties of the oil.

In general, miscible CO_2 displacement which is generally used for light, low-viscosity oils (25-48 API) involves the miscibility of supercritical CO_2 with the oil in all proportions forming a single phase fluid. This mixing causes the oil to swell, its viscosity and surface tension to reduce which consequently improves its ability to flow out of the reservoir. The CO_2's miscibility is strongly affected by pressure. The pressure in the reservoir needs to be higher than the minimum miscibility pressure (MMP) in the range of 10 to 15 MPa. At this pressure, the density of CO_2 is similar to that of the oil. Obviously, this pressure will depend on the composition of the oil and gravity, the reservoir temperature, and the purity of the CO_2 [39, 48]. The amount of oil recovered depends on factors such as the flow stability and

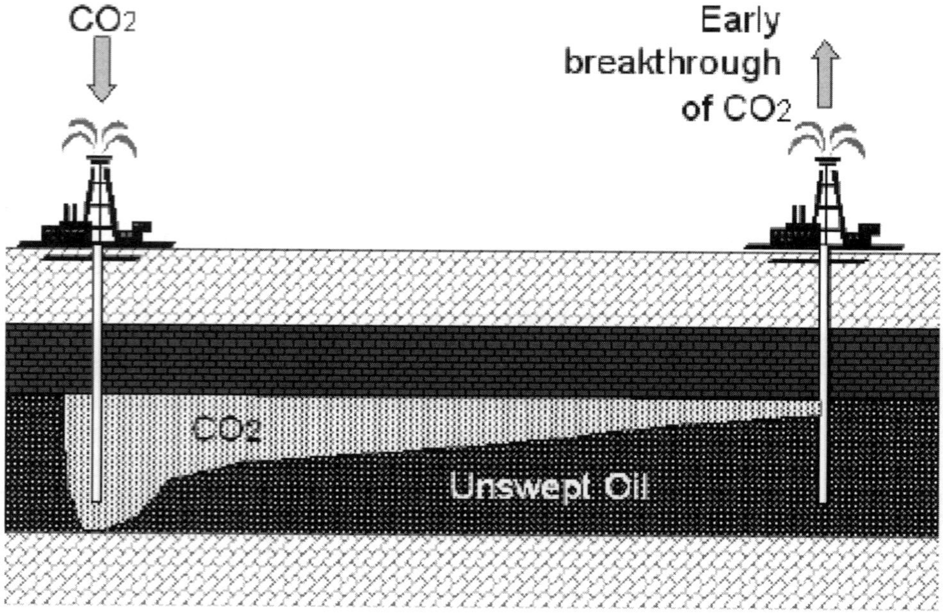

Figure 2-7. Early breakthrough of CO_2 due to unstable flow and gravity effects [39]

the avoidance of early CO_2 breakthrough as a result of unstable flow as shown in Figure 2-7. In order to prevent this from occurring, CO_2 is injected into the reservoir alternately with water (WAG). This involves injecting CO_2 first to swell the oil and increase flow followed by the injection of water to displace the oil bank toward the injection well. Figure 2-8 illustrates this process. The concurrent flow of water and CO_2 reduces the occurrence of viscous fingering due to the existence of multiple phases. The injected CO_2 that comes out with the

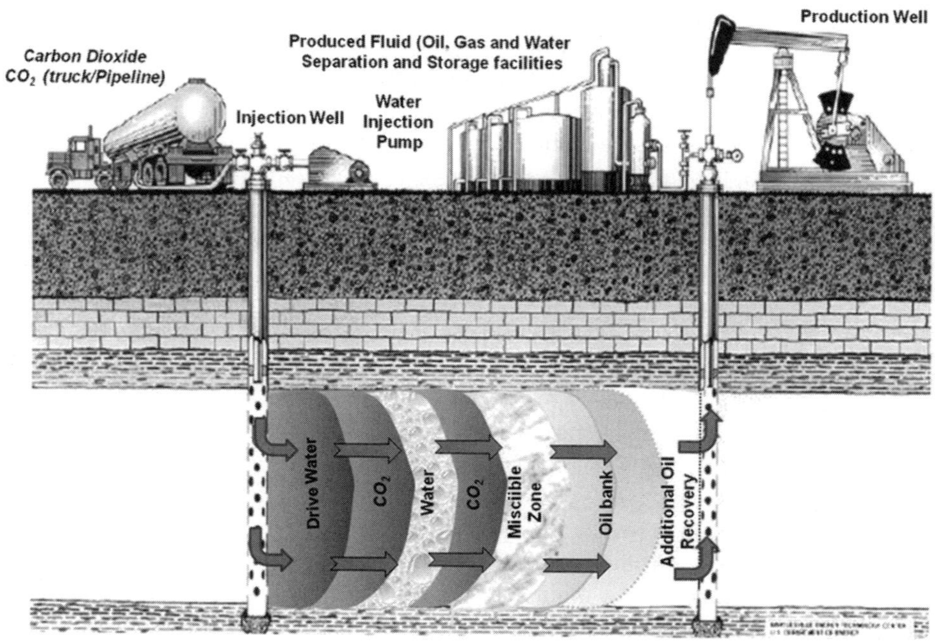

Figure 2-8. Miscible oil recovery using the WAG technique [39]

oil is separated, compressed, and injected back into the reservoir with new CO_2. The amount of CO_2 retained depends on the characteristics of the reservoir with typical ranges of 40% to 67% [37, 40].

Majority of the miscible projects exists in the South Western USA, in the Permian Basin of Western Texas and east of New Mexico. Most of these projects use high purity naturally occurring CO_2 [39]. Only a few projects utilize anthropogenic CO_2 from fertilizer and gas processing plants such as the Weyburn project in Canada which uses CO_2 from a coal gasification plant and other smaller projects such as the Rangely and Shannon ridge EOR fields [37, 39].

Immiscible oil recovery using CO_2 is used in reservoirs which have low pressures (i.e., the MMP is not reached). Although the CO_2 is unable to dissolve completely in the oil due to the unfavourable low pressure, a small amount of CO_2 is still soluble in the oil and can therefore cause some swelling, where swelling is defined as the volume of an oil–CO_2 mixture divided by the CO_2 free oil volume at the same pressure and temperature [44]. The role of CO_2 here is to increase or maintain the reservoir pressure. Only a limited number of projects use the immiscible recovery method where water is less favorable due to geological conditions. The CO_2 is injected in the gravity stable gas injection (GSGI) although WAG is also a possibility. The GSGI is involve in injecting the CO_2 at the crest of the reservoir filling the pore volume of the reservoir rock which in turn creates a gas cap above the oil. This gas cap pushes the oil downward and toward the rim of the reservoirs in which the production wells are located. Figure 2-9 shows a schematic representation of this technique. The presence of water in the reservoir is not favorable for this type of recovery as it inhibits oil flow. Only one large-scale immiscible project using CO_2 exists, the Bati Raman Oil Fields on the Turkish–Iraqi border in which the field contains very heavy oil with low gravity [49].

Immiscible recovery has been applied to only a few projects worldwide as it is less favorable compared to miscible recovery which has been applied to numerous oil recovery projects. Other than economics, immiscible recovery is less favored due the significant amounts of CO_2 needed and the construction of new wells. Immiscible recovery, on the other hand, is able to utilize the same equipment as water flooding (secondary recovery). Immiscible recovery requires a longer duration (approximately 10 years) compared to miscible recovery (approximately 1-5 years) before the well begins producing additional oil. Also, immiscible techniques need to be implemented to the whole reservoir, whereas

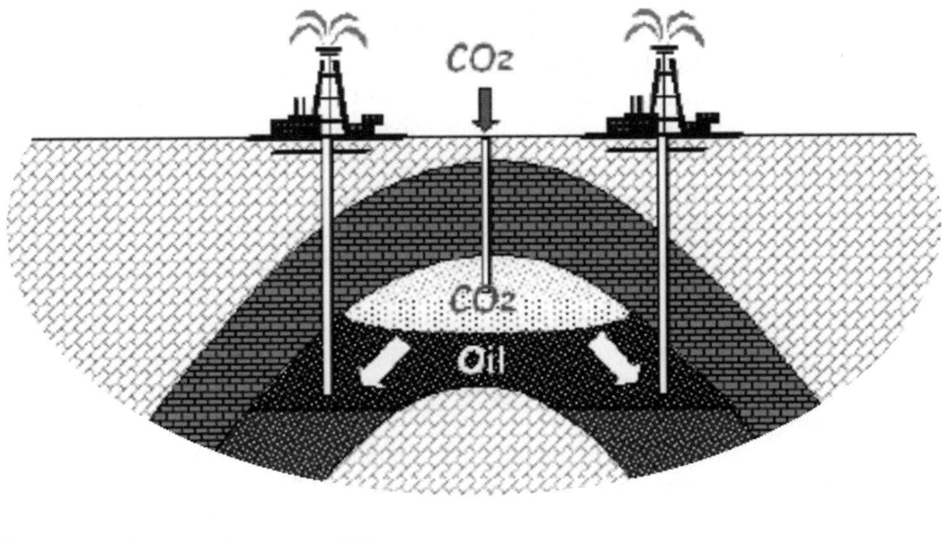

Figure 2-9. Immiscible recovery [39]

TABLE 2-7. Comparison of miscible and immiscible recovery using CO_2 [39]

	Miscible	Immiscible
Project duration	Short (<20 years)	Long (min. 10 years)
Project start r	Before or after water flooding is completed	After water flooding is completed
Oil extraction	Early (~1–3 years)	Late (>5–8 years)
Scale of project	Smaller	Larger
Recovery mechanism	Complex	Simple
CO_2 recycling	Unavoidable	Avoidable
Oil recovery potential	Lower (4–12% OOIP)	Higher (~18% OOIP)
CO_2 storage potential	Lower (0.3 t/bbl)	Higher (up to 1 t/bbl)
Experience	Significant	Little

miscible recovery can be implemented on a much smaller scale on certain parts of the reservoir. Table 2-7 shows the advantages and disadvantages of both means of oil recovery.

Given the potential environmental benefits of carbon capture and storage, immiscible recovery may prove to be somewhat more beneficial compared to miscible recovery due to the large volumes of CO_2 that can be stored in the reservoir. Also, CO_2 breakthrough is very much less likely to happen in immiscible projects favoring the permanent storage of CO_2 but is unavoidable in miscible recovery.

2.4.3 Enhanced Gas Recovery

Enhanced gas recovery using CO_2 is also considered as another feasible recovery method. It has been estimated that up to 95% gas in place can be recovered. This means that even gas reservoirs that have been depleted contain significant amounts of natural gas that can be recovered via EGR [50]. Studies [51–53] have shown that the injection of CO_2 into depleted gas reservoirs can recover additional natural gas and at the same time store CO_2 [51]. It has been estimated that the capacity of these reservoirs to store carbon is 140 GT worldwide [49]. The production of natural gas can then be used to offset the cost of CO_2 injection and storage [51–53]. The concept of EGR is based on the injection of CO_2 at a distance from the production well which would repressurize the reservoir, thus producing additional natural gas. There are currently only a few pilot scale EGR projects, such as the K12B project in the Netherlands [11]. One of the main concerns is that the injected CO_2 will rapidly mix with the natural gas, thus degrading the quality of the recovered natural gas [51]. Techniques to reduce the mixing of CO_2 and natural gas are currently being researched and show promising results [50, 51, 54]. One of the strategies proposed is taking advantage of the high density of CO_2 compared to natural gas and injecting the CO_2 deep into the reservoir and then extracting natural gas form a higher level in the reservoir [51].

Figure 2-10 shows an example of an enhanced gas recovery and storage of carbon dioxide from a power plant. The potential advantages of using depleted gas reservoirs for CO_2 storage are the benefits associated with gaining an additional amount of natural gas through EGR. Another advantage is that gas reservoirs have also demonstrated structural integrity due to their containment of natural gas over a long period of time, and it has also been characterized and well understood due to previous production history. Finally, the existing infrastructures, such as wells and pipelines, could be used in assisting EGR and CO_2 storage.

A number of technical issues will need to be addressed due to uncertainties that exist in certain areas of EOR using anthropogenic CO_2. These include the effect of impurities in the anthropogenic CO_2 mixtures, the feasibility of using existing pipeline infrastructure, and the applicable design codes that would be needed. As previously mentioned, impurities will

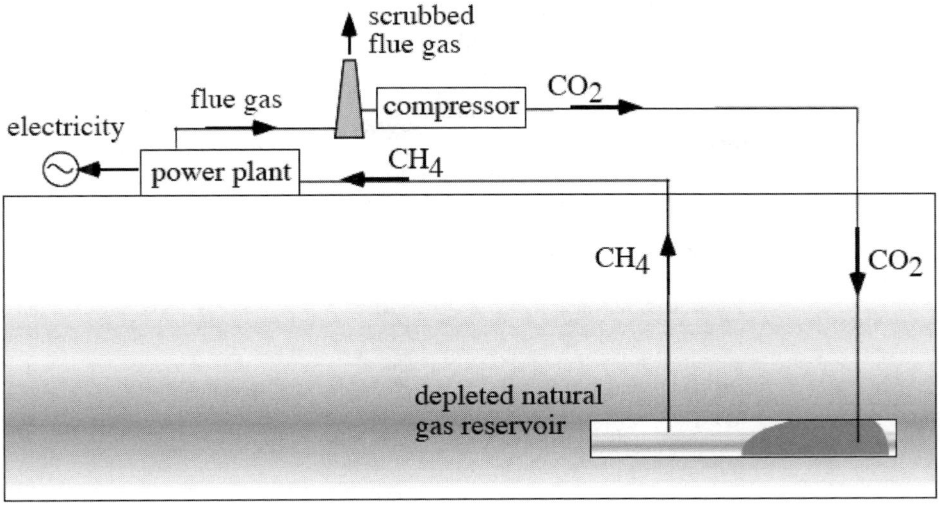

Figure 2-10. Schematic representation of EGR and CO_2 storage

affect the storage capacity and technical feasibility for both EOR and storage of CO_2 due to the effect on the trapping mechanisms of the aquifer and also the Minimum Miscibility Pressures (MMP) for EOR [20].

Therefore, taking into consideration the impact of different components on the storage of the CO_2 stream either in saline aquifers, depleted oil and gas fields or the use in EOR applications, a specification will need to be drawn up in order to satisfy the requirements of transport and storage. An economic analysis in determining the optimum option will need to be based on a trade off between the cleanup cost after the capture process and transportation and storage costs.

2.4.4 Enhanced Coal Bed Methane

Sequestration of CO_2 in coal is another means of potentially providing a market-based solution for carbon capture and storage. Enhanced coal bed methane involves injecting CO_2 into a coal bed which in turn would occupy pore space and also adsorb or bond with the carbon in the coal itself [55]. The affinity of coal to adsorb gaseous CO_2 is much higher than that of methane which results in the freeing up of methane which has a lower affinity to coal [11].

The geologic factor that effect coal bed methane producibility and CO_2 sequestration potential are shown in Table 2-8, some of which are interrelated with each other [56]. For example, the quality of coal in the formation will in turn determine the gas content. This plays a significant role in the productibility of coal bed methane and also the sorption capacity which is important in terms of sorption potential of the reservoir.

CO_2 has been used in ECBM applications worldwide, for example the Allison project North of New Mexico in the San Juan Basin where the CO_2 was sourced from the McElmo Dome in Colorado via the Kinder Morgan pipeline, in the Alberta Basin, Canada [57] and also the Black Warrior Basin in Alabama [56]. The storage of CO_2 in coal seams will depend on the future fate of the coal in the formation, in particular shallow coal, where the potential for mining and continued storage may result in a conflict [11].

Geological storage projects would depend on three components:

1. The Capture process,
2. The mode of transportation to the storage site,
3. Injection and storage in a reservoir.

TABLE 2-8. Similarities and differences in the screening criteria for assessing coal bed methane predictability and CO_2 sequestration potential I Modified from (Paschin et al [56] and IPCC [11])

Coal Bed Methane Producibility	CO_2 Sequestration Potential
Hydrodynamics – For example, reservoir pressure which determines the gas sorption capacity and water chemistry which also plays a part in the ability to dewater the formation and subsequent disposal of the water	**Hydrodynamics** – For example, reservoir pressure which determines the gas sorption capacity and water chemistry which also plays a part in the ability to dewater the formation and subsequent disposal of the water
Gas content – For example, thermogenic vs. biogenic and proportion of other gases determines the ease and quantity of a which methane can be produced	**Sorption capacity** – For example, pressures and temperatures, the rank (from low volatile bituminous to high-volatile bituminous), grade, and maceral composition of coal
Coal quality and content – Important in terms of amount of methane produced (volumetric ratio of adsorbable CO_2:CH_4 ranges from as low as one for mature coals such as anthracite, to ten or more for younger, immature coals such as lignite)	**Coal quality** – Important in terms in terms of the amount of methane extracted and the remaining pore space suitable to be taken up for CO_2 (volumetric ratio of adsorbable CO_2:CH_4 ranges from as low as one for mature coals such as anthracite, to ten or more for younger, immature coals such as lignite)
Stratigraphy – Important in terms of the location of the coal resources – For example in the upper or lower part of the formations, thickness and coal utilization as these are primary determinants of the completion techniques that are chosen – Homogeneous and confined coal seam(s) that are laterally continuous and vertically isolated are desired – Suitability of coal geometry (a few, thick seams are preferred rather than multiple, thin seams)	**Stratigraphy** – Important in terms of the location of the coal resources – For example, in the upper or lower part of the formations, thickness and coal utilization as these are primary determinants of the completion techniques that are chosen – Homogeneous and confined coal seam(s) that are laterally continuous and vertically isolated are desired – Suitability of coal geometry (a few, thick seams are preferred rather than multiple, thin seams)
Structural geology – Relates to fold geometry, fault geometry, and fracture architecture – Minimal faulting and folding is desired	**Structural geology** – Relates to fold geometry, fault geometry, and fracture architecture – Minimal faulting and folding is desired
Permeability – Generally decreases with increasing depth – Low permeability increases difficulty in injection of CO_2	**Permeability and seal integrity** – Generally decreases with increasing depth – Low permeability increases difficulty in injection of CO_2 – Seal integrity determines if the CO_2 will continue to remain in the ground long after the ECBM operation has been decommissioned

One of the most overlooked aspect of CCS is CO_2 transportation via pipeline due to the current experience with CO_2 pipelines in the USA. While, a lot can be learned from this experience, it must be emphasized that anthropogenic CO_2 differs from CO_2 from naturally occurring sources which have high purity. This is due to the capture process itself which introduces impurities into the CO_2 stream. Hence, the chapters that follow will be focusing on pipeline transportation of CO_2 and the implications of transporting anthropogenic CO_2.

2.5 PIPELINE TRANSPORTATION SIGNIFICANCE

The options for transportation of CO_2 and CO_2 stream with impurities are fully described in Chapter 5 with a focus on pipeline transportation and system design needs. However, the following highlights the reasons for and experience with transportation of CO_2 by pipelines globally.

Pipelines are one of the most cost-effective and safest options for transcontinental transportation of energy. It is therefore the most widely used transport mode for hydrocarbons in the oil and gas industry. Similarly, pipeline transportation of CO_2 does exist although primarily for EOR purposes. It is, therefore, considered to be the most viable option for the transport of anthropogenic CO_2 for CCS.

2.5.1 Existing Experience with CO_2 Pipeline Transportation

There are currently 2400 km of CO_2 pipelines in operation. In addition, there are about 700 km of CO_2 distribution lines that branch from the trunk line. Therefore, in total, there are about 3100 km of pipelines worldwide with a capacity of transporting 44 Mt per year of CO_2 [9]. The sources of the CO_2 transported can be divided into two categories which are CO_2 from naturally occurring sources and from anthropogenic sources. Tables 2-9 and 2-10 show the details of some operational CO_2 pipelines transporting natural and anthropogenic sources of CO_2.

The two main carbon capture and storage projects ongoing in the North Sea are the Snohvit and Sleipner projects both of which are for storage purposes. The Sleipner project, which has been operated by Statoil since 1996, does not require transportation of the captured CO_2 as the CO_2 is separated from produced natural gas on the platform and stored in a deep saline aquifer formation (Utsira Sand) about 1000 m beneath the seabed underneath the platform [26]. The Snøvhit project is the only project for which an offshore pipeline exists, and it has recently been commissioned in 2008. The pipeline is 203.2 (8 in.) in diameter and extends 150 km from the LNG plant, where the CO_2 is captured and transported for storage in the Tubåen formation [58, 59].

The following section provides further details of the current worldwide experience with CO_2 pipelines indicating where it may be relevant or transferable to the establishment of pipeline networks that have both onshore and offshore components.

2.5.1.1 Transport of CO_2 from Natural Sources

The main sources of natural CO_2 in the US are the McElmo and Sheep Mountain Domes in Colorado, the Bravo Dome in New Mexico and the Jackson Dome in Mississippi [60]. CO_2 from these sources is transported via pipeline networks to oil fields in Texas and Louisiana, where it is used for EOR. Outside of the US, the main pipeline transporting natural CO_2 is the Bati Raman pipeline in Turkey. Details of these pipelines are provided in Table 2-9.

2.5.1.2 Transport of Anthropological Sources of CO_2

The main industrial sources of CO_2 in the USA are located in Texas, Oklahoma, Wyoming, and North Dakota. As with natural sources of CO_2, the anthropogenic CO_2 in the USA is used exclusively for EOR projects. Three particular projects are highlighted for discussion to draw out the existing experience with anthropogenic CO_2 transport, Canyon Reef

TABLE 2-9. Some examples of existing long distance CO_2 pipelines transporting natural sources of CO_2 (modified from Gale and Davison [61], Seevam et al [62], $KMCO_2$ [8] and Denbury [3])

Pipeline	Location	CO_2 Capacity (Mt/y)	Length (km)	MAOP (bar)	Source	Year
Cortez	USA	19.3	808	186	McElmo Dome	1984
Sheep Mountain	USA	9.5	660	132	Sheep Mountain	1983
Bravo	USA	7.3	350	165	Bravo Dome	1984
Central Basin Pipeline	USA	20	278	170	Denver City Hub	1985
North East Jackson Dome (NEJD)	USA	–	295	152	Jackson Dome	1986
Bati Raman	Turkey	1.1	90	170	Dodan Field	1983

Carriers (CRC), Weyburn, and LaBarge. Table 2-10 shows the details of some of the pipelines transporting anthropogenic CO_2.

The first CO_2 pipeline was built by Danberry in 1963 for pure CO_2 followed by CRC (between 1970 and 1972) to transport anthropogenically produced CO_2 from gasification plants in Texas to the SACROC field. In 1988, this pipeline was connected to the Central Basin pipeline and the Comanche Creek pipelines. The Central Basin Pipeline receives CO_2 from the Cortez, Sheep Mountain and Bravo pipelines, which are transporting CO_2 from the natural sources located in Colorado and New Mexico, and delivers it from Denver City to West Texas. Therefore, this pipeline is transporting a mix of anthropogenic and natural sources of CO_2.

The Weyburn oilfield in Saskatchewan, Canada, utilizes CO_2 from a synfuels plant, which is producing methane from coal in North Dakota, for EOR. The CO_2 is transported via a 328 km pipeline to the Weyburn field for EOR. The Weyburn pipeline transports CO_2 at a higher pressure (about 186 bar) than any other pipeline. The Weyburn pipeline is interesting as it crosses a national border, and therefore, there were numerous government agencies involved in its construction and development. There were also right of way issues to acquire a corridor for building the pipeline. It should also be noted that the pipeline route passes through mostly unpopulated areas, therefore, a broad right of way (ROW), which is defined as a corridor of land at which a pipeline operator has the right to conduct activities in accordance with an agreement with the land owner, was possible (Table 2-9).

The Labarge project in Wyoming is an example of the transport of both natural and anthropogenic CO_2 in one facility. The LaBarge field was originally developed to recover and sell methane [63]. The composition of the raw feed gas from the field is 66% CO_2, 22% CH_4, 7% N_2, 4.5% H_2S, and 0.5% He. Initially, the field was not thought to be viable as the CH_4 had low-energy content. However, with a gas processing facility, CO_2, S, and He can be extracted and sold in addition to the CH_4, making the plant economical. The raw feed gas stream is dehydrated and piped to the Labarge plant along a 64 km, 711.2 mm (28 in.) pipeline. The CO_2 is extracted from the stream and then compressed for pipeline transportation for EOR in the Rocky Mountain area (e.g., via the Bairoil pipeline).

Currently, all the operating pipelines transport CO_2 at high pressures ranging from 132 to 204 bar. Therefore, the CO_2 is transported in the supercritical or dense phase. The routing of the CO_2 pipelines in the United States is mostly through dessert terrain with low population densities [64]. However, dense phase transport raises important regulatory issues as current pipeline regulations do not completely consider transport in this phase. This may be of particular concern as onshore sections of a pipeline may run closer to population centres thus posing a greater risk to the public. For shorter distances, it may be beneficial to

TABLE 2-10. Some examples of existing long distance CO_2 pipelines transporting anthropogenic sources of CO_2 (modified from Gale and Davison [61] and Seevam et al [62])

Pipeline	Location	CO_2 Capacity (Mt/y)	Length (km)	MAOP (bar)	Source	Year
Canyon Reef Carriers	USA	5.2	225	140	Gasification Plant	1972
Bairoil	USA	8.3	180	–	La Barge Gas Plant	1986
Val Verde	USA	2.5	130	140	Val Verde Natural Gas Plant	1998
Weyburn	USA/Canada	5	328	186/204	Gas Plant	2000
Snøhvit	Statoil	0.7	153	150	Liquefied Natural Gas	2008

transport the CO_2 in the gas phase which are at much lower pressures than dense phase or supercritical CO_2. Examples of this type of transport include the pipelining of CO_2 for use in greenhouses in Holland [65].

2.6 CO_2 TRANSPORTATION FOR CCS — A BACKGROUND

This section provides a background to pipeline transportation of CO_2 containing impurities, however, as indicated previously, Chapters 5 and 7 actual design condition, material requirement, routing, etc.

A- New Pipelines

Carbon capture and storage projects differ significantly from past experience as they will be focusing on anthropogenic sources from major polluters such as fossil fuel power plants. The distances between the CO_2 sources and sinks will determine the length of the pipelines which could potentially involve both long distance onshore and offshore pipelines. Currently operating CO_2 pipelines are mostly high pressure, onshore, and routed through areas of low population density. Early work on CO_2 pipelines was conducted in the late 1970s and 1980s due to the development of EOR in the USA using CO_2 flooding techniques [66–69]. During this time, gas phase transport was considered undesirable for long distance transportation of CO_2 because of the low density (and consequent requirement for a large diameter pipe), throughput implications for EOR and high-pressure drops particularly in hilly terrains [Meissner [70]; 8–10]).

It has also been acknowledged that the design of CO_2 pipelines for EOR in the USA is conservative with excessive investment costs [5]. This is because similar techniques to natural gas cannot be employed as there are key differences between designing supercritical CO_2 pipelines and natural gas pipelines [3–5, 69]. In other words, many of the general design rules for natural gas are not applicable to CO_2 pipelines [5]. This is due to the non-linear behavior of supercritical CO_2 within the conventional natural gas pipeline operating regions and therefore any linear approximations used in natural gas pipeline design will yield inaccurate results [5, 69]. The non-uniform or unusual properties of CO_2 are associated with the location of the critical point in its phase diagram which is in close proximity to its pipeline operating conditions. This non-uniform behavior is further accentuated by the presence of impurities occurring in anthropogenic CO_2 [5, 66, 67, 71]. Refer to Chapter 5 for further details.

Fossil fuel power plants will produce CO_2 with varying combinations of impurities depending on the capture technology used. It is desirable to design a transport system that is capable of carrying flows from all capture technologies retaining as many of the impurities as is allowable from environmental, social, and economic constraints. The transported CO_2 for CCS could be at pressures close to the critical point in which conditions it is most prone to discontinuous behavior [72]. The type, combination, and quantity of impurities in the CO_2 can have significant effects [64] in terms of pipeline flow assurance and design aspects. This is due to the dramatic changes they induce in physical properties, such as the phase equilibrium, critical point, density, viscosity, and compressibility of CO_2 [5, 67, 73, 74]. The physical properties of anthropogenic CO_2 will have a direct effect on the development of pipeline regulations, pipeline design hydraulics, property equations, such as equations of state, operation, maintenance, routing, and economic costing. There is little published work concerned with transporting CO_2 with impurities, such as SO_X, NO_X, and Argon, all of which can be present in the CO_2 captured for CCS. This is important because the effects of these impurities in a CO_2 pipeline needs to be understood as power plant or anthropogenic CO_2 could potentially contain these impurities.

B- Use of Existing Pipelines

Another anthropogenic CO_2 transport option being evaluated in some parts of the world is the re-use of existing pipeline infrastructure. This option could provide an alternative for pipelines that are due to be decommissioned or removed from service. However, re-using infrastructure would involve additional requirements to those already necessary for new pipelines, such as up rating and change of use requirements, integrity assessment, capacity requirements, fracture control, and also the availability of the pipeline infrastructure.

Note: The transportation of anthropogenic CO_2 or CO_2 containing impurities will be discussed technically in Chapter 5. Requirement for pipe materials is detailed in Chapter 7.

REFERENCES

[1] Pascala, S. and Socolow, R., 2004, "Stabilisation Wedges: Solving the Climate Problem for the Next 50 Years with Current Technologies," Science, vol 305.

[2] HSE, 2011, General Hazards of CO_2- EET Programme Work Stream, Accessed June 2011 http://www.hse.gov.uk/carboncapture/carbondioxide.htm.

[3] Denbury, 2008, Personal Communication, Connell Rader, Operations Manager Denbury Resources Inc.

[4] Kinder Morgan CO_2, 2008, Personal communication, Marco A Roncal, Pipeline Engineer.

[5] Farris, C., 1983, "Unusual Design Factors for Supercritical CO_2 pipeline" Energy Progress, vol. 3, pp. 150–158.

[6] Sulzer Patent, 1948, Verfahren zur Erzeugung von Arbeit aus Warme Swiss Patent 599. 1948.

[7] Dorstal, V., Driscoll, M. J. and Heizlar, P., 2004, "A Supercritical Carbon Dioxide Cycle for Next Generation Nuclear Reactors," PhD Thesis, Department of Nuclear Energy, Massachusetts Institute of Technology, USA.

[8] West, J. M., 1974, "Design and Operation of a Supercritical CO_2 Pipeline-Compression System, SACROC unit, Scurry County, Texas," Society of Petroleum Engineers Permian Basin Oil and Gas Recovery Conference, Texas, USA.

[9] Zhang, Z. X., Massarato P. and Rudolph, V., 2006, "Optimization of Pipeline Transport for CO_2 sequestration," Energy Conversion and Management, Vol. 47: pp. 702–715.

[10] Mohitpour, M., Golshan, H. and Murray, A., 2003, "Pipeline Design and Construction," 1st edition, ASME publication, USA.

[11] IPCC, 2005 "IPCC Special Report on Carbon Dioxide Capture and Storage," Working Group III of the Intergovernmental Panel on Climate Change, 2005, Eds. Metz, B., Davidson, O., de Coninck, H. C., Loos, M. and Meyer, L. A., Cambridge University Press, Cambridge, United Kingdom and New York, NY, USA.

[12] DTI, 2003, "Review of the Feasibility of Carbon Dioxide Capture and Storage in the UK," Cleaner Fossil Fuels Programme, report number: URN 03/1261 Department of Trade and Industry, UK.

[13] IEA, 2003 "Potential for Improvement in Gasification Combined Cycle Power Generation with CO_2 Capture," Report Number PH4/19, International Energy Agency Green House Gas R&D Programme.

[14] IEA, 2004 "Improvement in Power Generation with Post Combustion Capture of CO_2," Report no. PH4/33, International Energy Agency.

[15] Hendriks, C., 1994, "Carbon Dioxide Removal from Coal Fired Power Plants," PhD Thesis, Utrecht University, Netherlands.

[16] Reimer, P. W. F. and Ormerod, W. G., 1995, "International Perspectives and the Results of Carbon Dioxide Capture, Disposal and Utilisation Studies," Energy Conversion Management, vol 36 (6–9), pp. 813–818.

[17] DTI, 2005, "Optimising CO_2 capture from Pulverized Coal Plants-Post Combustion Capture with Amine Solvents," Carbon Abatement Technologies Programme, report no. DTI/Pub URN 05/1575, Imperial College London, UK.

[18] Drage, T. C. and Arenillas, A., 2006, "Comparison of Pre and Post Combustion CO_2 Adsorbent Technologies," 8th Green House Gas Technologies Conference (GHGT-8), Norway.

[19] Rao, A. B. and Rubin, E. S., 2002, "A Technical, Economic and Environmental Assessment of Amine based CO_2 Capture Technology for Power Plant Green House Gas Control," Journal of Environmental Science and Technology, Vol. 36, pp. 4467–4475.

[20] IEA, 2004(b), "Impact of Impurities on CO_2 Capture, Transport and Storage," Report no. PH4/32, study conducted by SNC Laevulin, August 2004.

[21] White, V., Allan, R. and Miller, E., 2006, "Purification of Oxyfuel Derived CO_2 from Sequestration or EOR," 8th Green House Gas Technology Conference (GHGT-8).

[22] Anheden, M., Andersson, A., Bernstone, C., Eriksson, S., Yan, Liljemark, S. and Wall, C., 2005, "CO_2 Quality Requirement for a System with CO_2 Capture, Transport and Storage," Proceedings of the 7th Greenhouse Gas Technologies Conference GHGT-7, Vancouver, vol 2, pp. 2559–2563.

[23] de Visser, E. and Hendriks, C., 2007, "Towards Hydrogen and Electricity Production with Carbon Capture and Storage-DYNAMIS Quality Recommendation," DYNAMIS, Project No.: 019672.

[24] Oosterkamp, A. and Ramsen, J., 2008, "State of the art Overview of CO_2 Pipeline Transport with Relevance to Offshore Pipelines," Polytec Report no. POL-O-2007-138-A, Polytec, Norway.

[25] Hendriks, C., Hagedoorn, S. and Warmenhoven, H., 2007, "Transportation of Carbon Dioxide and Organisational Issues of CCS in the Netherlands," Report prepared by Ecofys for EnergieNed, the Ministry of Economic Affairs and the Ministry of Housing, Spatial Planning and the Environment.

[26] Dolf-Gielen, J. P. and Podkánski, J., 2004, "Energy Technology Analysis — Prospects for CO_2 Capture and Storage" International Energy Agency and Organisation for Economic Co-Operation and Development, Paris, France.

[27] IEA, 2001 "Putting Carbon Back into the Ground," International Energy Agency Green House Gas R&D Programme,website: http://www.ieagreen.org.uk/putcback.pdf.

[28] Race, J. M., Seevam, P. N. and Downie, M. J., 2006, "Challenges for Offshore Transport of Anthropogenic Carbon Dioxide," 26th International Conference on Offshore Mechanics and Arctic Engineering, California, USA.

[29] Holloway, S., Bentham, M. S., Kirk, K. L. and Angel, M., 2006, "Carbon Capture and its Storage - An Integrated assessment," Chapter 2, (eds.) Shackley S, Gough C, Ashgate Publishing; Aldershot, UK.

[30] Pearce, J. M., Holloway, S., Wacker, H., Nelis, M. K., Rochelle, C. and Bateman, K., 1996, "Natural Occurrence as Analogues for Geological Disposal of Carbon Dioxide," Energy Conversion Management, Vol. 37(6–8), pp. 1123–1128.

[31] DTI and BGS, 2006, "Industrial Carbon Dioxide Emissions and Carbon Dioxide Storage potential in the UK," Report No, COAL R308 DTI/Pub URN 06/2027.

[32] Solomon, S., 2007, "Carbon Dioxide Storage: Geological Security and Environmental Issues — Case Study on the Sleipner Gas field in Norway," Bellona Report.

[33] Steeneveldt, R., Berger, B. and Torp, T. A., 2006, "CO_2 Capture and Storage: Closing the Knowing and Doing Gap," Trans IChemE, Part A, Chemical Engineering Research and Design, Vol 84(A9), pp. 739–763.

[34] Kovsek, A. R., 2002, "Screening Criteria for CO_2 Storage in Oil Reservoirs," Journal of Petroleum Science and Technology, Vol. 20(7–8), pp. 841–866.

[35] Drake, L. P., 1978, "Fundamentals of Reservoir Engineering," Elsevier, Amsterdam, The Netherlands.

[36] Bachu, S., W. D., Gunter, E.H. and Perkins, 1994, "Aquifer Disposal of CO_2: Hydrodynamic and Mineral Trapping," Energy Conversion and Management, Vol. 35(4), pp. 269–279.

[37] Gozalpour, F., Ren, S. R. and Tohido, B., 2005, "CO_2 EOR and Storage in Oil Reservoirs," Journal of Oil and Gas Science and Technology, Vol 60(3), pp. 537–546.

[38] Holt, T., Jensen, J. L., and Linderberg, E., 1995, "Underground Storage of CO_2 in Aquifers and Oil reservoirs," Energy Conversion and Management, Vol 36(6–9), pp. 535–538.

[39] Tzimas, E., Georgakaki., A., Garcia Cortes, C. and Peteves, S., D., 2005, "Enhanced Oil Recovery using Carbon Dioxide in the European Energy System," Rep. No. EUR 21895 EN, DG JRC, Institute for Energy, Petten, The Netherlands.

[40] Bondor, P. L., 1992, "Application of Carbon Dioxide in Enhanced Oil Recovery," Energy Conversion and Management, Vol. 33(5), pp. 579–586.

[41] Martin, F., D., and Taber, J., J., 1992, "Carbon Dioxide Flooding," Journal of Petroleum Technology, Vol. 44(4), pp. 396–400.

[42] Moritis, G., 2003, "CO_2 Sequestration Adds New Dimension to Oil, Gas Production," Oil and Gas Journal, Vol. 101(9), pp. 71–83.

[43] Goodyear, S. G., Hawkyard, I. R., Masters, J. H. K., Woods, C. L., and Jayasekera, A., J., 2003, "Subsurface Issues for CO_2 Flooding of UKCS reservoirs," Transaction of IChemE, Vol. 81, Part A.

[44] Sankur, V., Emmanuel, A. S. and Di Julio, S. S., 1986, "Recovery of Heavy Oils by CO_2 Injection," Energy Progress, Vol. 6(2), pp. 91–96.

[45] Mangalsingh, D. and Jagai, T., 1996, "A Laboratory Investigation of the Carbon Dioxide Immiscible Process," SPE Latin America/Caribbean Petroleum Engineering Conference, Port of Spain, Trinidad.

[46] Klins, M., A., 1984, "Carbon Dioxide Flooding," D. Reidel Publishing Co., Boston, MA, USA.

[47] Klins, M., A. and S. M. Farouq Ali, 1982, "Heavy oil production by carbon dioxide injection," Journal of Canadian Petroleum Technology, Vol. 21(5), pp. 64–72.

[48] Metcalfe, R., S., 1982, "Effects of Impurities on Minimum Miscibility Pressures and Minimum Enrichment Levels for CO_2 and Rich Gas Displacements." SPE Journal, Vol. 22 (2), pp. 219–225.

[49] IEA, 2000, "Barriers to overcome in implementation of CO_2 capture and storage — Storage in disused oil and gas field," Report no. PH3/22, International Energy Agency.

[50] Oldenburg, C. M. and Benson, S. M., 2002, "CO_2 Injection for Enhanced Gas Production and Carbon Sequestration," Society of Petroleum Engineers SPE-74367, SPE International Petroleum Conference and Exhibition in Mexico, Villahermosa, Mexico.

[51] Oldenburg, C. M., Pruess, K. and Benson, S. M., 2001, "Process modelling of CO_2 injection into natural gas reservoirs for carbon sequestration and enhanced gas recovery," Energy and Fuels, Vol. 15, pp. 293–298.

[52] Van der Burgt, M. J., Cantle, J. and Boutkan, V. K., 1992, "Carbon Dioxide Disposal from Coal-Based IGCC's in Depleted Gas Fields." Energy Conversion and Management, Vol. 33(5–8), pp. 603–610.

[53] Blok, K., Williams, R. H., Katofsky, R. E. and Hendriks, C. A., 1997, "Hydrogen Production from Natural Gas — Sequestration of Recovered CO_2 in Depleted Gas Wells and Enhanced Natural Gas Recovery," Journal of Energy, Vol. 22(2,3), pp. 161–168.

[54] Mamora, D. D. and Seo, J. G., 2002, "Enhanced Gas Production by Carbon Dioxide Sequestration in Depleted Gas Reservoirs," SPE Annual Technical Conference and Exhibition, San Antonio, Texas, USA.

[55] Wickstrom, L. H., Slucher, E. R., Baranoski, M. T. and Mullett, D. J., 2008, "Geologic Assessment of the Burger Power Plant and Surrounding Vicinity for Potential," Open-File Report 2008-1, Ohio Department of Natural Resources Division of Geological Survey, USA.

[56] Paschin, J. C., Groshong, R. H. and Caroll, R. E., 2001, "Enhanced Coalbed Methane Recovery Through Sequestration of Carbon Dioxide: Potential for a Market-Based Environmental," NETL 1st National Conference on Carbon Sequestration, USA.

[57] Gunter, W. D., Mavor., M. J. and Robinson, J. R., 2005, "CO_2 storage and enhanced methane production: field testing at Fenn-Big Valley, Alberta, Canada, with application," Proceedings of the 7th International Conference on Greenhouse Gas Control Technologies (GHGT-7), Vol. 1, pp. 413–422, Vancouver, Canada.

[58] Statoil Hydro, 2007, website: http://www.statoilhydro.com/en/technologyinnovation/protectingtheenvironment/carboncaptureandstorage/pages/researchanddevelopment.aspx.

[59] Oil and Gas, 2008, "Statoil Hydro starts CO_2 Injections from Snohvit," Uchenna Izundu, International Editor Oil and Gas Journal, April 2008.

[60] Stern, N., 2006, "STERN REVIEW: The Economics of Climate Change," Final report published 30/10/06 by HM Treasury Department, website:http://www.hmtreasury.gov.uk/independent_reviews/stern_review_economics_climate_change/sternreview_index.cfm.

[61] Gale, J., and Davison, J., 2004, "Transmission of CO_2-Safety and Economic Considerations" Energy, Vol. 29 (9–10), pp. 1319–2004.

[62] Seevam, P. N., Race, J. M. and Downie, M., J., 2007, "Carbon Dioxide Pipelines for Sequestration in the UK – Engineering Gap Analysis," Proceedings of the conference on Transmission of CO_2, H_2 and Biogas: Exploring New Uses for Natural Gas Pipelines, Amsterdam, Netherlands.

[63] Hunter, J. K.and Bryan, L. A., 1987, "LaBarge Project: Availability of CO_2 for Tertiary Projects," Journal of Petroleum Technology, Vol. 39(11), pp. 1407–1410.

[64] Barrie, J., Brown, K., Hatcher, P. R. and Schelhase, H. U., 2005, "Carbon Dioxide Pipeline: A Preliminary Review of Design and Risks," 7th International Conference on Greenhouse Gas Control Technologies (GHGT-7), Vancouver, Canada.
[65] Van Bermum, E., 2009, Website on OCAP — Organic Carbon Dioxide for the Assimilation of Plants: http://blog.mariposítas.org/horticulture/ocap-organic-carbon-dioxide-for-assimilation-of-plants.
[66] Hein, M., 1986, "Pipeline design Model Addresses CO_2's challenging behaviour," Oil and Gas Journal, June 1986, pp. 71–75.
[67] Wallace, C. B., 1985, "Drying supercritical CO_2 Demands Care," Oil and Gas Journal, vol 83(25), pp. 98–99.
[68] George, B. A. and Heiman, B. S., 1986, "Design and Operating Experience of Amoco's Bravo Dome Facility," Energy Progress, Vol 6(4), pp. 219–221.
[69] Wallace, C. B., 2008, Personal communication, private consultant for Kinder Morgan.
[70] Meissner, R., E., 1980, "Purifying CO_2 for use in Enhanced Oil Recovery," World Oil, pp. 189–198.
[71] King, G. G., 1982a, "Here Are Key Design Considerations for CO_2 Pipelines." Oil and Gas Journal, Vol 80(39), p. 219.
[72] George, B. A., 1982, "Solving Unusual Design Problems with Equation of State," Proceedings of the 61st Annual Convention, Gas Processors Association, Dallas, Texas.
[73] Mohitpour, M., Golshan, H. and Murray, A., 2007, "Pipeline Design and Construction," 3rd edition, ASME Publication, USA.
[74] Recht, D. L., 1984, "Design Consideration for Carbon Dioxide Pipelines," Pipeline Gas Journal, Vol. 61(3), pp. 53–54.

Chapter 3

ANTHROPOGENIC CO_2 PROPERTIES

3.1 ANTHROPOGENIC CO_2 COMPOSITION

This chapter highlights and explores compositional quality scenarios of anthropogenic CO_2 for the different capture technologies and pipeline transportation. Anthropogenic CO_2 originates largely from power plants which employ the post-combustion, pre-combustion and oxyfuel capture technologies. The levels of impurities present are driven by the economic, environmental and safety considerations for both capture technology and pipeline transportation. While this chapter highlights both globally, Chapter 5 details pipeline related issues with respect to impact of impurities on pipeline design and specifications.

Whether the source is natural or anthropogenic, the CO_2 stream will always contain impurities. It has been established that anthropogenic CO_2 will likely contain more and different impurities compared to natural sources of CO_2. Their influence on the properties of CO_2 will have important implications for the safety, regulation, design, and operation of CO_2 pipelines. Therefore, this chapter focuses on the current compositional specification for currently operating pipelines and provides an overview of current knowledge of the CO_2 quality from the capture plants.

The CO_2 used for EOR demands relatively high purity [1], and therefore must adhere to a stringent specification for both miscible and immiscible oil recoveries. Miscible recovery can be used when the reservoir pressure is large enough and occurs when the injected CO_2 mixes with the oil such that the interfacial tension between the oil and the CO_2 disappears and the capillary number becomes infinite. This technique is generally used for reservoir depths >1200 m and oil lighter than 22 API gravity. Immiscible recovery is when the injected CO_2 remains physically distinct from the oil, causing the oil to swell, thus reducing the density and viscosity, increasing permeability, increasing pressure buildup, and improving the mobility. This is less efficient than miscible recovery and is used for very low-pressure reservoirs and/or when the oil gravity is relatively high [2].

By and large, CO_2 used for EOR demands relatively high purity due to the minimum miscibility pressure (MMP) of CO_2 with crude oil. The MMP is one of the important parameters related to miscible recovery of oil, which is defined as the minimum pressure at which CO_2 can be fully miscible with oil [3]. It is dependent on the composition of the crude oil, purity of the CO_2, and also reservoir conditions. Hence, a miscible CO_2 displacement technique can only be implemented when CO_2 can be injected at a pressure higher than that of the MMP, which in turn must be lower than the reservoir pressure. It is therefore essential to limit impurities such as N_2, H_2, Sulphur, SO_x, NO_x, and CH_4 which raise the MMP [4] and is therefore undesirable for EOR. However, H_2S increases the miscibility of CO_2 in oil, and, therefore, it could be seen as a desirable impurity in the CO_2 stream. It should also be recognized that there is an economic incentive to remove impurities for EOR, owing to the revenue stream from the quality of the produced oil, which would not be present for storage applications.

Most of the current oil miscible displacement projects using CO_2 are located in the southwestern USA, in the Permian basin of western Texas and eastern New Mexico. Other projects have been developed in the Rocky Mountain region, the Gulf of Mexico, and the Midwest. Most of these projects utilize naturally occurring CO_2 from high-pressure high-purity geological deposits. For example, the McElmo dome in Colorado contains over 283 million m^3 of CO_2 at a pressure of about 13.5 MPa [4, 5]. Only a small number of projects use CO_2 captured from anthropogenic sources, such as waste streams from gas processing and fertilizer plants. An example of this is the Weyburn EOR project which uses anthropogenic CO_2 with 96.8% purity, with 1.1% H_2S, and 1.3% light hydrocarbons [6]. The major impurities, such as nitrogen, carbon monoxide, water vapor, and hydrogen, are removed as they increase the MMP and pressure drop along the pipeline. However, H_2S and the light hydrocarbons are known to lower the MMP and, hence, increase oil recovery [7].

3.2 CO_2 QUALITY SCENARIOS FOR CURRENTLY OPERATING CO_2 PIPELINES

The CO_2 stream compositional quality for current pipelines are also described in Chapter 5. Chapter 5 describes the interrelationship between pipeline design and CO_2 stream product quality and as well as CO_2 stream specification for sweet/sour, point to point, and network pipelines. This chapter, however, explores the current as well as various anticipated compositional specifications proposed/anticipated by the industry globally and explains impact of impurities in a CO_2 stream and their impact on the ultimate usage of the stream, e.g., EOR or general sequestration.

The CO_2 mixture quality requirements specified for the current CO_2 pipelines are dictated by the effects of impurities on the EOR process as well as by economic, safety, and hydraulic considerations. The Kinder Morgan ($KMCO_2$) CO_2 quality specification [8] is the specification defined for all supercritical CO_2 pipelines owned and operated by $KMCO_2$ The adoption of a universal specification is possible because the majority of the CO_2 transported by $KMCO_2$ is from naturally occurring sources which are relatively pure and fairly consistent in composition. The main issue faced by $KMCO_2$ is the high water content at the source and therefore, the CO_2 requires dehydration to prevent corrosion and hydrate formation. The nitrogen and hydrocarbon content in the CO_2 are also carefully controlled as these components can affect the MMP of the CO_2 used for EOR operations.

The Denbury CO_2 company pipeline specification for CO_2 stream is also based on a quality requirement for EOR. The very low concentration of H_2S (0.0015%) in the Denbury specification takes into account the potential for sulphide stress cracking [9]. The specification for the Canyon Reef Carriers (CRC) pipeline, as defined in IPCC [7], is similar that of $KMCO_2$ [10] as shown in Table 3-1. It should be noted that the CRC pipeline is operated by Kinder Morgan ($KMCO_2$). This pipeline carries both natural and anthropogenic sources of CO_2, but, as with the $KMCO_2$ and Denbury specifications, the allowable impurity levels are defined by the end use, which is for EOR applications.

3.2.1 "Actual" Compositions of CO_2 in Operational Pipelines

It is interesting to compare the above specifications with actual CO_2 mixture compositions of transported. For example, the Cortez pipeline, operated by $KMCO_2$ and the NEJD pipeline, operated by Denbury, are operating at CO_2 concentrations well above those required by the specifications shown in Table 3-2.

It may also be that the Weyburn pipeline is transporting significantly more H_2S than the other pipelines. As mentioned, the presence of H_2S is desirable for EOR applications. In addition, as CO_2 is odorless, the presence of H_2S acts as a mercaptan which can be beneficial as it can be used to detect the presence of leaks in the pipeline due to its pronounced odor.

TABLE 3-1. Operating pipeline CO_2 stream quality specifications (%mol) with the published units highlighted in bold [7, 9, 10]

	KMCO$_2$	Denbury	CRC Pipeline
CO_2	>95%	>95%	>95%
N_2	<4%	<4%	<4%
Hydrocarbons	<5%	5%	<5%
H_2O	0.0638% **(30 lb/MMscf)**	0.047% **(257 ppmwt)**	<0.064% **(0.48 g/m^3)**
O_2	<0.00136% **(10 ppmwt)**	–	<0.00136% **(10 ppmwt)**
H_2S	0.00127–0.0254% **(10–200 ppmwt)**	0.0015%	<0.19% **(1500 ppmwt)**
Glycol	4.01×10^{-6}% **(0.3 gal/MMscf)**	–	$<2.5 \times 10^{-5}$ % **(4×10^{-5} L/m^3)**

Note: Conversions from the published units to percentage have been carried out and are based on the density at standard conditions of 1 bar and temperature of 15°C. Note: Both volume percent and mol percent are interchangeable at constant pressure and temperature for gases.

3.3 PROJECTED CO_2 PURITY SPECIFICATIONS FOR THE DIFFERENT CAPTURE TECHNOLOGIES

The amount and type of impurities in the CO_2 stream captured from power plants are dependent on the capture technology, the fuel source, and also economic considerations. Therefore, one method by which an impurity specification can be derived is by a detailed study of the process route to identify the levels and types of impurities that could be present in the captured CO_2.

It was mentioned in the previous chapter that there are three main process routes for capturing CO_2 from power plant, post-combustion capture, pre-combustion capture or integrated gasification combined cycle (IGCC), and oxyfuel. Within any of these process routes, there are then different technologies by which the CO_2 is captured, such as separation with solvents/sorbents, separation with membranes, and cryogenic separation. The choice of technology depends on parameters, such as the volume, concentration and pressure of the CO_2 that is to be captured.

3.3.1 IPCC Specification Scenario

The IPCC special report [7] identified the composition ranges detailed in Table 3-3 for CO_2 captured from coal-fired or gas-fired power stations, taking into account the process route used.

TABLE 3-2. Composition of CO_2 in currently operational pipelines with the published units in bold font ([9, 11] with permission)

Pipeline/Composition	Canyon Reef Carriers	Central Basin	Cortez	Weyburn	NEJD
CO_2	95	98.5	98.5	96	99.4
CH_4	5	0.2	0.14	0.7	0.3
N_2	<0.5	1.3	1.3	<0.047 **(300 ppmwt)**	0.3
H_2S	**0.0126 (100 ppmwt)**	–	–	0.9	–
C_2+	–	–	–	2.3	–
CO	–	–	–	0.1	–
O_2	–	–	–	<0.0068 **(50 ppmwt)**	–
Source	Anthropogenic	Natural	Natural	Anthropogenic	Natural

Note: Conversions from the published units to percentage have been carried out and are based on the density at standard conditions of 1 bar and temperature of 15°C. Note: Both volume percent and mol percent are interchangeable at constant pressure and temperature for gases.

It indicates that the CO_2 captured from a post-combustion solvent scrubbing process contains the lowest concentrations of impurities, in particular the sulphur compounds SO_2 and H_2S. In fact, for post-combustion solvent absorption processes, it is essential that sulphur is removed from the feed gas prior to the scrubbing unit as it poisons the solvent and therefore causes excessive solvent loss. The IPCC report suggests that for oxyfuel and pre-combustion capture, leaving sulphur compounds in the CO_2 product stream could be economically beneficial as it will reduce the cost of capture. However, it should be noted that the allowable levels of these compounds may be restricted by other factors such as environmental legislation or pipeline hydraulic constraints. The proportion of sulphur compounds is also directly proportional to the sulphur content of the coal. For the IPCC analysis, the sulphur content was assumed to be 0.86%, therefore, using coal with higher sulphur content will increase the proportion of sulphur compounds.

The oxyfuel case, as shown in Table 3-3, assumes that the CO_2 has passed through a cryogenic purification process to remove some of the N_2, Ar, O_2, and NO_x. If this unit were to be removed, then the levels of these impurities would be increased but the cost of capture would be reduced. Conversely, if a distillation process was included in the cryogenic unit, then purity levels of 99.99% could be achieved, but at an increased cost.

It is emphasised that the IPCC composition assumes that water has been removed to very low levels to avoid corrosion and hydrate formation. The report also identified that trace heavy metals could be present in the capture stream, although there is not much literature published on the amounts or effects of these trace metals.

3.3.2 ENCAP Specification Scenario

As part of the EU Framework 6 European Enhanced Capture of CO_2 (ENCAP) [12] project, the power company Vattenfall have analyzed the potential compositions of the CO_2 stream that could result from post-combustion, pre-combustion, and oxyfuel capture [12]. The results are presented in Table 3-4.

The main focus of this analysis was coal-fired power plants burning German lignite coal and, as described in the previous section, the presence of sulphur in the fuel increases the amount of SO_2 and H_2S in the pre-combustion and oxyfuel capture stream. This is denoted by the "CO_2+ H_2S" and "CO_2+SO_2" columns and the highlighted cells in Table 3-4. If it is assumed that the sulphur products are not removed, then the levels of SO_2 and H_2S become higher in the CO_2 product stream.

As with IPCC, the Vattenfall analysis also assumes that all the water has been removed from the CO_2 and that mercury and metals will be present in trace amounts.

Vattenfall's quality requirement is based on the principle that excessive purification of the CO_2 has to be avoided in order for CCS to be economically viable [12] This scenario

TABLE 3-3. Predicted composition of CO_2 from power plant capture [7]

Coal-fired Power Plants	Component	Coal Fired (%)	Gas Fired (%)
	SO_2	<0.01	<0.01
	NO_x	<0.01	<0.01
Post-Combustion Capture	$N_2/Ar/O_2$	0.01	0.01
	H_2S	0.01–0.6	<0.01
Pre-Combustion Capture (IGCC)	H_2	0.8–2.0	1
	CO	0.03–0.4	0.04
	CH_4	0.01	2
	$N_2/Ar/O_2$	0.03–0.6	1.3
	SO_2	0.5	<0.01
Oxyfuel	NO_x	0.01	<0.01
	$N_2/Ar/O_2$	3.7	4.1

TABLE 3-4. Possible CO_2 composition for the three capture technologies evaluated by Vattenfall (adapted from Anheden et al [12])

	Oxyfuel		Post-combustion	Pre-combustion	
	CO_2/SO_2	CO_2+SO_2	MEA Absorption	CO_2/H_2S	CO_2+H_2S
Component			**Concentration (%)**		
H_2O	0.14	0.14	0.14	0.14	0.14
CO_2	91	90	99.8	97.8	95.6
SO_2	0.076	1.5	0.001	–	–
NO	0.25	0.24	0.002	unknown	unknown
H_2S	–	–	–	0.01	2.3
HCI	trace	trace	trace	trace	trace
HF	trace	trace	trace	trace	trace
CO	unknown	unknown	0.001	1.7	1.7
Ar	5.7	5.6	0.021	0.050	0.049
O_2	1.6	1.6	0.003	unknown	unknown
N_2	0.061	0.60	0.021	0.030	0.030
H_2	–	–	–	1.7	1.7
CH_4	–	–	–	0.035	0.035
Hydrocarbons	–	–	0.003	unknown	unknown
HCN	trace	trace	–	<0.0005	<0.0005
COS	–	–	–	<0.0005	<0.0005
NH_3	trace	trace	unknown	0.003	0.003

is only valid if the legislative regime permits of the transport and storage of the resultant impure CO_2 stream and the cost of cleaning the stream are higher than the cost of transportation. For example, if the level and types of impurities in the CO_2 stream change the hydraulic behavior such that re-pressurization is required along the pipeline route, then the cost of transportation will increase.

3.3.3 Oosterkamp and Ramsen Specification

The Oosterkamp and Ramsen quality specification [13] is based on compiled ranges of impurities published by IEA [14, 15] and White et al [16] for post- and pre-combustion capture. The composition for the oxyfuel stream is derived from both IEA [14] and White et al [16], with additional analysis conducted using the NIST thermo physical properties of fluid systems database (NIST, 2006). As a result of this analysis, the maximum ranges of impurities that could be present are shown in Table 3-5. The levels indicated in this Table, along with those of the IPCCs and ENCAPs, are without any further purification after capture.

3.4 PROJECTED CO_2 PIPELINE SPECIFICATIONS SCENARIO FOR CCS TRANSPORT

The CO_2 specifications reviewed in the previous sections (IPCC, ENCAP and Oosterkamp & Ramsen) are derived from detailed analyses of the capture processes used, identifying the potential impurities that could be present, taking into account the fuel type and economic considerations. The present section, however, describes published specifications from the Dynamis EU project [17, 18] and an Ecofys study [19] where the focus has been on defining the specification for CO_2 quality required for pipeline transportation.

3.4.1 Dynamis Specification

The Dynamis project is an integrated European project investigating the production of hydrogen and electricity from fossil fuels with CO_2 capture and permanent storage. The

TABLE 3-5. CO2 quality specification in percentage composition in terms of maximum ranges (adapted from Oosterkamp and Ramsen [13]) with the values in bold converted from ppmv to volume or mol percent

Component	Post-combustion (IEA [14]) (%)	Pre-combustion (IEAGHG, 2004) (White et al [16]) (%)	Oxyfuel (%)
CO_2	>99	>95.6	>90
CH_4	**<0.01**	**<0.035**	–
N_2	<0.17	<0.6	<7
H_2S	Trace	<3.4	trace
C_2+	**<0.01**	<0.01	–
CO	**<0.001**	<0.4	trace
O_2	<0.01	Trace	<3
NO_x	**<0.005**	–	<0.25
SO_x	**<0.001**	–	<2.5
H_2	Trace	<3	Trace
Ar	Trace	<0.05	<5

Note: Both volume and mole percent are interchangeable for a gas.

Dynamis transport specification (Table 3-6) is based on the ENCAP specification for pre- and post-combustion capture applied to a process for co-production of hydrogen and electricity.

The oxyfuel capture technology was not considered and therefore careful consideration needs to be taken when applying this specification to oxyfuel capture technology or to other types of capture processes. In the Dynamis specification, the ENCAP specification has been modified to take account of safety and toxicity limits, in the event of a release from the pipeline; infrastructure durability, in terms of the need to avoid free water formation, hydrate formation, and corrosion; and transport efficiency [18].

The effects of safety and toxicity have been taken into account by the change in the allowable levels of H_2S and CO which are determined by using the short-term exposure limits (STEL) for these compounds to set a maximum concentration in the CO_2 stream, allowing for air dilution effects during a pipeline release. The Dynamis study [18] also indicated that followup work will be carried out to provide recommendations in terms of quantity for impurities, such as SO_x NO_x and O_2. The allowable water content was increased from the ENCAP specification, which has been justified by the fact that the revised limit is still below the water saturation level under the pipeline operating conditions considered, allow-

TABLE 3-6. Pipeline specifications (converted to %) proposed by the Dynamis [17,18] and Ecofys project [19]

	DYNAMIS		
	Storage	EOR	ECOFYS
CO_2	>95.5		>95
H_2O	<0.05		<0.05
SO_x	0.01		Not critical
NO_x	0.01		Not critical
H_2S	0.02		<0.02
CO	0.2		<0.2
H_2			
Ar		<4	
N_2	<4		<4
O_2		0.01- 0.1	
CH_4		<2	

ing for the effects of other impurities on the solubility of water in CO_2, particularly H_2S and CH_4 as well as the risk of hydrate formation.

3.4.2 Ecofys Specification Scenario

A Dutch study on CO_2 transport, performed by Ecofys [19], took a similar approach to the Dynamis project [17,18], except that the case study was based on potential impurities from coal-fired power plants [19].

In contrast to the Dynamis specification, the Ecofys specification does not set a limit for SO_x and NO_x as these are not considered to be critical impurities, particularly, in the case of SO_2, if no free water is present. The total of the non-condensable components, such as N_2, O_2, H_2, CH_4, and Ar, is limited to <4% in this specification. This limit is primarily based on the assumption that these gases will occupy a finite portion of the pipelines transport capacity and therefore increase the pipeline investment and compression energy consumption.

3.5 WATER CONTENT IN CURRENTLY OPERATIONAL CO_2 PIPELINES

In the presence of water, highly corrosive carbonic acid is formed, and it has been reported that carbon steel can corrode at rates of more than 10 mm/year in wet pure CO_2 [20, 21]. Therefore, in order to prevent corrosion in transportation pipelines, it is necessary either to dry the product prior to transportation to prevent the formation of free water or to specify a more corrosion resistant material. Levels of 2.88×10^{-4} to 4.8×10^{-4} kg/st.m^3 of water in the pipeline are accepted by the industry for CO_2 transmission in carbon steel pipelines [22]. For hydrocarbon pipelines, a requirement to dry the gas to 50 ppm of water is often used to ensure no free water is in the pipeline. In terms of free water, a number of specifications have been published. Existing CO_2 pipeline transportation systems are reviewed in terms of operational parameters, particularly drying specifications by Heggum et al [23]. Based on specific calculations of water solubility for a selected case, it was found that the most stringent drying requirements (e.g., 50 ppm proposed (not operational) for Hammerfest LNG) may be relaxed to 600 ppm, which is the present Kinder Morgan specification.

The specification of an acceptable level of water in the pipeline is dependent on the solubility of water in the fluid at the operating temperature and pressure. The solubility limit for water in pure CO_2, at supercritical conditions is up to160 lb/MMscf (0.0026 kg/m^3), and it has been determined that corrosion of carbon steel will not occur if the water content is below 60% saturation (i.e., 0.0015 kg/m^3) [24]. There is no general consensus for an acceptable level of water in gaseous CO_2 pipelines as there is insufficient experience with transporting CO_2 via pipeline in the gas phase. The reduced solubility of water in CO_2 at low temperatures and pressures which increases the amount of free water present could possibly be mitigated by imposing 60% saturation limit on the amount of water present. This would then be part of a corrosion control strategy (Renfro, 1979). Indeed, some of the current experience with two-phase CO_2 (CO_2 gas and dense phase CO_2) from naturally occurring sources (i.e., CO_2 domes/reservoirs) underground which are transported to the CO_2 processing facilities, has its gas phase dried in the glycol contactors (separators) to water levels of 1.28×10^{-4} kg/m^3 to 1.44×10^{-4} kg/m^3 [25]. Table 3-7 reviews some of the published water specifications for operating supercritical CO_2 pipelines. It can be seen that they are all well below the 60% saturation limit in order to ensure that corrosion does not occur. In fact, in many cases, the actual water content achieved in the pipeline is below the specification. For example, CO_2 from the McElmo Dome source, which is transported via the Cortez pipeline, is supplied to a pipeline specification of 30 lb/MMscf (0.00048 kg/m^3), although in fact a water content of around 15 lb/MMscf (0.00024 kg/m^3) is actually recorded [20].

The drying requirement specified in the Bati Raman pipeline was based on the specification for drying natural gas and also the design specification used in the CRC pipeline [26]. As previously mentioned, the drying specification for natural gas is likely to be too conservative when applied to supercritical CO_2. This conclusion was also drawn by Heggum et al [23] when commenting on the water specification for the Snøvhit pipeline which is based on the drying requirements for natural gas. They conclude that a "maximum of 600 ppm water may be a suitable and sufficient requirement for supercritical CO_2 pipelines."

The most stringent water specification in the literature is 20 ppm, for the Weyburn pipeline [7]. It is not clear if this is a requirement or whether it is the nature of the process itself, which happens to produce 20 ppm of water, or the fact that the Weyburn pipeline has a high H_2S content. It is interesting to note that the presence of H_2S may actually increase the solubility of water in the CO_2 and therefore have a beneficial effect [17].

The DNV recommended practice for the design and operation of CO_2 pipelines (DNV RP J 202) does not provide a specific limit on maximum water content but states that the maximum water content should be determined on a case-by-case basis taking into account the normal operation pressure envelope, the pipeline shut in pressure combined with the minimum ambient temperature, and the pipeline depressurization scenario. Nevertheless, this recommended practice has provided some indicative limits on maximum water content for CO_2 pipelines based on other operational experience, such as Kinder Morgan and Statoil for the Snohvit pipeline, similar to those shown in Table 3-7 and also those quoted on the literature (i.e., the DYNAMIS project specification and Mohitpour et al [25]). The DNV recommended practice also mentions that the limit on the water content should consider the potential for hydrate formation and the potential of forming hydrates from other non-condensables components in the CO_2 product stream. However, no specification on maximum water content is mentioned. The main concern for hydrates comes from a flow assurance perspective, where the formation of a hydrate in the pipeline could cause blockage and restrict flow, potentially resulting in a safety incident. Additionally, hydrates could

TABLE 3-7. Operating pipeline quality specifications NB. The boxes shaded in are the values quoted in the literature, the remaining figures quoted are conversions to allow comparison

Pipeline/Company Specification	Water Content			Reference
	lb/MMSCF	kg/m^3	ppmv	
CRC Pipeline	30	0.00048	638.32	IPCC [7]
$KMCO_2$	30	0.00048	638.32	Kinder Morgan [10]
Cortez Pipeline	30	0.00048	638.32	Wallace [20]
Sheep Mountain	30	0.00048	638.32	Watson and Siekkinen [28]
Denbury	30	0.00048	638.32	Denbury [9]
Bravo Pipeline	25	0.0004	532.15	Najera [24]
Bati Raman	4	0.00006	85.06	Kantar and Connor [29]
CRC Pipeline	2.31	0.0000377	50	Gill [26]
Snøvhit (proposed)	2.31	0.0000377	50	Heggum et al [23]
Weyburn	0.94	0.000015	20	DTI [27]

Note: The conversions to ppmv are based on standard conditions of 1 bar and temperature of 15°C which corresponds to a CO_2 density of 1.842 kg/m^3. This is based on the oil and gas pipeline design convention. It is also cautioned that the literature does not generally state whether the specification in ppm is by volume or mass and therefore these conversions should only be used as a guide, reference always being made to the original literature source. The unit used by Gill [26], Heggum et al [23], and DTI [27] is assumed to be in ppmv although it is not stated in the original literature.

potentially form during or restart of a pipeline if the de watering procedure is not carried out thoroughly [30].

In conclusion, it should be emphasized that it is not the water content in the CO_2 fluid that needs to be controlled but the amount of (free) water dropping out (i.e., insoluble in the carbon dioxide). The solubility of water is controlled by the pressure and temperature as well as impurities. In the case of transport, the pressures and temperatures are maintained at prescribed levels; therefore consideration has to be given to the amount and type of impurities. This can become even more accentuated in a pipeline network where multiple sources of CO_2 using different capture technologies may exist.

3.5.1 Solubility of Water in Anthropogenic CO_2

All of the operational CO_2 transmission pipelines are manufactured from plain carbon steel which is essentially non-corrosive in pure dry CO_2. The level to which the CO_2 stream has to be dried is dependent on the water solubility limit, which, in turn, is dependent on the pressure, temperature, and the level of impurities. The solubility of water in pure CO_2 is well understood and documented as a function of temperature and pressure, the effect of impurities on the solubility of water in supercritical CO_2 is less well researched. At a constant temperature, the solubility of water in pure CO_2 in the gas phase decreases with increasing pressure. However, in the liquid phase, the solubility increases with increasing pressure. This behavior results in a turning point in water solubility as a consequence of the phase change from gas to liquid. This minimum represents the optimal operating conditions at which to remove water. These trends of water solubility in the gaseous and liquid phase are shown in Figure 3-2.

At a constant pressure, the solubility of water in both the liquid phase and gas phase increases with increasing temperature for pure CO_2. This is illustrated by the experimental data of King et al [31, 32] and also by the calculations conducted by Austegard et al [33] and cited by de Visser and Hendriks [17] as presented in Figure 3-1 and Figure 3-2. Similar results have been reported by Seiersten [34]. The significance of this behavior of CO_2 is that for pipelines operating at low temperatures and pressures, the water content would have to be more tightly controlled. This has been supported by corrosion experiments conducted by Seiersten [34] whereby a higher corrosion rate of carbon steel was observed in low-pressure CO_2 compared to high-pressure CO_2, for the same amount of water levels present at a constant temperature. Consequently, pipelines operating in the gas phase would require more stringent allowable water content specifications than those operating in the liquid or dense phase. For example, a CO_2 pipeline built in colder climates at which ground temperatures could reach 5°C, water solubility, or more accurately, insolubility could pose a problem in which water drop out or free water could be present at higher levels.

At pressures below 40 bar, it is shown from Figure 3-2 that at temperatures below 4°C, the solubility of water which is directly related to the dew point of water is about 375 ppmv which is more than half of the maximum water specifications of 638 ppmv for operating CO_2 pipeline given in Table 3-7. Therefore, low-pressure (below 40 bar) gaseous CO_2 at ground temperatures of 5°C would require more stringent water specifications than supercritical CO_2 to prevent free water dropout in the pipeline. A similar concern was expressed for offshore pipelines where a proposed water specification 500 ppmv may pose a problem if the pressure in the pipeline goes below 40 bar at the North Sea seabed temperature of 4°C [18]. Therefore, a potential solution could be the application of a 60% saturation limit on the minimum water solubility at the minimum ground temperature. As shown in Figure 3-2, the minimum solubility of water at 4°C is 375 ppmv and therefore a 60% saturation limit would impose a water specification of 225 ppmv at pressures below 40 bar. Generally, for CO_2, the saturated water content is lower at pressure ranges of 40 to 80 bar compared to

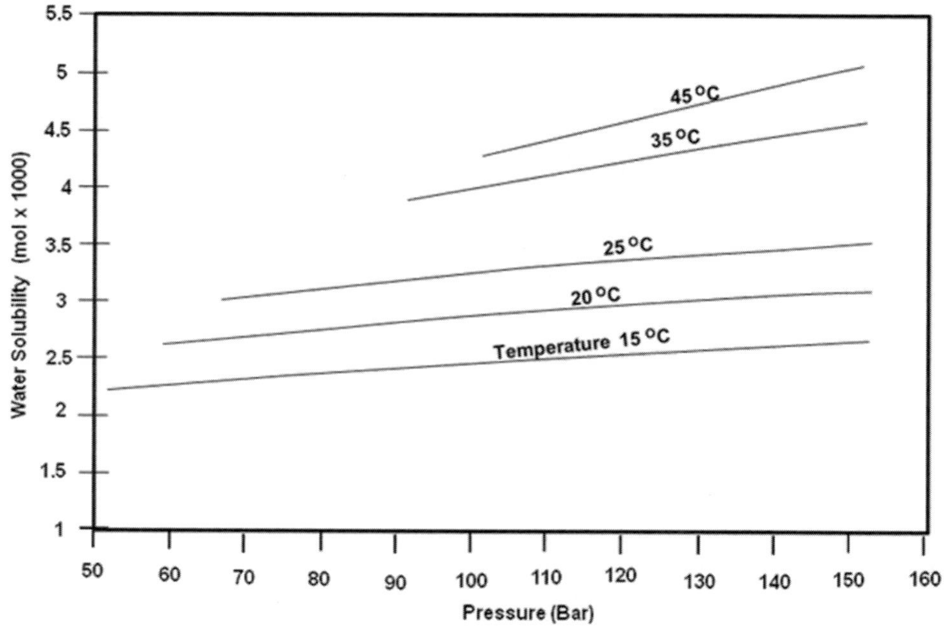

Figure 3-1. Experimental data for the solubility of water in pure CO_2 at a range of pressures and temperatures (reproduced from King et al [31]) with a power law fit to the data

pressures above 100 bar. Above the critical point of CO_2, the solubility of water increases more dramatically with increasing pressures and temperatures.

It should be noted that the minimum in the water saturation behavior shown for CO_2 is not exhibited by natural gas over the same range of temperatures and pressures [35, 36]. This is illustrated in Figure 3-3. For natural gas, the solubility of water decreases with increasing pressure over the full pressure range. For example, at 83 bar and a dew point of −6.6°C, natural gas can have approximately 18 times less water saturation compared to pure CO_2 at the same conditions. In natural gas systems, water solubility decreases with

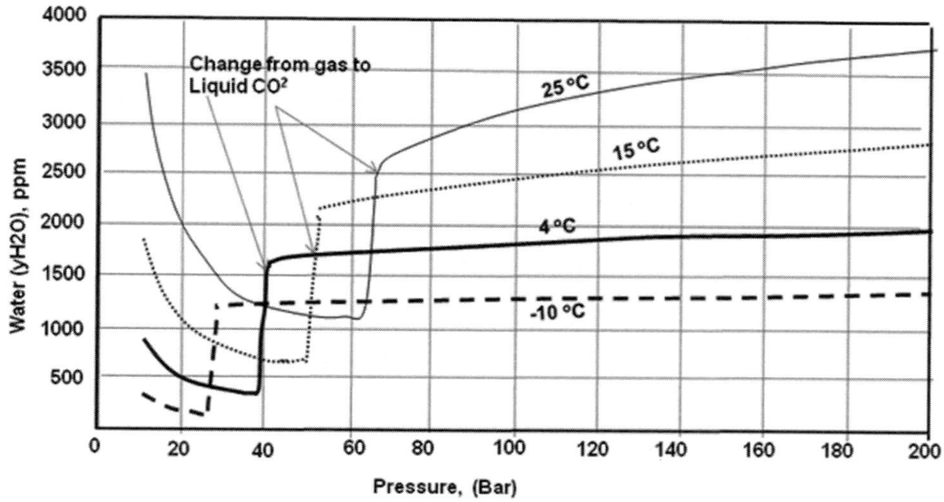

Figure 3-2. Solubility of water in pure CO_2 as a function of pressure and temperature (Austegard et al [33] as cited by de Visser and Hendriks [17])

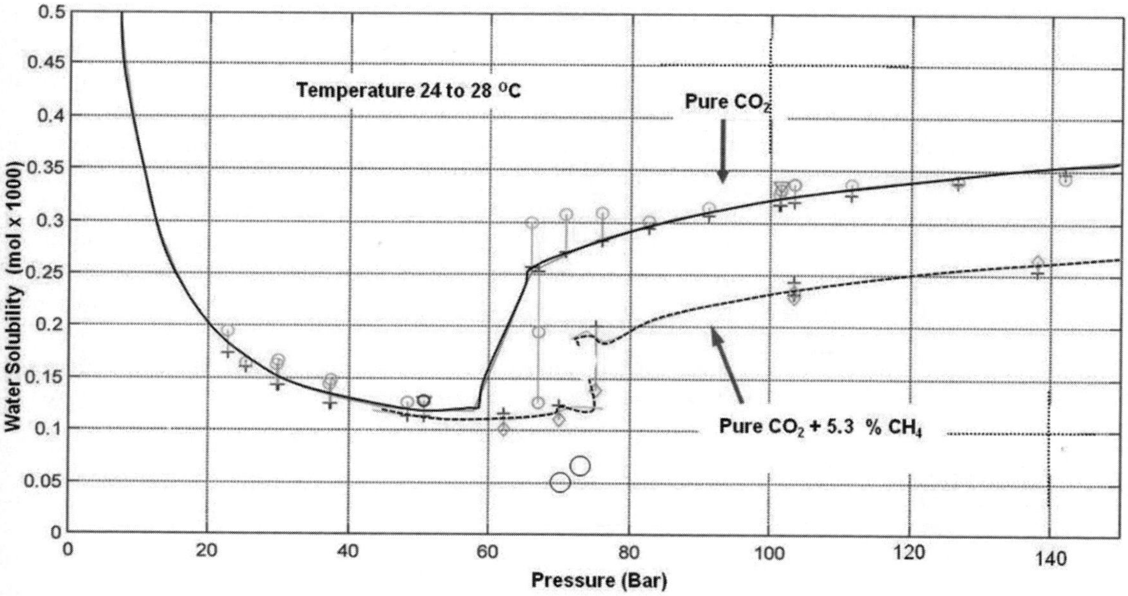

Figure 3-3. Solubility of water in pure CO_2, pure CH_4 and a mixture of CO_2 and 5.31% CH_4 at 25°C [32]

increasing pressure which is in contrast with supercritical CO_2 [20]. Dehydration specifications for supercritical CO_2 pipelines based on natural gas specifications would therefore be overly conservative.

As detailed in the previous section, the solubility of water in pure CO_2 is well understood and documented as a function of temperature and pressure. However, the effect of impurities on the solubility of water in CO_2 containing impurities is less well researched although there is some limited data available on the CO_2–H_2O–CH_4 system and the CO_2–H_2O–H_2S system. Indeed the DNV RP J202 recommended practice for the design and operation of CO_2 pipelines acknowledges that there is limited knowledge of water solubility of CO_2 containing impurities and it goes on to states that indicative levels of solubility of water in pure CO_2 should not be used as a representative specification level in a stream of CO_2 containing impurities [30].

A recent experimental study by Seiersten and Kongshaug [21] on the CO_2–H_2O–CH_4 system concluded that the water solubility in the liquid and supercritical phase decreases with increasing CH_4 content and therefore the risk of forming free water increases. This conclusion is in agreement with the results and calculations quoted by Heggum et al [23] and Austegard et al [33] and also shown in Figure 3-3. This data also indicates that the minimum in the solubility versus pressure plot shown by pure CO_2 is still exhibited by the CO_2–CH_4 mixture. On this basis it is interesting to note that the specification for the Canyon Reef Pipeline, which contains 5% CH_4 in the CO_2 stream is one of the most stringent used in the industry.

In contrast, a study conducted as part of the Dynamis project [18] showed that for the CO_2–H_2O–H_2S system at 4°C and 100 bar, adding H_2S increased the solubility of water. However, it was shown that at the low levels of H_2S expected in the CO_2 stream, around 200 ppm, this effect was negligible. It is emphasised that this study was based on calculations and no experimental data was generated. The consequence of these observations is that the drying specification for CO_2 containing impurities will be dependent on the type and quantity of impurities present. It is evident that some impurities may increase the solubility of water whilst others will reduce it. Therefore experiments are required to determine the solubility of water in CO_2 containing the expected impurities from a specific capture process or technology.

The consequence of these observations is that the drying specification for CO_2 containing impurities will be dependent on the type and quantity of impurities present. It is evident that some impurities may increase the solubility of water whilst others will reduce it. Therefore experiments are required to determine the solubility of water in CO_2 containing the expected impurities from power station capture.

By and large, impurities have the potential of decreasing the solubility of water in supercritical CO_2 and options such as pipeline material, dehydration and clean up of the flue gas needs to be considered. The presence of water and oxygen could also initiate certain reactions between impurities. An example of such a process is the lead chamber process whereby at high pressures, SO_2 is oxidized by O_2 to SO_3 which then reacts with water forming H_2S which is highly corrosive [37].

3.6 ANTHROPOGENIC CO_2 FOR CCS

This review of specifications has indicated that, at the present time, there is no general consensus on pipeline quality CO_2 specification. The specifications reviewed in the literature are based on health and safety limits, as well as potential hydraulic limitations [18]. However, it is unclear if any hydraulic analysis has been carried out in determining the effect of the components on pipeline transport. The review has also identified some of the potential barriers to being able to achieve it, most specifically economic constraints. In principle it is possible to remove impurities to very low levels but, for storage purposes (as opposed to EOR), there is no economic incentive to achieve these low levels. In addition, it has been shown that the type and quantity of impurities will change for different fuel types and capture processes. Although the impact of these impurities can be considered at an individual level, none of the specifications take into account the interaction of impurities (other than water) and their potential effects on pipeline hydraulic and integrity.

The anthropogenic CO_2 specification for water will depend on the phase of transport and also the lowest pressure and temperature that could be encountered during normal operation. However, this may change during an upset condition where free water maybe present. The presence of free water can be exacerbated by the presence of impurities such as CH_4 which reduce the solubility limit of water in CO_2.

3.7 CONCLUDING REMARKS

Compositional quality scenarios of anthropogenic CO_2 for the different capture technologies are highlighted. Anthropogenic CO_2 originates largely from power plants which employ the post-combustion, pre-combustion and oxyfuel capture technologies.

Potential CO_2 stream compositional scenarios are explored from two perspectives, the capture technology and the pipeline. In either scenario, the types of impurities that could be present are dependent on the capture technology, process used and type and composition of the fuel. However, the actual levels of impurities present will be driven by the economic, environmental and safety considerations. There is currently no CO_2 pipeline specification which is relevant for all capture technologies, refer to Chapter 5 for details.

The potential specification scenarios reviewed indicate that the type and quantity of impurities present will be different from those transported in currently operational CO_2 pipelines. For example impurities such as SO_x, NO_x and Ar have never been transported before in CO_2 pipelines. These are further detailed in Chapter 5.

The water specifications or level of dryness for CO_2 pipelines reviewed indicated industry accepted levels of water for CO_2 pipelines range between 2.0×10^{-5} to 4.8×10^{-4} kg/m^3, well below the accepted limit. Further it may be - concluded, however, that specification scenarios based on the drying requirements for natural gas may not be appropriate and may be overly conservative for supercritical CO_2 as the saturation behavior is different. Indeed recent work on specifications for transporting anthropogenic CO_2 indicate that under certain circumstances a water specification of 2.0×10^{-5} kg/m^3 would be adequate to prevent corrosion and hydrate formation. However, for pipelines operating in the gas phase, the solubility of water is reduced and even more stringent drying may be required. Indeed it may be that, depending on the level of impurities, the water specification for natural gas may be appropriate for gas phase CO_2.

REFERENCES

[1] Sankur, V., Emmanuel, A. S. and Di Julio, S. S., 1986, "Recovery of Heavy Oils by CO_2 Injection," Energy Progress, Vol 6(2), pp. 91–96.

[2] Lake, L. W. and Walsh, M. P., 2008, "Enhanced Oil Recovery (EOR) Field Data Literature Search," Technical Report prepared for: the Danish North Sea Partner, Danish Energy Agency and Mærsk Olie og Gas AS, Department of Petroleum and Geosystems Engineering University of Texas at Austin, Texas, USA.

[3] Tzimas et al., 2005, "Enhanced Oil Recovery using Carbon Dioxide in the European Energy System". EUR 21895 EN Vol. Petten, The Netherlands: Institute for Energy, 2005. Print.

[4] Goodyear, S. G., Hawkyard, I. R., Masters, J. H. K., Woods, C. L., and Jayasekera, A. J., 2003, "Subsurface Issues for CO_2 Flooding of UKCS reservoirs," Transaction of IChemE, vol 81, Part A.

[5] IEA, 2000, "Barriers to overcome in implementation of CO_2 capture and storage — Storage in disused oil and gas field," Report no. PH3/22, International Energy Agency.

[6] Perry, M. and Eliason, D., 2004, "CO_2 Recovery and Sequestration at Dakota Gasification Company,": http://www.gasification.org/Docs/2004_Papers/11ELIA.pdf.

[7] IPCC, 2005, "IPCC Special Report on Carbon Dioxide Capture and Storage," Working Group III of the Intergovernmental Panel on Climate Change, 2005, Eds. Metz, B., Davidson, O., de Coninck, H. C., Loos, M. and Meyer, L. A., Cambridge University Press, Cambridge, United Kingdom and New York, NY, USA.

[8] Kinder Morgan CO_2 (KMCO_2), 2008, "Personal communication," Marco A Roncal, Pipeline Engineer.

[9] Denbury, 2008, "Personal Communication," Connell Rader, Operations Manager Denbury Resources Inc. used with permission.

[10] Kinder Morgan CO_2, (KMCO_2), 2006, "Presentation: Kinder Morgan" CAPP Investment symposium, Canada (June 13, 2006).

[11] Race, J. M., Seevam, P. N. and Downie, M. J., 2007, "Challenges for Offshore Transport of Anthropogenic Carbon Dioxide," 26th international Conference on Offshore Mechanics and Arctic Engineering, California, USA.

[12] Anheden, M., Andersson, A., Bernstone, C., Eriksson, S., Yan, Liljemark, S. and Wall, C., 2005, "CO_2 Quality Requirement for a System with CO_2 Capture, Transport and Storage," Proceedings of the 7th Greenhouse Gas Technologies Conference GHGT-7, Vancouver, vol 2, pp. 2559–2563.

[13] Oosterkamp, A. and Ramsen, J., 2008, "State of the art Overview of CO_2 Pipeline Transport with Relevance to Offshore Pipelines," Polytec Report no. POL-O-2007-138-A, January 2008.

[14] IEA 2004, "Improvement in Power Generation with Post Combustion Capture of CO_2," Report no. PH4/33, International Energy Agency.

[15] IEA 2004, "Impact of Impurities on CO_2 Capture, Transport and Storage," Report no. PH4/32, study conducted by SNC Laevulin, August 2004.

[16] White, V., Allan, R. and Miller, E. 2006, "Purification of Oxyfuel Derived CO_2 from Sequestration or EOR," 8th Green House Gas Technology conference (GHGT-8).

[17] de Visser, E. and Hendriks, C., 2007, "Towards Hydrogen and Electricity Production with Carbon Capture and Storage- DYNAMIS Quality Recommendation," DYNAMIS, Project No.: 019672.

[18] de Visser, E., Hendriks, C., Barrio, M., Mølnvik, M. J., de Koeijer, G., Liljemark, S. and Le Gallo, Y., 2008, "Dynamis CO_2 Quality Recommendation," International Journal of Greenhouse Gas Control, Vol. 2, pp. 478–484.

[19] Hendriks, C., Hagedoorn, S. and Warmenhoven, H., 2007, "Transportation of Carbon Dioxide and Organisational Issues of CCS in the Netherlands," Report prepared by Ecofys for EnergieNed, the Ministry of Economic Affairs and the Ministry of Housing, Spatial Planning and the Environment.

[20] Wallace, C. B., 1985, "Drying supercritical CO_2 Demands Care," Oil and Gas Journal, Vol. 83(25), pp. 98–99.

[21] Seiersten M. and Kongshaug K. O., 2005, "Materials Selection for Capture, Compression, Transport and Injection of CO_2", Carbon Dioxide Capture for Storage in Deep Geologic Formations," edited by D. C. Thomas and S. M. Benson, Elsevier Ltd., vol. 2, pp. 937–953.

[22] Mohitpour, M., Golshan, H. and Murray, A., 2003, "Pipeline Design and Construction," 1st edition, ASME publication, USA.

[23] Heggum, G. T., Weydahl, R. Mo, Mølnvik, M. and Austergaard, A., 2005, "CO_2 Conditioning and Transportation Carbon Dioxide Capture for Storage in Deep Geologic Formations," D. C. Thomas and S. M. Benson, Elsevier Ltd. vol 2, pp. 925–936.

[24] Najera, G., 1986, "Maintenance Techniques Proven on CO_2 Line," Oil and Gas Journal, vol. 84(46), pp. 55–57.

[25] Mohitpour, M., Golshan, H. and Murray, A., 2007, "Pipeline Design and Construction," 3rd edition, ASME publication, USA.

[26] Gill, T. E., 1985, "Canyon Reef Carriers, Inc. CO_2 Pipeline: Description and 12 Years Operation," 8th Annual Pipeline Engineering Symposium, Dallas, TX, USA, ASME.

[27] DTI, 2003, "Review of the Feasibility of Carbon Dioxide Capture and Storage in the UK," Cleaner Fossil Fuels Programme, report number: URN 03/1261 Department of Trade and Industry, UK.

[28] Watson, K. D. and Siekkinen, D. G., 1986, "Sheep Mountain CO_2 Project Development and Operation," Energy Progress, vol. 6(3), p. 155.

[29] Kantar, K. and Connor, T. E., 1984, "Turkey's Largest Oil Field Poised for CO_2 Immiscible EOR Project," Oil and Gas Journal, vol. 82(48), pp. 95–98.

[30] DNV, 2010, "Recommended Practice DNV-RP-J202: Design and Operation of CO2 Pipelines," Det Norske Veristas, April 2010 Edition.

[31] King, M., B. Mubarak, A. Kim, J. D. & Bott, T. R., 1992, "The Mutual Solubilities of Water with Supercritical and Liquid Carbon Dioxide," The Journal of Supercritical Fluids, Vol.(5), pp. 296–302.

[32] King, G. G., 1982a. "Here Are Key Design Considerations for CO_2 Pipelines," Oil and Gas Journal, Vol. 80(39), p. 219.

[33] Austegard, A, Solbraa, E., de Koeijer, G. and Mølnvik, M. J., 2006, "Thermodynamic Models for Calculating Mutual Solubilities in H_2O–CO_2–CH_4 Mixtures", Chemical Engineering Research and Design (ChERD), Part A. Special Issue: Carbon Capture and Storage, Vol. 84(A9), September 2006, pp. 781–794.

[34] Seiersten, M., 2001, "Material Selection for Separation, Transportation and Disposal of CO_2" Proceedings of Corrosion 2001, Houston, Texas, USA.

[35] Recht, D. L., 1987, "Carbon Dioxide Pipeline Design Considerations," ASME Pipeline Engineering Symposium ETCE, Dallas, Texas.

[36] McCain, W. D. Jr., 1990, "The Properties of Petroleum Fluids," 2nd edition, PennWell Publishing.

[37] Kiefer, D. M., 2001, "Sulphuric Acid: Pumping up the volume," American Chemical Society, http://pubs.acs.org/subscribe/journals/tcaw/10/i09/html/09chemch.html.

Chapter 4

OVERVIEW OF EOS APPLICABLE TO CO_2 WITH IMPURITIES

4.1 OVERVIEW OF EOS APPLICABLE TO CO_2 WITH IMPURITIES

The physical and transport properties of natural and anthropogenic CO_2 are important in the hydraulic process calculations in order to provide accurate material and energy balances, heat transfer, mass transfer, and pressure loss calculations [1]. They are also important in addressing design issues (such as fracture control properties of material, refer to Chapter 7). Therefore, in order to design, operate, and maintain a CO_2 pipeline or pipeline network, these properties need to be considered and understood. It should be noted that there is an important difference between calculating phase equilibrium based on mixture compositions and calculating typical physical, thermodynamic, or transport properties of the mixture of known composition. In the latter case (i.e., a pipeline), the properties of a mixture as a whole are of interest, whereas in the former, the partial properties of the individual components that make up the mixture and how those properties affect the mixture as a whole are important.

To a certain extent, the physical and transport properties of CO_2 have been modeled. Most of the equations embodied in these models are semi-empirical in nature and have their range of validity. The equation of state (EOS) can be used to determine physical properties, such as phase behavior, density, compressibility, and other thermodynamic properties related to pressure, volume/flow, and temperature. Additional correlations (e.g., viscosity correlations), constants, and interaction parameters are sometimes used along with the equation of state. The calculated physical and thermodynamic properties will be the inputs into pipeline hydraulic modeling equations, such as the flow and heat transfer equations, which in turn are used to calculate pressure and temperature profiles along the length of the pipeline. Figure 4-1 shows an overview of the relationship between the components of physical property modeling and that of the pipeline hydraulic modeling.

4.2 PHYSICAL PROPERTIES — EQUATIONS OF STATE

The physical and thermodynamic properties of a pure substance or mixtures are modeled using semi-empirical equations that best describe the characteristics of the fluid. The validity of these equations will depend on the assumptions and conditions under which they were developed. Equations of state (EOS) are used in determining densities, compressibilities and phase behavior, and are used directly in the calculation of entropy, enthalpy, and heat capacity, along with the fluid's calorimetric properties. The viscosity and thermal conductivity of a fluid may be determined indirectly from an equation of state in the sense that terms that are dependent on pressure, density (or specific volume), and temperature are derived from the equation of state. Figure 4-2 shows the relationship between the various thermophysical and transport parameters and their interdependency. It can be observed that

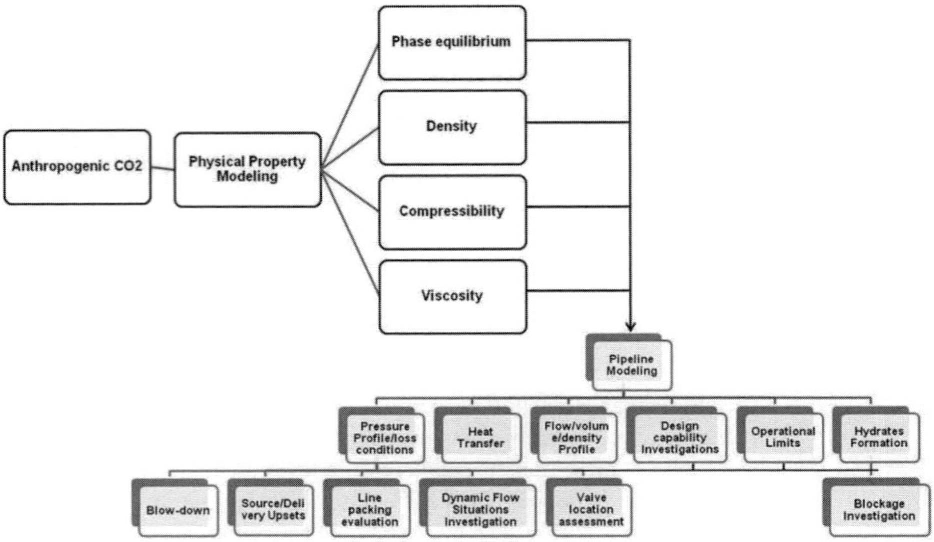

Figure 4-1. Schematic diagram illustrating the relationship between the physical properties and transport properties of a fluid

density is the central linking property between the thermodynamic properties and the fluid transport properties.

The equation of state (EOS) is used to generate a phase diagram which can be used to determine the phase or phases of a fluid (vapor, liquid, solid, dense, or supercritical) at particular pressures, temperatures, enthalpies, and densities. The compressibility factor Z is also a derivative of the equation of state parameters, such as pressure, temperature, and density, which are also used directly or indirectly in determining thermodynamic and transport properties.

The phase equilibrium determination for a CO_2 stream-containing impurities involves calculations of multiphase equilibrium algorithm of CO_2 mixture components. This is interfaced with thermodynamic models (through equation of states (EOS) that utilize physical property information.

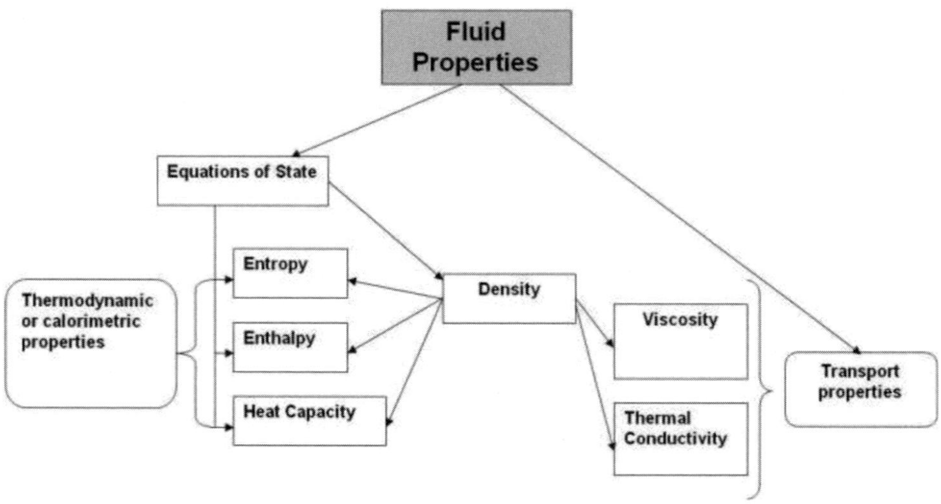

Figure 4-2. Inter-relationship between the various fluid properties

There are several equation of state (EOS) correlations that have been utilized in predicting CO_2 stream phase diagrams. These are Peng–Robinson (PR), Soave–Redlich–Kwong (SRK), Benedict–Webb–Rubin–Starling (BWRS), and GERG-2008. Although references to the Span and Wagner equation of state will be made in relation to the validation of pure CO_2 properties, the overall validation and comparisons will be evaluated with the PR, SRK and BWRS, and GERG-2008 equations of state for the purpose of standardization as the Span and Wagner equation of state (EOS) can only be used for pure CO_2. The evaluation was based on a review of literature and to some extent comparison with available experimental data. Transport properties, such as viscosity, are calculated using the densities determined from the selected equations of state. Thermodynamic properties, such as enthalpy, entropy, and internal energy, are derivatives from these basic properties and also from the available empirical data.

4.3 OVERVIEW OF THE PHYSICAL PROPERTY MODELS

The equilibrium calculations consist of a multiple phase equilibrium algorithm that is interfaced with thermodynamic models described by equations of state, which is also interfaced with physical property databanks. The variables involved in flash calculations are pressure, temperature, density, enthalpy, entropy, internal energy, and mol fraction of a particular phase, whereby two of the mentioned variables need to be defined. The flash calculations are based on the thermodynamic principles that the fugacity, the temperature, and the pressure of all phases are equal and also that the solution satisfies the set variable constraints. The flash calculation then allows the determination of the number and type of phases present and the respective composition and properties of these phases.

The phase equilibrium calculations are computed using a selected equation of state. Flash calculations to determine the number of phases, molar amount, and composition of each phase are performed at given temperatures and pressures. Flash calculation detailed methodology can be found in a number of references [2, 3]. Since the number of phases at equilibrium is not known in advance, stability analyses are carried out using the Gibbs free energy to define equilibrium states. The equilibrium state for a closed system at a given pressure and temperature is defined when the Gibbs free energy is at a minimum with respect to all possible changes [2]. This is used as a part of a convergence criterion to determine the equilibrium parameters. These in turn will produce the vapor–liquid equilibrium (VLE) line on the phase diagram.

The Gibbs free energy (G) is a function of pressure, temperature, and the number of moles of each component. It can be expressed as:

$$\Delta G = \Delta H - T\Delta S \qquad (4-1)$$

in which, for a spontaneous change, the Gibbs free energy must decrease, as is consistent with a spontaneous process where enthalpy (H) will decrease and entropy (S) increase, and (T) is the temperature. Stability analysis is an important step in vapor–liquid equilibrium (VLE) flash calculations and the numerical methods used are based on Gibbs tangent plane criterion. This method used by Multiflash to calculate phase equilibrium using a given equation of state is well documented [4, 5].

Phase properties were calculated over a range of fluid conditions. Fluid property tables are generated at these given conditions and subsequently used to calculate pressure and temperature profiles or other varying parameters along the pipeline. The output from these calculations has been used to compute the pipeline hydraulic conditions.

4.4 EQUATIONS OF STATE (EOS)

4.4.1 Ideal Gas Law and Van der Waals EOS

Physical properties of a fluid depend on the behavior of the molecules of a substance. The first gas law proposed by Charles and Gay-Lussac combined with Avogadro's hypothesis was in the form of

$$PV = NRT \tag{4-2}$$

where:
P Absolute pressure of the gas (Pa)
V Molar volume of the gas (m^3)
N Number of moles
R Universal gas constant (8.314472 $m^3 \cdot Pa \cdot K^{-1} \cdot mol^{-1}$)
T Absolute temperature (K)

It is also known as an ideal gas law and is used for gases at low pressures and high temperatures. This equation can be thought as the simplest form of an equation of state. It basically describes the relationship between the state variables and provides the state of a fluid under given physical conditions of pressure and temperature. Accurate prediction of fluid properties requires that suitable equations of state (EOS) are used appropriately to their respective working ranges.

The limitations of the ideal gas law on pressures and temperatures, and the simplification in terms of compositional effects, led to the development of improved equations of states. The Van der Waals equation of state developed using the principle of corresponding states was introduced following the ideal gas law [6, 7]. The Van der Waals equation of state was the first to consider intermolecular forces. Equation 4-3 shows the Van der Waals equation of state.

$$\left(P + \frac{a}{V^2}\right)(V - b) = RT \tag{4-3}$$

The coefficients a and b, or commonly referred to as the binary interaction parameters, represent the attractive and repulsive forces which account for non-ideal behavior of gases.

4.4.2 Span and Wagner EOS

The reference equation of state for pure CO_2 is Span and Wagner [8]. It is an empirical equation of state, but based on the form of a fundamental equation. It covers the fluid region (i.e., the gas and liquid states) from the triple point temperature, 216.692 K to 1100 K, and pressures up to 800 MPa. The Span and Wagner equation of state has 42 coefficients, which is reported to represent almost all of the reliable experimental data for pure CO_2 in the homogeneous region (i.e., single phase) and in liquid–vapor equilibrium states (i.e., the liquid–vapor phase boundary) to within their experimental uncertainty. The region of interest to pipelines, the uncertainty in the prediction of density, is reported to not exceed ±0.05%. The equation of state was developed from experimental data for the thermal and caloric properties of CO_2 in the single-phase region and for liquid–vapor, solid–vapor, and solid–liquid phase equilibria. It is also reported that the experimental data on the density of CO_2 over a range of temperatures and pressures in the single-phase region consists of a total of 5508 data points (from 59 papers, the earliest published in 1903).

4.4.3 Cubic Equations of State

The first modification in the Van der Waals equation of state came about in 1949 by Redlich–Kwong [9] in which the temperature dependence of the attraction parameter was established. Soave [10] then introduced the acentric factor and reduced temperature in the temperature-dependent attraction term where the reduced property is expressed as the fraction of the critical property (normalisation) $\frac{T}{T_c} = T_r$ where T is temperature and subscripts c and r denote critical and reduced. The Peng–Robinson (PR) equation of state has a close relation to the SRK equation in that it addresses the latter's weakness in the critical region of fluids and its inaccurate liquid density prediction [11].

The Peng–Robinson (PR), Soave–Redlich–Kwong (SRK), and the Benedict–Webb–Rubin–Starling (BWRS) equations are analytical equations of state, which means that an algebraic relationship exists between pressure, temperature, and molar volume [3]. The PR and SRK equations of state are classified as cubic equations of state as the polynomial expansion of the equation results in a cubic volume term. The BWRS, however, adds 5th and 6th power exponential density terms and is therefore more computationally complex. However, it is applicable to wider ranges of pressures and temperatures compared to the cubic equations of state. The general cubic equation of state is given by Eq. 4-4 [12].

$$P = \frac{RT}{V-b} - \frac{a}{(V^2 + ubV + \omega b^2)} \qquad (4-4)$$

The corresponding constants for the Peng–Robinson are

$$u = 2$$

$$\omega = -1$$

$$a = 0.45724\frac{R^2T_c^2}{P_c}\left[1 + (0.37464 + 1.54226\omega - 0.26992\omega^2)\left(1 - \sqrt{\frac{T}{T_c}}\right)\right]^2$$

$$b = 0.07780\frac{RT_c}{P_c}$$

The corresponding constants for the SRK are

$$u - 1$$

$$\omega = -1$$

$$a = \frac{0.42748R^2T_c^2}{P_c}\left[1 + (0.48 + 1.574\omega - 0.176\omega^2)\left(1 - \sqrt{\frac{T}{T_c}}\right)\right]^2$$

$$b = \frac{0.08664RT_c}{P_c}$$

where the subscript c indicates critical temperature and pressures, and ω is the acentric factor.

The a and b coefficients can be defined using different parameters, such as critical temperature and pressure, acentric factors, saturated liquid volume, or vapor pressure [3, 12, 13]. Equation 0-4 reduces to the Van der Waals equation when $u = 0$ and $\omega = 0$. The determination of a, b and ω are based on fit parameters obtained from empirical data of hydrocarbon vapor pressure. It should be noted that variation of the numerical constants due to fitting with other data sets, and also variation in the temperature dependence parameter, do exist [14].

Equations of state are developed from vapor pressure data of pure compounds, mainly normal alkanes and some non-hydrocarbon gases. They are first developed for gases, and then extended to liquids and also solids (i.e., wax and asphaltenes). Therefore, there are uncertainties associated in predicting the behavior of fluids with high densities, liquids, and supercritical fluid phases. As a result of this, validation of these equations is required especially with dense phase and supercritical fluids.

The two constant equations of state such as the Peng–Robinson (PR) and Soave–Redlich–Kwong (SRK) will predict densities well [15]. As mentioned earlier, modifications in the fit parameters can be carried out in order to extend their functionality and address the limitations. A comparison between the SRK and PR was performed, and it was found that the PR predicted more accurate liquid density values for a mixture of isobutene and CO_2 at pressures between 40.8 bar and 272.1 bar within a temperature range of 38°C to 237°C [12]. These pressures and temperatures can be considered as an indication of the range of validity of the PR and SRK equation of state.

Mixing Rule for Mixtures: The above expressions for the cubic equations of state are for pure components. In order for them to be used for mixtures, mixing rules need to be employed. There are a number of different mixing rules available for different component systems and equations of state. In general, mixing rules are obtained by averaging pure component constants to obtain constants that characterize the mixture and also by using empirical correlations which have resulted from the comparison of experimental and calculated data. The equation of state mixture parameters have to be determined separately for mixtures of different compositions. Given that this type of experimental data is scarce, it is therefore necessary to reduce and correlate the limited data available to make the most accurate interpolation and extrapolation [3]. This is usually done to obtain the binary interaction parameters of the components that make up the mixture.

Generally, the a and b parameters for pure components in the PR and SRK equations are replaced as follows for mixtures. These are the most common mixture rules used, and it should be noted that many more complex mixing rules are also available resulting from both empirical and theoretical studies [3].

$$a_{\text{mix}} = \sum_i \sum_j y_i y_j (a_i a_j)^{\frac{1}{2}} (1 - \bar{k}_{ij}) \tag{4-5}$$

$$b_{\text{mix}} = \sum_i y_i b_i \tag{4-6}$$

The value of $\bar{k}_{ij}$, the binary interaction parameter (BIP), is determined using experimental data. It should be noted that the parameters a_{mix} (which is determined by $\bar{k}_{ij}$), and b_{mix} play an important role in determining the phase properties of CO_2 mixtures. This has been shown in a study conducted by Li and Yan [16] using available vapor–liquid equilibrium data for binary mixtures of CO_2 at particular temperatures and pressures. Binary mixtures of CO_2 with H_2S, Ar, N_2, O_2, SO_2 were analyzed, and it was concluded that different equations of state provide different levels of accuracy for varying binary mixtures for different values of $\bar{k}_{ij}$. However, the value of the BIP had no effect on the saturated vapor pressure prediction of the EOS used, as only the vapor composition prediction was improved.

4.4.4 Benedict–Webb–Rubin–Starling Equation of State

The Benedict–Webb–Rubin–Starling (BWRS) equation of state is an example of virial equation of state and is a power series in density. The coefficients of a virial expansion can be calculated from a theoretical model of the inter-molecular potential energy of the gas molecules. It was developed for light gases and hydrocarbons. It is an eleven-parameter, non-cubic, equation of state, and hence there are eleven parameters to be defined for each component. Mixing rules are defined in terms of a single binary interaction parameter for each binary mixture. Starling [17] conducted multi-property regression analysis to determine the 11 parameters for each of fifteen substances (light gases and hydrocarbons) applied to this EOS. The binary interaction parameters were determined from experimental data on liquid–vapor equilibrium. Starling also conducted a simultaneous multi-property regression analysis to define a generalized BWRS equation of state, where the 11 parameters were defined in terms of the critical temperature, critical density, molecular weight, and acentric factor of the pure substance. It is believed that the generalized BWRS equation of state can be applied to substances other than the fifteen substances in Starling [17].

Lin and Hopke [18] and Hopke and Lin [19] conducted a simultaneous multi-property regression analysis on pure component, binary mixture, and multi-component mixture data to develop an optimized set of pure component parameters and binary interaction parameters for the BWRS equation of state. The substances considered were methane, ethane, propane, iso-butane, n-butane, iso-pentane, n-pentane, nitrogen, and carbon dioxide. A generalized correlation to determine the parameters for heavy hydrocarbon fractions (hexane and above) was also developed.

As such, the BWRS is a more complex equation of state as it is an equation with eleven coefficients requiring extensive data regression and therefore also requiring an increased amount of computing time [3, 20]. The range of validity of this equation of state is between temperatures of 10°C and 50°C and pressure ranges of 15 bar to 600 bar for pure CO_2 [15, 17]. It is worth noting that although pressure and temperatures in the critical region are within the validated pressure and temperature ranges, the region or points around the critical point have not been included in the validation. Hence, extrapolation by the equation of state into the supercritical region is still necessary.

The coefficients in the equations of state are modified according to such factors as the type of fluid, composition, available experimental data, pressure, and temperature ranges of interest. They are determined either via extrapolation into regions where experimental data are unavailable or via data fitting exercises with experimental data. For example, the BWRS coefficients have been extensively tested for hydrocarbons and coal-derived compounds and therefore would be used for calculating the properties of such fluids [3]. A general drawback of the BWRS EOS is the complexity of the equation in terms of solving parameters as it is not a simple cubic equation of state. Also, the interaction parameters must be available for all the system components [21].

4.4.5 GERG-2008 Equation of State

The Groupe Européen de Recherches Gazières (GERG) equations of state [22, 23] allow accurate modeling of thermodynamic properties in a pressure and temperature range suitable for the transport of CO_2. The equation was designed to have a simple structure with a relatively small number of terms, giving high calculation speeds and high accuracy results over a wide range of pressures and temperatures (including around the critical region). Phase envelope calculations are at least as accurate as cubic equations of state.

TABLE 4-1. 18 Components applicable in the GERG-2008 Equation of State (EOS)

Carbon Dioxide	*Hydrogen Sulphide	n-Pentane
Methane	Water	Isopentane
Nitrogen	Helium	n-Hexane
Hydrogen	Ethane	n-Heptane
Oxygen	Propane	n-Octane
Argon	n-Butane	*n-Nonane
Carbon monoxide	Isobutane	*n-Decane

GERG in 2004 allows the use of 18 components (shown in Table 4-1), while in GERG 2008 version has an extra 3 components shown by (*) in Table 4-1.

Similar to most modern equations of state, the GERG equation is based on the molar Helmholtz energy, $a\ (\rho, T)$. In terms of the dimensionless Helmholtz energy = a/kT, the GERG equations of state have the basic structure

$$a(\delta,\tau,x) = a^{o}(\rho,T,x) + \sum_{i=1}^{N} x_i a_{0i}^{r}(\delta,\tau) + \Delta a^{r}(\delta,\tau,x) \qquad (4-7)$$

where

$$\delta = \frac{\rho}{\rho_r(x)}, \tau = \frac{T_r(x)}{T}$$

and ρ_r and T_r are the reduced density and inversely reduced temperature, respectively. For the 2004 version, $N = 1,...,$ 18 for the 18 components, and for the 2008 version, $N = 1,...,$ 21. All thermodynamic properties can be deduced from derivatives of Eq. 4-7. Each term of Eq. 4-7 has a similar form allowing for integrated rapid calculations. The first term of Eq. 4-7 takes into account the ideal gas behavior, the second term takes into account the pure substance behavior for each component, and the third term is the departure function.

Each component, $1,...,N$ is described by a pure substance equation of state (the second term of Eq. 4-7), these are shown in Table 4-2. The pure substance equations of state were chosen because of their mathematical simplicity and similar structure, so that the whole process can be optimized in an integrated manner. New equations of state were created by GERG to describe the pure substance behavior of hydrogen, carbon monoxide, water, and helium [22].

The departure function and reducing functions describe the multi-component behavior of mixtures. The reducing function matches to empirical density and temperature data, while the departure function uses empirical data for binary mixtures. No ternary or higher mixture data are used because of the poor quality and relative sparseness of the data. For poor-quality binary data, no departure function is used.

The departure function takes the form

$$\Delta a^{r}(\delta,\tau,x) + \sum_{i=j+1}^{N} \sum_{j=1}^{N-1} \Delta a_{ij}^{r}(\delta,\tau,x),$$

which captures the multi-component behavior.

TABLE 4-2. Pure substance equation of state included in the GERG equation of state

Substance	Equation of State
Carbon Dioxide	Klimeck [24]
Methane	Klimeck [24]
Nitrogen	Klimeck [24]
Ethane	Klimeck [24]
Hydrogen	GERG 2004
Oxygen	Span and Wagner [25]
Carbon Monoxide	GERG 2004
Water	GERG 2004
Argon	Span and Wagner [25]
Helium	GERG 2004
Propane	Span and Wagner [25]
n-Butane	Span and Wagner [25]
Isobutane	Span and Wagner [25]
n-Pentane	Span and Wagner [25]
Isopentane	GERG 2004
n-Hexane	Span and Wagner [25]
n-Heptane	Span and Wagner [25]
n-Octane	Span and Wagner [25]

Note: No reference equations are included for hydrogen sulphide, n-nonane and n-decane as GERG 2008, yet to be published.

Phase equilbria can be calculated in the usual way by finding the minimum of the Gibbs free energy. However, phase equilibria can also be calculated directly from the Helmholtz energy, which has the advantage of not requiring the use of iterative calculations to find densities from the equation of state.

4.5 EVALUATION OF EOS FOR ANTHROPOGENIC CO_2 MIXTURES

There is no consensus in the literature regarding the equation of state that should be used for the design of CO_2 pipelines. McCullough [26], King [27], and Farris [28] all use the Benedict–Webb–Rubin–Starling (BWRS) equation of state for their calculations, whereas Zhang [29] used the Peng–Robinson (PR) equation with Boston Mathias modifications and Hein [30, 31] used the PR equation and also Soave–Redlich–Kwong (SRK). The DNV RP J202 [32] recommended practice for the design and operation of CO_2 pipelines [32] recommend using the Peng–Robinson equation of state for a composition containing overwhelmingly CO_2. However, it also states that for CO_2 mixtures containing significant levels of impurities, the equation of state used should be tuned using experimental data or, as a minimum, it should be validated against experimental data to evaluate the level of uncertainty in its calculation. The recommended practice also cautions the use of an equation of state close to the critical point as this point is associated with non-linearities. The equation of state should also be validated with experimental data for water solubility/drop out and hydrate calculations for the project specific CO_2 stream.

Li and Yan [33] has conducted a comparative study of all of these equations of state with the exception of the Span and Wagner EOS and concludes that the selection of the equation of state may have a significant impact on the pipeline design, although without experimental data, it was not possible to identify the most accurate equation of state to use. Indeed, King [27] recommends that the equation of state selected should be based on reliable experimental data and the validation of the equation of state for a given impurity combination is necessary. He, therefore, concludes that it is advisable to validate the equation of state being used for the particular impurities under consideration using bubble-point experiments. This has not been done for the impurities expected from power station capture of CO_2. A comparison of the output from these equations with available experimental data from the literature will be carried out in this chapter. It should be noted that the optimum equation of state may differ for pure CO_2 and CO_2 containing impurities. This aspect will also be investigated.

The equations of state used for modeling anthropogenic CO_2 in this chapter were evaluated based on available experimental data in the published literature. It is impossible to carry out experiments for all the physical properties required for the entire range of power plant impurities and thermodynamic states of the mixtures [34]. Therefore, complete phase equilibrium validation of CO_2 with power plant impurities at pressures and temperatures relevant to pipeline operation was not possible.

The phase properties of pure CO_2 were first evaluated by comparing experimental data with the Span and Wagner, PR, SRK and BWRS equation of state. It may be noted that the Span and Wagner equation of state is complex and only applicable to pure CO_2. The phase properties of this equation were evaluated using published results of the equation for pure CO_2 [8]. shows a comparison of the equations of state in predicting the vapor–liquid equilibrium line of pure CO_2. It can be seen that all the equations of state evaluated give good agreement with experimental data [35, 36] for pure CO_2 except for the BWRS equation of state which produces higher saturated vapor pressures along with a higher critical pressure and a lower critical temperature. Also, there is the question as to whether this equation with its eleven coefficients can be extrapolated as reliably as EOS with two coefficients like the Peng–Robinson and Soave–Redlich–Kwong equations. These complexities of the BWRS, such as the high number of constants and binary interaction parameters, have limited its use in commercial simulation packages [15].

Figure 4-3 shows the values of critical temperature and pressure predicted by the equations of state and that measured from the experiments of Duschek et al [35, 36]. Other than the BWRS equation of state, the critical temperatures and pressures measured experimentally and calculated using the equations of state have little variation. The BWRS gives the worst prediction of these parameters when compared to the experimental data. The best or

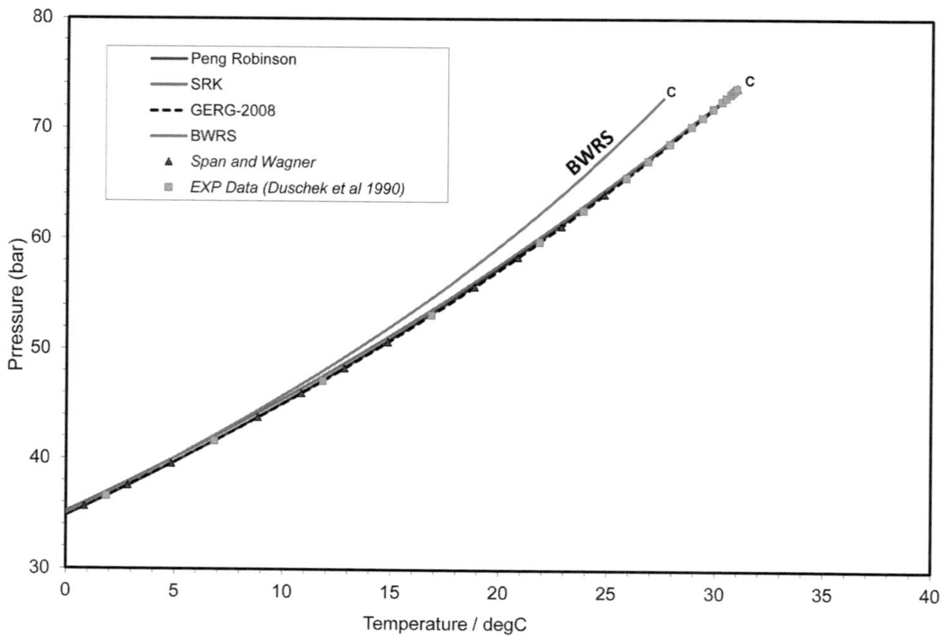

Figure 4-3. Comparison of the equations of state with experimental data for pure CO_2

TABLE 4-3. Comparison of the critical pressures and temperatures of pure CO_2 calculated by the equations of state and experimentally measured values

Equation of State	Critical Temperature (Tc)	Critical Pressure (Pc)	% Deviation from Experimental Data	
			Tc	Pc
PR	30.86	73.57	−0.29	−0.19
SRK	30.81	73.51	−0.45	−0.27
BWRS	27.56	72.82	−10.95	−1.21
GERG-2008	30.97	73.76	100.06	100.07
Span & Wagner	30.98	73.77	0.10	0.08
Exp (Duscheck et al)	30.95	73.71	0.00	0.00

most accurate prediction of critical temperatures and pressures is from the Span and Wagner equation of state followed by the Peng–Robinson (PR) with deviations of −0.09°C and −0.14 bar, respectively. It is worth emphasizing that the Span and Wagner and PR EOS for pure CO_2 give very similar results in terms of both the critical temperature and also the vapor–liquid equilibrium line.

The experimental data for CO_2 mixtures are scarce and insufficient to completely validate the phase modeling presented here. It is also deemed difficult to obtain data relevant to the proposed CO_2 pipeline operating region. Most of the experiments are well away from the critical point either at very high pressures and temperatures which are well into the supercritical region or very low pressures and temperatures. The binary or ternary CO_2 mixture data available in the literature are with low concentrations of CO_2 whereby the CO_2 is an impurity and not the main component of the mixture [21, 37]. This is because most of the data available have been developed for hydrocarbon processing systems in which CO_2 is an impurity.

An equation of state can be validated using a number of fluid properties obtained via experiments (Figure 4-4). These properties include vapor–liquid equilibrium (VLE) data, pressure–volume–temperature (PVT) data, pressure–temperature–density (P Tρ) data, saturated density, compressibility factor, enthalpy, and heat capacity measurements. This is because such properties are directly predicted by an equation of state. Other parameters, such as viscosity and thermal conductivity, are predicted using separate correlations although some input maybe required from the parameters predicted by the equation of state.

The validity of an equation of state will essentially be limited by the composition, temperature, and pressure ranges used in its development and evaluation [16, 21], although most analytical equations of state can provide reasonably good results beyond this range, and there are validity checks that can be carried out on the extrapolation behavior of the equations of state [8, 21]. The region around the critical point for CO_2 has a number of discontinuities and uncertainties associated with it. Therefore, systematic deviation from experimental data can occur [35]. These uncertainties and discontinuities are further exacerbated by the presence of impurities [28, 38–40]. The current CO_2 pipeline transportation operations are carried out in the vicinity of the critical region and therefore accurate thermo physical calculations are necessary to ensure the safety and efficiency of the pipeline [8, 27, 28]. However, it should be emphasized that calculated results must be treated cautiously in the critical region due to the potential limitations and also discontinuous behavior of most equations of state in representing the critical region of fluid mixtures accurately especially when the molecular weights of the components differ markedly [20].

A preliminary assessment of the most appropriate EOS for the modeling of an impure CO_2 capture stream has been conducted using experimental data for a binary acid gas system CO_2–H_2S [41], as this contains at least one of the expected impurities in the captured CO_2 stream.

The experimental data for two different concentrations of H_2S is modeled with the PR, BWRS, and SRK equation of state, and the results are presented in Figure 4-4. The Span

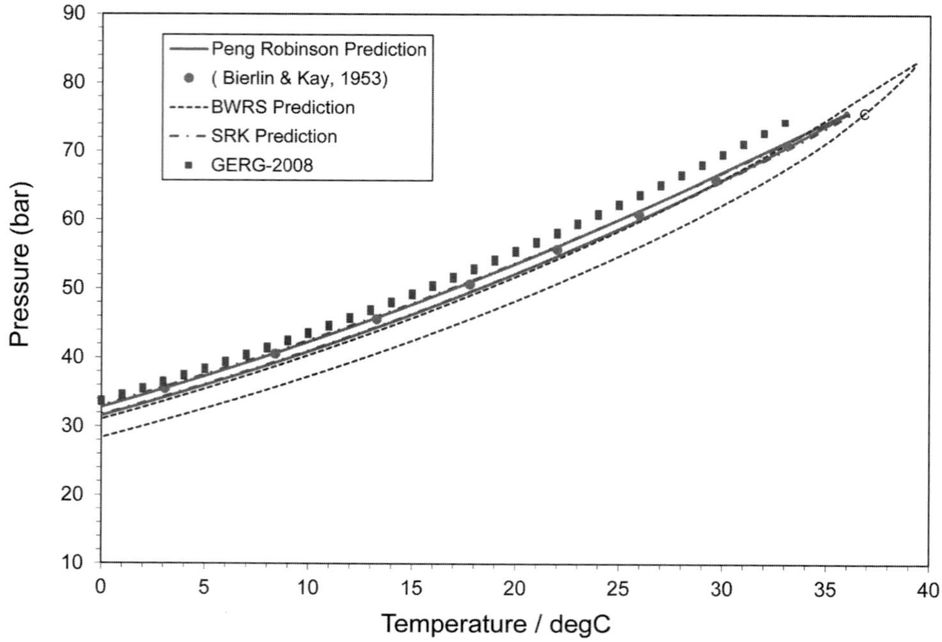

Comparison of results for the 90.02%CO_2-9.98%H_2S system

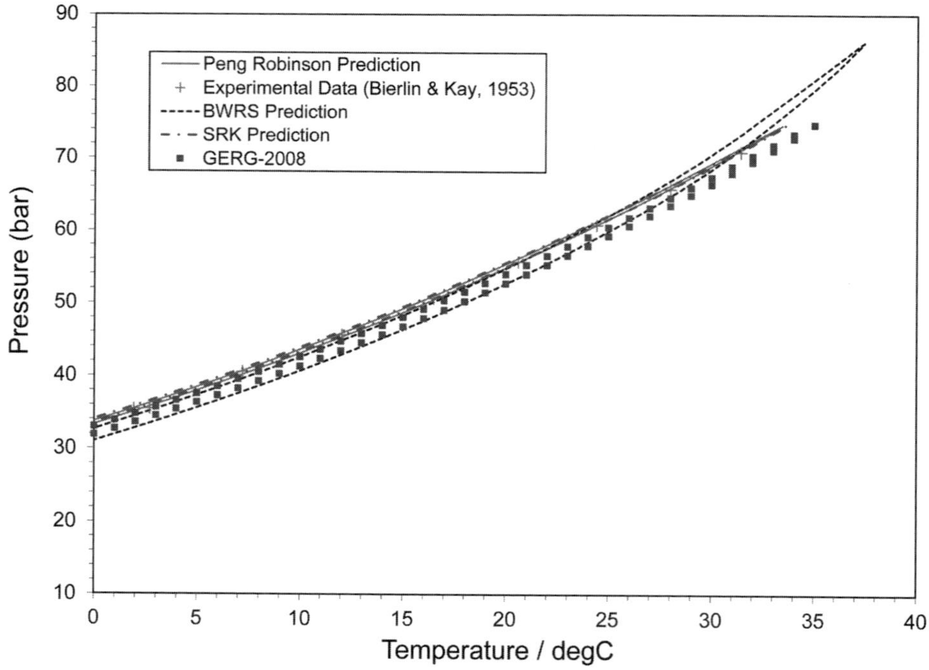

Comparison of results for the 82.92%CO_2-17.08%H_2S system

Figure 4-4. Comparison of experimental phase equilibrium data for the CO_2–H_2S system with phase envelopes generated using the PR, BWRS, and SRK and GERG-2008

and Wagner EOS was not evaluated here as it is not applicable for CO_2 mixtures. Although both the PR and SRK equation of state provide a good fit to the experimental data, close inspection indicates that the PR EOS is slightly more accurate over both sets of experiments. Therefore, on the basis of the literature review and the results of the previous comparisons with experimental and published data, the PR EOS [12] has been selected for the physical property analysis carried out in this chapter.

4.6 EFFECT OF IMPURITIES ON THE PHASE PROPERTIES OF CO_2

The phase behavior governing the physical properties of CO_2 changes when impurities are introduced into the system. For efficient pipeline transport, it is essential that the CO_2 remains in the dense phase. Therefore, the effect of impurities on the critical temperature, critical pressure, and also the phase envelope itself needs to be understood. This is because these parameters determine the area of the dense phase and supercritical region which in turn plays a part in determining the pipelines operating region. In addition, two-phase flow in a pipeline designed and specified for single-phase transport should be avoided if possible due to the potential for damage to the pipeline and its facilities. It is therefore desirable to study the effects of impurities on the potential to form two-phase regions in the phase diagram to ensure that the pipeline is operated outside these areas.

Figure 4-5 shows the vapor–liquid equilibrium line for pure CO_2. The addition of impurities to the CO_2 product stream causes a change in the resulting phase envelope. The addition of impurities into pure CO_2 has two main effects on the phase envelope, one is to change the critical pressure and temperature, and the second is to open up a two-phase envelope or region. This effect of impurities in a binary system with CO_2 is shown in Figure 4-5 from the phase envelopes generated for some of the impurities identified as a result of the power plant capture process. It can be observed that some impurity combinations tend to cause a large increase of the phase envelope or two-phase area; H_2, N_2, and NO_2 open up large two-phase regions, whereas the two-phase region formed with the addition of H_2S is almost negligible [42].

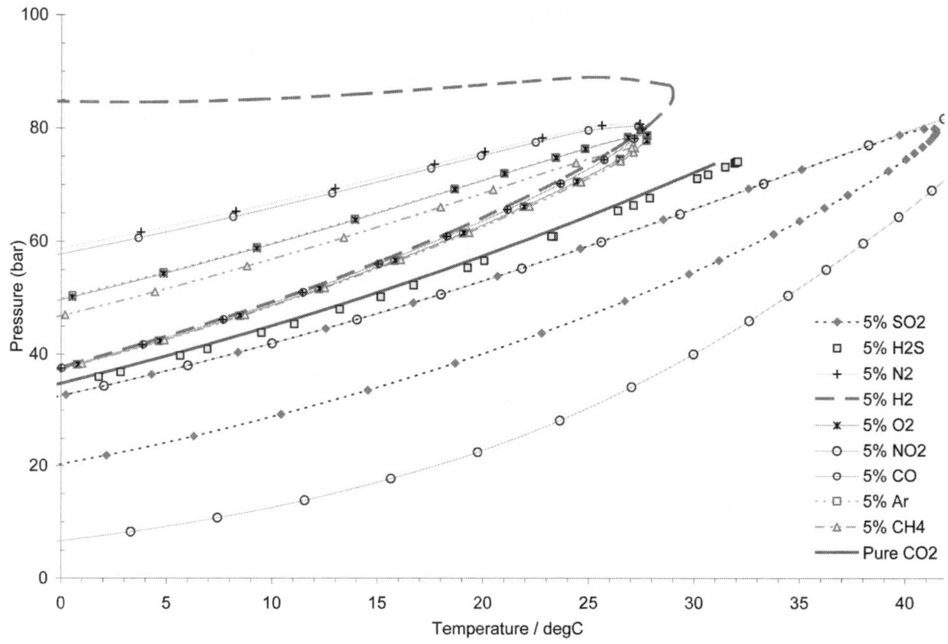

Figure 4-5. Phase diagram for CO_2 containing 5% impurity

It can be seen from Figure 4-5 and the corresponding two-phase areas calculated in Table 4-4, that the binary mixture of 95% CO_2 and 5% NO_2 has the largest two-phase area of 1075.92 unit^2, followed by that of 5% H_2 and 5% SO_2. It can also be observed that components, such as NO_2 and SO_2 with molecular weights greater than CO_2, form a phase envelope that expands below the VLE line of pure CO_2 when in a binary combination with 95% CO_2 and 5% of the component. The opposite is true for lighter components with a smaller molecular weight than CO_2 such as H_2 whereby the phase envelope forms above the VLE line of pure CO_2. However, for the binary mixture phase envelopes that form above the VLE line, the 95%CO_2–5% H_2 has the largest two-phase area of 820.76 unit^2.

The effect of increasing percentages of impurities on the critical temperature and pressure are shown in Figure 4-6 and Figure 4-7, whereby an increase in critical pressure and a decrease in critical temperature are observed, with the exception of NO_2 and SO_2 in which the critical temperature rises with increasing proportions of these impurities. The different combinations of impurities have different effects on the phase properties of CO_2, and the general trend is always to raise the critical pressure. However, the presence of SO_2, NO_2, and H_2S raises both the critical temperature and pressure of the CO_2 mixture, but the addition of H_2, N_2, O_2, CO, Ar, and CH_4 only raises the critical pressure but lowers the critical temperature. The changes in the mixtures critical properties are related to the critical properties of the individual component [3] shown in Table 4-4. It can be observed that both the critical pressure and temperature of the individual components NO_2, SO_2, and H_2S are greater than that of pure CO_2, which is related to the resulting increase in both the critical temperature and pressure the binary mixtures Table 4-3.

In general, the critical pressures for the binary mixtures are proportional to the calculated two-phase area as shown in Table 4-4. Therefore, in descending order of phase envelope area which is proportional to the increase in critical pressure, H_2 has the highest critical pressure and also the largest two-phase area, which is then followed by N_2, CO, O_2, Ar, and CH_4 with the exception of H_2S, which does not form a phase envelope and therefore has the lowest critical pressure compared to the other binary mixture. Once again, the binary combination of CO_2 involving NO_2 and SO_2 does not exhibit similar behavior which can be attributed to the higher molecular weight of these components compared to CO_2, which results in a phase envelope forming below that of the pure CO_2 vapor equlibrium line (VLE) (Figure 4-8).

It is interesting to observe that the critical temperature of pure H_2, N_2, O_2, CO, Ar, and CH_4 are all below 0°C and therefore very much lower than that of pure CO_2, which in turn results in a lowered overall critical temperature for the binary mixture of CO_2 containing these individual components. This is because the critical temperature of a mixture is a linear function of composition and the critical temperature of individual components in the mixture. Hence, the critical temperature of a binary mixture of CO_2 with other components will decrease or increase based on the critical temperature of the pertinent component.

TABLE 4-4. Calculated two-phase area for 5% impurity combination with 95% CO_2 and properties of individual components in the anthropogenic CO_2

Component X	Molecular Weight	Critical Temperature (°C)	Critical Pressure (bar)	Two-phase Area for 95% CO2– 5%X Combination
SO_2	64	157.7	78.8	470.82
H_2S	34	100.1	89.4	0
N_2	28	−147	33.9	397.91
H_2	2	−240	13	848.39
O_2	32	−118.6	50.4	252.43
NO_2	46	157.9	101	1082.99
CO	28	−140.25	35	377.57
Ar	40	−122.4	48.7	255.64
CH_4	16	−82.75	46	179.88
CO_2	44	31	74.1	0

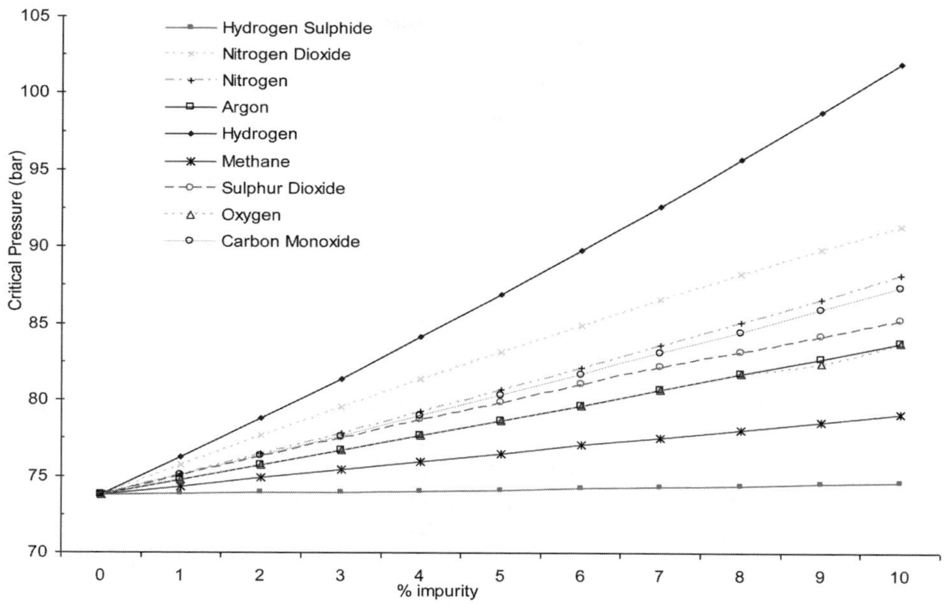

Figure 4-6. Change in critical pressure with percentage of impurity

However, this is not the case for the critical pressure as it is highly non-linear with concentration and the critical pressure of individual components. The critical pressure of a mixture is also a function of acentricity of constituent molecules, hence, the acentric factor (related to shape of molecules) is also used in critical pressure calculations in addition to other properties. Based on this non-linearity, the critical pressure of a binary mixture of CO_2 with other components not only depends on the critical pressure of the constituents but also on the acentric factor and the critical temperature. This highly non-linear function with

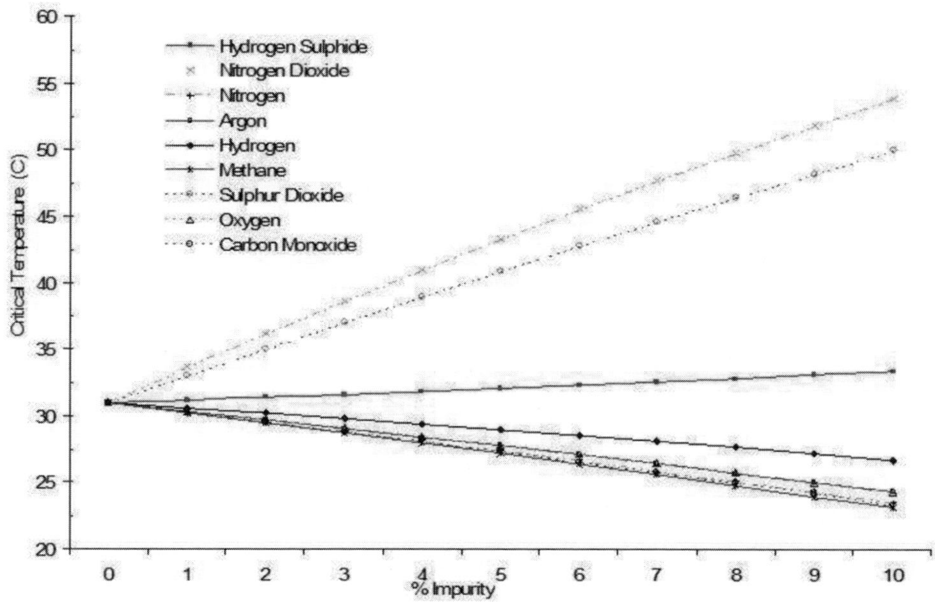

Figure 4-7. Change in critical temperature with percentage of impurity

concentration makes it difficult to discuss the trend of the critical pressure of the binary CO_2 mixtures which is in contrast to the critical temperature of a mixture.

It is worth emphasizing that other than the individual critical properties of a component, properties such as intermolecular forces, molecule shapes (acentricity), and polarity will play a part in determining the phase properties of the mixture. These factors become more significant for multi-component mixtures of anthropogenic CO_2. Therefore, the type, combination, and quantity of the impurity components will play a significant part in determining the physical behavior of anthropogenic CO_2.

In order to illustrate the effect of the type and combination of impurities, on the phase behavior of the anthropogenic CO_2, binary and ternary systems involving NO_2, as shown in Figure 4-8, were selected as an example due to the contrasting phase behavior of NO_2 compared to the other components in a binary mixture with CO_2. The ternary system 90% CO_2, 5% N_2, and 5% NO_2 shows an even more significant effect on the phase envelope in terms of two-phase area, the critical pressure, and temperature, compared to a binary mixture of 95% CO_2 and 5% NO_2. The phase envelope has now extended to both below and above the VLE line of pure CO_2, thus reducing the operating region for both single-phase gas and liquid. The two-phase area for the 95%CO_2–5%N_2 also increase with the addition of 5% CH4 (90%CO_2–5%N_2–5%CH_4); however, the two-phase area here is less than that of the 90% CO_2, 5% N_2, and 5% NO_2 mixture. In general, the size of the two-phase area of the ternary system to a certain extent reflects the phase behavior of the individual components in a binary combination with CO_2.

In the examples shown, an increase in the critical pressure and a decrease in critical temperature which reduces the supercritical area are observed when decreasing the percentage of CO_2 with the addition of a third component to the binary systems as shown in Figure 4-8. The increase in the critical temperature and pressure decreases the supercritical area thus reducing the optimum pipeline operating region. The reverse is true for the decrease in the critical point. Therefore, it has been demonstrated that the type, amount, and combination of the impurities play a crucial role in the phase properties of the CO_2 stream. However, further research, which is beyond the scope of this book, is required in understanding the effect of molecular forces and molecule shapes on the phase behavior of the anthropogenic mixture.

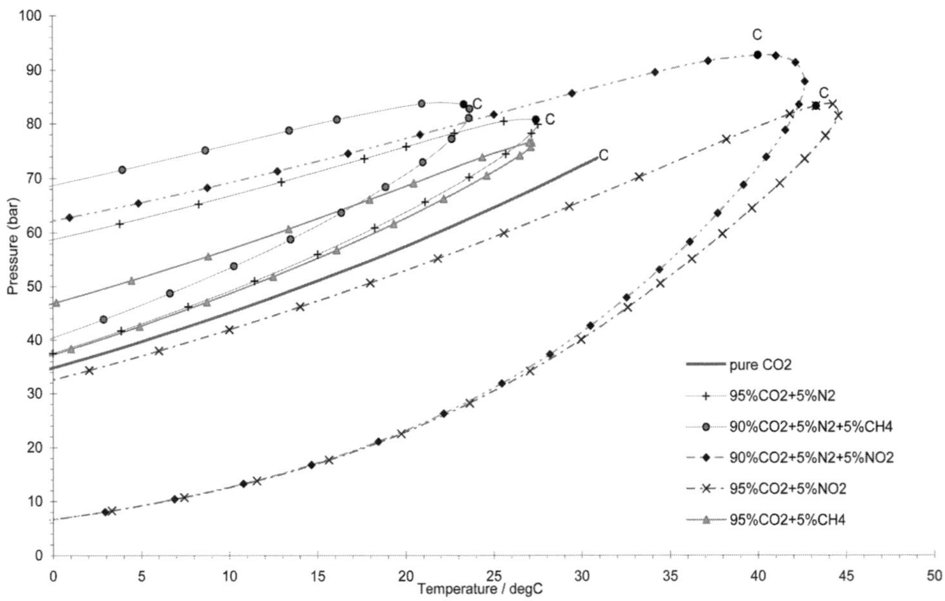

Figure 4-8. Phase diagram for CO_2 in binary and ternary combinations

4.7 OPERATING ENVELOPES OF A CO_2 PIPELINE

The operating and design aspects of the pipeline and associated equipment are determined by the physical and thermodynamic properties of CO_2 and contained impurities. Current operating practice for long distance CO_2 pipelines is to maintain the pressure well above the critical pressure as it is deemed the most efficient way to transport CO_2 in terms of throughput and it minimizes energy losses and also the number of pumps required [29]. However, for shorter distances and low volumes, it may be beneficial to transport CO_2 in its gaseous phase [29, 43, 44]. The captured and subsequently transported anthropogenic CO_2 will contain a certain amount of impurities which will have an effect on the thermo physical properties of CO_2, namely, on the area of two-phase flow and the critical point of the mixture. However, anthropogenic CO_2 will contain a combination of impurities due to the different capture technologies proposed, refer to Chapter 2 for details.

Current research has tackled the problem of defining a CO_2 quality specification scenario for anthropogenic CO_2 from power plants from two different perspectives/options. These are the following:

1. Taking the capture process as the starting point and analyzing the impurities that could be present, and
2. taking the pipeline and defining the levels of impurities that could be safely transported from both a health, environmental, and technical point of view.

Nevertheless, the second option is technically determined by the maximum allowable operating pressure of the pipeline which is in turn determined by the delivery pressure, a safety factor determined by the regulators/codes, for anthropogenic CO_2.

This chapter l focus will be on the quality specifications reviewed in Chapter 3 for the three different capture technologies with the aim to define a potential "best" and "worst" scenarios for CO_2 quality from the capture plant followed by analyzing the physical properties of the selected "best" and "worst" cases. These examples would provide a general indication of the composition and trends associated with physical and hydraulic properties that could be expected from an anthropogenic CO_2 capture stream.

The "best" and "worst" cases were determined based on the size of the two-phase area and the critical points of each anthropogenic CO_2 quality scenario. Subsequently, the effect of the physical properties, such as viscosity, density, compressibility, thermal conductivity, and enthalpy of the anthropogenic CO_2 mixture on the whole system, is explored. The effect of the currently defined water specification for CO_2 pipelines will also be analyzed and discussed.

4.8 PHASE BEHAVIOR OF ANTHROPOGENIC CO_2 QUALITY SCENARIOS FOR CCS

Table 4-5 shows a summary comparison of the impurities present for all the anthropogenic CO_2 gases reviewed in Chapter 3 including the critical pressure and size of the two-phase area for each of these systems. The CO_2 concentration has been determined by assuming that all of the other impurities are present at the maximum limit specified and calculating the percentage of CO_2 as the remainder, ensuring that the CO_2 minimum specification is still met. However, for the Oosterkamp and Ramsen specification scenario [37], this process results in a lower than specified CO_2 concentration. Therefore, in this case, the CO_2 concentration has been set at the minimum specified and the concentration of the remaining impurities have been adjusted in the proportions indicated in the specification.

As might be expected, due to the high percentage of CO_2 present in the mixture, the phase envelopes produced by the post-combustion specification are similar to that of pure CO_2 with critical pressures ranging from 73.82 bar to 74.04 bar and critical temperatures ranging from 30.85°C to 31.02°C and also the absence of a two-phase envelope. The "best" case for the post-combustion capture technology, based on the literature review, was the IPCC specification scenario with a critical pressure of 73.82 bar and the "worst" case for the post-combustion capture technology was the Oosterkamp and Ramsen [37], with a critical pressure of 74.04 bar. Based on the criteria defined previously, the "worst" case within the pre-combustion capture scenario was the IPCC [45] minimum purity specification scenario and the "best" case was the maximum purity IPCC [45] compositional stream.

A similar evaluation criteria for determining the "best" and "worst" cases for the oxy-fuel compositional specification indicate that the Oosterkamp and Ramsen [37], "Coal and Gas" scenario was the "worst" case and the IPCC gas compositional scenario was the "best" case for the oxyfuel specifications reviewed. The Oosterkamp and Ramsen [37] oxy-fuel compositional specification scenario has a critical pressure of 86.69 bar and a critical temperature of 28.53°C with the largest two-phase area. The IPCC gas "best" case oxyfuel compositional specification has a critical pressure of 78.39 bar and a critical temperature of 28.34°C.

The overall "best" and "worst" cases among all the reviewed capture technologies were also determined using the same criteria which was based on critical pressures and phase envelope areas. This in turn forms part of the analysis that is to follow and consequently defines the "best" and "worst" cases. In accordance to these criteria, the "best" case impurity specification was the post-combustion specification published by IPCC [45] and the "worst" case specification was the oxyfuel specification of Oosterkamp and Ramsen [37] which are highlighted in Table 4-5. It can be seen both from Table 4-5 and Figure 4-9 that the area of two-phase flow increases with the critical pressure of the anthropogenic CO_2 specifications.

From Table 4-5, it can be inferred that, of the three capture technologies, post-combustion capture produces CO_2 streams with critical pressures which are very close to that of pure CO_2. However, there is significantly more variation in the critical pressures for the published specifications for pre-combustion and oxyfuel capture. The selected "worst" case for the pre-combustion technology which is the IPCC [45] minimum purity specification results in a larger two-phase area and higher critical pressure than the other pre-combustion specifications, even though it contains a higher percentage of CO_2. This can be attributed to the higher proportion of H_2 (2%) present in the IPCC [45] minimum purity specification compared to the other pre-combustion quality specification reviewed.

The "worst" case specification from the oxyfuel technology which is also the overall "worst" case specification contains both NO_X (0.14%) and SO_X (1.41%) which are assumed to be NO_2 and SO_2 for the purpose of the analysis, both of which as previously shown result in large two-phase areas below the vapor–liquid equilibrium (VLE) line of pure CO_2 and high critical pressures when present in binary combination with CO_2. However, the amounts of these components present are well below 5%, and the higher percentages of the lighter components such as N_2 (3.94%), O_2 (1.69%), and Argon (2.82%) have influenced the phase behavior in terms of ensuring that the resulting phase envelope is above the vapor–liquid equilibrium (VLE) line of pure CO_2 as shown in Figure 4-9. Indeed, the ENCAP oxyfuel specifications did not contain NO_X and therefore did not result in a higher two-phase area or critical pressure than the "worst" case specification eventhough similar amounts of O_2 and higher amounts of Argon were present. Nevertheless, the ENCAP (CO_2 + H_2S) specification scenario containing 1.5% of SO_X has resulted in the second largest two-phase area and critical pressure among the quality specifications reviewed as shown in Figure 4-9. Hence, it can be concluded that for the oxyfuel specifications reviewed, the presence of relatively large amounts of N_2 along with the presence of both SO_X and NO_X

TABLE 4-5. Comparison of CO_2 quality specifications

Component		Post-Combustion				Pre-combustion						Oxyfuel					Pipeline Specification	
		IPCC		ENCAP	O&R†	IPCC			ENCAP		O&R†	IPCC		ENCAP		O&R†	Dynamis	Ecofys
						Coal			Coal					Coal				
	Unit	Coal	Gas	Coal	Coal & Gas	Max. Purity	Min. Purity	Gas	CO_2/ H_2S	CO_2+ H_2S	Coal & Gas	Coal	Gas	CO_2/ SO_2	CO_2+ SO_2	Coal & Gas	Post-/Pre-combustion	Coal
CO_2	vol%	99.97	99.97	99.95	99.79	99.14	96.39	95.66	98.00	95.72	95.6	95.80	95.87	90.76	90.46	90.00	95.76	95.78
CH_4	vol%	–	–	–	0.0100	0.0100	0.01	2.0000	0.0350	0.0350	0.0206	–	–	–	–	–	0.8000**	0.8000**
N_2	vol%	0.0033*	0.0033*	0.0210	0.1700	0.0033*	0.2000*	0.4300	0.0300	0.0300	0.3522	1.2300*	1.3700*	0.6100	0.6000	3.9400	0.8000**	0.8000**
H_2S	vol%	–	–	–	Trace	0.0100	0.6000	0.0100	0.0100	2.3000	1.9960	–	–	–	--	Trace	0.0200	0.0200
C2+	vol%	–	–	0.0030	0.0100	–	–	–	–	–	0.0587	–	–	–	--	–	–	–
CO	vol%	–	–	0.0010	0.0010	0.0300	0.4000	0.0400	0.1700	0.1700	0.2348	–	–	–	--	Trace	0.2000	0.2000
O_2	vol%	0.0033*	0.0033*	0.0030	0.0100	0.0033*	0.2000	0.4300	--	--	Trace	1.2300*	1.3700*	1.6000	1.6000	1.6900	0.8000**	0.8000**
NO_X	vol%	0.0100	0.0100	0.0020	0.0050	–	–	–	–	–	–	0.0100	0.0100	--	--	0.1400	0.0100	–
SO_X	vol%	0.0100	0.0100	0.0010	0.0010	–	–	–	–	–	–	0.5000	0.0100	0.0760	1.5000	1.4100	0.0100	–
H_2	vol%	–	–	–	Trace	0.8000	2.0000	1.0000	1.7000	1.7000	1.7612	–	–	0.2500	0.2400	Trace	0.8000**	0.8000**
Ar	vol%	0.0033*	0.0033*	0.0210	Trace	0.0033*	0.2000	0.4300	0.0500	0.0490	0.0294	1.2300*	1.3700*	6.7000	5.6000	2.8200	0.8000**	0.8000**
HCN	vol%	–	–	–	–	–	–	–	0.0005	0.0005	N/A	–	–	–	–	–	–	–
COS	vol%	–	–	–	–	–	–	–	0.0005	0.0005	–	–	–	–	–	–	–	–
P_{crit}	bar	73.82	73.82	73.83	74.04	75.81	80.08	78.9	78.38	78.53	79.13	78.62	78.39	83.77	84.90	86.69	79.24	79.19
Two-phase area	$Unit^2$	Critical pressure used to determine best and worst case due to the absence of a phase envelope.				149.33	422.46	322.57	324.30	332.77	370.87	284.32	279.83	430.03	604.00	719.46	331.23	329.54

Notes: * The limit for N_2, O_2 and Ar in the IPCC specification was quoted as a total for all three impurities in combination. For the purposes of comparison, it is assumed that the impurities are present in equal proportion. ** The Dynamis and Ecofys specification limits the total of non-condensable gases to 4%. As for the IPCC specification, it is assumed that these impurities are present in equal proportion.

† O&R refers to the specification defined by Oosterkamp and Ramsen [37].

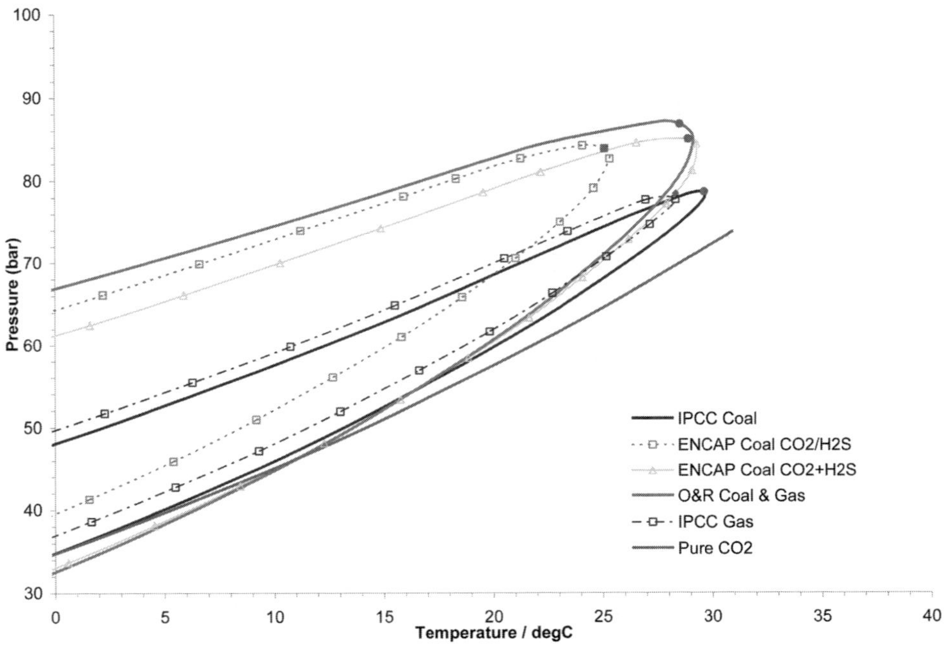

Figure 4-9. Phase diagram comparisons for the CO_2 quality specifications of the oxyfuel capture technology

has resulted in the largest two-phase area and critical pressure which was then defined as the "worst" case specification.

It is worth noting that the highest percentage of CO_2 with much smaller amounts of SO_X and NO_X in the post-combustion compositional scenario had insignificant effects on the phase properties for the post-combustion specifications which have similar phase behavior to that of pure CO_2. The pipeline compositional scenario recommended by Dynamis and Ecofys ranks as an intermediate between the "best" and the "worst" case specification and is based on the pre-combustion capture technology. Both specifications have also assumed that the total non-condensable (H_2, Ar, O_2, and N_2) are less than 4%.

Nevertheless, this assumption could result in erroneous assumptions of critical pressure and two-phase area as it was previously shown that the presence of these impurities in binary combination with CO_2 could result in significant differences in two-phase area and critical pressures. For example, the presence of 4% H_2 and absence of other non-condensable components results in larger two-phase areas and critical pressure than the presence of 4% of Ar, N_2, or O_2, respectively. It is acknowledged that other indicators of "best" and "worst" case quality specifications could include the effect of the impurity combinations on enhanced oil recovery (EOR), compressor power, economic, and safety considerations for the following discusses the hydraulic of these effects.

4.9 CO_2 WATER CONTENT SPECIFICATIONS

The anthropogenic CO_2 specifications scenarios detailed in Section 4.7 excluded water as the allowable water content are restricted due to the possibility of corrosion. Therefore, the amount of water present in the captured CO_2 stream was deemed too small to have any significant effect on pipeline hydraulics. In order to demonstrate the effect of water content on the phase behavior or bubble and dew point of the anthropogenic CO_2, a small amount

TABLE 4-6. The modified "worst" case specification to include the supercritical (0.064 vol%) and gaseous phase (0.0225 vol%) water specifications

Component	Worst case specification with 640 ppmv (0.064 vol%) water	Worst case specification with 225 ppmv (0.0225 vol%) water
CO_2	89.936	89.978
H_2O	0.0640	0.0225
NO_X	0.1400	0.1400
SO_X	1.4100	1.4100
Ar	2.8200	2.8200
N_2	3.9400	3.9400
O_2	1.6900	1.6900

of water was introduced into the CO_2 stream and the mixture flashed at different constant temperatures to obtain the dew point and bubble point pressure which are related to respective points on the phase envelope. The modified "worst" case composition to include water is shown in Table 4-6 whereby the 640 ppmv (0.064 vol%) water content based on existing specifications for CO_2 pipelines was included for high pressure dense or supercritical phase CO_2 transport, whereas the 225 ppmv (0.025 vol%) water specification was included for low-pressure gaseous transport, as per the recommended 60% saturation limit of water in CO_2 pipelines [46, 47].

At pipeline operating temperature and pressures, it is difficult to determine the phase boundaries for systems containing water as when a two-phase system is always formed — water will be present in both the liquid and gaseous phases of CO_2, and at the same time, CO_2 is present in the liquid and gaseous phase of the water. This is due to the complex thermodynamic behavior of strongly associating polar molecules, such as water, which can lead to unstable phase calculations [48, 49]. This has further been attributed to the lack of optimized binary interaction parameters for polar molecules, such as water due to the scarcity of experimental data [3]. In acid-gas CO_2 applications, it has been shown that the presence of an aqueous phase is difficult to model with the cubic equation of state (refer to Section 4.3.2.1) as it is difficult to determine the phase boundary in the presence of water and, therefore, research is currently being carried out in the area to address this [50]. However, it is possible to perform dew and bubble point flashing at given pressures and temperatures using cubic equations of state for very low concentration of water whereby the water does not drop out or become insoluble as the saturation limit is reached [51].

The "worst" case specification and pure CO_2 containing 0.064 vol% of water, and also in the absence of water, were flashed at different temperatures to obtain the resulting bubble and dew point pressures and the results are presented in Figure 4-10. It can be observed that the bubble point pressure limit for the "worst" case specification does not change in the presence of 0.064 vol% water. However, the dew point pressure limit shows deviation when 0.064 vol% water is present in the "worst" case specification at temperatures of 5°C and 10°C. Similar behavior was also observed in pure CO_2 containing 0.064 vol% of water.

As previously shown by the solubility curves of water in pure CO_2 (Chapter 3), the amount of water soluble in CO_2 reduces with decreasing pressure and temperature, whereby the solubility of water in the lower pressure gas phase, at temperatures below 15°C, is lower than that of the dense or supercritical phase of CO_2, thus causing water to condense out of the gas phase. Hence, a similar scenario of 0.064 vol% of water content for the supercritical phase should not be considered for the gaseous phase CO_2 below approximately 40 bar when low temperatures are anticipated. It is also worth noting that the dew points for pure CO_2 and the "worst" case specification at 5°C, 10°C, and 15°C are the same, which highlights that at these pressures and temperatures water will condense for both the "worst" case and the pure CO_2, containing 0.064 vol% water. Nevertheless, as previously mentioned,

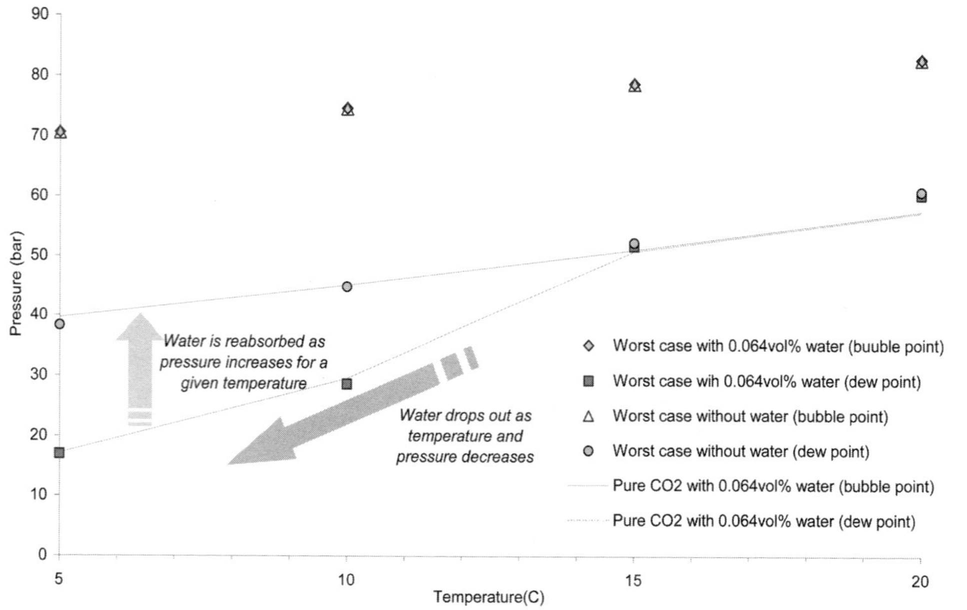

Figure 4-10. Bubble and dew point flashing of the single branch "worst" case specification with 0.0064 vol% of water based on the supercritical pipeline specifications for water content

certain impurities, such as CH_4, which lowers the solubility limit of water, will increase the amount of water condensed or dropping out in the gaseous phase.

A similar analysis was conducted for the single branch "worst" case specification using the lower gaseous phase water specification of 0.0225 vol% and the results are shown in Figure 4-11. In contrast to Figure 4-10, the dew point pressures for the "worst" case specification containing 0.0225 vol% water remains the same even when water is not present at all pressures and temperatures as no condensation was observed at these lower pressures

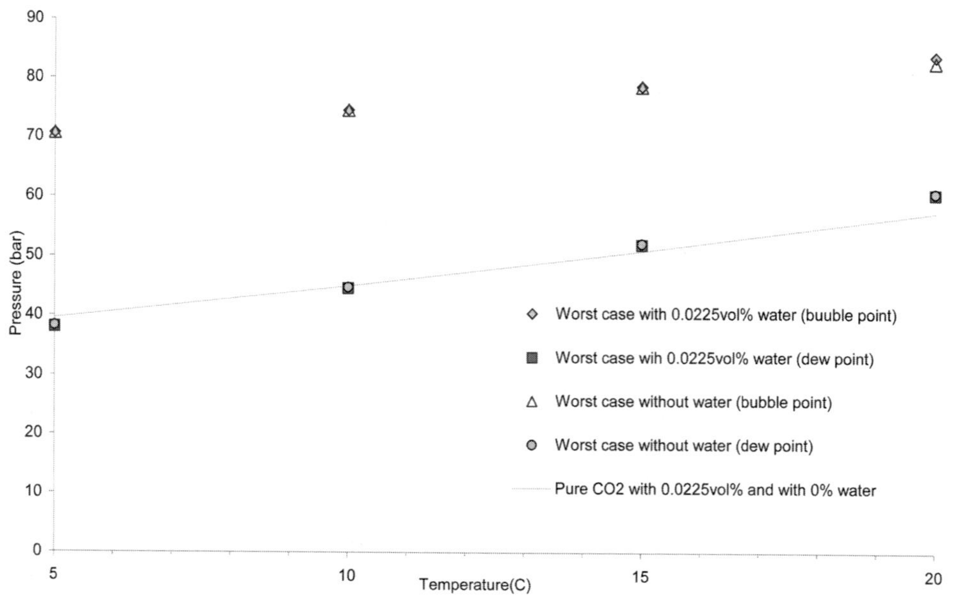

Figure 4-11. Bubble and dew point flashing of the single branch "worst" case specification with 0.00225 vol% of water based on gaseous pipeline specifications for water content

and temperatures. This trend was also similar to that of pure CO_2 containing 0.0225 vol% of water, which leads to the conclusion that the water content for the gaseous phase CO_2 does not have a significant effect on the phase behavior for both the "worst" case composition and pure CO_2 and that further work may be necessary to determine more accurately the phase behavior of anthropogenic CO_2 containing water.

Given that these low water content have no effect on the bubble and dew point of the "worst" case specification, the phase behavior and physical property analyses that follow will not consider water content as a component in anthropogenic CO_2 as the currently recommended water content has been shown not to have any significant effect on the phase behavior.

4.10 PHASE BEHAVIOR OF ANTHROPOGENIC CO_2

A phase envelope comparison of the "best" and "worst" case scenario (with no water content) and that of pure CO_2 is shown in Figure 4-12. From the previous comparison of the critical pressure and calculation two-phase area, the "best" case compositional specification has a vapor–liquid equilibrium line similar to pure CO_2 whereas the "worst" single branch case has the largest two-phase region with the highest critical pressure. The "worst" single branch case has a critical pressure of 12.9 bar above the critical pressure of pure CO_2.

The increase in the two-phase area for any given CO_2 compositional stream is highly dependent on the type, quantity, and combination of impurities. The effect of some individual impurities in binary combination with CO_2 has been discussed previously in Section 4.7. On the basis of the results and analysis discussed in Section 4.7, the large two-phase area for the "worst" case scenario was mainly attributed to the presence of impurities, such as N_2, SO_X, and NO_X which tend to increase the two-phase area even in a binary combination.

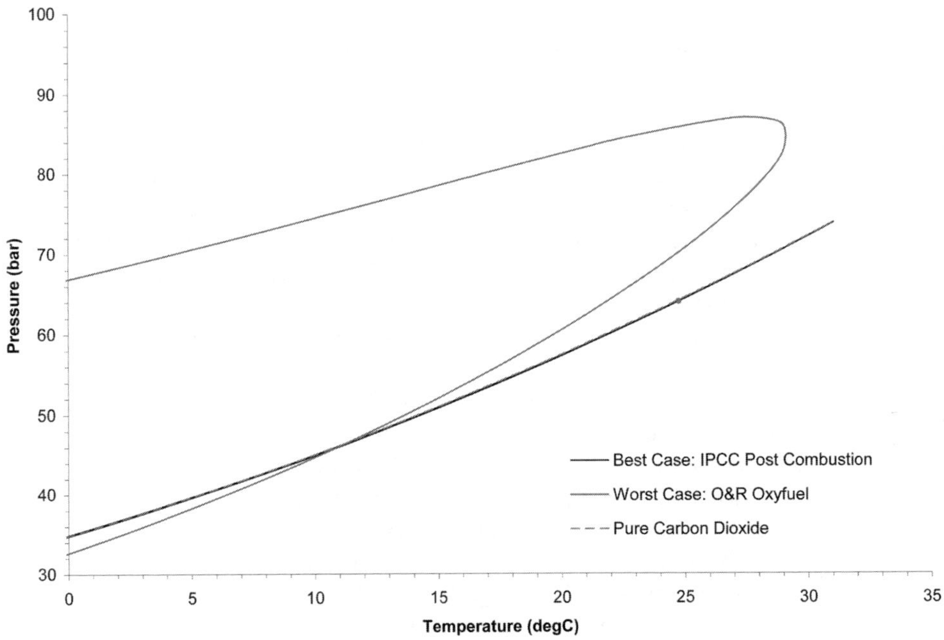

Figure 4-12. Phase envelopes for the different quality specifications of CO_2

It should be clarified that the resultant phase envelope for a network case (where different compositional streams are introduced at different locations into the pipeline) could differ if the compositional specification of each point source changes as a result of changes that could occur in the capture or combustion process or the type of fuel used at the power station. This phase envelope could also be worse (larger) than that of the individual phase envelopes generated due to the capture technology employed at the source.

The properties evaluated in this chapter will have an effect on the pipeline network capacity, pressure and temperature profile, flow characteristics on hilly terrain, design limits, compression or pumping power requirements, and also in determining the safety requirements in terms of pipeline design.

4.11 EFFECT OF IMPURITIES ON CO_2 STORAGE, ENHANCED OIL AND GAS RECOVERY

Impurities affect almost all aspects of CO_2 capture, transport, and storage. However, while the emphasis of this book will be on pipeline transportation, it is recognized that in order to develop or determine quality specifications for CO_2 transport, it is imperative to understand the possible contaminants that could be present in the captured CO_2 and the effect of impurities on the CO_2 sinks, such as aquifers and reservoirs. These effects along with the requirements for pipeline transport can then form a basis of a single CO_2 quality specification satisfying all technical requirements of capture transport and storage. Contaminants, such as SO_X, NO_X, and H_2S, may require classification in terms of health and safety. Depending on whether the CO_2 is stored or used for EOR, regulations may differ in their approach.

The effect of impurities on deep saline formations or aquifers has yet to be studied in detail. It is expected that issues associated with impurities in depleted oil and gas reservoirs will be somewhat similar to that experienced by aquifers [52]. The effect of impurities on saline aquifers will greatly depend on their specific chemical and geological characteristic [53].

CO_2 storage in saline aquifers is dependent on two main trapping mechanisms, which are solubility (dissolution) and mineral trapping (precipitation). It involves the interaction of CO_2 with formation water. The presence of impurities will affect the trapping mechanisms' reaction kinetics in the formation [54]. Leaching of heavy metals from reservoir rock by contaminants, such as SO_2 and O_2, is possible and could lead to permeability changes and subsequently plugging of the aquifer [53]. The presence of significant amount of H_2S should not pose a problem in terms of reactions that could occur in the reservoir. However, the presence of H_2S with SO_2 which can significantly lower the pH of the reservoir is undesirable. This is because the extremely low pH could potentially cause the loss of porosity due to the formation of anhydrites [54]. It should be noted that more research is necessary in this area as this could also lead to potential cost saving in terms of co-storage of CO_2 with other contaminants [45].

In enhanced oil recovery (EOR) operations using CO_2 flooding, the impurities have more effect on the miscible recovery processes compared to immiscible EOR. The main effect impurities have on immiscible recovery is that it takes up volume space which can actually be occupied by the CO_2, this in turn increases the specific storage cost of CO_2 [55]. Viscosity improvements which are achieved through CO_2 solubility in the oil phase are better for heavier oils via the immiscible process where oil swelling has proven to be better for lighter oils through the miscible process [56, 57]. Although it is worth mentioning that some swelling does occur during immiscible recovery due to a small degree of CO_2–oil miscibility. It has also been found that pure CO_2 is more effective in reducing viscosity due to its higher solubility at a given pressure. The presence of N_2 reduces this solubility in continuous CO_2 and low water alternating gas or WAG injection. This has proven to be detrimental to CO_2 flood performance in the immiscible recovery processes [56].

In terms of miscible recovery, the presence of impurities in the CO_2 stream has a relatively more significant effect in terms of the Minimum Miscibility Pressure (MMP). The MMP is important in determining the number of economically feasible miscible projects. Therefore, the effect of impurities on this pressure is of great importance. It is also important to be able to determine an optimum CO_2 stream with a minimum possible range of impurities that can be injected with the assurance that the impurities present will be of minor concern. Impurities that reduce the minimum miscibility pressure would be more acceptable than those that increase it. This is because a low MMP will enable the CO_2 to mix with the fluid at a low pressure compared to a high MMP whereby a higher pressure is necessary at reservoir conditions [55]. Results from tests [58] have shown that the presence of CH_4 (C1) and H_2 raises the MMP where H_2S and ethane (C_2H_6) lower the MMP. Hydrocarbon impurities with carbon numbers higher than 2 (i.e. C2+) are more effective in decreasing the MMP compared to H_2S and ethane (C_2H_6). However, the presence of H_2S with methane reduces the MMP when compared to a CO_2 stream with methane only. It should be noted that the amount of reduction in MMP will vary from one case to another and that higher temperatures promote a decrease in minimum miscibility pressure (MMP) [58]. Figure 4-13 shows the proportion of increase or decrease in MMP for a given type and amount of each impurity [55, 59].

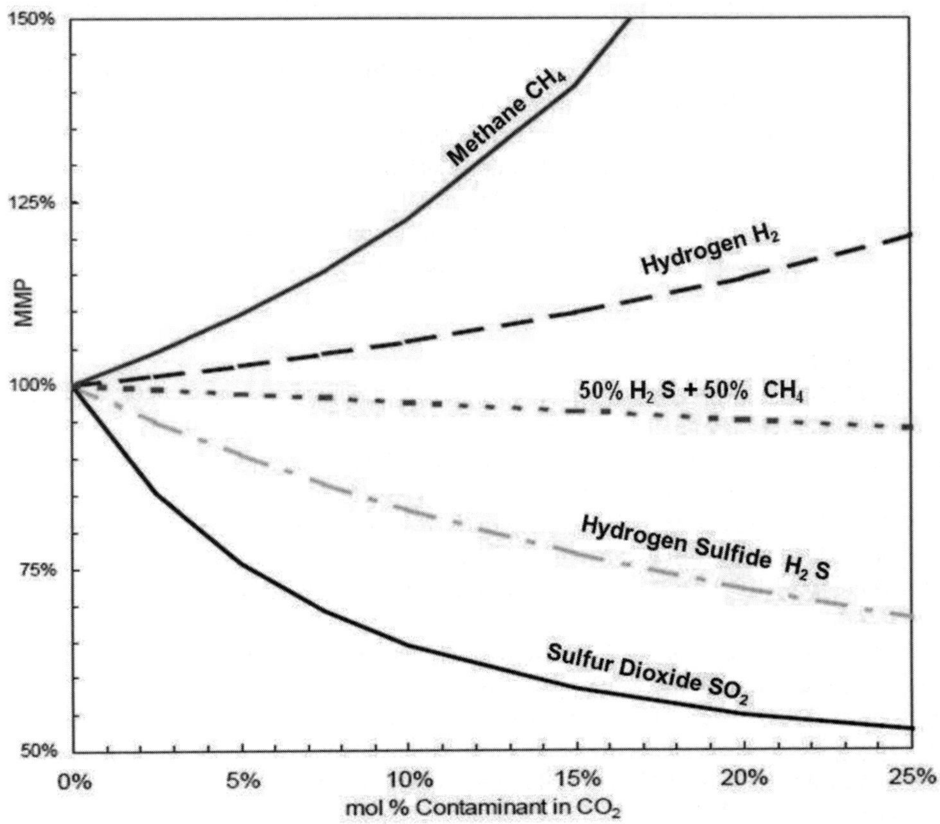

Figure 4-13. The effect of impurities on the Minimum Miscibility Pressure MMP in relation to EOR [55]

The presence of N_2, O_2, and CO also increases the MMP and a combined total of 5% of these impurities could introduce major drawbacks in EOR applications [55, 59, 60]. Figure 4-13 also shows that the presence of SO_2 [61] decreases the MMP quite significantly where studies have shown that the presence of NO_X can decrease CO_2 miscibility with oil and therefore reduce oil recovery.

Impurities also pose other concerns for both miscible and immiscible recovery. These include corrosion, injection pressure, clogging of injection equipment, turning existing "sweet" field "sour," oil quality, and so on. For example, O_2 could be quite detrimental in the sense that it could react exothermically with the oil in the reservoir and increase the viscosity of the oil. O_2 is also known to increase the MMP and in the presence of free water and accelerate the corrosion of production and process equipment [45, 55]. Corrosion of downhole equipment in the presence of SO_2 and H_2S is also a probable threat. Water in combination with CO_2 forms very corrosive fluid and has been highlighted as a potential operational problem for CO_2 floods. However, the industry has addressed this by dehydration for continuous CO_2 injection floods and the use of stainless steel and/or plastic-coated tubing for water-alternating CO_2 floods. Chemical corrosion inhibitors are also used in production wells for some cases [55, 62]. The presence of H_2S could turn normally "sweet" fields into sour ones, thus increasing cost and safety concerns. The deposition of elemental sulphur due to the injection of SO_2 into a field already containing H_2S could potentially reduce the quality of the oil.

Finally, CO_2 is injected into the reservoir in the supercritical or dense phase [56–58, 63], the presence of impurities, such as CO, N_2, O_2, Ar, H_2, NO_2, and SO_2 in a given quantity and combination, could increase the requirement for injection pressure due to changes in critical pressures and consequently the power requirement [64]. In terms of reservoir rock reactions, in aspects such as permeability and porosity issues, similar issues seen by aquifers could occur depending on the reservoir geological characteristics.

In summary, the impurities will affect CO_2 transportation in terms of design codes, health and safety, hydraulics, routing, and cost. Therefore, taking into consideration the impact of impurities on the storage of CO_2 either in saline aquifers, depleted oil and gas fields, or the use in EOR applications, a specification will need to be drawn up in order to satisfy the requirements of transport and storage. An economic analysis in determining the optimum option must thus be carried out on balancing the cleanup cost at the capture side versus transportation and storage cost.

REFERENCES

[1] McAllister, E. W. (2005), "Pipeline Rules of Thumb Handbook: A Manual of Quick, Accurate Solutions to Everyday Pipeline Engineering Problems," 7th edition, Gulf Professional Publishing.

[2] Pedersen, K. S. and Christensen, P. L. (2006), "Phase Behaviour of Petroleum Reservoir Fluids," Publisher Taylor and Francis group, London, UK.

[3] Reid, R. C., Prausnitz, J. M. and Poling, B. E. (1987), "The Properties of Gases and Liquids," 4th edition, McGraw-Hill Book Company., USA.

[4] Michelsen, M. L. (1982a), "The Isothermal Phase Problem–Part 1 Stability," Journal of Fluid Phase Equilibria, vol 9, pp. 1–19.

[5] Michelsen, M. L. (1982b), "The Isothermal Phase Problem–Part 2 Phase Split Calculation," Journal of Fluid Phase Equilibria, vol 9, pp. 21–40.

[6] Waals, J. D. (1912), "The Law of Corresponding States for Different Substances," Proceedings of the Royal Netherlands Academy of Arts and Science (KNAW), 1913, pp. 971–981.

[7] Guggenheim, E. A. (1945), "The Principle of Corresponding States," Journal of Chemical Physics, vol (13)7, pp. 253–261.

[8] Span, R. and Wagner, W. (1996), "A New Equation of State for Carbon Dioxide Covering the Fluid Region from the Triple Point Temperature to 1100K at Pressures up to 800 MPa," Journal of Physical Chemistry, vol 25(6).

[9] Redlich, O. and Kwong, J. N. S. (1949), "On the Thermodynamics Solutions. V: An equation of state, fugacities of gaseous solutions," Chemical Reviews, vol 44 (1), pp. 233–234.

[10] Soave, G. S. (1972), "Equilibrium Constants from a Modified Redlich Kwong equation of State," Chemical Engineering Science, vol 27, pp. 1197–1203.

[11] Raal, J. D. and Muhlbauer, A. L. (1997), "Fluid Phase Equilibria: Measurement and Computation," Taylor and Francis, USA.

[12] Peng, D., Y. and Robinson, D. B. (1976), "A Near Two Constant Equation of State," Journal of Fundamental Ind. Eng. Chem., vol 15(1), pp. 59–64.

[13] Brule, M. R. and Starling, K. E. (1984), "Thermophysical Properties of Complex Systems: Applications of Multiproperty Analysis," Journal of Industrial Chemical Process Development., vol 23(4).

[14] Modisette, D. L. (2000), "Equation of State tutorial," Pipeline Simulation Interest Group PSIG) website:http://egasgrid.si.fr.atosorigin.com/egasgrid/publications/D32_attached%20documents/PSIG/key_docs/200108_equation%20of%20state%20tutorial_modisette.pdf.

[15] George, B. A. (1982), "Solving Unusual Design Problems with Equation of State," Proceedings of the 61st Annual Convention, Gas Processors Association, Dallas, Texas.

[16] Li, H. and Yan, J. (2009), "Evaluating Cubic Equations of State for Calculation of vapor–liquid equilibrium of CO_2 and CO_2 Mixtures for CO_2 capture and Storage Processes," Journal of Applied Energy, vol 86(6), pp. 826–836.

[17] Starling, K. E. (1973), "Fluid Thermodynamic Properties for Light Petroleum Systems," Gulf Publishing Company.

[18] Lin, C. J. and Hopke, S. W. (1973), "Application of the BWRS Equation to Methane, Ethane, Propane and Nitrogen systems, 75th National AIChE Meeting, American Institute of Chemical Engineers, New Orleans, Louisiana, USA.

[19] Hopke, S. W. and Lin, C. J. (1974), "Application of BWRS Equation to Natural Gas Systems," 76th National AIChE Meeting, American Institute of Chemical Engineers, Tulsa, Oklahoma, USA.

[20] Moorwood, T. (2009), Personal communication, Senior Consultant, Infochem UK ltd., UK, used with permission.

[21] Manning, F. S. and Thompson, R. E. (1991), "Oilfield Processing of Petroleum Volume One: Natural Gas," PennWell Publishing Company, Oklahoma, USA.

[22] Kunz, O., Klimeck, R., Wagner, W. and Jaeschke, M. (2007), "The GERG-2004 Wide-Range Equation of State for Natural Gases and Other Mixtures," GERG Technical Monograph 15(2007), website: http://www.gerg.info/publications/tm/tm15_04.pdf.

[23] Kunz, O. and Wagner, W. (2011), "The GERG-2008 wide-range equation of state for natural gases and other mixtures: An expansion of GERG-2004," To be submitted to J. Chem. Eng. Data (2011). (not yet published).

[24] Klimeck, R. (2000), "Entwicklung einer Fundamentalgleichung für Erdgase für das Gas- und Flüssigkeitsgebiet sowie das Phasengleichgewicht." Dissertation, Fakultät für Maschinenbau, Ruhr-Universität Bochum.

[25] Span, R. and Wagner, W. (2003), "Equations of state for technical applications. II. Results for Nonpolar fluids," Int. J. Thermophysics, vol 24, pp. 41–109.

[26] McCullough, D. E. (1986), "Central Basin Pipeline: A CO_2 System in West Texas," Energy Progress, vol 6(4), pp. 230–240.

[27] King, G. G. (1982), "Here Are Key Design Considerations for CO_2 Pipelines." Oil and Gas Journal, vol 80(39), p. 219.

[28] Farris, C. (1983), "Unusual Design Factors for Supercritical CO_2 pipeline" Energy Progress, vol. 3, pp. 150–158.

[29] Zhang, Z. X., Massarato P. and Rudolph, V. (2006), "Optimization of Pipeline Transport for CO_2 sequestration," Energy Conversion and Management, vol 47, pp. 702–715.

[30] Hein, M. A. (1985), "Rigorous and Approximate Methods for CO_2 Pipeline Analysis in Facilities, Pipelines and Measurements: A Workbook for Engineers," 41st Petroleum Mechanical Engineering Workshop and Conference, Kansas City, MO, USA.

[31] Hein, M. (1986),"Pipeline design Model Addresses CO_2's challenging behaviour," Oil and Gas Journal, June 1986, pp. 71–75.

[32] DNV (2010), "Recommended Practice DNV-RP-J202: Design and Operation of CO_2 Pipelines," Det Norske Veristas, April 2010 Edition.

[33] Li, H. and Yan, J. (2006), "Comparative Study of Equations of State for Predicting Phase Equilibrium and Volume Properties of CO_2 and CO_2 Mixtures," 8th International Conference on Greenhouse Gas Control Technologies (GHGT-8), Trondheim, Norway.

[34] Nieto de Castro, C. A. and Wakeham, W. A. (1992), "The Prediction of Transport Properties of Fluid," Fluid Phase Equilibria, vol 79, pp. 265–276.

[35] Duschek, W., Kleinrahm, R. and Wagner, W. (1990a), "Measurement and Correlation of the Pressure, Density, Temperature Relation of Carbon Dioxide I. The homogenous Gas and Liquid Regions in the Temperature range from 217K to 340K at Pressures up to 9 MPa," Journal of Chemical Thermodynamics, vol 22, pp. 827–840.

[36] Duschek, W., Kleinrahm, R. and Wagner, W. (1990b), "Measurement and Correlation of the (pressure, density, temperature) relation of Carbon Dioxide II. Saturated-liquid and Saturated-Vapour Densities and the Pressure along the Coexistence Curve," Journal of Chemical Thermodynamics, vol 22, pp. 841–864.

[37] Oosterkamp, A. and Ramsen, J. (2008), "State of the art Overview of CO_2 Pipeline Transport with Relevance to Offshore Pipelines," Polytec Report no. POL-O-2007-138-A, Polytec, Norway.

[38] Recht, D. L. (1987), "Carbon Dioxide Pipeline Design Considerations," ASME Pipeline Engineering Symposium ETCE, Dallas, Texas.

[39] Ely, J. F., Haynes, W. M. and Bain, B. C. (1989), "Isochoric ($p,V_m.T$) measurements on CO_2 and on ($0.982CO_2 + 0.018N_2$) from 250 to 330K at pressures to 35 MPa," Journal of Chemical Thermodynamics, vol 21, pp. 879–894.

[40] Levelt Sengers, J. M. H., Morrison, G. and Chang, R. F. (1983), "Critical Behaviour in Fluid and Fluid Mixtures," Fluid Phase Equilibria, vol 14, pp. 19–44.

[41] Bierlein, J. A. and Kay, W. B. (1953), "Phase Equilibrium Properties of System Carbon Dioxide – Hydrogen Sulphide," Journal of Industrial and Engineering Chemistry, vol 45(3), pp. 618–624.

[42] Seevam, P. N., Race, J. M., Downie, M. J. and Hopkins, P. (2008), "Transporting the Next Generation of CO_2 for Carbon, Capture and Storage: The Impact of Impurities on Supercritical CO_2 Pipelines," 7th International Pipeline Conference (IPC 2008), Calgary, Alberta, Canada.

[43] West, J. M. (1974), "Design and Operation of a Supercritical CO_2 Pipeline-Compression System, SACROC unit, Scurry County, Texas," Society of Petroleum Engineers Permian Basin Oil and Gas Recovery Conference, Texas, USA.

[44] Fradet, A., Saysset, S., Odru, P., Broutin, P., Ruer, J. and Bonnissel, M. (2007). "Technical and Economic Assessment of CO_2 Transportation for CCS Purposes," Journal of Pipeline Engineering, vol. 6(3), pp. 173–180.

[45] IPCC 2005, "IPCC Special Report on Carbon Dioxide Capture and Storage," Working Group III of the Intergovernmental Panel on Climate Change, 2005, Eds. Metz, B., Davidson, O., de Coninck, H. C., Loos, M. and Meyer, L. A., Cambridge University Press, Cambridge, United Kingdom and New York, NY, USA.

[46] Najera, G. (1986), "Maintenance Techniques Proven on CO_2 Line," Oil and Gas Journal, vol 84(46), pp. 55–57.

[47] Renfro, J. J. (1979), "Sheep Mountain CO_2 Production Facilities- A Conceptual Design," Proceedings of the Production Operation Symposium of the Society of Petroleum Engineers of AIME, Oklahoma City, USA.

[48] Yokozeki, A. (2004), "Solid–liquid–vapour phases of water and water–carbon dioxide mixtures using a simple analytical equation of state," Fluid Phase Equilibria, vol 222–223 (2004), pp. 55–66.

[49] Martinez. S. A. and Hall, K. R. (2007), "Phase equilibria in water containing binary systems from molecular based equations of state," Fluid Phase Equilibria, vol 254(1–2), pp. 112–125.

[50] Caroll, J. J. (2006), "Compression and High Pressure Pipeline Transportation of Carbon Dioxide Rich Fluids," 1st Regional Symposium on Carbon Management – Carbon Management Challenges and Opportunities for Petroleum Industry, Canada.

[51] Fatosh (2009), Personal communication, Fluid Specialist, BP, Houston, USA.
[52] IEA 2004, "Improvement in Power Generation with Post Combustion Capture of CO_2," Report no. PH4/33, International Energy Agency.
[53] Gunter, W. D., Perkins, E., H. and Hutcheon, I. (2000), "Aquifer Disposal of Acid Gases: Modelling of Water-rock Reactions for Trapping of Acid Wastes," Journal of Applied Geochemistry, vol 15(8), pp. 1085–1095.
[54] Knauss, K. G., Johnson, J. W. and Steefel, C. I. (2005), "Evaluation of the impact of CO_2, co-contaminant gas, aqueous fluid and reservoir rock interactions on the geologic sequestration of CO_2," Journal of Chemical Geology, Elsevier, vol 217(3–4), pp. 339–350.
[55] IEA 2004(b), "Impact of Impurities on CO_2 Capture, Transport and Storage," Report no. PH4/32, study conducted by SNC Laevulin, August 2004.
[56] Sankur, V., Emmanuel, A. S. and Di Julio, S. S. (1986), "Recovery of Heavy Oils by CO_2 Injection," Energy Progress, vol 6(2), pp. 91–96.
[57] Kovsek, A. R. (2002), "Screening Criteria for CO2 Storage in Oil Reservoirs," Journal of Petroleum Science and Technology, 20(7&8), pp. 841–866.
[58] Metcalfe, R. S. (1982), "Effects of Impurities on Minimum Miscibility Pressures and Minimum Enrichment Levels for CO_2 and Rich Gas Displacements" SPE Journal, vol 22(2), pp. 219–225.
[59] Alston, R. B., Kokolis, G. P. and James, C. F. (1985), "CO_2 Minimum Miscibility Pressure: A Correlation for Impure CO_2 Streams and Live Oil Systems," Society of Petroleum Engineers Journal, vol 25(2), pp. 268–274.
[60] Sebastian, H. M., Wenger, R. S. and Renner, T. A. (1985), "Correlation of Minimum Miscibility Pressure for Impure CO_2 Streams," Journal of Petroleum Technology, vol 37(12), pp. 2076–2082.
[61] Bryant, S. and Lake, L. (2005), Effect of impurities on subsurface CO_2 storage processes, Carbon Dioxide Capture for Storage in Deep Geologic Formations - Results from the CO_2 Capture Project, v. 2: Geologic Storage of Carbon Dioxide with Monitoring and Verification, S.M. Benson (ed.), Elsevier, London, pp. 983–998.
[62] Bondor, P. L. (1992) "Application of Carbon Dioxide in Enhanced Oil Recovery," Energy Conversion and Management, vol 33(5), pp. 579–586.
[63] Goodyear, S. G., Hawkyard, I. R., Masters, J. H. K., Woods, C. L. and Jayasekera, A. J. (2003), "Subsurface Issues for CO_2 Flooding of UKCS reservoirs," Transaction of IChemE, vol 81, Part A.
[64] Seevam, P. N., Race, J. M. and Downie, M. J. (2007), "Carbon Dioxide Pipelines for Sequestration in the UK – Engineering Gap Analysis," Proceedings of the conference on Transmission of CO_2, H_2 and Biogas: Exploring New Uses for Natural Gas Pipelines, Amsterdam, Netherlands.

Chapter 5

PIPELINE TRANSPORTATION OF CO_2

CO_2 injection techniques for enhanced oil recovery (EOR) have been growing steadily since early 1970s [1]. In 2002, in the USA, the number of CO_2 injection projects exceeded the number of thermal recovery. The following are reasons for this increase and the continuing trend:

- CO_2 remains a dense fluid over much of the range of reservoirs pressures and temperatures.
- CO_2 is miscible or partially miscible with many hydrocarbon components of crude oil at reservoir conditions.
- Dense CO_2 has low solubility in water compared to oil.
- CO_2 at high pressures has higher viscosity comparing to other gases used in gas flooding.
- There are a lot of naturally occurring CO_2 resources near many oil fields.
- CO_2 sequestration has environmental benefits by reducing greenhouse gas effects.

CO_2 is a very compressible gas and at high pressures forms a phase whose density is close to that of liquid and with a higher viscosity comparing to other gases (although it remains low compared to that of liquids). Dense phase CO_2 has the ability to extract hydrocarbon components from oil more easily than if it were in the gaseous phase. CO_2 has also been used as cushion gas in gas injection and recovery processes.

5.1 ESTABLISHING DESIGN PARAMETERS

For the purpose of pipeline transportation system design, the following data is generally required and if unavailable must be rationally assumed:

- Pipeline route and profile;
- Pipeline flow and flow build up;
- CO_2 mixture properties (level of impurities) as well as viscosity and density at several temperatures and pressure ranges;
- Maximum inlet and operating as well as minimum delivery pressures;
- Inlet, maximum and minimum operating temperatures;
- Pipe material/grade;
- Pipe diameter and wall thickness and roughness;
- Depth of cover;
- Ground/environment temperature, conductivity, specific heat and density or otherwise the overall heat transfer film coefficient;
- Piping pressure loss through stations (compression, pumping, and measurement stations);
- Applicable codes, standards and regulations.

Transporting carbon dioxide from natural occurring sources via pipeline is commonly undertaken in the USA (Refer to Chapter 1) since early 1970. Pipelines are used to transport CO_2 in liquid state to oil fields where it is injected under pressure into depleted oil fields to enhance the recovery of oil. In this enhanced oil recovery (EOR), the injection of carbon dioxide pays for itself by helping recover more fossil fuels that would otherwise remain underground. It can also maintain reservoir pressure, thus enhancing the oil recovery process. However, transportation of CO_2 gas contaminated with impurities (H_2S, CO, SO_x, NO_x, etc.) is relatively new. The first example of such pipelines is the Weyburn (on-streamed in year 2000) that transports CO_2 captured from flu stacks of a gasification plant in North Dakota, USA to oil fields in Weyburn, Saskatchewan, Canada (Refer to Chapter 1). The pressure of the CO_2 leaving the plant is 17 MPa (2500 psig) while the delivery pressure at Weyburn is 14.8 MPa (2,175 psig). The delivery pressure matches the injection pressure requirement into the depleted oil reservoirs.

The delivery pressures of impure CO_2 are relatively high in comparison with normal pipeline transportation of other fluids, such as natural gas and oil.

5.1.1 Natural Gas Versus CO_2 Transportation

Natural Gas (NG) is a mixture of hydrocarbon gases, predominantly methane with lesser parts of ethane, propane, butane, and pentane, plus some impurities. The molecular weight of NG is typically between 16 and 18, with a density that ranges between 0.552 and 0.621 kg/m^3 [2] and a viscosity between 0.0104 and 0.0108 centipoises [3] at normal temperature and pressure (NTP).

The phase envelopes for typical sweet natural gas and a CO_2 stream containing impurities are shown in Figure 5-1.

As can be observed from Figure 5-1, CO_2 mixtures have a critical pressure and a critical temperature that are significantly higher than those of natural gas. In practical terms, this signifies that, in order to maintain a single-phase fluid, CO_2 gas pipelines must be operated at higher pressure and/or temperature than the conditions needed for the transporta-

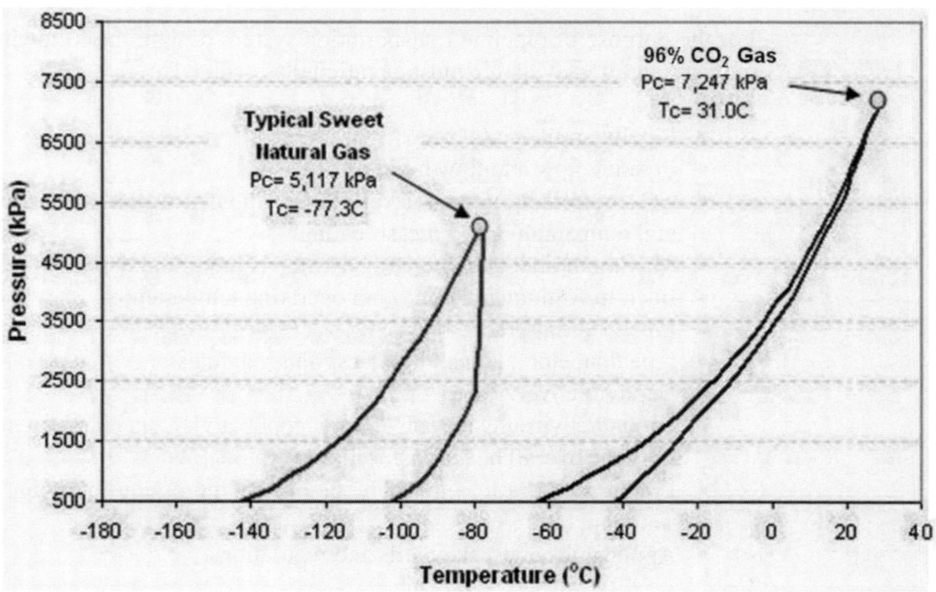

Figure 5-1. Phase envelope comparison, natural gas versus CO_2

tion of natural gas. In general, natural gas and CO_2 have different operating environments, mechanical design requirements, and safety issues as summarized below:

- Natural gas: typically transported by pipelines at 7–10 MPa (1000–1440 psi), no phase change.
- CO_2: transported in supercritical phase at 15–17 MPa (2200–2465 psi), 40–60°C.
- Different pipeline fracture control properties (CO_2 has higher vapor pressure).
- CO_2 and impurities (e.g., H_2S) — dispersion considerations associated with leak/rupture or blow-down.
- Unlike natural gas CO_2 is non-flammable.
- CO_2 pipelines are considered low risk facilities, but regulators consistently use the language of "hazardous materials and carbon dioxide" requiring rigorous design and inspection.
- Natural gas systems are not usually suitable for high pressure CO_2 transmission.

The above issues must be addressed for CO_2 pipelines independent of information available from other fluid pipeline sources.

5.2 ESTABLISHING CO_2 PIPELINE PRESSURE

An important aspect of CO_2 pipelines (besides knowing the gas composition) is establishing the pipeline design and operating pressures usually based on the delivery pressure requirements. The delivery pressure is generally determined by customer requirement usage for example for injection into oil and gas reservoirs or aquifers for performance maintenance/enhancement or for storage purposes.

5.2.1 CO_2 Density/Compressibility

5.2.1.1 Density–Depth Relationship

For sequestration purposes, CO_2 is generally injected to depths well over 1000 m (3280 ft). As injection depth is increased, higher injection pressure is required. At greater depths, CO_2 increases in density and becomes a supercritical fluid below 0.8 km (Figure 5-2, [4]). This figure also illustrates the difference in volume between carbon dioxide as a gas at the surface and as a supercritical fluid below 800 m underground. The numbers in this figure show the volume of CO_2 at each depth compared to a volume of 100 at the surface, which indicates how carbon dioxide can be compressed for more efficient sequestration purposes [5]. Supercritical fluids take up much less space than gases and diffuse better than either gases or ordinary liquids through the tiny pore spaces in storage rocks.

As indicated in Figure 5-2, at a depth greater than approximately 1000 m, an increase of depth will not change the CO_2 density as significantly as for the first 800 m of depth. At a depth of 1500 m the CO_2 density will remain almost constant [4]. However, CO_2 density varies as a function of surface temperature and subsurface geothermal conditions (cold versus hot) at reservoir depth (Figure 5-3) in a sedimentary basin [6].

Because of the geothermal gradient and typical depth of storage reservoirs, CO_2 will be normally in supercritical condition by virtue of temperature and possibly pressure depending on depth and geothermal state. The density of CO_2 changes drastically around its critical point of 31°C and 73.8 bars.

Injection basins fall into two geothermal regions: cold and hot. The extent of the geothermal condition of a sedimentary basin has a significant effect on the behavior of CO_2 and thus volume or pressure required for the reservoir yield or maintenance purposes [7]. Higher pressure/volume will be required in hot basins (high geothermal gradient, say

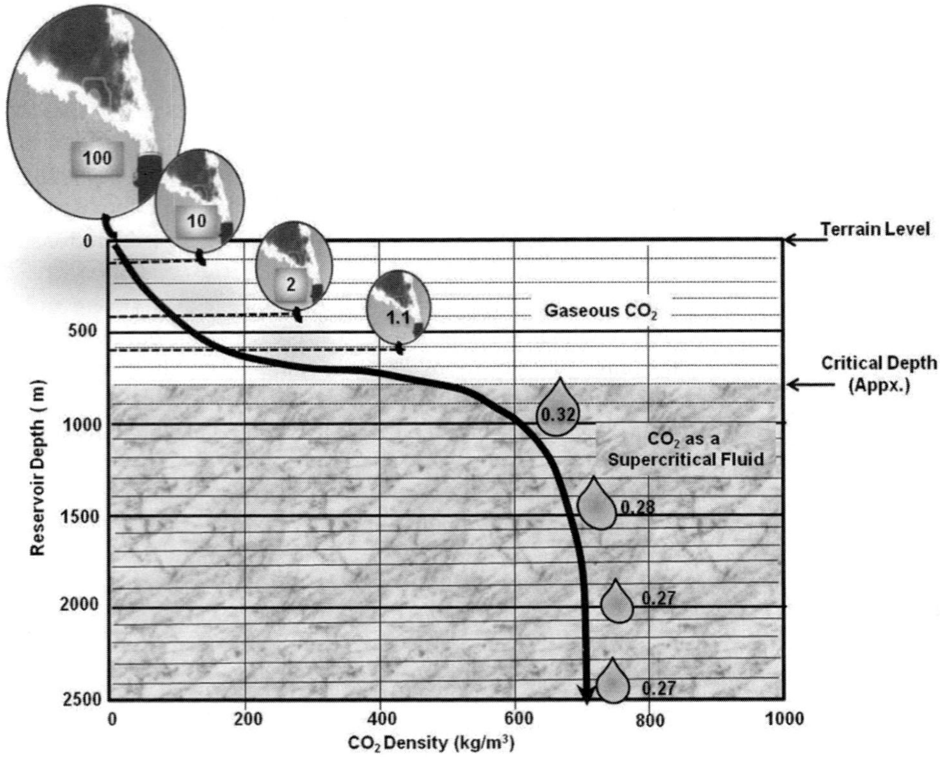

Figure 5-2. Compressibility comparison of CO_2 at different injection depths for a geothermal gradient of 25°C/km [1, 5]

50°C/km) than in a cold basin (low geothermal gradient, say 25°C/km). CO_2 density is lower in hot basin than in a cold basin at the same depth. Therefore, more compression will be required to either maintain higher reservoir pressure or more flow to achieve higher compressibility for reservoir flow maintenance in hotter than colder basins. This difference in behavior of CO_2 (in cold and hot basins) will determine the optimum injection depth, selection of injection basin locations, and compression requirements [6].

The geothermal state and the required pressure for a sedimentary basin would decide the optimum ranking of depth for which the capacity of CO_2 storage is optimized. Therefore, basins with low geothermal gradients (<25°C/km) are preferred since such basins can permit the injection of CO_2 in supercritical conditions at lower depths (less and lower injection pressure and/or volume) than hot basins. Hot basins (with geothermal gradient 50°C/km) are less suitable for storage of CO_2 due to lower storage capacity capability in terms of mass of CO_2 at the same pressure. However, the optimum injection depth into a cold basin is between 800 and 1000 m while for a hot basin it is between 1500 and 2000 m [7].

5.2.1.2 Compressibility

The presence of impurities such as sulfur dioxide(SO_2), hydrogen sulfide (H_2S), methane (CH_4), etc affect the volume in geological basins as such impurities affect the overall properties of the mixture by lower critical points and affecting the phase diagram. For example, if CO_2 contains CH_4 (methane), its density would be reduced significantly. This not only affects pipeline design but also storage capability. Also depending on requirement for storage; EOR, Gas Cushioning, Sequestration (coal seams or aquifers), it would affect storage capability.

The compressibility of pure CO_2 versus CO_2 mixture under pipeline conditions is indicated in Figure 5-4. Over the range of typical conditions shown in Figure 5-4, the density

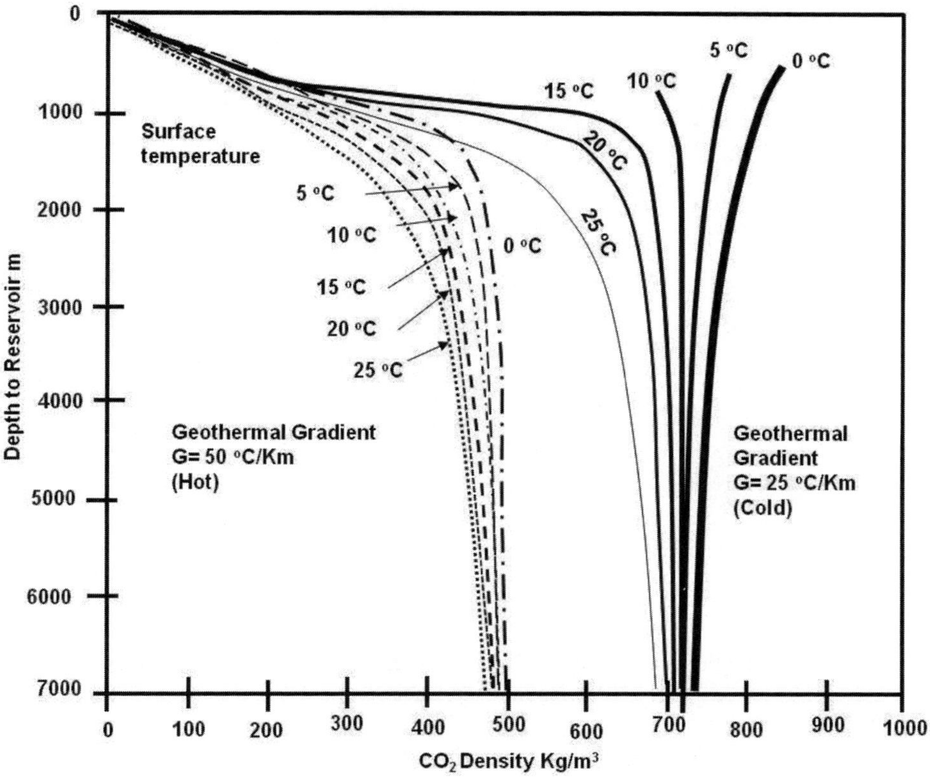

Figure 5-3. CO_2 density variation as a function of geothermal gradient condition, surface temperature and depth to reservoir [6]

of CO_2 varies between approximately 700 kg/m^3 and 1000 kg/m^3. For the same pressure, less compressibility would be expected for CO_2 with impurities. This in turn would affect the pipeline flow, i.e., the volume requirement to maintain reservoir pressure maintenance or sequestration requirement.

As can be observed from Figure 5-4, the compressibility of CO_2 is non-linear in the range of pressures common for pipeline transport and is highly sensitive to any impurities, such as hydrogen sulfide (H_2S) or methane (CH_4). Figure 5-4 also shows that there is a significant difference between the compressibility of pure CO_2 and CO_2 with impurities (e.g., 10% H_2S (by volume). To mitigate difficulties in design and operation, generally CO_2 pipelines are operated at pressures greater than 8.6–10 MPa or dense phase conditions where the sharp changes in compressibility of CO_2 can be avoided across a range of temperatures that may be encountered in the pipeline system operations [8].

5.2.2 Reservoir Pressure Versus Pipeline Delivery Pressure

Pipeline delivery pressure is based on the requirement for knowing the injection pressure. Determining the operating pressure at the top of the well leading to underground reservoirs requires consideration of the pressure at the bottom of the well to force CO_2 into the injection zone, the pressure increase in the pipe due to the height of the CO_2 column, and the pressure loss due to flow in the pipe.

The reported rate of pressure rise with depth in most reservoirs ranges from 0.105 to 0.124 bar/m (0.464 to 0.548 psi/ft) with a few sites having gradients as high as 0.23 bar/m (1.02 psi/ft) [9, 10].

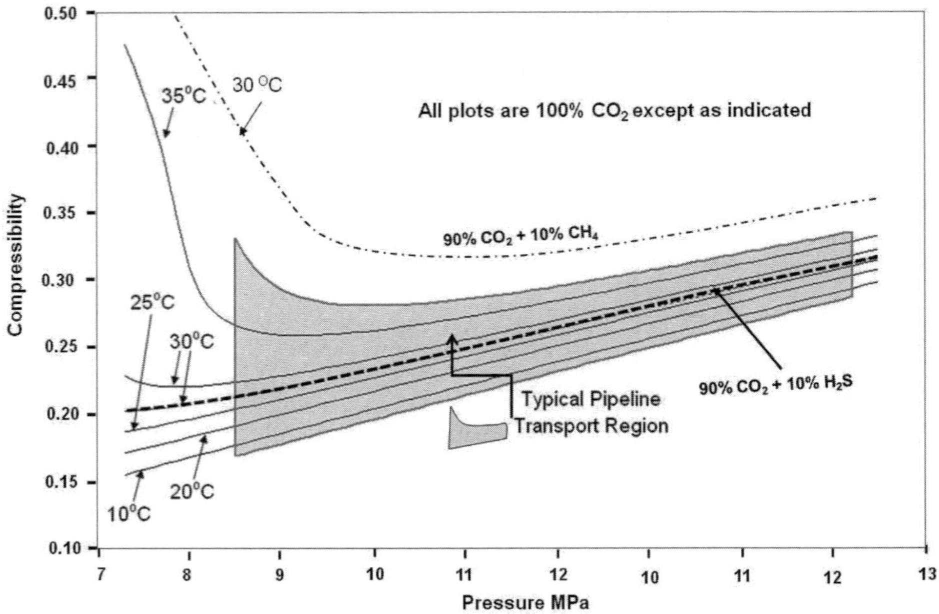

Figure 5-4. Pure and impure CO_2 compressibility-pressure variation and pipeline operating range (Prediction: Peng-Robinson Equation of State [8, 11]

Moving the CO_2 into a below ground reservoir (e.g., aquifers) requires raising the CO_2 pressure sufficiently above the in situ pressure to provide a driving force. However, this pressure must not be so high as to risk hydro-fracturing the injection interval. Typically, the CO_2 injection pressure is about 9% to 18% above the in situ bottom pressure [9]. This will approximately account for the pressure loss in the injection/extraction well tubing and other fittings.

Pressure caused by the weight of the column of CO_2 in the injection tubing provides some of the required pressure. As can be deduced from Figure 5-2, this pressure contribution is a function of the density of the CO_2 at the pressure and temperature conditions in the injection tubing. Additionally, the basin's geothermal conditions and the extent of the impurities in CO_2 will also be contributors as indicated previously. The results of typical calculations to determine the well head pressure for various depths for pure CO_2 are shown in Table 5-1.

Pressures in this table have been calculated on the basis of the vapor pressure of CO_2 + hydrostatic pressure column between the reservoir and the wellhead. Thus, at 1000 m depth, the required injection pressure will be 7.5 (MPa) + [1000m depth × 600 Kg/m^3/100000] = 7.5 + 6.0 = 13.7 MPa. Adding on about 9% loss, the injection pressure will be = 14.7 MPa (Column 4 of Table 5-1). It may be noted that the CO_2 density was taken from Figure 5-2 above. See also Chapter 1, Figure 1-25 for typical CO_2 density for pipeline transportation. In this example minimum vapor pressure for CO_2 is assumed to be 7.5 bar.

TABLE 5-1. Estimated injection using pressure for cold thermal basin [9]

Depth (m [ft])	CO_2 Pressure at Well Head (MPa, [psi])	Specific Gravity at Well Head	CO_2 Pressure at Injection Point (MPa, [psi])	Specific Gravity at Injection Point
1000 [3281]	7.50 [1090]	0.71	14.7 [2140]	0.77
2000 [6562]	12.8 [1860]	0.83	29.5 [4278]	0.87
3000 [9843]	18.7 [2710]	0.83	44.2 [6417]	0.91

However, while the bottom hole injection pressure (BHIP) is frequently estimated by assuming it is equal to the hydrostatic pressure exerted by a static column of fluid, it is only an approximation and valid at low injection flow rates, as the loss of pressure due to friction between the flowing fluid and the surrounding reservoir environment (soil and rocks) counteracts the pressure increase due to hydrostatic pressure. As a consequence for a high capacity reservoir, pressure losses due to friction will not be negligible and must be considered. Moreover, heat transfer between the surrounding rock (i.e., geothermal condition of the basin) and the wellbore also affects the pressure and temperature distribution in the reservoir.

A typical vertical injection well completion is indicated in Figure 5-5. The CO_2 from the capture pipeline flows from the surface to the reservoir through the production tubing and enters the reservoir through perforations in the production casing.

The rate at which CO_2 can be injected into a reservoir, e.g., an oil reservoir, is determined by the injectivity of the system. Injectivity (I) is defined as the injection rate (q) divided by the excess pressure above reservoir equilibrium pressure driving injection [12].

$$I = q / (p_{wb} - p_e) \qquad (5-1)$$

Thus, given an injection rate, q, and reservoir equilibrium pressure (vapor pressure), p_e, the pressure at the well bottom, p_{wb}, can be calculated if the injectivity is known.

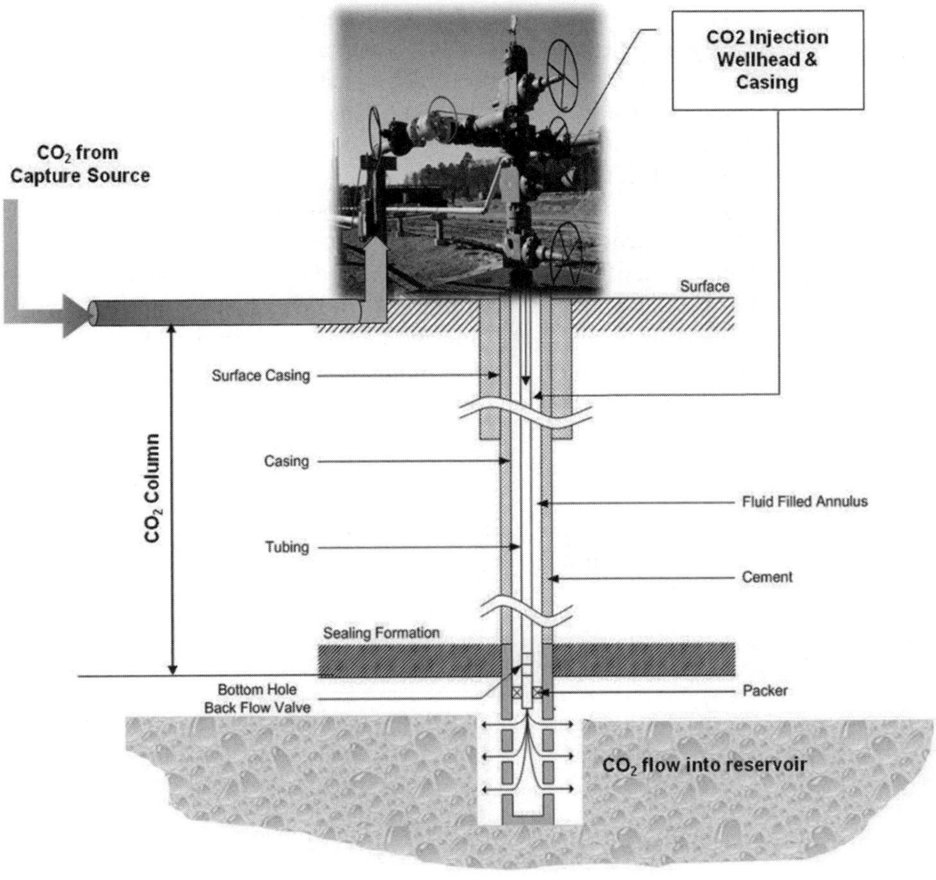

Figure 5-5. A typical CO_2 injection well completion (from McCoy [8])

Assuming that the pressure gradient (ΔP) in the wellbore is a linear function of depth L and, thus, that the Bottom Hole Injection Pressure (BHIP) can be written as [8]:

$$p_{wb} = (\Delta P / \Delta L)\, L + p_{\text{wh}} \quad (5-2)$$

where, p_{wb} is the BHIP pressure, p_{wh} is the wellhead pressure, and L is the wellbore length.

It may be noted that the above Eq. 5-2 indicates the pressure gradient in the well is linear with depth; however, it should be expected that it becomes non-linear with increase in injection flow rates. The above equation holds to flow rates < 1 Mt CO_2/year at < 30°C and p_{wh} >8 MPa for a 76 mm (3 in.) inside diameter tubing.

The following equations can provide a better estimate of pressure drop in a wellhead tubing [8]:

$$\Delta P / \Delta L = -7.008 \text{ x } 10^{-4}\, q^2 + 1.812 \text{ x } 10^2 \log[\,(P_{wh})^2] \quad (5-3)$$

Coefficients in Eq. 5-3 indicate that the wellbore pressure gradient is negatively correlated with the volumetric flow rate, and positively correlated with the square of the base-ten logarithm of pressure. In addition, wellhead temperature influences the pressure gradient through its effect on volumetric flow rate (affected by the density of CO_2 and the geothermal gradient.

5.2.3 Injection/Withdrawal Rate and Pressure Requirements

While injection pressure calculations (referred to in Section 5.2.2) determine pipe delivery pressure at the injection site, the actual injection/withdrawal rates and thus reservoir pressure requirement may impose additional limitations/operating conditions for the pipeline leading to the well site injection piping. This may require additional control facilities as explained below:

5.2.3.1 Injection/Withdrawal

An example is the injection of CO_2 into hydrocarbon/fossil fuel depleted reservoirs. One of the attractions of disposing CO_2 in such reservoirs (besides EOR) is that underground withdrawal of fossil fuels can be balanced by CO_2 storage in the subsurface, i.e., possibility that the rate of withdrawal of fuel can be offset by CO_2 injection into a reservoir. However, as it is shown below, the amount of CO_2 sequestered would not necessarily be equal to the amount that is emitted from burning of fossil fuels [13].

Fossil fuels contain carbon and hydrogen. When they are burned, carbon dioxide and water are produced. If instead of air, pure oxygen is used, the combustion of natural gas, oil, and coal is as follows [13]:

- **Natural gas:** $CH_4 + 2O_2 \rightarrow CO_2 + 2H_2O$
- **Fuel oil:** $2CH_2 + 3O_2 \rightarrow 2CO_2 + 2H_2O$
- **Coal:** $C + O_2 \rightarrow CO_2$

The material balance by weight is:

- **Natural gas:** 1 kg CH_4 + 4 kg $O_2 \rightarrow$ 2.75 kg CO_2 + 2.25 kg H_2O
- **Fuel oil:** 1 kg CH_2 + 3.43 kg $O_2 \rightarrow$ 3.14 kg CO_2 + 1.29 kg H_2O
- **Coal:** 1 kg C + 2.67 kg $O_2 \rightarrow$ 3.67 kg CO_2

Assuming that there is no oxygen and water present and that the densities of natural gas, oil, and CO_2 under average reservoir conditions are 150, 800 and 700 kg/m^3, respectively, the material balance by underground volume for oil and gas is:

- **Natural gas:** 1 m^3 CH_4 0.59 m^3 CO_2
- **Fuel oil:** 1 m^3 CH_2 2.75 m^3 CO_2

As can be inferred from above, it can be expected that the underground volume of the natural gas can be replaced by the CO_2 produced by the burning of the gas. On the other hand, the burning of oil produces much more CO_2 than the underground volume of oil removed, i.e., the volume of oil removed is much smaller than the volume that would be occupied by the CO_2 produced.

A lot of power stations are, however, coal powered. The combustion of coal produces more CO_2 that can be stored in the space the coal came from. To dispose of the CO_2 permanently, it has to be sequestered within depleted hydrocarbon reservoirs or a trap on an aquifer directly analogous to the hydrocarbon reservoir. Such an underground reservoir would need a sealed trap to prevent the CO_2 from percolating upwards through the water and reaching the surface because supercritical, "liquid" CO_2 is lighter than water and tends to force its way up through the water-saturated rocks.

5.2.3.2 Operational Pressure Requirement

Pure CO_2 density and viscosity at different temperatures are shown in Figure 5-6 and Figure 5-7, respectively. As indicated in Figure 5-6, the critical temperature of CO_2 is 31.1°C and its critical pressure is 7.38 MPa. Also, as can be seen from Figures 5-6 and 5-7, at temperatures below the critical temperature (31.1°C), CO_2 density and viscosity increase rapidly when gaseous CO_2 is pressurized to obtain liquid state. At temperatures above the critical point, however, the CO_2 gas/CO_2 liquid border is not defined clearly [14].

Additionally, Figure 5-8 has been included to relay and understand pure CO_2 density variation in liquid and gaseous conditions. For comparison purposes, variations of densities and viscosities of pure carbon dioxide (CO_2), methane (CH_4), and nitrogen (N_2) with pressure and at 40°C are shown in the same diagram in Figure 5-9.

Considering Figures 5-6 and 5-7 and as shown previously the underground temperature and pressure below a depth of about 800–850 m (in a geo-thermally cold basin) can be such that CO_2 will be in its supercritical state. Under these conditions, no distinction can

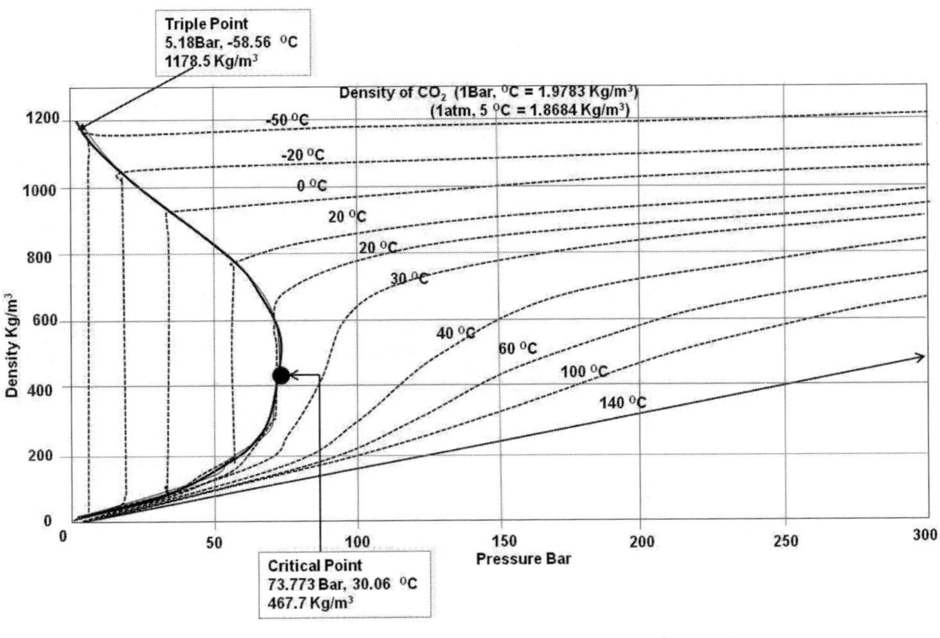

Figure 5-6. Pure CO_2 density variation with pressure and temperature (reproduced from Ref [13])

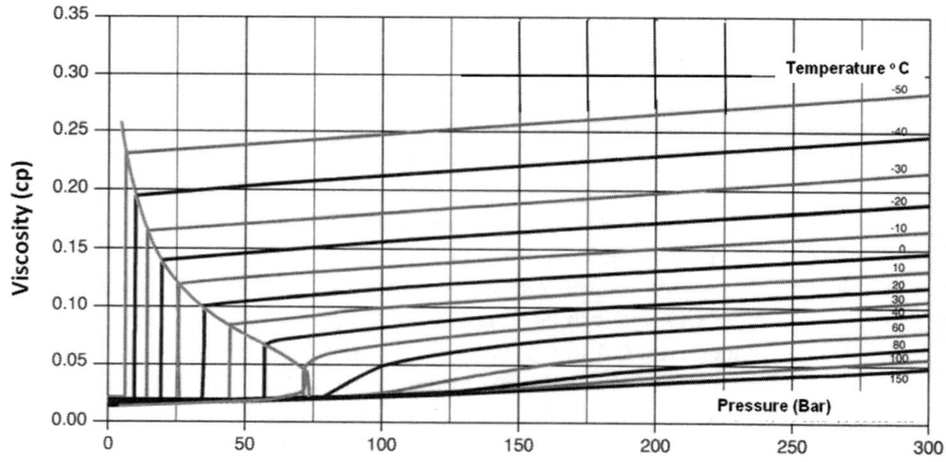

Figure 5-7. Pure CO_2 viscosity variation with pressure and temperature ([13])

be made between liquid and vapor and the CO_2 acts as a gas-like compressible fluid that takes the shape of its container and fills it but has a liquid-like density. At higher depths and particularly in geo-thermally hot basins, the CO_2 is a gas and its density is too low to store large volumes economically [13].

One of the main differences between supercritical fluids such as CO_2 and conventional fluids is their density and compressibility. Such a variation is compared in Figure 5-9. Conventional fluids in the liquid phase require very large pressures to change the density, whereas for supercritical fluids, very significant changes in density can be achieved by comparatively small pressure and/or temperature changes, particularly around the critical point (Figure 5-6).

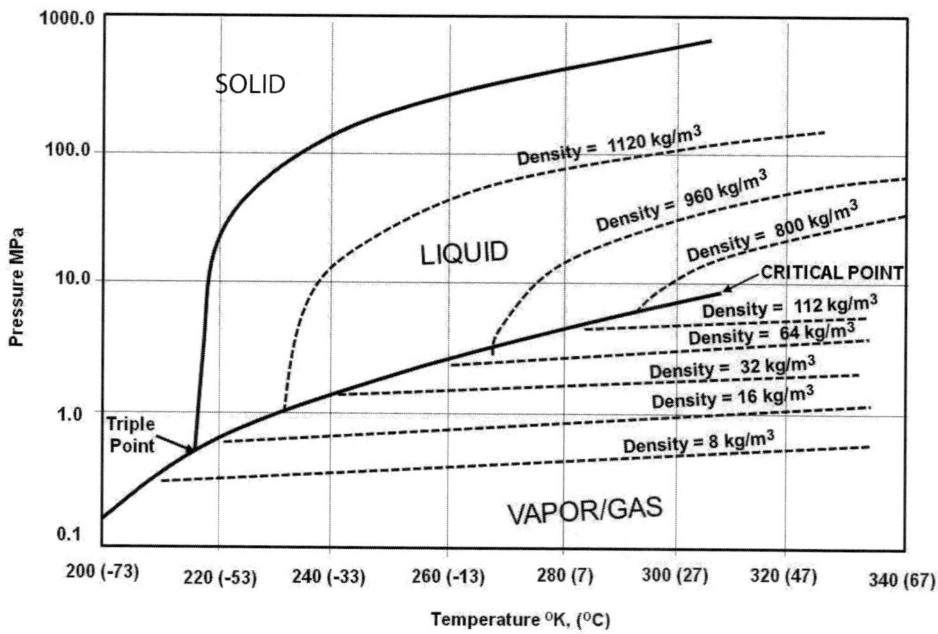

Figure 5-8. Variation of density in pure CO_2 under various conditions in phase diagram (from [15], from NIST REFPROP)

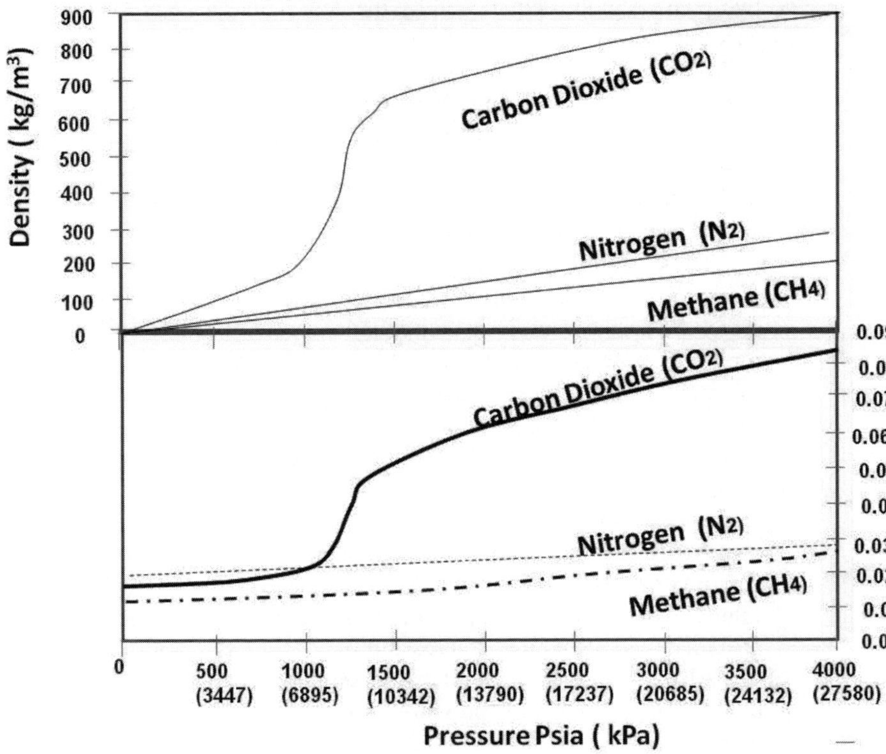

Figure 5-9. Densities and viscosities of (CO_2), CH_4, and N_2 with pressure at 40°C

Thus, CO_2 mixtures with different kinds and amounts of impurities can provide a very wide range of properties (Figure 5-10) which can in some cases significantly affect the outcome of their transmission and subsequent storage. Figure 5-10 indicates as example of the effect of the level of impurities in a CO_2 stream phase diagram. It shows the phase diagram for a mixture of 96% mol CO_2 and 4% mole nitrogen, a mixture of 91% CO_2 and 9% nitrogen as well as pure (100% CO_2). From this figure, it may be noted that the critical pressures of both $CO_2 + N_2$ mixtures are higher than for pure CO_2. Additionally, a

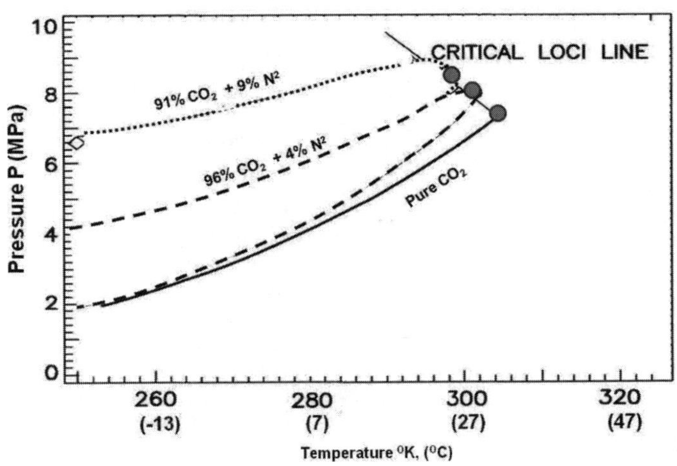

Figure 5-10. Typical phase diagram for a CO_2 stream with impurities ($CO_2 + N_2$) [16]

two-phase region (demarcated by the dew point curve at the bottom and a bubble point curve at the top) can be seen for the mixture. Figure 5-10 also illustrates experimental data all as reported by Verma et al [16].

5.2.3.2.1 Effect of Impurities on Operational Pressure

In general, the delivery of CO_2 by pipelines is for injection into deep reservoirs. As previously indicated, in-ground temperature varies with depth and geothermal conditions of the basin into which injection takes place (cold basin versus hot basin). As shown, a typical geothermal gradient is about 25°C/km for a cold basin.

Thus, in any injection options, considerations must be given to the required pipeline pressure (at the injection point) meeting the reservoir depth consideration which in turn affect the density, compressibility, and viscosity of the CO_2 mixture at the required depth for the purpose of sequestration. While magnitude of injection pressures can be predicted reasonably for pure CO_2, such prediction is more complicated when dealing with CO_2-containing impurities, particularly for operational purposes [17]. An example quoted from Oldenburg [18] for using CO_2 mixture for natural gas cushioning purposes will indicate this complication when using CO_2-containing impurities for pressure maintenance purposes as explained below:

Figure 5-11 indicates typical gas density (ρ) and compressibility (Z) of various CO_2–CH_4 mixtures as a function of pressure (up to 200 bar) at 40°C [18]. It may be noted from Figure 5-11 that for pure methane (CH_4) compressibility factor (Z) varies from 1.0 to approximately 0.85, while compressibility (Z) for CO_2 varies from 1.0 to less than 0.3.

Also from Figure 5-11, it can be observed that the density (ρ) of pure CO_2 increases by a factor of 5 when the pressure changes from 60 to 130 bar, whereas pure CH_4 density increases by just over a factor of 2. If CO_2 is used as a cushion gas within the pressure range of 60–130 bar, it would allow significantly larger quantities of working gas to be injected with less increase in pressure relative to using a cushion gas consisting of native gas or inert gas (e.g., nitrogen, N_2).

Furthermore, when the working gas is withdrawn and the reservoir pressure decreases, there will be a corresponding larger gas drive due to the rapid decrease in density (i.e., increase in volume) of the CO_2 cushion gas.

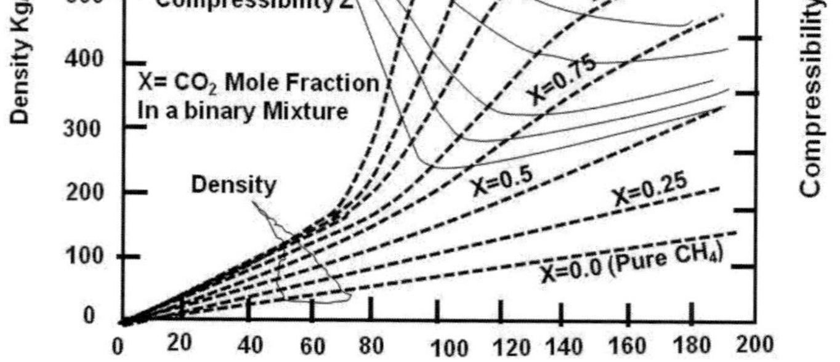

Figure 5-11. Density and compressibility value of CO_2–CH_4 mixtures at 40°C. ([18], Courtesy of American Chemical Society)

Figure 5-12 indicates the density (ρ) and viscosity (μ) of CO_2-CH_4 mixtures at various pressures as a function of impure CO_2 composition at 40°C. Although the density of supercritical CO_2 approaches that of liquid water, the viscosity is always gas-like, i.e., on the order of 10^{-5} Pa s. Hence, dense CO_2 mixture will have the density of a liquid, but viscosity of gas. This affects the mobility of the two gases as much as they affect the design of the transmission pipeline.

The relative mobility of the two gases, CO_2 and CH_4, can be explained and is controlled by the following relationship:

$$M = k_r / \mu \tag{5-4}$$

M is the mobility ratio, k_r is gas relative permeability, and μ is gas viscosity

As indicated in Figure 5-12, the mobility of CH_4 is larger than the mobility of CO_2 due to the lower viscosity of CH_4. This affects the operation of the reservoir as it makes the interface between CH_4 and CO_2 unstable as working gas is injected against the CO_2 cushion gas.

However, the opposite situation will arise during the working gas withdrawal cycle as relatively viscous CO_2 pushes against CH_4.

The coefficient multiplying the pressure gradient to give the mass flux of gas in a porous medium is ρMk, k = permeability of the medium [18].

Figure 5-12 indicates a strong similarity of trends in the variations in the density and viscosity of the CO_2 and CH_4 and their mixtures. This correlation will make the flow term ρMk relatively constant as a function of pressure for the two gases. In this case, the compensating effects of density and viscosity imply that the displacement processes between the two gases will not change significantly as pressure changes.

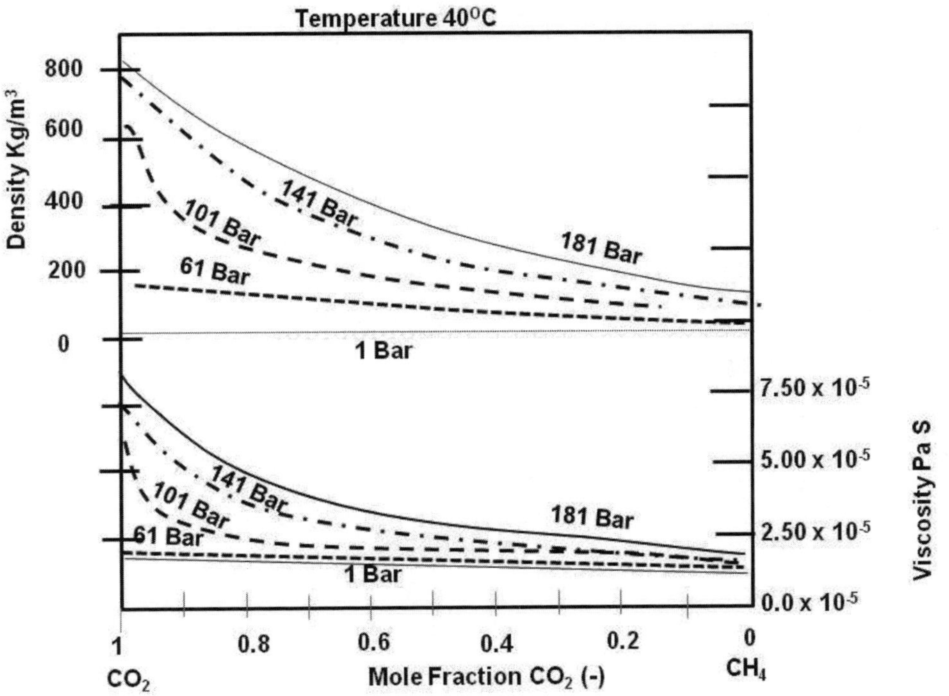

Figure 5-12. Density and viscosity of CO_2–CH_4 mixtures at various pressures and at T — 40°C ([18], Courtesy of American Chemical Society)

Therefore, consideration must be given to the purposes of storage of anthropologic CO_2 and establishing the operational pressure for both sequestration and pipeline transmission as these will affect equipment/facility needs.

5.3 CO_2 PRODUCT QUALITY FOR PIPELINE TRANSPORTATION

CO_2 properties from capture points of view are discussed and well represented in Chapter 3. This section will discuss the subject matter from pipeline transportation point of view.

Since 1970s, the oil and gas industry has produced and safely transported more than 11 TCF of CO_2 from seven naturally occurring sources in the USA, 1.2 TCF of which came from sources that otherwise would have been vented. In the USA well over 5000 km (3100 miles) of CO_2 mainline pipeline systems are operational, transport about 80,000 tons of CO_2/day, and have led to production of over 1.2 billion barrels of incremental oil (Figure 5-13). There are secured operating practices for the following [19]:

- Corrosion/integrity management,
- Measurement and quality management,
- Material metallurgies, compatibility to elastomers,
- Separation, dehydration and hydrocarbon extraction,
- Compression/pumping,
- Injection and production well completion and operation.

Figure 5-13. Current naturally occurring CO_2 sources and mainline system, USA

There are also other CO_2 pipelines outside the USA [20]. These pipelines are dedicated for enhanced oil recovery and are as follows:

- Raman (EOR) operated by Turkish Petroleum and
- Snohvit operated by Statoil Hydro in Norway (Offshore EOR and CCS).

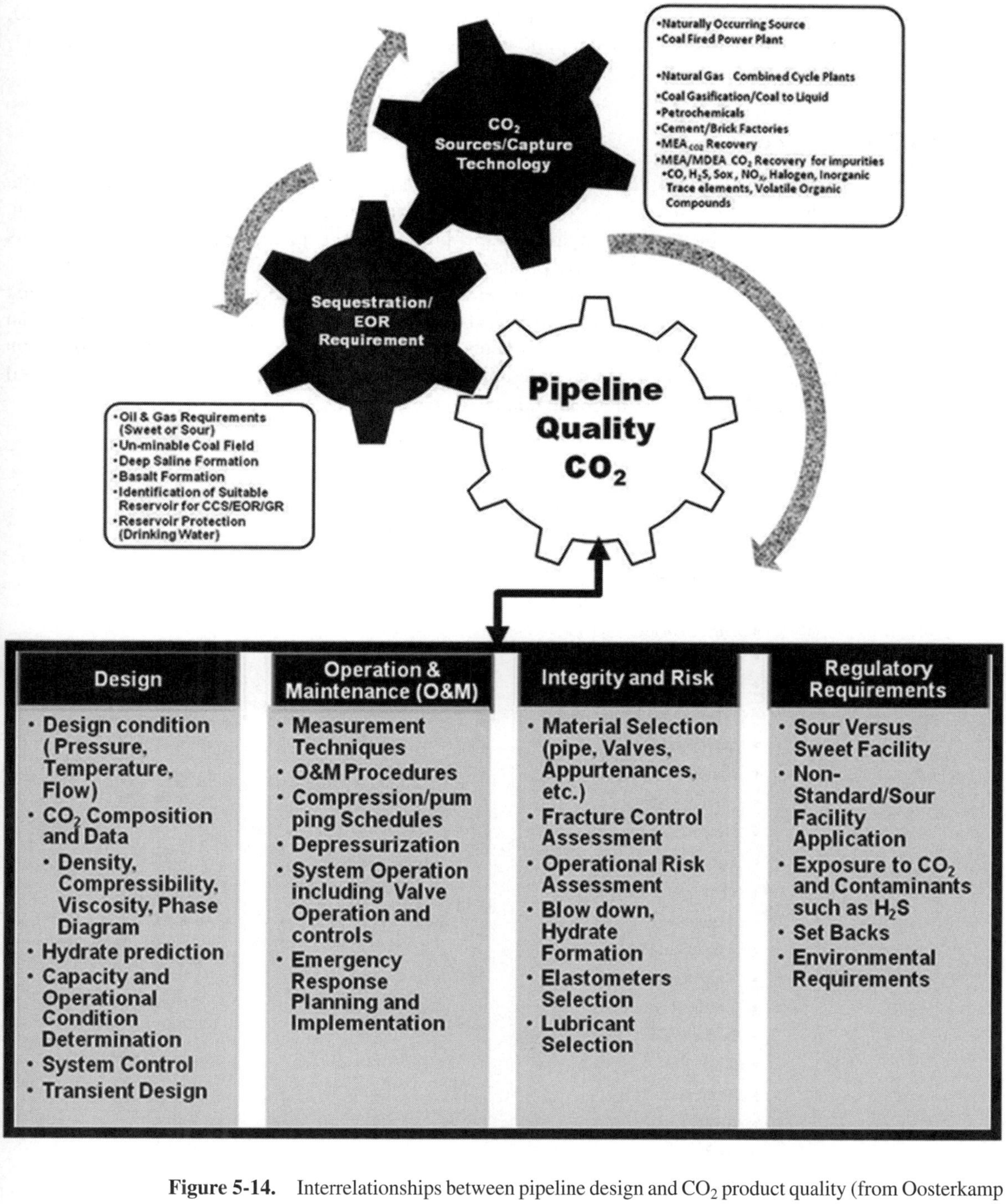

Figure 5-14. Interrelationships between pipeline design and CO_2 product quality (from Oosterkamp [21])

However, crucial to the success of a CO_2 pipeline design and relevant hydraulics, general mechanical design and material selection, is having realistic CO_2 properties. As such properties will drive pipeline system design and its operational conditions.

Compositional properties of CO_2 all influence aspects of pipeline transportation system design, operation, and maintenance. The relation of CO_2 pipeline design and operation/ maintenance issues as related to CO_2 product specifications is illustrated in Figure 5-14.

The requirements for the source(s) and the market(s) for the CO_2 need to be known and are complementary. This then sets the product range of characteristics and leads to the development of the product specification. From there, CO_2 pipelines can be specified, designed, publicly approved, built, operated, and maintained for the life of the entire CO_2 infrastructure.

5.3.1 Sources of CO_2

As indicated in Chapter 1, CO_2 can be found in both naturally occurring sources and from industrial processes (such as power generation, natural gas processing, ethanol production, cement manufacturing and pulp and paper production). Typical flue gas stream constituents from coal- and gas-fired power plants are indicated in Table 5-2 [22].

CO_2 from industrial sources is removed from bulk streams such as flue gases using various CO_2 capture technologies. CO_2 from these "sources," depending on its quality, can then be used for applications such as enhanced oil recovery (EOR), food-grade carbonation for beverages or can be sequestered in permanent storage. A majority of projects planned for CO_2 capture and storage are related to coal-fired power plants [23].

TABLE 5-2. Typical power plant flue gas stream constituents (Source: CO_2 Capture Project [22])

Flue Gas Constituent	Coal-Fired (% vol)	Gas-Fired CCGT (% vol)
• Carbon dioxide (CO_2)	14	4
• Carbon Monoxise (CO)	?	?
• Nitrous oxides (NO_x)	80–85	85–95
• Sulphur oxides (SO_x)	1–3	<0.1
• Carbon monoxide	1–3	1–3
• Hydrogen sulphide (H_2S)*	?	?
• Elemental oxygen (O_2)	1–3	1–3
• Elemental nitrogen (N)	< 1%	< 1%
• Mercury (Hg)	~60,000 mg / Nm^3	trace
• Arsenic (As)	~5,000 mg / Nm^3	trace
• Selenium (Se)	1,000 mg /Nm^3	trace
• Hydrogen Chloride (HCl)	trace	–
• Argon (Ar)	trace	trace
• Vanadium (V)	trace	trace
• Cadmium (Cd)	trace	trace
• Lead (Pb)	trace	trace
• Nickel (Ni)	trace	trace
• Zinc (Zn)	trace	trace
• Manganese (Mg)	trace	trace
• Molybdenum (Mo)	trace	trace
• Moisture (H_2O)	1–2	1–2
• Particulates/dust	1	–

*may be present in reducing conditions under poor operating conditions

The type of processing and cleanup required to meet end-user requirements may vary depending on the levels of contaminant acceptable. The following basic processes, as detailed in the CCS Guidelines, may be used to treat raw CO_2 product to meet specifications from power plant sources.

5.3.2 CO_2 Grade for Pipeline Transportation

Typically, the following "Quality Specifications" are stipulated by the industry for pipeline transportation of CO_2 from naturally occurring sources [19]:

- **Minimum % CO_2** ninety-five mole percent (95 mol%), typical values are about 98.45% to 98.72 %,
- **Water Content** to contain no free water, and no more than thirty (30) pounds of water per MMSCF (480 mg/m^3) in the vapor phase,
- **CO_2 Stream** generally is expected not to contain more than:
  - **H_2S** twenty (20) parts per million, by weight,
  - **Sulfur** thirty-five (35) parts per million, by weight (total)
  - **Nitrogen** four mole percent (4 mol%)
  - **Oxygen** ten (10) parts per million, by weight
  - **Glycol** 0.3 gallons of glycol per MMcf (0.04 l/m^3), at no time glycol must be present in a liquid state at the pressure and temperature conditions of the pipeline
- **Operating Temperature** CO_2 stream temperature is generally kept below 50°C
- **Hydrocarbons** CO_2 stream is not to contain more than five mole percent (5%) of hydrocarbons with the dew point below (−20°F, −29°C).

The above conditions are based on data from several CO_2 transportation pipelines operational in the USA and Canada (see Figure 5-13). These CO_2 pipelines are intended to either:

- Connect a dedicated source of CO_2 to a single end user or to multiple deliveries or
- Connect multiple sources of CO_2 (Natural and Anthropogenic) to multiple delivery points through interconnecting pipelines (Permian Basin).

5.3.2.1 CO_2 Product Quality

There are currently no standard compositional requirements (e.g., moisture or co-constituents) for the transport and geologic storage of CO_2. Typically, specifications are included in the contract language between the pipeline and source and pipeline and end user.

However, a number of guidelines are available that provide potential product specifications for CO_2 pipelines [20]. The guidelines are developed with stakeholder participation involving business, government regulators, and academia and provide a comprehensive review and series of recommendations regarding CCS.

Potential CO_2 stream specification scenarios for pipelines transportation are indicated in Table 5-3. It may be noted that Type I CO_2 pipelines are designated useful for a single-use CCS project (i.e., source to sink). Type II pipelines are the most common CO_2 pipelines mostly in use in the USA and are typical of EOR projects mostly carry CO_2 from naturally occurring sources. An example of the Type III pipeline is the Weyburn CO_2 Pipeline. It may be noted that it would be difficult to integrate Type II and Type III pipeline systems due to carrying CO_2 streams of different compositional impurities. Type III CO_2 pipeline should be considered as sour service because of carrying large quantities of entrained H_2S.

TABLE 5-3. Types of CO_2 product quality for pipeline transportation [20]

Parameter	Type I	Type II	Type III
CO_2, % by volume	>95%	>95%	>96%
H_2S, ppmbw*	<10	<20	<10,000
Sulphur, ppmbw*	<35	<30	–
Total hydrocarbons, % by volume	<5	<5	–
CH_4, % by volume			<0.7
$C_2{}^+$ hydrocarbons, ppm			<23,000
CO, ppm			<1,000
N_2, % by volume/weight	<4	<4	<300 ppm
O_2, ppm by weight/ volume	<10	<10	<50
H_2O, lb/mmcf** (mg/m^3) or ppm by volume	<25**(400 mg/m^3)	<30*(480 mg/m^3)	<20***(320 mg/m^3)

Notes: C_2 = Hydrocarbon; CH_4 = methane; CO = carbon monoxide; CO_2 = carbon dioxide; H_2O = water; H_2S = hydrogen sulfide; **mmcf = millions of cubic feet; N_2 = nitrogen; ppm = parts per million; O_2 = oxygen; *ppmbw = ppm by weight

Type I = Single Pipeline (Source to Sink), Type II = Current CO_2 Pipeline from Natural Sources and Type III = Pipeline Transporting CO_2 with Impurities from flue stacks

In comparison is the European Union (EU) proposed specification for CO_2 pipelines. The specification focuses on pre combustion and post combustion processes for CCS prevalent in Europe (Table 5-4).

The following notes are applicable to Table 5-4 [24, 25]:

- A minimum of 95% CO_2 is generally required for EOR operations to ensure miscibility of CO_2 into the oil at reservoir temperature and pressure in the appropriate geothermal basin.
- H_2 in the gas stream affects pipeline capacity, compression, EOR miscibility, pipe fracture propagation behavior and has value in its energy content.
- Acid Gas EOR and Storage components range from: 95% H_2S / 5% CO_2 to 1% H_2S / 98% CO_2.
- Limits set to meet Health and Safety requirements. However, the recommended 2000 ppm CO content recommended by DYNAMS is above the 1200 ppm Immediately Dangerous to Life and Health (IDLH) limit recognized in some countries, e.g., in Canada.
- Non condensable gases include N_2, H_2, Ar.

TABLE 5-4. CO_2 pipeline quality stream recommendations [24]

Component	Concentration	Limitation
H_2O	500 ppm	Technical: below solubility limit of H_2O in CO_2. No significant cross effect of H_2O and H_2S, cross effect of H_2O and CH_4 is significant but within limits for water solubility
H_2S	200 ppm	Health and safety considerations
CO	2000 ppm	Health and safety considerations
O_2	Aquifer < 4 vol%, EOR 100–1000 ppm	Technical: range for EOR, because of lack of practical experiments on effects of O_2 underground.
CH_4	Aquifer < 4 vol%, EOR < 2 vol%	
N_2	< 4 vol % (all non-condensable gasses)	
Ar	< 4 vol % (all non-condensable gasses)	
H_2	< 4 vol % (all non-condensable gasses)	Further reduction of H_2 is recommended because of its energy content
SO_x	100 ppm	Health & safety considerations
NO_x	100 ppm	Health & safety considerations
CO_2	>95.5%	Balanced with other compounds in CO_2

- CO_2 gas temperature limit is a function of pipeline coating and applicable rating only.
- Volatile inorganic trace elements contained in the coal (As, Be, Cd, Cr, Co, Hg, Mn, Ni, Pb, Sb, Se, Ba, B, Mo, V, and others) which may be present in coal-fired power generator flue gasses.
- Organic compounds that result from incomplete combustion of coal or natural gas.

A review of existing pipelines requirement for transportation of CO_2 indicates the following key considerations:

- Networked pipelines and point-to-point systems must be considered separately as they have differing product quality requirements to match source(s) to end user(s);
- Contaminants such as hydrogen, higher levels of H_2S, inorganic trace elements, CO, NO_x and SO_x all may lead to increased capital and/or operating costs to the pipeline infrastructure, as well as potentially increased regulatory oversight. For example, inorganic trace elements may lead to stricter pipeline abandonment requirements and increased cost when the infrastructure has met its life expectancy;
- There are no standard specifications presently used in North America and Europe to manage a broad spectrum of end-use requirements, such as EOR and CCS; and
- The removal of impurities provides flexibility to a backbone pipeline system network as it will allow greater variety of supply (source) and demand (sink) to be accommodated. For example, some EOR operations may not want CO_2 containing heavy metals or high H_2 or H_2S content in the supply stream.

5.3.3 CO_2 Stream Quality Scenarios — Network and Point to Point System

Network System: A CO_2 gathering and transportation network system must have key contaminants determined. This includes inorganic trace elements such as Hg and gaseous impurities such as CO, H_2, NO_x, SO_x. Permissible H_2S levels could be relatively high, depending on end use, safety, regulatory, and design requirements. An example of a CO_2 specification for a network pipeline can potentially be similar to the Type III (Table 5-3, [20]) which is found in the Weyburn pipeline with adaptation for H_2S. Refer to Table 5-5 below for sweet or sour networks.

Point to Point (Source to Sink) System: is generally tailored to meet the end user and pipeline transportation requirements. Essentially, a Type I (Table 5-3, [20]) specification could form the basis of any point-to-point system, with specific modifications as required. For example, an EOR use may require that more processing be performed upstream to ensure that inorganic trace or other undesirable elements are eliminated or reduced to an acceptable limit from the feedstock. Carbon capture, on the other hand, may have much higher allowable concentrations of contaminants.

Table 5-5 provides a summary of CO_2 stream quality specification for a network (backbone type) as well as a point to point (source to sink) pipeline systems. Table 5-6 summarizes the reasons for limitations provided.

In both types of systems, it is important to evaluate thoroughly the opportunity and possible limits that may be required to ensure a good match between source and demand while ensuring the pipeline project is economical and safe. Once the particular case is well understood, negotiation between parties would occur and the process would move forward through the regulatory and public consultation framework.

TABLE 5-5. CO_2 Stream specification for network (sweet and sour) and point to point pipeline

Component	Concentration		
	(Sweet Network)	(Sour Network)	(Point to Point)
Caron dioxide, CO_2	>95.5%	>95.5%	to >95%
Water, H_2O	<30 lbs/MMscf (480 mg/m^3)	<20 ppm	<5 to <30 lb/MMscf (<80 to <480 mg/m^3)
Hydrogen sulphide, H_2S	<16 ppm	<10,000 ppm	<16 ppm to < 95%
Total sulfur	<30 ppm	Not Required	< 30 ppm to N/S
Carbon monoxide, CO	<1000 ppm	<1000 ppm	TBD
Oxygen, O_2	<10 ppm	<10 ppm	<10 ppm to < 0.4%
Methane, CH_4	Aquifer < 4 vol%, EOR < 2 vol%	<0.7 %	<0.5% to 4%
Nitrogen, N_2	< 4 vol % (all non condensable gasses)	<23,000 ppm	<0.5% to 4%
Argon, Ar			<0.5% to 4%
Hydrogen, H_2	< 4 vol % (all non condensable gasses)	<300 ppm	TBD
Sulphur oxide, SO_x	<100 ppm	TBD	100 ppm
Nitrogen oxides, NO_x	<100 ppm	<100 ppm	100 ppm
Hazardous organic compounds	Below detectable limits	<100 ppm	TBD
Inorganic trace elements	Below detectable limits	TBD	TBD

5.4 OTHER DESIGN PARAMETERS

5.4.1 Pipe Roughness

The following pipe roughness is generally used by the pipeline industry when undertaking hydraulics design calculations [26, 27].

Pipe Condition	Absolute Roughness k [mm]
• New clean bare steel	0.0127–0.02
• welded used, but cleaned steel	0.15
• Scraper burnished pipe	0.0076–0.0127
• Internally coated pipe	0.0051–0.0076
• steel, after 2 years of atmospheric exposure	0.045–0.0508
• Sand/shot blasted steel	0.005–0.0076

The relative roughness n is often used for hydraulics calculation and is given by the ratio of the absolute roughness k and the pipe diameter D

$$n = k/D \qquad (5-5)$$

5.4.2 Depth of Cover/Buried Depth

Typically, CO_2 pipeline are buried within trench, at a minimum of about 0.9 m, (distance from top of pipe to the surface) [4].

5.4.3 Design Lifetime

A review of available technical papers as reported by Wildbolz [24] and industry practice indicates a design lifetime of 25-30 years for CO_2 pipeline [28, 29].

TABLE 5-6. Basis of limitations for CO_2 stream specification (refer to Table 5.5)

Component	Limitations/Comments (Sweet Network)	(Sour Network)	(Point to Point)
Caron dioxide, CO_2	A minimum of 95% CO_2 is required for EOR operations to ensure miscibility of CO_2 into the oil at reservoir temperature and pressure		Acid Gas EOR and Storage components range from: 95% H_2S/5% CO_2 to 1% H_2S/98% CO_2
Water, H_2O	<60% water saturation level in the gas stream, refer to WRI [23]		Technical: below solubility limit of H_2O in CO_2. No significant cross effect of H_2O and H_2S, cross effect of H_2O and CH_4 are significant but within limits for water solubility
Hydrogen suphide, H_2S	Health and safety considerations, refer to WRI [23]		Acid Gas EOR and Storage components range from: 95% H_2S/5% CO_2 to 1% H_2S/98% CO_2
Total sulfur	Based on WRI [23] "CCS Guideline, Guidelines for Carbon Dioxide Capture, Transport, and Storage"		Guidelines/Quality Standards for Transportation of Gas Containing Mixed Corrosive Constituents, Pipeline Research Council International Inc, February 2004
Carbon monoxide, CO	Based on WRI [23] "CCS Guideline, Guidelines for Carbon Dioxide Capture, Transport, and Storage"		Health and safety considerations limit CO concentrations
Oxygen, O_2	Based on Guidelines/Quality Standards for Transportation of Gas Containing Mixed Corrosive Constituents [25].		Stakeholder negotiations are required to determine limits or acceptability
Methane, CH_4		Based on WRI [23] "CCS Guideline, Guidelines for Carbon Dioxide Capture, Transport, and Storage"	Including all Hydrocarbons, C_2 plus
Nitrogen, N_2			Total inert materials <4%
Argon, Ar			Total inert materials <4%
Hydrogen, H_2	H_2 in the gas stream affects capacity, compression, EOR miscibility, pipe fracture arrest requirements and has value in its energy content		H_2 in the gas stream affects capacity, compression, EOR miscibility, fracture control properties and energy content
Sulphur oxide, SO_x	Health and safety considerations		
Nitrogen oxides, NO_x	Health and safety considerations		
Hazardous organic compounds	Health and safety considerations	Stakeholder negotiations are required to determine limits or acceptability	
Inorganic trace elements		Stakeholder negotiations are required to determine limits or acceptability	

A 30 years lifetime for both design and economic considerations is generally adopted by the industry. However, it may be noted that the industry has avoided the transportation of wet CO_2 in carbon steel pipe due to the high corrosion rate caused by the acid character when CO_2 and water come into contact.

5.4.4 Transportation Velocity Limit

The industry considers a typical velocity in a pipeline transporting CO_2 to be limited between 1 and 5 m/s [4, 30]. Hydrocarbon pipeline transportation is limited to the maximum of the range to mostly avoid/limit and mitigate static electricity generation during pipeline transportation and economically optimum facilities (i.e., pipe diameter versus compression). However, the same range can also be adapted for CO_2 transportation because the range limits pressure drop in the pipeline, the higher the limit the shorter the distances that will be required between pumping/compression facilities thus, affecting the economics of the pipeline transportation.

5.4.5 Corrosion

The presence of free water (i.e., water that is not dissolved) in the CO_2 stream may cause corrosion of the pipeline steels and/or lead to hydrate formation in the pipeline. All CO_2 transmission pipelines currently in operation are carbon steel, which is essentially non-corrosive in pure CO_2. Trace water dissolved in the CO_2 stream is not a significant problem. However, in the presence of *free* water, highly corrosive carbonic acid is formed.

At a constant temperature, the solubility of water in gaseous CO_2 drops with increasing pressure until it reaches a minimum as it approaches the boundary of the CO_2 liquid/vapor envelope, as shown in Figure 3-3 (Chapter 3). As the solubility of the water in CO_2 falls, it is more likely to form the free water phase.

Hydrates are solid compounds with similar properties to ice; consequently, they can block the pipeline and plug piping and appurtenance such as heat exchangers [31]. In order for hydrates to form in a CO_2 pipelines, there must be the required combination of pressure and temperature and a sufficient amount of water present. Under CO_2 pipeline operating pressures, and with certain water contents, it would be possible for hydrates to form at around 10–11°C and therefore there is a requirement to control water levels [32].

On the other hand, CO_2 streams from flue gases (from flue stacks) include impurities and free water from the combustion of fossil fuel. There is a consensus that this water must be removed prior to pipeline transportation, and there exist a number of effective drying technologies and processes [31].

The industry, however, through transportation agreements, limits water content in pipeline transportation CO_2 to 480 mg/m^3 (refer to Table 5-5).

Therefore, generally, it is very unlikely that there would be any free H_2O in the CO_2 stream because of a strict contractual requirement between parties. This is because CO2 is an acid gas, and it will react with water to form carbonic acid (Figure 5-15).

As indicated in Figure 5-16 in a wet CO_2 environment, free water (H_2O) reacts with carbon dioxide (CO_2) and forms carbonic acid (H_2CO_3) as hydrate. This hydration can dissociate into bicarbonate (affecting the pH of the CO_2 stream) to become alkaline solution ($CO_3{}^{2-}$ + H^+ (i.e., carbonates and hydrogen)).

Carbonic acid in a aqueous solution (CO_2 and free water) releases electrons which in turn give up bicarbonate thus affecting the pH content of the CO_2 stream. The electrons release free hydrogen reacting with steel (Fe) to create ferrous carbonate ($FeCO_3$) thus causing (general and localized) corrosion in the steel pipeline.

Carbonic acid is corrosive to carbon steel pipes and causes damage referred to as sweet gas corrosion. If this corrosion is not caught as soon as possible, the life of the pipeline will

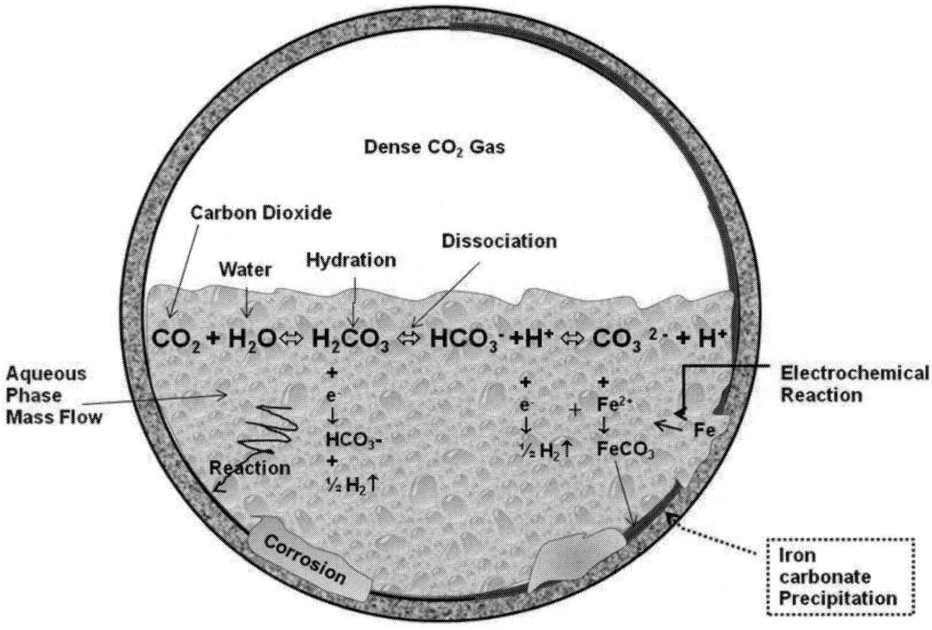

Figure 5-15. CO_2 corrosion mechanisms (from Anderson [33])

be greatly reduced. US DOT CFR 49 part 195 the DOT has set additional corrosion requirements for pipeline transportation of CO_2 that are generally followed.

Corrosion allowance is not usually considered and accounted for in the design of CO_2 pipelines because corrosion is very low for pipeline transportation operating with dry supercritical CO_2. However, impurities (depending on their level), if not removed from flue gas streams, may have corrosive effect on equipment in pipeline transportation and on the storage reservoir. In storage, corrosiveness may either enhance or weaken the trapping and sealing of CO_2 in specific geologic formations. Furthermore, health, environment, and safety issues will need to be addressed if leakage of the CO_2 and the impurities occurs.

Should wet CO_2 stream (outside the pipeline grades) be transported, proper precautions for prevention of internal pipeline corrosion should also be considered. Because an acid gas is being transported through the pipeline, corrosion must be checked biannually at intervals no greater than 7.5 months. Corrosion inhibition is not generally used in CO_2 pipelines and neither is the use of internal coating to reduce corrosion. The industry has accepted frequent pigging, to mitigate any corrosion reactions [27].

It has been noted through literature search that corrosion rate in carbon steel (such API 5L X60) amounts to less than 0.5 μm/year at a temperature of 22°C, a pressure of 140 bar and 1000 ppm (0.1%) H_2S [4]. For an AISI 1080 steel a corrosion rate of about 0.01 mm/yr (at 90–120 bar and 160–180°C for 200 days has been indicated [34] and [26]. It has also been reported that a relative humidity less than 60% does not corrode the type carbon steels typically used for CO_2 pipelines. It may be noted that AISI 1080 is a Standard grade *Carbon Steel*. It is composed of (in weight percentage) 0.75–0.88% carbon (C), 0.60–0.90% manganese (Mn), 0.04%(max) phosphorus (P), 0.05% (max) sulfur (S), and the base metal iron (Fe). Other designations of AISI 1080 carbon steel include UNS G10800 and AISI 1080.

However, in the presence of *free* water, highly corrosive carbonic acid is formed, and it has been reported that carbon steel can corrode at rates of more than 1.0 mm/year in wet pure CO_2 [31].

5.5 TYPICAL PIPELINE SYSTEMS DESIGN

For efficient pipeline transportation, supercritical CO_2 will be needed over the entire distance of the pipeline [35].

CO_2 is typically received from a flue stack source (such as a power plant, oil upgrading site, or another industrial site). The CO_2 is also generally received at atmospheric pressure. Once captured from the source, the CO_2 needs to be purified of impurities that can cause potential pipeline and equipment corrosion, operational problems, and/or mechanical damage. Impurities as indicated previously may include, but are not limited to, nitrogen, hydrogen, hydrogen sulphide, methane and heavier hydrocarbons, and water. Once the impurities are removed to acceptable level (refer to Table 5-5), the CO_2 is compressed to the required inlet pressure of the pipeline. The appropriate maximum operating pressure will depend on the balance between compression and pipe costs, which will be determined by the final design hydraulics [27]. Compression is generally achieved using integrally geared compressors similar to those used for urea operation. This is because of high compression ratio required to raise the pressure from atmospheric condition to the required pipeline pressure for transporting CO_2 in the supercritical phase. Thus, the discharge pressure from the receipt compression into the pipeline is typically likely to be in the range of 10 MPa (1440 psig) to 20 MPa (2900 psig).

Once the CO_2 is compressed, it will need to be cooled to a temperature suitable to be transported in a pipeline (i.e., < 60°C). Depending on the flow and pipeline temperature required, this is typically performed by air coolers and/or water chillers. Once the CO_2 is cooled to a suitable temperature to transport, it will be sent to the pipeline.

Once the CO_2 enters the pipeline, it will travel until the pressure approaches the minimum acceptable level. For pure CO_2 in the supercritical phase, the appropriate minimum pressure would likely be about 8 MPa (1160 psig), as this is approximately 10% above the critical pressure. By this point, the CO_2 will likely have cooled to the dense phase (due to the heat transfer that occurs between the pipeline and the cooler surrounding soil) and will behave as a liquid. Thus, if it is required, the mixture can be pumped (using a centrifugal pump) back to the maximum operating pressure of the pipeline or to a suitable pressure determined based on economic feasibility and/or delivery requirements.

5.5.1 Pipeline Configuration

Typical CO_2 pipeline facilities from a (power plant) flue stack through to separation, compression, transportation, pumping and an injection well head and injection well piping configuration is schematically shown in Figure 5-16.

Typical pipeline system from the receipt of initial compression to injection facilities depicting pipeline pumping, valving, traps, and metering facilities (as usually deemed by a pipeline operating company) is shown in Figure 5-17.

A large compressor system is usually required to compress the source CO_2, (leaving the point source at atmospheric pressure condition). Depending as to what level the pressure has to be raised to, a number of stages will be required for a sequential increase in pressure to a final pressure for pipeline transportation/application. Figure 5-18 provides an example of a four (4) stage compression system. Compression stages depend on manufacture and type of equipment (reciprocating versus centrifugal) and generally inter-stage air-cooling is also utilized for the design specifications. This cooling process will allow the compressors to operate at a lower temperature therefore increasing the working life and also the overall efficiency of the compression system cooling will also allow the exit temperature of CO_2 to be reduced which will help with the economics of the pipeline materials and coating protections system.

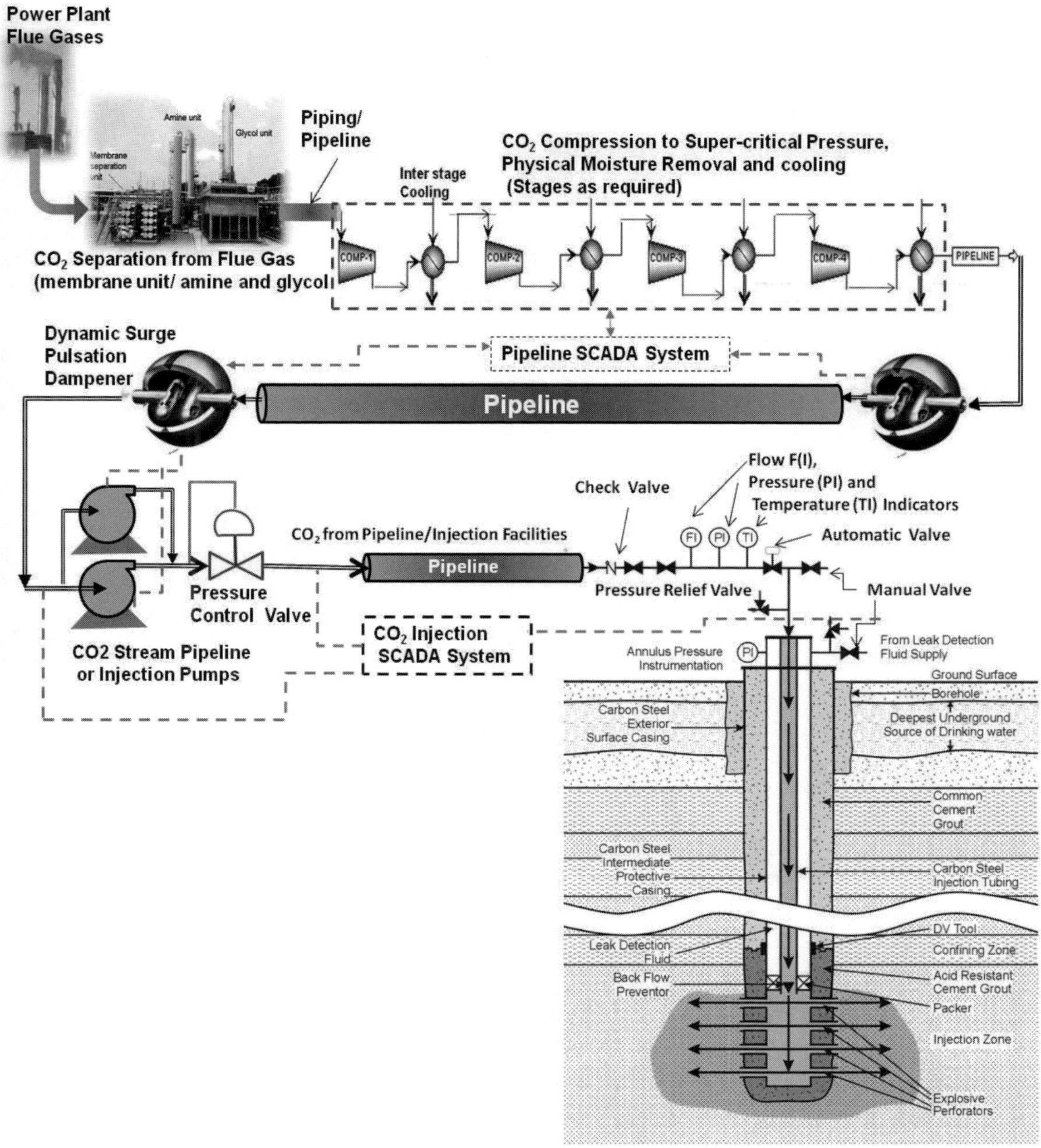

Figure 5-16. Schematic of a typical impure CO_2 gathering, pipeline transportation (processing, compression, and injection) configuration

5.5.2 CO_2 Pipeline Compression

There are a number of choices for CO_2 compression that depend on the manufacturer as well as system compression requirements [36]. Within the centrifugal compressor markets, there are two technologies, namely, in-line (between bearings) centrifugals and integral-gear centrifugals [37]. However, there are several manufacturers that market the compressors that are or could be used in CO_2 compression service for Carbon Capture and Storage (CCS) projects. These manufacturers that produce very large compressors are as follows:

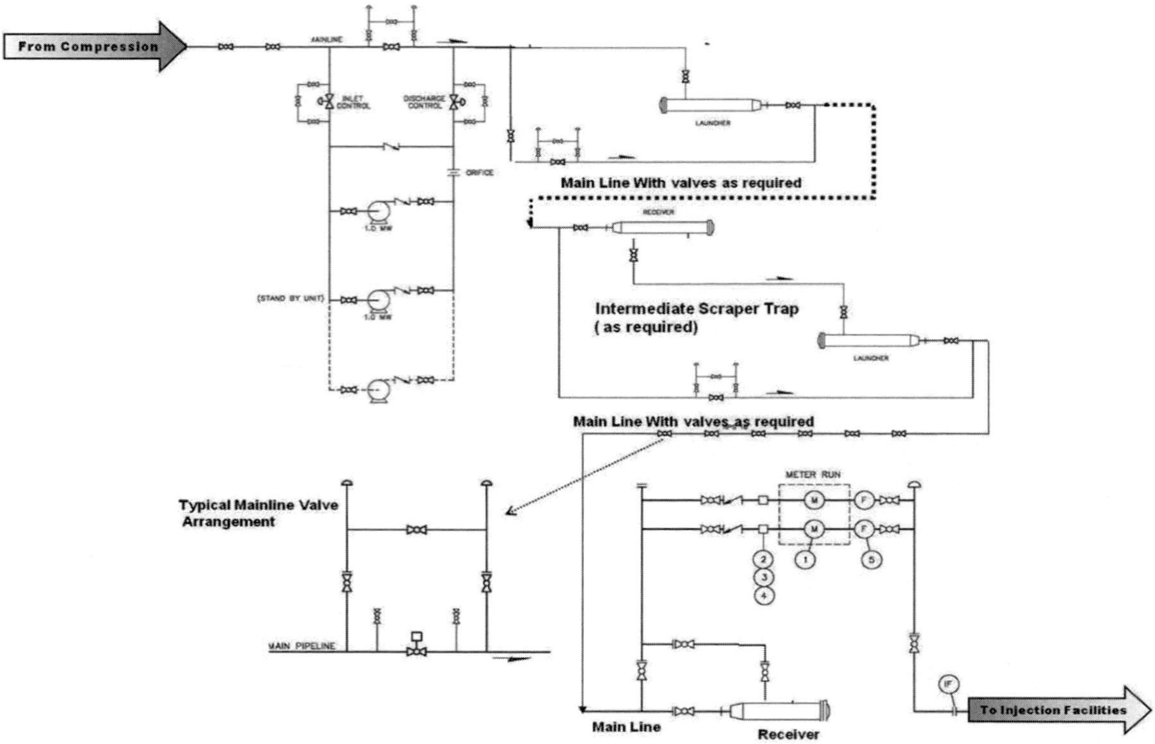

Figure 5-17. Schematic of a typical impure CO_2 pipeline transportation configuration

- Integrally geared compression units (ManTurbo),
- Hitachi Inline Process CO_2 Compressor,
- Ramgen Discrete Drive LP-Stage,
- Dresser Rand Datum Compressor,
- General Electric (GE),
- Seimens.

Compression systems and configuration choices for pipelines transporting CO_2 containing impurities are extensively dealt with in Chapter 8.

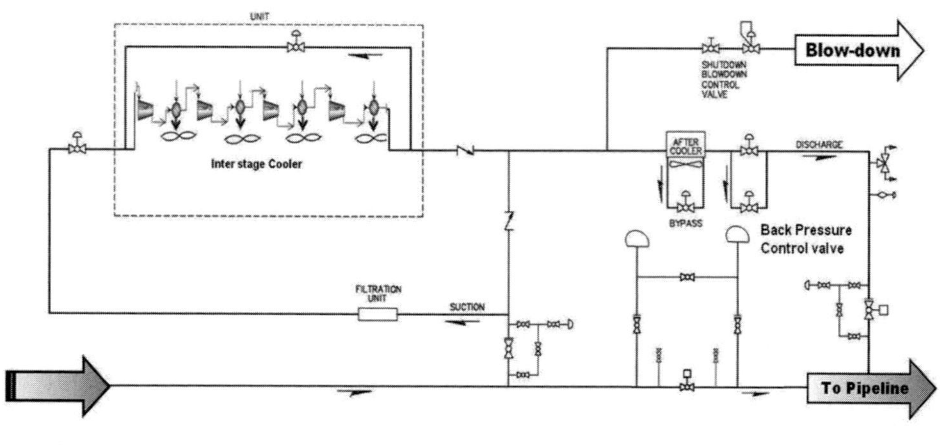

Figure 5-18. Typical impure CO_2 pipeline compression system

5.6 MEASUREMENT FACILITIES LAYOUT

CO_2 is usually metered with orifice meters, turbine or Coriolis meters (refer to Chapter 8 and Mohitpour et al [27] for metering types). A custody transfer meter station is generally installed at the outlet of the compressor station prior to entering the pipeline as well as at the delivery points from the pipeline, see Figure 5-18. Details of all measuring units are provided in Chapter 8. General measurement methods for pipeline control are outlined in Chapter 10, Section 10.5.

This measurement station would also be used for monitoring pressure, temperature, CO_2 gas quality (CO_2, H_2S, H_2O, O_2, etc. as required) and recording flows. Depending on the design flow range, two meter runs or more may be required for the range ability and back-up measurements when maintenance and calibration is required. Economics will dictate the type of metering required. A typical receipt meter station isometric is shown in Figure 5-19.

A delivery station is also required. Metering would be required at each delivery station if there were more than one delivery point. Typical multiple run delivery measurement station including a common header, meter proving facility, and meter runs with strainers, etc. for a multi-injection system is shown in Figure 5-20 for multi-injection systems. All meter runs have a vent/relief and connections to vents must take place underneath the meter piping. In this way, liquid CO_2 disposal can be ensured during any relief operation. Meter runs typically must use ball valves, with 100% blocking capability and a double seating arrangement (EDPM 85 or Viton seats) hard metal surfaces to 4 RMS and exposed parts must be chrome plated and anodized.

The delivery station generally includes regulators/control valves to maintain a back pressure on the pipeline and protect the downstream systems from overpressure. Pressure regulation would be located downstream of the meters.

Coriolis meters are generally preferred as they have many advantages over orifice and turbine meters. They are mass flow and density meters that do not have any movable parts, do not require venting to change an orifice plate, and are simple to install. Pressure and temperature measurements are not required to operate this device. A gas chromatograph is

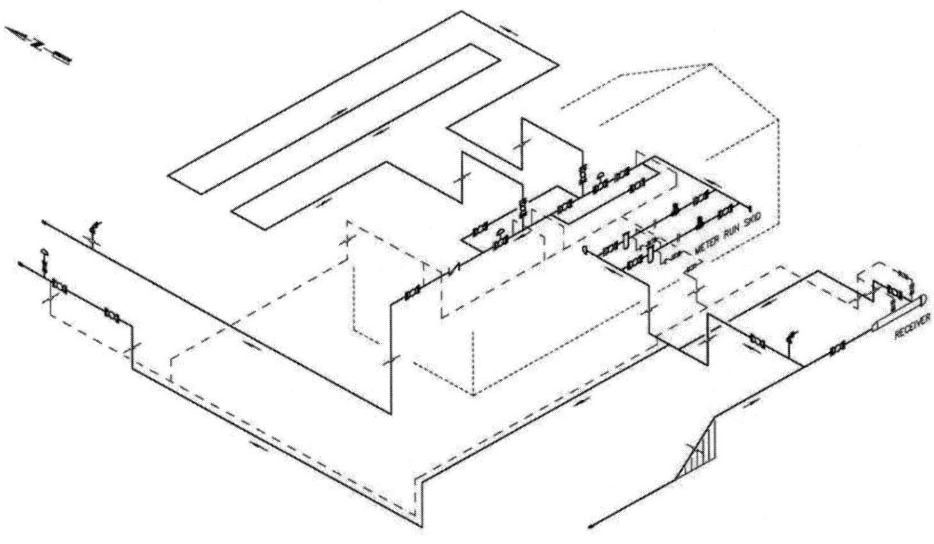

Figure 5-19. Typical CO_2 meter station piping arrangement

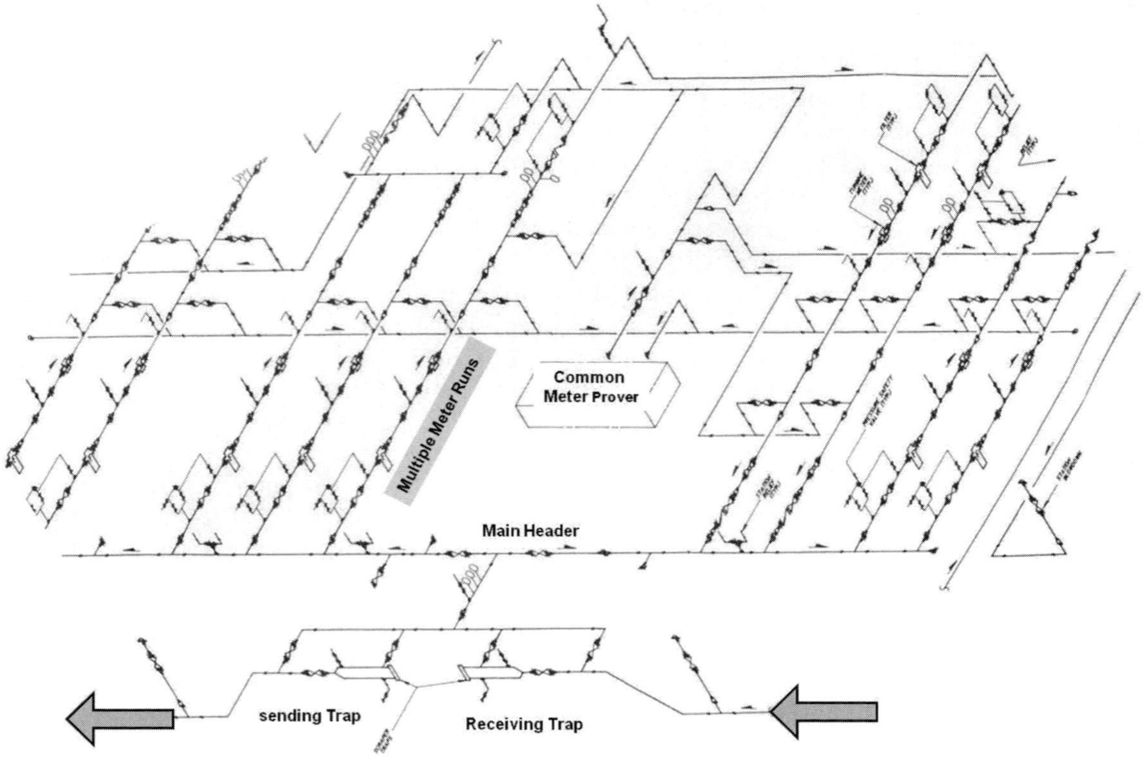

Figure 5-20. Typical multi-run CO_2 measurement station with trap facilities

not required for mass flow measurement. The expected accuracy for turbine and Coriolis meters are expected to be about ±0.1%.

Using the same type of measurement for the receipt and delivery stations would provide the repeatability required to monitor the system mass balance. This is important for system control and leak detection. It must be noted that because of measurement, the accurate measurement of density is extremely important. Densitomers are used for this purpose with repeatability better than 1% (calibration error about 0.05%).

Moisture analyzers are also required (to assure moisture content between 20 and 30 lb/MMSCF (350–450 mg/m^3).

5.7 PIPELINE VALVES AND BLOW-DOWNS

In order to provide for safety, block valves are typically required to be placed along the pipeline about every 15 km (10 miles) maximum, per regulation such as US DOT CFR regulations. This prevents loss of pressure integrity of the entire pipeline, whereby sections can be isolated, limiting the amount of CO_2 released into the atmosphere. The ability of CO_2 to collect in depressions and other low-lying areas near the pipeline presents a significant hazard if leaks continue undetected. Typical mainline valve arrangement is shown in Figures 5-17 and 5-18.

Mainline valve assemblies are required to be installed at regular intervals to sectionalize the pipeline (Figure 5-21). Mainline valves are required to isolate sections of lines for maintenance and emergencies. Valve spacing in general should be performed based on an engineering assessment that considers all of the relevant factors. However, Canadian Standard, CSA Z662 Table 5-7 provides default values that can be used; for other than Class 1

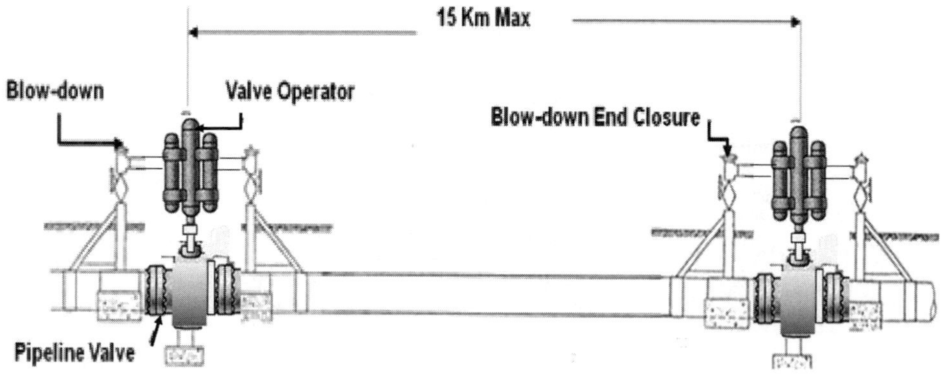

Figure 5-21. Typical CO_2 pipeline mainline sectionalizing valve and components

locations, the maximum spacing for CO_2 pipelines is 15 km. This distance may need to be shorter depending on access, blow-down time requirements, gas dispersion behavior from a blow-down operation, the effect on inhabitants close to blow-downs and the requirements of codes which generally require for "HVP and carbon dioxide pipelines, sectionalizing valves be located outside cities, towns, and villages, at the transition from Class 1 to a higher class location."

Mainline valves are typically ball valves as under high differential pressures they would respond to operation more easily than gate valves. Similar to meter run valves, these valves must also have 100% blocking capability and similar seat finishing arrangements.

Blow-down time for a section of a pipeline (depending on pipeline and blow-down size) could be up to 3 hours; longer times may be dictated by considerations of temperature changes within the pipeline. Dispersion model studies should be done to help plan the block valve locations and blow-down times, as well as to assist in developing emergency response plans. Further details on these issues are provided in Chapter 11 "Operation and Maintenance Requirements."

Blow-down valves are typically full-bore ball. The end closures (as shown in Figure 5-22) or flanges on the blow-down outlet should be smaller than the bore of the blow-down valves to ensure that the full pressure and temperature drop occurs across the assembly outlet to atmosphere, rather than across the blow-down valve.

The valve assemblies, as shown in the typical block valve (Figure 5-22), will include two blow-down stacks and a bypass between them to help with loading and pressuring sections of line (also refer to Figure 5-18). Mainline valves should also be full bore ball valves.

Valve operators would be included with the mainline valve for remote operation as part of the pipeline control system. This can also form part of the line break and leak detection system. Pressure should be monitored at each block valve.

The valve assemblies should be shop prefabricated, pre-tested and coated before field installation. The mainline valves are generally buried with the operators located above ground. The assemblies are typically shop hydrotested to 1.5× design pressure.

5.8 PIPELINE CONTROLS

Pipeline control is important for providing continuous monitoring of the pipeline system. The pipeline system must be able to respond quickly to emergency situations. It should monitor key parameters such as the operating temperature and pressure of the pipeline, as well as the flow rate of supercritical CO_2 gas through the pipeline system. The control system

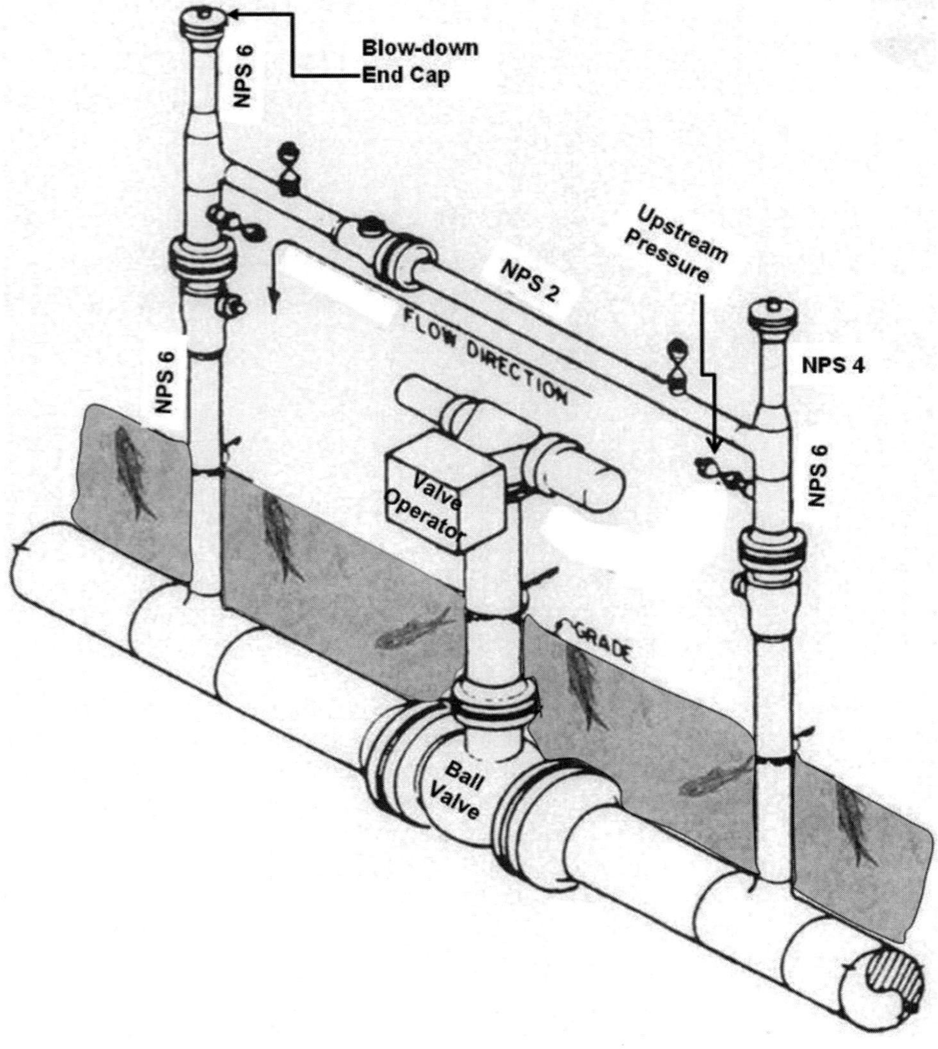

Figure 5-22. Type CO_2 pipeline stream blow-down configuration

should be able to detect leaks and also provide predictions on the fatigue of the system. Refer to Chapter 10 for further details.

A method for controlling a CO_2 stream with impurities in a pipeline thus may comprise pressure and flow controllers and indicators and well as temperature indications at frequent intervals along the pipeline in order to maintain the stream as a single-phase liquid or supercritical fluid stream in the pipeline.

Controls can include compositional analyzer, dew point controller, and simulation software generating a phase analysis of the stream in order to set and maintain operating inlet and delivery pressures and temperatures as per preset criteria for avoiding a two-phase situation [16]. For further details refer to Chapter 10, Section 10.21.

Another concern in pipeline design is the formation of hydrates in the pipeline. These can build up in the pipeline and cause the flow rate to drop or block off the pipe. With low percentage of water and methane in the system, hydrate formation should not be a major concern but it must be addressed.

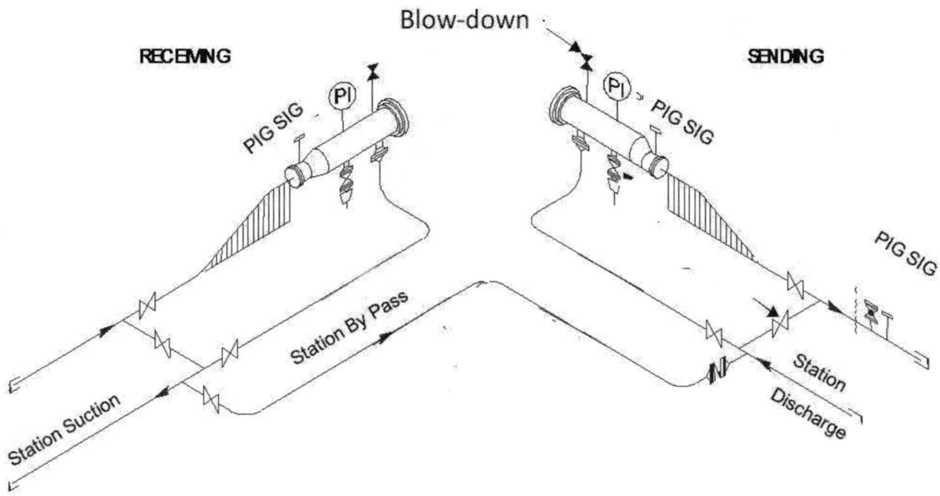

Figure 5-23. Typical CO_2 pipeline pig launching and receiving piping configuration

Scrapers or "pigs" can be used to clean out the inside of the pipe. Pig launchers and receivers are generally installed between stations (Figure 5-23) in order to speed up the pigging process and make shut down as short as possible. Injecting inhibitors such as methanol or salt solutions during the pigging process can minimize the hydrate formation, refer to Figure 5-17 for pipeline location of traps.

5.9 PIPING AND COMPONENTS MATERIALS

5.9.1 Pipe and Appurtenances

Table 5-7 indicates the summary of current materials generally used for CO_2 mainline pipes and components [27]. For pipeline material selection and fracture control, refer to Chapter 7.

5.9.2 Injection Well

A summary of the materials commonly used for individual CO_2 injection well components is listed in Table 5-8 below [38]:

It is indicated that complementing the material as listed in the table has been the development of operational and safety practices by the industry, including:

- Use of corrosion protection of the casing strings via impressed and passive currents and chemically inhibited (oxygen, biocide, corrosion inhibitor) fluid in the casing-tubing annulus;
- Use of special procedures for handling and installing the production tubing;
- To provide gas tight seals between adjacent tubing joints and eliminate coating or liner damage,
- Use of tubing and casing leak detection methods and repair techniques, using both resin and cement squeeze technologies as well as insertion of fiberglass and steel liners,
- Formulation and implementation of criteria unique to siting wells in or near populated areas incorporating: fencing, monitoring and atmospheric dispersion monitoring elements to protect public safety.

TABLE 5-7. Summary of materials for pipeline components

Description	Material
Pipe	API 5L X65-X70 (current experience),
Flanges	API 16.5, usually 900# flange rating
Pipeline Welding	Welds are preheated and special welding rods such as 8018 C3 electrodes and a special welding procedure are used.
Pipeline Coating and Cathodic Protection (C.P.)	A fusion bonded epoxy or polyurethane coating with full cathodic protection is utilized. Meter and control facilities usually provide the status of the C.P. If alarmed, checks are made of the C.P. system.
Mainline valves	Ball type (with globe valves only on blowdown)
Relief Valves	Anderson Greenwood or equivalent is used and valves are tested with a nitrogen pilot. Valves must seal under high CO_2 pressure. Pressure relief arrangement must be installed on all exposed piping.
Valve Operators	EIM and Limitorque operators or equivalent are installed on all valves that use electric power.
Valve Seat	Usually Viton as CO_2 disintegrates Teflon. Other elastomers include. EPDM with 85 durometer hardness. Attention needs to be paid to the method of manufacturing elastomers
Valve seats	4 RMS recommended for best surface finish for valve seats. Dense chrome plating is best for hard seats. Anodized aluminum should be used if in contact with CO_2.
Valve Sealant/Lubricant	Chemflow or equivalent. Petroleum based react with CO_2 and decompose. Also sealants like Valtec 80 not recommended as it hardens in contact with CO_2.
Valve Gaskets	Flexitallic or equivalent.
Instrument Fittings	Swagelok

5.10 CODES/STANDARDS AND REGULATIONS

5.10.1 CO_2 Injection Well Technology

The following documents are a sampling of those currently in use by the industry [38]:

API Specifications and Recommended Practices for Well and Field Piping

Spec 5/CT ISO 11960	Specifications for casing and tubing
Bull 5C2	Performance properties of casing tubing and drill pipe
Spec 5L	Specification for line pipe
Spec 5LD	CRA Clad or Lined Steel Pipe
Spec 6A	Specifications for wellhead and christmas tree

TABLE 5-8. Materials for CO_2 Injection Wells, Meyer [39]

Component	Material
Upstream Metering and Piping Runs	316 SS, Fiberglass
Christmas Tree (Trim)	316 SS, Nickel, Monel
Valve Packing and Seals	Teflon, Nylon
Wellhead (Trim)	316SS, Nickel, Monel
Tubing Hanger	316 SS, Incoloy
Tubing	GRE lined carbon steel, IPC carbon steel, CRA
Tubing Joint Seals	Seal ring (GRE),Coated threads and collars (IPC)
ON/OFF Tool, Profile Nipple	Nickel plated wetted parts, 316 SS
Packers	Internally coated hardened rubber of 80–90 durometer strength (Buna-N), Nickel plated wetted parts
Cements and Cement Additives	API cements and/or acid resistant specialty cements and additives

Equipment

Spec 6D/ISO 14313	Specifications for pipeline valves
Bull 6J	Testing of oilfield elastomers
RP 10B-2 through 5	Testing well cements
Spec 10A/ISO 10426-1	Specifications for cements and materials for well cementing
TR 10TR1 C	Cement sheath evaluation
RP65 Part 1	Cementing shallow water flows in deep water wells
Spec 11D1/ISO 14310	Petroleum and natural gas industries — downhole

Equipment — Packers and Bridge Plugs

Spec 15HR	High pressure fiberglass line pipe
Spec 15LR	Low pressure fiberglass line pipe
RP 15TL4	Care and use of fiberglass tubulars
RP 90	Annular casing pressure management for offshore
Wells	

Supplementing these documents are those related to corrosion control (NACE), mechanical properties of materials (ASME), and the design experience of a vast cadre of professionals.

From a design point of view, a well is characterized as consisting of two basic elements, namely:

1. The wellbore, that is, the penetration into the earth consisting of casing, cement, and wellhead whose purpose is containment of reservoir pressure and isolation of groundwater resources from contamination, and,
2. the mechanical completion equipment consisting of valves, tubulars, and packers used to inject fluids into or to produce fluids from a formation.

5.10.2 Pipeline and Appurtenances

While these codes are referred to in Chapter 1 the following elaborates on their applicability and associated standards applications;

European Standards including BS EN series supported by other published documents (such as the British Standards PD series) provide a sound basis for the design of pipelines including CO_2 pipelines [38]; These codes as generally listed in Chapter 1 are as follows:

- BS PD 8010 2004 Parts 1 — Steel pipelines on land and 2-subsea pipelines.
- BS EN 14161: 2003 — Petroleum and natural gas industries. pipeline transportation Systems.
- Institute of Petroleum Pipeline Code IP6 (not widely used).
- DNV OS-F101 - Submarine pipeline systems (2007).
- DNV RP, Cathodic protection of pipelines.
- Pipelines safety regulations 1996 (PSR).
- DNV-RP-J202, 2010 Design and operation of CO_2 pipelines.

Codes IP6, BS EN 14161, BS PD 8010 and DNV OS-F101 are all applicable to pipelines transporting CO_2; the last three categorizing it as a non-flammable, non-toxic fluid which is gaseous at ambient temperature and pressure. IP6 also treats CO_2 as a gas.

DNV-RP-J202 document provides guidance and sets out criteria for the concept development, design, construction and operation of steel pipelines for the transportation of CO_2. It is written to be a supplement to existing pipeline standards and is applicable to both onshore and offshore pipelines.

However, it may be noted that because of largest concentration of CO_2 pipelines and years of experience, the US and Canadian codes and regulatory framework govern carbon dioxide pipelines.

As indicated previously in Chapter 1, design, construction, and operation of CO_2 pipelines is essentially under the same framework of standards for other pipelines while some additional requirements are stipulated. Regulatory framework in the USA is under US DOT Code of Federal regulation Title 49 parts:

- Part 195 "(Transportation of Hazardous Liquids By Pipelines Systems" and
- Part 192 " Transportation of Natural and Other Gas By Pipeline : Minimum Federal Safety Standards"

In Canada Canadian Standard Association CSA Z662-07 "Oil and Gas Pipeline Systems" apply.

CFR 192 is only applicable if CO_2 is transported in gaseous condition but not in supercritical state where CO_2 will have density of a liquid and viscosity of a gas

As a part of US regulation CFR title 49 Part 195, a number of established codes referred to the operation of CO_2 pipelines and facilities. These codes are referred to in Figure 5-24 below:

ASME pipeline codes, particularly B31.4 and B31.8, are relevant in the design of such pipelines, integrity maintenance and also the design of plant/facilities piping systems.

Other relevant codes are follows:

- API Pipelines Codes for Oil and Gas Services,
- SSPC, dealing with safety protection coatings in pipelines, at:
- ASNT SNT-TC-1A-2006, dealing with welders qualifications and tests of pipelines,
- ISO 8501-1—In 2007, updated version of the standard measure for the visual evaluation of rust and purity levels of non-coated steel. Also known as the "Swedish

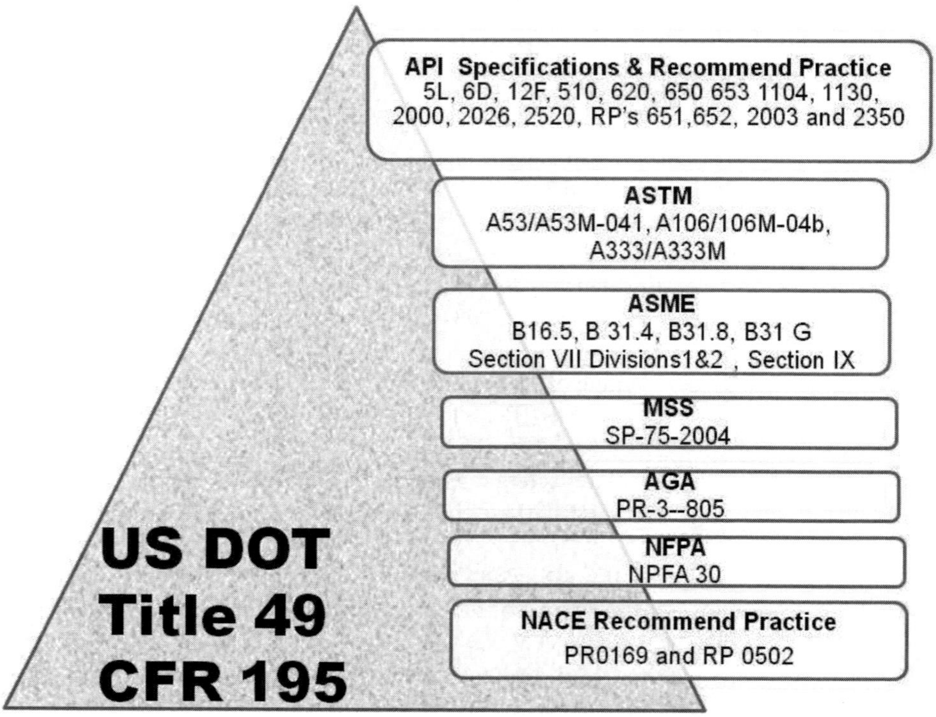

Figure 5-24. Prescribed standards and codes under US DOT CFR Title 49 part 195

steel blasting grades" (Sa). Contains high-quality color photographs for estimating the rust levels and purity levels after cleaning manually or by machine (also with blasting). Target audience: Blasting and painting companies and their clients, paint technicians, advice bureaux, and industrial coating businesses.

- ASTM A370—10 Standard Test Methods and Definitions for Mechanical Testing of Steel Products
- ASTM D4984—06 Standard Test Method for Carbon Dioxide in Natural Gas Using Length-of-Stain Detector Tubes
- AGA—American Gas Association, AGA PR-3-805, "A Modified Criterion for Evaluating the Remaining Strength of Corroded Pipe."
- AWS D1.1-2006, Structural Welding Code—Steel,
- OSHA Hazards Associated with De-Watering of Pipelines,
- U.S. Department of Transportation (DOT)—PIPELINE AND HAZARDOUS MATERIALS SAFETY ADMINSTRATION—PHMSA, which deals with safety problems in pipelines.

5.11 PIPELINE ROUTING

CO_2 pipeline best practices include but are not limited to selecting pipeline routes and facilities siting that reduce the probability of accumulation resulting from leakage, rupture, or injection well failure.

Routing of CO_2 pipelines thus must consider leakage, rupture, and blow-down scenarios such that pipelines are located in areas with the least potential impact on the environment. For example, low-lying areas (where CO_2 may collect), sensitive wildlife habitats and location of nearby residents and communities should be considered.

Route selection must therefore be considered an integral component of design, safety, environmental and construction planning. Selection of an appropriate route is a key to the success of any pipeline project. All pipeline projects encounter numerous challenges by the stakehold-

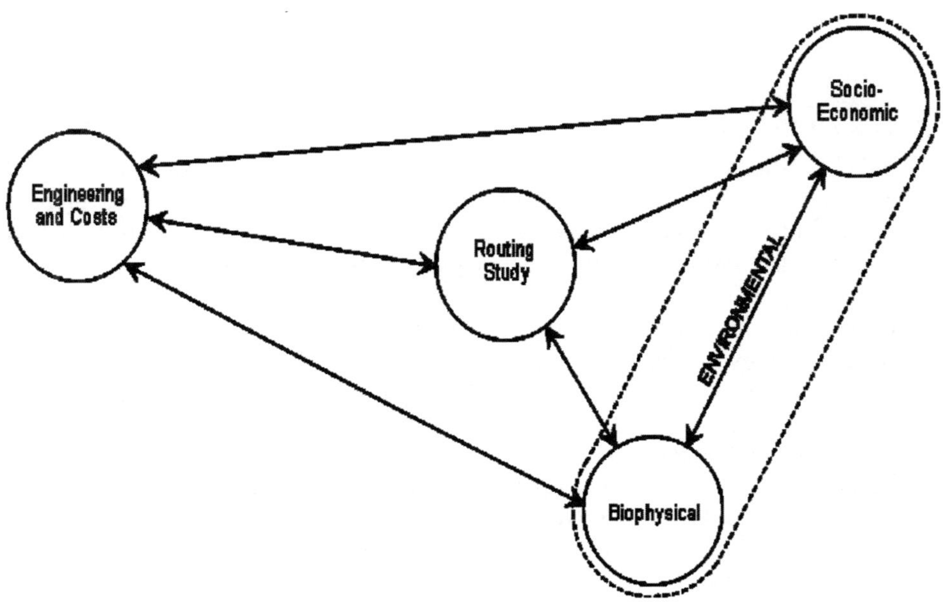

Figure 5-25. Major considerations during the CO_2 pipeline route selection process

ers—landowners, leaseholders, environmental and other government and regulatory agencies. Although the stakeholders are motivated by a variety of issues and concerns, almost all of their objections will be focused on the route, particularly if the project entails a large-diameter, high-pressure pipeline. Potential routing issues can be addressed through the following steps:

- Preliminary route evaluation;
- Detailed assessment (considering adjacent communities, proximity to critical wild life habitat, vegetation, water features, historical and archaeological sites, and physical terrain features); and
- Route design.

The route design must consider all engineering, cost, socio-economic factors and biophysical aspects of the pipeline right of way (Figure 5-25, [27]. The route design must consider existing and new corridors, geo-hazards (landslides, seismic zones, settlement and subsidence, buoyancy conditions, etc.), as well as how a CO_2 pipeline will impact the adjoining biophysical environment. Terrain profile is very important, in that design needs to avoid low spots where CO_2 can accumulate in the event of unforeseen leak/rupture.

Risk to adjacent communities and human health effects as a consequence of constructing and operation of the CO_2 pipeline must be considered when selecting the route. Proper design must assure minimal impact in these situations. The following must be addressed:

- Economic feasibility of proposed route;
- Potential impacts (environmental, socio-economic, and commercial impacts);
- Public concerns and consultation issues;
- Pipeline and facilities locations and land right acquisition, including access; and
- Safety and design aspects of pipeline operation.

For facilities siting, best practices would involve selecting a site with sufficient ventilation (e.g., open areas) to prevent accumulation. An additional measure to reduce risk could include adding chemical odorants, like those added to natural gas, which help in detecting leaks especially around more populous areas. This technique has had a positive impact on leak detection at the Weyburn facility and its supplying pipeline [40].

REFERENCES

[1] NETL (National Energy Technology Laboratory), 2008 "Oil & Natural Gas Technology-Report on Technology on In-Situ Gas Generation to Recover Residual Oil Reserves," prepared by New Mexico Institute of Mining and Technology, July 20 *DOE Award No.: DE-FC26-05NT15478.*

[2] Rejoy, A. and Jaffret, C., 1997 " Natural gas, Production and Transport" Published by Institut Francais Du Petrol.

[3] GPSA (Gas Processor Supply Association), 1994 "Engineering Data Book."

[4] IPCC (Intergovernmental Panel on Climate Change), 2005 "Carbon Dioxide Capture and Storage" Special Report on Prepared by Working Group III of the Intergovernmental Panel on Climate Change [Metz, B., O. Davidson, H. C. de Coninck, M. Loos, and L. A. Meyer (eds.)]. Cambridge University Press Camb, UK and NY, USA.

[5] Thomson, G., 2009 *"Burying Carbon Dioxide in Underground Saline Aquifers: Political Folly or Climate Change Fix?"* A Report prepared for the Program on Water Issues, University of Toronto, Munk Centre for International Studies, Sept. http://beta.images.theglobeandmail.com/archive/00242/Munk_Centre_Paper_242701a.pdf.

[6] Bachu, S., 2003 "Screening and ranking of sedimentary basins for sequestration of CO_2 in geological media in response to climate change," *Environmental Geology,* 44, 277–289.

[7] Ruiz, C., 2008 "Almacenamiento geológico de CO2: Criterios de selección de emplazamientos," 9th CONAMA Cogress (Congreso Nacional de Medio Ambiente), Dec 1–5, MADRID. Palacio Municipal de Congresos del Campo de Las Naciones.
[8] McCoy, S. T., 2008 "The Economics of CO_2 Transport by Pipeline and Storage in Saline Aquifers and Oil Reservoirs" Ph.D Thesis, Carnegie Mellon University Pittsburgh, PA, April.
[9] Hendriks, C. F. and Blok, K., 1993 "Underground Storage of Carbon Dioxide" Energy Conversion and Management, 34 (9011) pp. 949–957.
[10] Hendriks, C. F., 1994 *"Carbon dioxide removal from coal fired power plant"* 1 ed.: Kluwer Academic Publishers.
[11] McCoy, S. T. and Rubin, E. S., 2008 "An engineering-economic model of pipeline transport of CO_2 with application to carbon capture and storage," International Journal of Greenhouse Gas Control, Vol 2, pp. 219–229.
[12] Craft, B. and Hawkins, M., 1959 "Applied Petroleum Reservoir Engineering." 1st ed. Prentice-Hall Chemical Engineering Series. Englewood Cliffs, NJ: Prentice-Hall. 437.
[13] van der Meer, B., 2005 "Carbon Dioxide Storage in Natural Gas Reservoirs," Oil & Gas Science and Technology – Rev. IFP, Vol. 60 (2005), No. 3, pp. 527–536.
[14] Lemmon, E. W., Huber, M. L. and McLinden, M. O. 2010 "NIST Standard Reference Database 23: Reference Fluid Thermodynamic and Transport Properties-REFPROP, Version 9.0, National Institute of Standards and Technology, Standard Reference Data Program, Gaithersburg. http://www.boulder.nist.gov/div838/theory/refprop/Frequently_asked_questions.htm#REFPROPReference.
[15] Span, R. and Wagner, W., 1996 "A New Equation of State for Carbon Dioxide Covering the Fluid Region from Triple-Point Temperature to 1100 K at Pressures up to 800 MPa," J. Phys. Chem. Ref. Data, Vol 25,#6, pp.1509–1597.
[16] Verma, S., Oakes, C. S. and Ramakrishnan, T. S., 2010 "Methods and Systems for Optimizing Carbon Dioxide Sequestration Operation," US patent application 2010/0287123 A1, Nov 11, http://www.freepatentsonline.com/20100287123.pdf.
[17] Brubacher, L., 2006 "Obtaining reliable data online—Phase diagram for carbon dioxide," Chem 13 News, November, http://www.chem13news.uwaterloo.ca/issues/342/342_nov_2006_pages_13.pdf.
[18] Oldenburg, C. M., 2003 "Carbon Dioxide as Cushion Gas for Natural Gas Storage" American Chemical Society, Jr of *Energy & Fuels Vol. 17, No 1,* pp. 240–246, http://www.lbl.gov/tt/publications/2570pub.pdf.
[19] Havens, K., 2008 "CO_2 Transportation," Kinder Morgan CO2 Company, Indiana CCTR Panel Meeting, Bloomington IN, June 5.
[20] Forbes, S., Verma, P., Curry, T. E., Friedmann, S. J. and Wade, S. M., 2008 " Guidelines for Carbon Dioxide Capture, Transport, and Storage" World Resources Institute (WRI). Washington, DC, October, http://www.wri.org/publication/ccs-guidelines.
[21] Oosterkamp, A., 2008 " CO_2 Pipeline Transmission-A Broad State Of The Art," Polytech, November 07. http://www.docstoc.com/docs/43753551/Offshore-vindkraft-og-Polytec.
[22] CO_2 Capture Project 2004 "Regulatory Treatment of CO_2 Impurities for CCS," Policy Position Paper. 24., http://energy-resource.net/ccp/downloads/report/CCP2_Principles_CO2_Impurities.pdf.
[23] WRI (World Resources Institute), 2008 "CCS Guidelines: Guidelines for Carbon Dioxide Capture, Transport, and Storage." Washington, DC.
[24] Dynamis, 2007 "Towards Hydrogen and Electricity Production with Carbon Dioxide Capture and Storage" Report # D 3.1.3 - CO2 Quality Recommendations Project, June., http://www.dynamis-hypogen.com/publications/Deliverables/Year_2/D3-1-3%20DYNAMIS%20CO2%20quality%20recommendations.pdf.
[25] Oosterkamp, A. 2008. "CO_2 Pipeline Transmission -a broad state of the art-." R&D Foundation Polytec, 2008 http://www.climit.no/frontend/files/CONTENT/CLIMIT/porsgrunn_polytec.pdf.
[26] Wildbolz, C., 2007 "Life Cycle Assessment of Selected Technologies for CO_2 Transport and Sequestration" Diploma thesis, Swiss Federal Institute of Technology Zurich, http://www.doka.ch/CCSDiplomaWildbolz07.pdf.
[27] Mohitpour, M., Golshan, H. and Murray, A., 2007 "Pipeline Design and Construction a practical Approach," ASME Press NY, USA.

[28] Johnson, N., Yang, C. and Odgen, J., 2006 "Build-Out Scenarios for Implementing a Regional Hydrogen Infrastructure" Proc. National Hydrogen Association (NHA) Annual Conference. Long Beach, CA, USA: http://hydrogen.its.ucdavis.edu/publications/pubpres/2006pub/johnsonnha/.

[29] McCollum, D. L. and Odgen, J. M., 2006 "Techno-Economic Models for Carbon Dioxide Compression, Transport, and Storage & Correlations for Estimating Carbon Dioxide Density and Viscosity" Institute of Transportation Studies, University of California, Davis. http://hydrogen.its.ucdavis.edu/publications/pubpres/2006pub/mccollum06/.

[30] BMU (Bundesministerium für Umelt, Naturschutz und Reaktorsicherheit) 2007 "Economic comparison of renewable energy technologies (RE) with Carbon Capture and Storage. Report by Fischedick, M., Esken, A., Pastowski, A., Schüwer, D., Supersberger, N., Nitsch, J., Viebahn, P., Bandi A., Zuberbühler, U., Edenhofer, O." Wuppertal Institut für Klima, Umwelt, Energie GmbH; DLR – Institut für Technische Thermodynamik; Zentrum für Sonnenenergie- und Wasserstoff-Forschung; Potsdam-Institut für Klimafolgenforschung. Wuppertal, Stuttgart, Potsdam. http://www.erneuerbareenergien.de/files/pdfs/allgemein/application/pdf/reccs_zusammenfassung.pdf

[31] IEA (International Energy Agency) 2010 "CO_2 pipeline Infrastructure: An analysis of global challenges and opportunities" Prepared by Element Energy Limited, Cambridge, UK, April 27. http://www.ccsassociation.org.uk/docs/2010/IEA%20Pipeline%20final%20report%20270410.pdf.

[32] Fradet, A., Saysset, S., Odru, P., Broutin, P., Ruer, J. and Bonnissel, M., 2007 "Technical and Economic Assessment of CO_2 Transportation for CCS Purposes," Journal of Pipeline Engineering, Vol. 6, #3, pp. 173–180.

[33] Anderson, S., 2008 "CO_2 Pipelines in the United States" Gassnova CO_2 Workshop Shell Global Solutions Presentation, Nov.

[34] Mazzotti, M., Boulouchos, K., Burg, J.-P., Dones, R., Hirschber, S., Jansohn, P. and Siegrist, M., 2007 "Carbon dioxide management in Swiss power generation (CARMA)" Presented at CCEM General Assembly, PSI on June 12. Mohitpour, M., Botros KK and Van Hardevels, T., 2008 "Pipeline Pumping and Compression Systems.

[35] Robertson, E. P., 2007 "Analysis of CO_2 Separation from Flue Gas, Pipeline Transportation, and Sequestration in Coal" Idaho National Laboratory (INL), Report # NL/EXT-08-13816, Prepared for the U.S. DOE, Fossil Energy, under Contract DE-AC07-05ID14517, http://www.inl.gov/technical publications/Documents/4010767.pdf.

[36] Gray, M. 2009 "Carbon Capture & Storage with Alstom's Chilled Ammonia Process at AEP's Mountaineer Plant" Alstom Presentation, July 27. http://www.rwe.com/web/cms/mediablob/de/289802/data/285934/1/rwe/innovationen/stromerzeugung/innovationszentrum-kohle/Mark-Gray-AEP-CCS-with-Alstoms-Chilled-Ammonia-Process-at-AEPs-Mountaineer-Plant-.pdf.

[37] Bovon, P. L. and Habel, R., 2007 "CO_2 Compression Challenges" presented on May 15, 2007 at the ASME Turbo Expo, Montreal, CO_2 Compression Panel.

[38] HSE (Heath and Safety Executives) 2008 "Pipeline design codes and standards for use in UK CO_2 Storage and Sequestration projects," Feb 19, Gas & pipeline Unit, http://www.hse.gov.uk/pipelines/resources/designcodes.htm. IEA International Energy Agency (IEA) 2004, "Prospects for CO_2 Capture and Storage." Paris Cedex, France: OECD/IEA.

[39] Meyer, J. P. 2007 "Summary of Carbon Dioxide Enhanced Oil Recovery (CO_2EOR) Injection Well Technology" Report Prepared for API, http://www.gwpc.org/e-library/documents/co2/API%20CO2%20Report.pdf.

[40] Gale, J. and Davison, J., 2002 "Transmission of CO_2 – Safety and Economic Considerations," IEA Greenhouse Gas R&D Programme, presented at the GHGT-6 Conference, Kyoto, Japan, October.

Chapter 6

CHARACTERISTICS OF DECOMPRESSION WAVE SPEED IN CO_2 CONTAINING IMPURITIES

6.1 INTRODUCTION

In most jurisdictions, one of the key elements in the design and operation of transportation pipelines carrying flammable or toxic gases or hazardous liquids is fracture control. CO_2 is explicitly categorized in many national standards and regulations as a fluid that requires measures to be taken to control fracture initiation and propagation. Materials for use in CO_2 pipelines can be specified to prevent fracture initiation and brittle fracture propagation in the same way as for gas pipelines, as is discussed in Chapter 7. It is the control of ductile fracture propagation that presents the greatest challenge. Because compressible fluids like CO_2 do not expand to atmospheric pressure quasi-instantaneously, as conventional liquids do, a high pressure can be maintained in the plane of a propagating ductile fracture. In order to specify materials that can prevent long ductile fracture propagation, it is thus necessary to understand the decompression behavior of the expanding fluid, so that the local pressure that is driving the fracture can be calculated. Supercritical CO_2 is known to be particularly challenging from the point of view of fracture control, since its decompression characteristics are such that a high pressure can be maintained at the crack front for a long time.

The determination of the decompression behavior requires accurate knowledge of the thermodynamic characteristics of the fluid (particularly the speed of sound and the phase boundary). The thermodynamic characteristics of pure CO_2 have been well understood for many years, but systematic study of impure CO_2 began more recently, as a result of growing interest in transportation from anthropogenic sources for the purpose of carbon capture and storage (CCS). Quantitative and validated understanding is still far from complete. However, the methods that need to be applied to determine decompression wave speed are well understood. This chapter will describe the background to the analysis methods and illustrate their use by examining the decompression behavior predicted by two different, widely used equations of state (EOS) for binary mixtures of CO_2 with a range of impurity species that might be present in carbon capture streams resulting from the use of different capture technologies. For each of the cases examined, the important aspects of the predicted decompression wave characteristics are reviewed, in particular initial speed of sound and saturation pressure (the pressure corresponding to phase boundary crossing during decompression). These are particularly significant in the determination of required material characteristics for fracture arrest.

6.2 BACKGROUND

Carbon capture and storage (CCS) is one of the technologies that have been proposed to reduce emissions of carbon dioxide (CO_2) to the atmosphere [1,2]. As an example, new coal-fired power stations are now being designed to either include CCS equipment or at least to be "CCS-ready." CCS will require the transportation of the CO_2 from the "capture" locations to the "storage" locations. For large-scale projects, pipelines appear to be the only viable transportation method, as has been discussed in Chapter 5. While CO_2 pipeline technology for enhance oil recovery purposes has been well developed and understood since the early 1970s, the infrastructure that will be required for large-scale CCS will require an extensive network of new pipelines designed specifically for the transportation of impure CO_2 streams. In addition, to determining the feasibility of converting existing gas or oil pipelines to CO_2 service, their characteristics will need to be re-assessed against the requirements imposed by the change in service.

As has been mentioned, fracture control is an important requirement for CO_2 pipelines in all jurisdictions. This is because CO_2 is considered a hazardous fluid, as a result of its dispersion behavior and asphyxiating characteristics [3] (see Chapter 12). It is thus important to minimize the probability of release incidents and to control the size and extent of any releases that do occur. Conventional approaches can be applied to the prevention of fracture initiation and brittle fracture propagation. However, supercritical CO_2 poses significant challenges from the point of view of fracture propagation control since its thermodynamic characteristics are such that they can result in a high fracture driving force being sustained for a long time [4–5]. On the other hand, positive limitation of fracture length is required in a number of jurisdictions. As an example, the Australian Pipeline Standard AS2885.1 requires any rupture to be arrested within the initiating pipe in high consequence areas and within five pipes for R1 and R2 locations. As was mentioned previously, the determination of the required toughness for the arrest of ductile fracture requires a knowledge of the decompression behavior of the contained fluid, which in turn requires accurate knowledge of its thermodynamic characteristics (particularly the speed of sound and the phase boundary) [6–12].

Worldwide, a significant number of CO_2 pipelines have been in service since the early 1970s. However, the majority of these transport relatively pure CO_2 from geological sources for use in enhanced oil recovery (EOR) projects. They operate in relatively remote, rural areas, and in some cases, the control of ductile fracture propagation is achieved by the use of crack arrestors, rather than intrinsic, pipe body arrest. Clearly, this approach cannot practically limit fracture length to a short distance; typical distances between crack arrestors are 300 to 500 m. Most pipelines intended for CCS applications will originate in, or on the outskirts of, populated areas, and will typically terminate in more remote regions where potential sequestration sites or EOR projects are located. As a result, the potential consequences of major releases close to the receipt end of a CCS pipeline can be much more significant than for a typical existing EOR pipeline, particularly since some carbon capture streams can contain highly toxic species such as H_2S.

In terms of the determination of decompression behavior for impure CO_2, models exist that, in principle, will allow the effects of some impurities to be evaluated. However, they have generally not been developed or validated for the kinds of compositions and operating conditions that will be of interest, and as a result, their predictions need to be treated with caution, as was pointed out by Seevam et al. [6].

As a result of these considerations, it can be concluded that, in the present state of knowledge, fracture propagation can only be reliably contained within a short distance, as may be required in high-consequence areas, by reducing the hoop stress to a low level such that propagating ductile fracture cannot occur. AS 2885.1 suggests 40% of SMYS as a suitable value. For high-volume pipelines designed for the supercritical regime, the required wall thickness may be completely impractical. As a result, it is viewed as vital to gather experimental data on the rapid decompression behavior of gas mixtures representative of different types of carbon capture technology, from initial conditions that might be expected in a typical, supercritical transportation pipeline. Such data will be directly applicable in the design of certain pipeline systems and will provide a basis for any revision of thermodynamic models that is found to be needed. Shock tubes are very valuable and cost-effective tools in obtaining this data and have been proven to generate very credible decompression wave speed data in the past [13–21]. Section 6.13 provides a description of one such facility, while Section 6.14 provides examples of how the shock tube has been utilized to generate credible experimental data for cases of rich natural gas mixtures (as examples), for which the only applicable experimental data had previously come from very expensive full-scale burst tests.

Cosham et al. [4,5] studied the effect of impurities on fracture control in CO_2 pipelines, and identified the following issues, based on existing thermodynamic models:

- Impurities with a lower critical temperature than CO_2, such as nitrogen, will increase the toughness required to arrest a ductile fracture. For example, they showed that N_2 has an adverse effect on the decompression curve. The addition of even relatively small amounts of N_2 causes a significant increase in the plateau pressure of the decompression wave speed curve.
- The impact of impurities on the speed of sound in CO_2 is not well quantified, due to the lack of experimental data for CO_2+impurity mixtures in the dense or supercritical state. There can be large deviations and uncertainties in predictions of the speed of sound by various EOS. The remainder of this chapter considers some aspects of this knowledge gap in detail, in order to indicate the work that is still needed to facilitate the safe and economical design and construction of long, high-pressure pipelines transporting CO_2 with impurities.
- The industry-standard software that is used to determine decompression behavior in hydrocarbon-based gases [22] cannot be applied to mixtures containing many of the species of interest for impure CO_2 pipelines, and is prone to instabilities when applied to CO_2-rich gases.

Seevam et al. [6] have studied the effect of a range of impurities, including representative carbon capture streams, on the phase envelope of CO_2. They indicate a range of effects, from virtually indiscernible to a dramatic increase in bubble-point pressures that would have serious implications for fracture control design. They chose to use the well-established Peng–Robinson (PR) EOS for this work, but warned that, in the absence of experimental validation, the results should be treated with great caution. Chemical compositions of several different capture technologies and naturally occurring streams were reviewed, and the main impurity elements were identified as N_2, O_2, CH_4, and higher alkanes, H_2, CO, Ar, and H_2S. Of course, not all of these would be present in any one stream received by a pipeline. As has been pointed out, the methods of determining decompression speed are theoretically well understood, but the performance of, and consistency between, different EOS is not. As an illustrative step, it will be instructive to compare the predictions of two widely used EOS, namely, the Peng–Robinson (PR) and GERG EOS, for pure CO_2 and for binary mixtures of CO_2 with each of the species mentioned above. The first of these EOS is widely used in the gas pipeline industry, while the second has been shown to be one of the most accurate in predicting the speed of sound and decompression behavior of a range of hydrocarbon-based mixtures [20]. Such an analysis will indicate where discrepancies exist, and, in many cases, their thermodynamic origins, as well as pointing out where experimental validation is most urgently required.

6.3 DESCRIPTION OF CASES INVOLVING CO_2 CONTAINING IMPURITIES

A total of 16 cases involving pure CO_2 and binary CO_2+impurity mixtures will be analyzed to show the effects of impurities on the decompression wave speed. The CO_2+impurity mixture compositions considered, along with the initial pressures and temperatures, are shown in Table 6-1. Note that the initial pressure for all of the cases was selected at 15 MPa-a, which represents the ANSI 900 pressure rating for pipeline equipment (2220 psig). Two temperature values were selected for each binary; 35°C, which represents typical discharge temperature for a station along the pipeline, and a lower temperature (5°C) to replicate conditions that might be encountered in pipelines for delivery to subsea injection facilities. One value of impurity mole% was selected for all binaries (3%), except for the CO_2+N_2 binary (5%)—see Table 6-1. Predictions of the decompression wave speed using the GERG-2004 and GERG-2008 equations of state (EOS) [23], as implemented in Aspen [24] and REFPROP [25], and the PR EOS [26], also as implemented in Aspen and REFPROP, are compared. It should be noted that, in some cases, one of the software packages used failed to converge on the isentropic decompression of the gas mixtures. Nevertheless, it was possible to get all of the EOS predictions of the decompression wave speed for all of the 16 cases using one software or another. If a software application did not converge in one of the cases, it should not be construed as a reflection of the performance of the EOS itself, but rather a reflection of how the EOS was implemented in this particular software for the specific mixture composition and for the specific initial pressure and temperature.

6.4 DECOMPRESSION WAVE SPEED IN PURE CO_2

Two pure CO_2 cases from different initial temperatures will be analyzed, as shown in Table 6-1. Figure 6-1 shows comparison between predictions of the decompression wave speed by GERG-2008 and PR EOS for Case 1. The initial pressure for this Case is 15 MPa-a and the temperature is 35°C. It is shown that predictions by the two EOS vary in two aspects. The first concerns the part of the decompression wave speed above the so-called plateau, which represents a phase transition as the isentrope along the decompression wave crosses the

TABLE 6-1. Compositions and initial conditions—pure CO_2 and binaries of CO_2+ impurities

	Component (mole %)								Pi	Ti
	N_2	O_2	CH_4	H_2	CO	Ar	H_2S	CO_2	MPa–a	°C
Case 1	–	–	–	–	–	–	–	100	15	35
Case 2	–	–	–	–	–	–	–	100	15	5
Case 3	5	–	–	–	–	–	–	95	15	35
Case 4	5	–	–	–	–	–	–	95	15	5
Case 5	–	3	–	–	–	–	–	97	15	35
Case 6	–	3	–	–	–	–	–	97	15	5
Case 7	–	–	3	–	–	–	–	97	15	35
Case 8	–	–	3	–	–	–	–	97	15	5
Case 9	–	–	–	3	–	–	–	97	15	35
Case 10	–	–	–	3	–	–	–	97	15	5
Case 11	–	–	–	–	3	–	–	97	15	35
Case 12	–	–	–	–	3	–	–	97	15	5
Case 13	–	–	–	–	–	3	–	97	15	35
Case 14	–	–	–	–	–	3	–	97	15	5
Case 15	–	–	–	–	–	–	3	97	15	35
Case 16	–	–	–	–	–	–	3	97	15	5

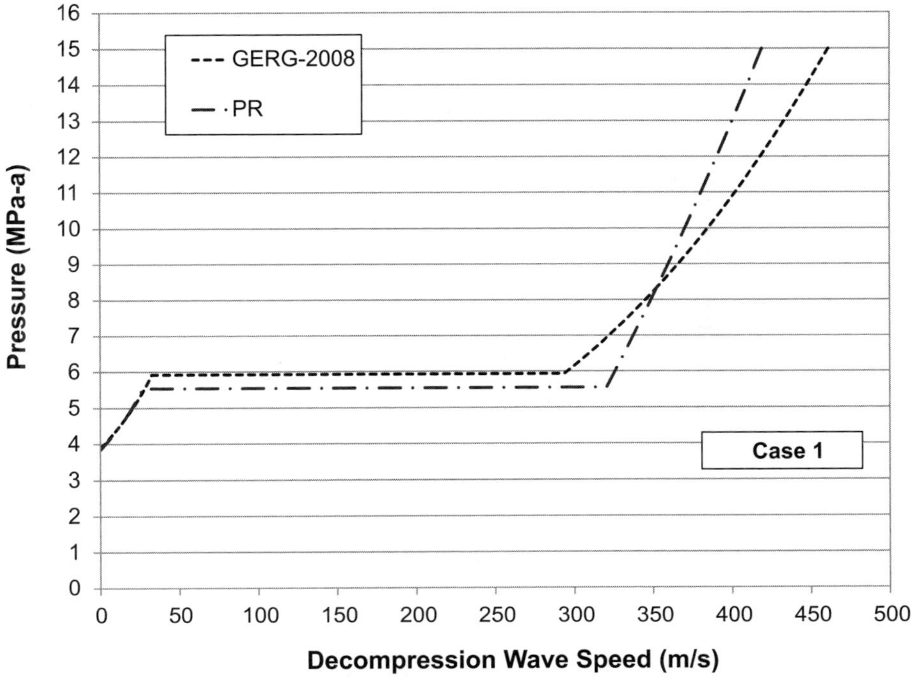

Figure 6-1. Prediction of the decompression wave speed based on GERG-2008 and PR EOS (Case 1)

equilibrium phase envelope (saturation line) of CO_2. At the initial pressure and temperature, prediction of the speed of sound by PR is much lower than that of GERG-2008, and this is reflected in the initial decompression wave speed. As the isentrope expands to a lower pressure (still above the saturation pressure, i.e., the 'plateau'), the PR prediction begins to show a higher estimation of the decompression wave speed than GERG-2008. In order to understand the main cause for these deviations, the isentropes in terms of pressure–density are plotted in Figure 6-2, based on both GERG-2008 and PR. From pressure–density profiles along the isentrope, it is possible to calculate the corresponding speed of sound (C), since:

$$C = \sqrt{\left(\frac{\partial P}{\partial \rho}\right)_s} \qquad (6-1)$$

Figure 6-3 shows the calculated speed of sound along the isentropes based on both GERG and PR for this test. It is clear that the difference in the decompression wave speed prediction by PR and GERG EOS is mostly attributable to the prediction of speed of sound. The contribution of the outflow velocity (U) is negligible in this part of the decompression wave speed profile, as shown in Figure 6-4. Recall that the decompression wave speed (W) is defined as:

$$W = C - U \qquad (6-2)$$

where U is the outflow velocity determined from:

$$U = -\int_{Pi}^{P} \frac{C\,d\rho}{\rho} = -\int_{Pi}^{P} \frac{dP}{C\rho} \qquad (6-3)$$

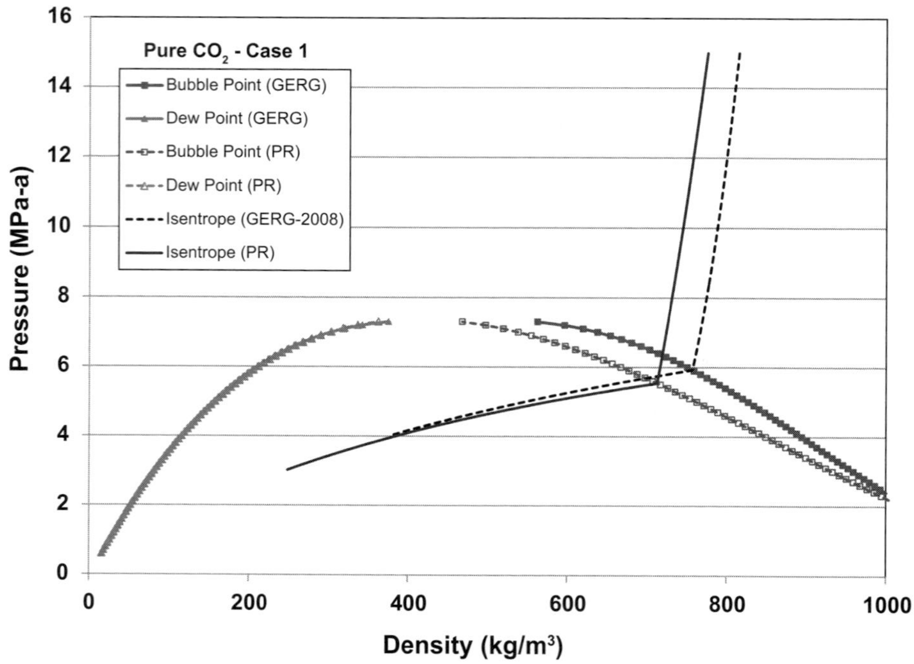

Figure 6-2. Pressure–density isentropes based on GERG-2008 and PR EOS (Case 1)

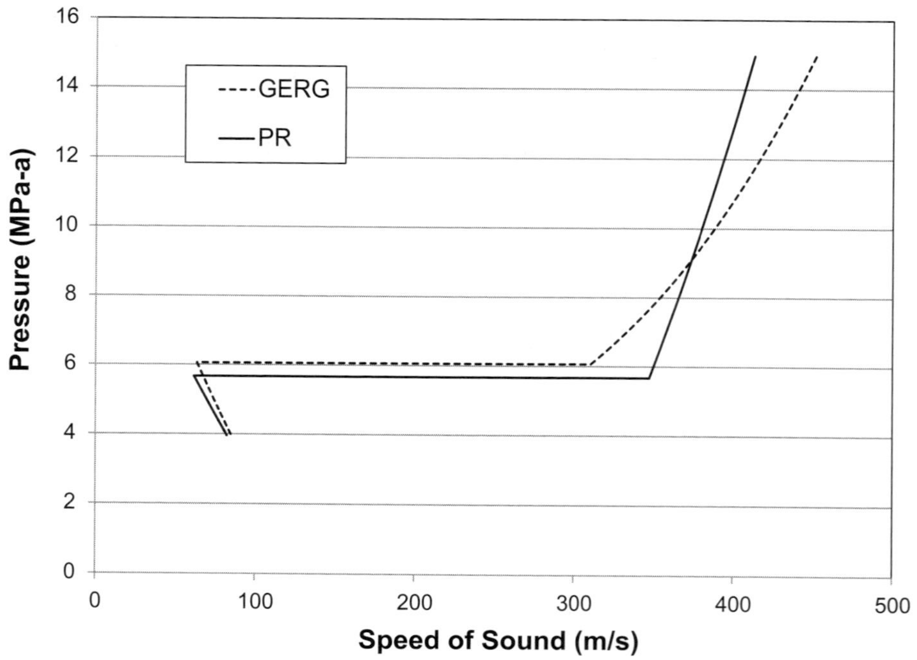

Figure 6-3. Speed of sound along the isentropes based on GERG-2008 and PR EOS (Case 1)

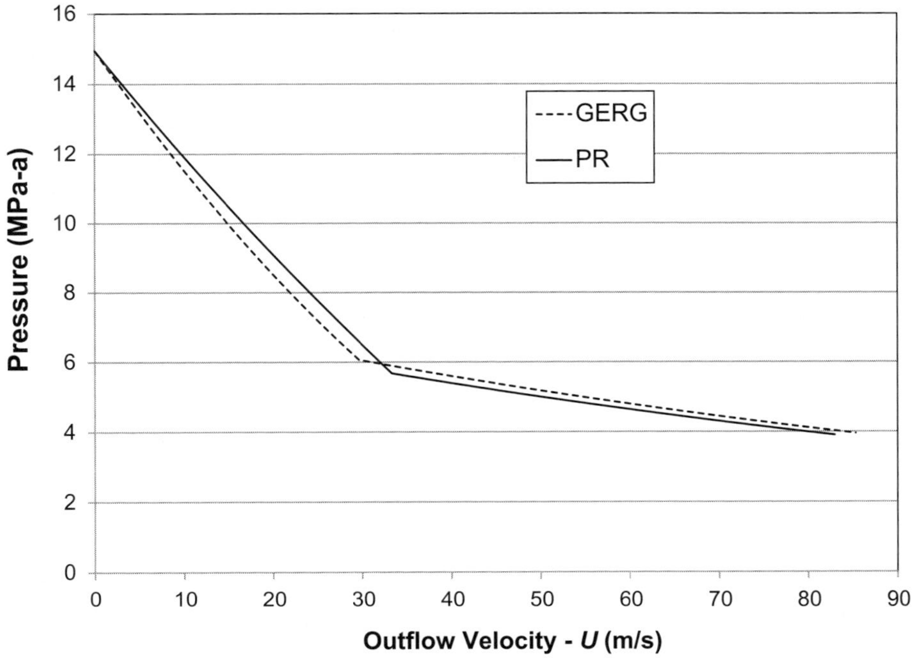

Figure 6-4. Outflow velocity along the isentropes based on GERG-2008 and PR EOS (Case 1)

where:

W - decompression wave speed
U - outflow velocity
P - decompressed pressure
Pi - initial pressure
ρ - local density behind the decompression wave

The second aspect of the difference between the predictions of W is the pressure corresponding to the plateau, which is the point where the isentrope crosses the phase envelope of pure CO_2 (Figure 6-5). It is shown that the phase crossing is at lower pressure based on the PR EOS than based on GERG-2008. This is also manifested in the pressure–temperature isentropes shown in Figure 6-5. As a result, the pressure corresponding to the plateau on the decompression wave speed profile is lower according to PR than GERG-2008. This will become an important aspect for determining the required fracture toughness to arrest ductile fracture in the design of a specific pipeline, as will be discussed in more detail in Section 6.12.

As will be shown later, and discussed in detail in Chapter 7, the pressure at isentrope phase crossing, which determines the plateau pressure of the decompression wave speed curve, is of significance for accurate determination of the required material toughness for given pipeline design parameters. In this context, the initial deviation in the speed of sound as predicted by the two EOS above may seem to be unimportant. However, it is the initial speed of sound and the decompression isentropic path that determine the point at which the decompression isentrope intersects the phase boundary, and they are thus of primary importance.

As Cosham and Eiber have pointed out [4,5], it is possible to determine the saturation pressure, for a given mixture and initial conditions, simply by calculating the initial entropy and comparing it with the phase boundary expressed in terms of pressure and entropy. This is exemplified in the results shown in Figures 6-6 and 6-7. Here, the isentropes, as predicted

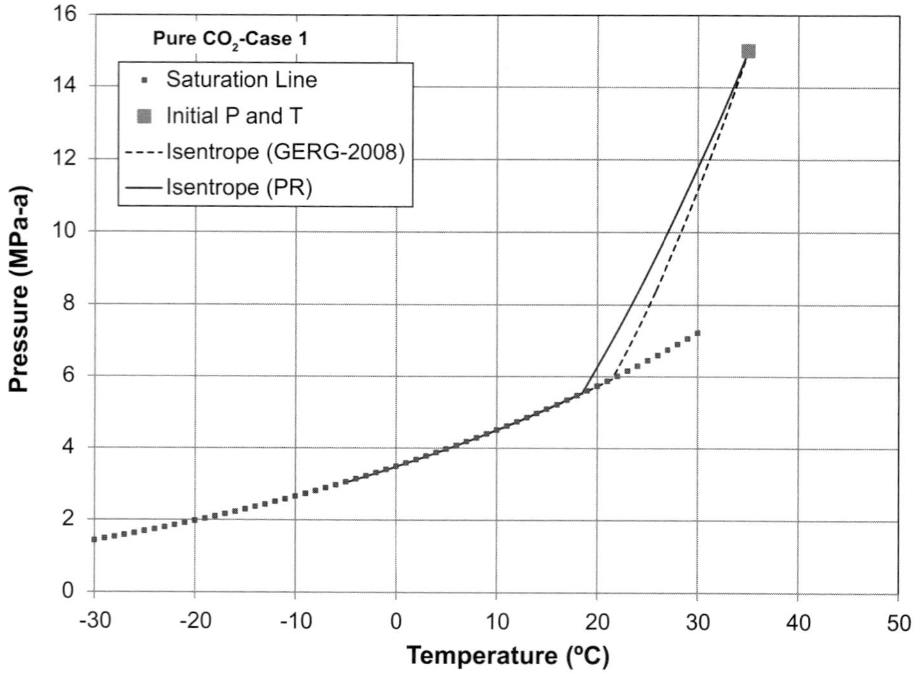

Figure 6-5. Pressure–temperature isentropes based on GERG-2008 and PR EOS for pure CO_2 (Case 1)

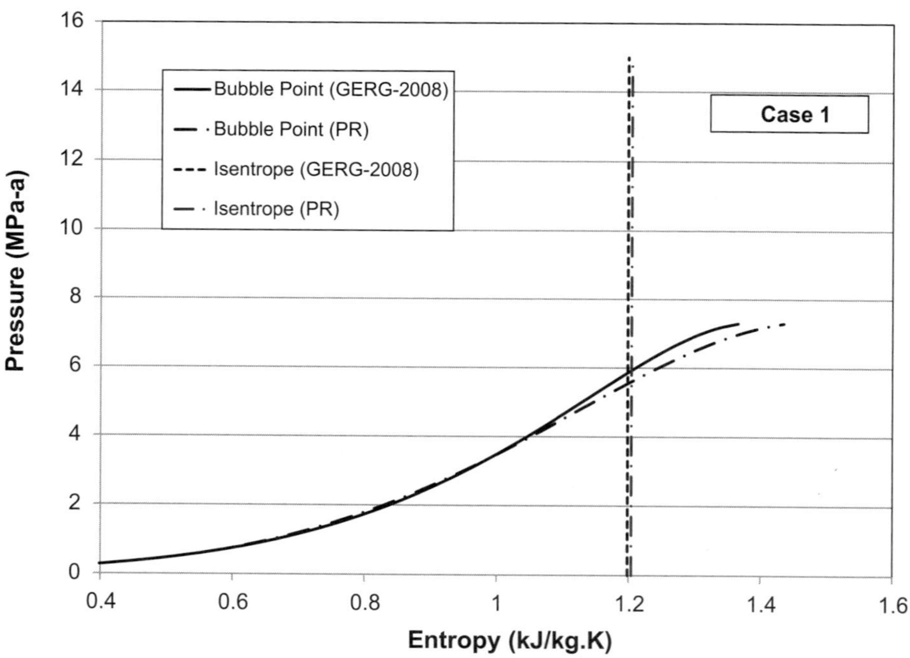

Figure 6-6. Isentrope intersection with phase boundaries based on GERG-2008 and PR EOS (Case 1)

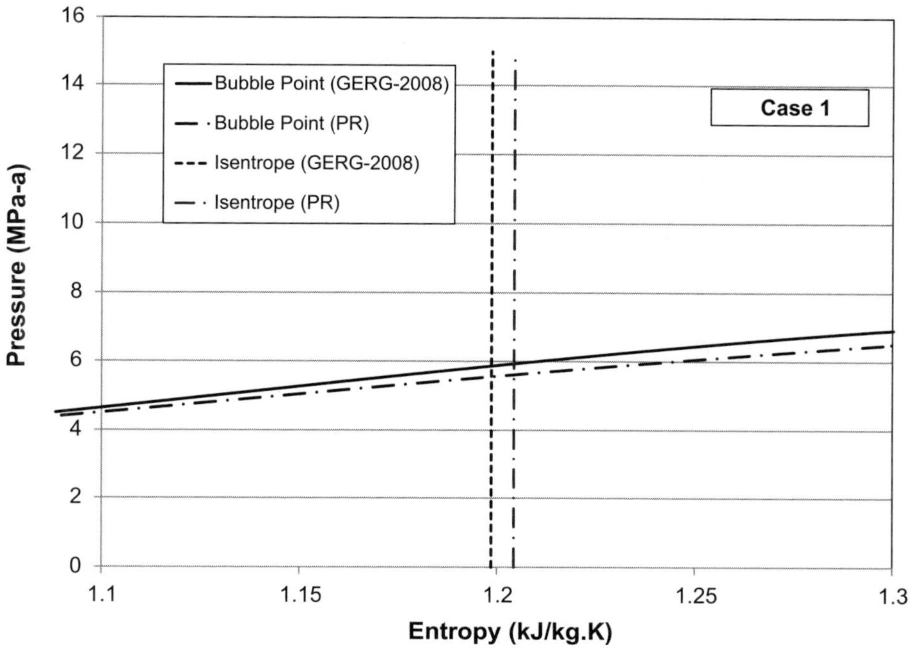

Figure 6-7. Isentrope intersection with phase boundaries based on GERG-2008 and PR EOS (Case 1) detail

by GERG-2008 and PR EOS, are shown on a pressure–entropy diagram. Note the corresponding intersections with the phase boundary of the two paths with the respective phase boundary. Therefore, an alternative and simpler procedure to determine the plateau pressure in the decompression wave speed is by matching the entropy at the initial conditions to the entropy at the phase boundary, and hence determining the saturation pressure. This, in turn, corresponds to the plateau and the required arrest pressure for the purpose of determining the Charpy energy for arrest (as will be seen later). Unless the software used to implement the EOS allows for the direct computation of phase diagrams in terms of entropy, however, this process may be more laborious than the calculation of the full decompression curve.

Case 2 has a lower temperature (5°C), but the same initial pressure (15 MPa-a). The results of the decompression wave speed (*W*) vs. pressure are shown in Figure 6-8 based on both GERG-2008 and PR EOS. Note again the significant deviation between the two EOS on the upper part of the decompression wave speed, which was caused by the significantly lower prediction of the speed of sound by PR EOS at the initial pressure and temperature. Because the initial temperature is lower than Case 1, the pressure difference between initial pressure and the saturation pressure at the isentrope intersection is higher. This leads to the plateau being developed at the tail end of the decompression wave speed profile and extends almost to zero speed at this pressure. Once again, the plateau based on PR is at a lower pressure than that based on GERG-2008 EOS. Figure 6-9 shows the pressure–density isentropes, which show that the slope based on PR EOS is lower than that which is based on GERG-2008. Figure 6-10 shows the same isentropes on a pressure–temperature diagram.

Figure 6-11 shows comparison of the isentropes predicted by both GERG-2008 and PR EOS, for Cases 1 and 2, on a pressure–temperature diagram. It is important to note the location of the initial condition in relation to the saturation curve of pure CO_2. One can easily determine what the corresponding plateau pressure will be for each case as predicted by both EOS and also the expected relative portion of the decompression wave speed above and below the plateau pressure.

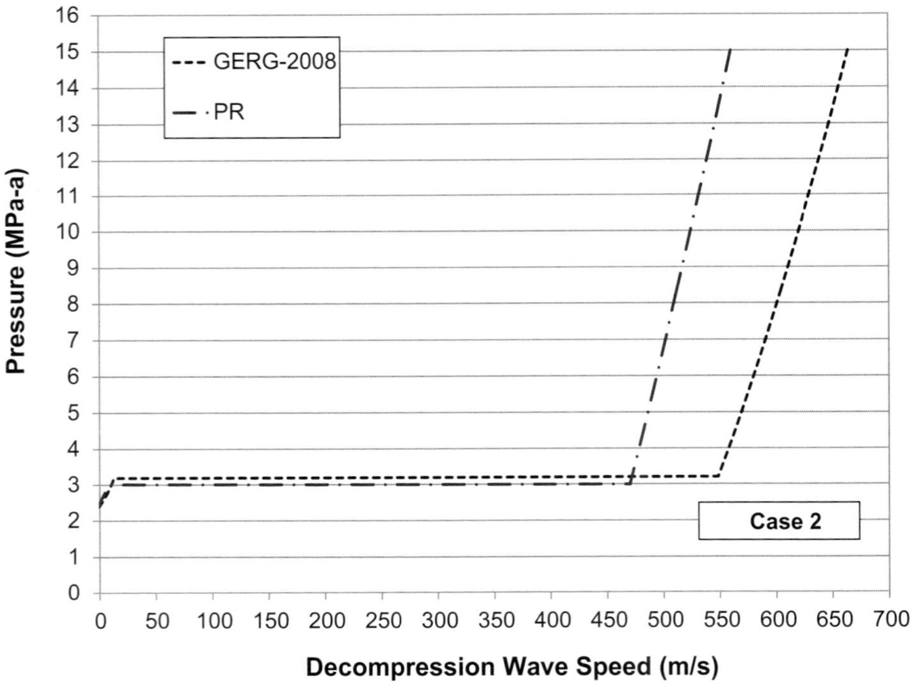

Figure 6-8. Decompression wave speed predictions based on GERG-2008 and PR EOS (Case 2)

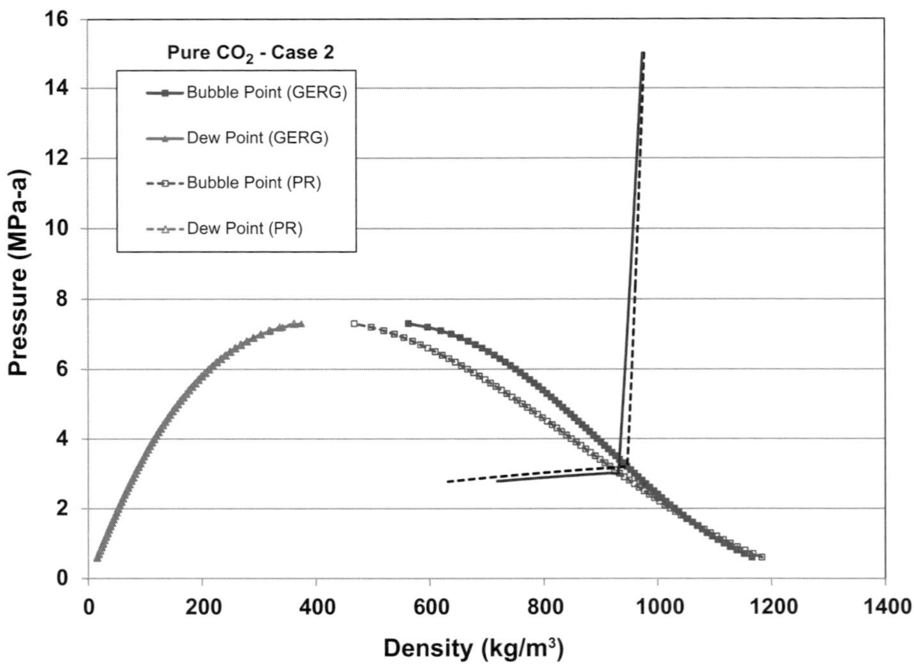

Figure 6-9. Pressure–density isentropes based on GERG-2008 and PR EOS of pure CO_2 (Case 2)

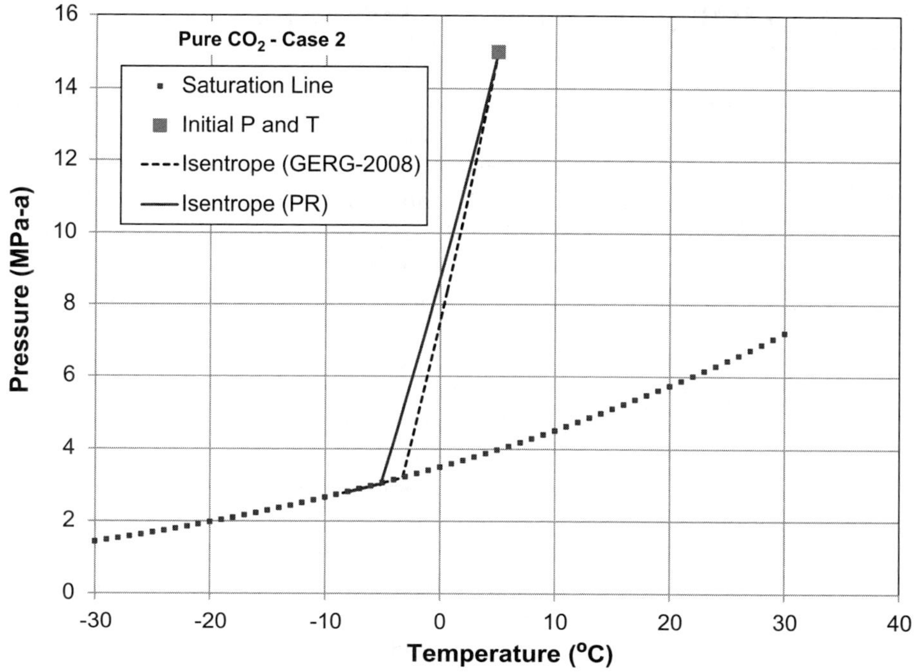

Figure 6-10. Pressure–temperature isentropes based on GERG-2008 and PR EOS of pure CO_2 (Case 2)

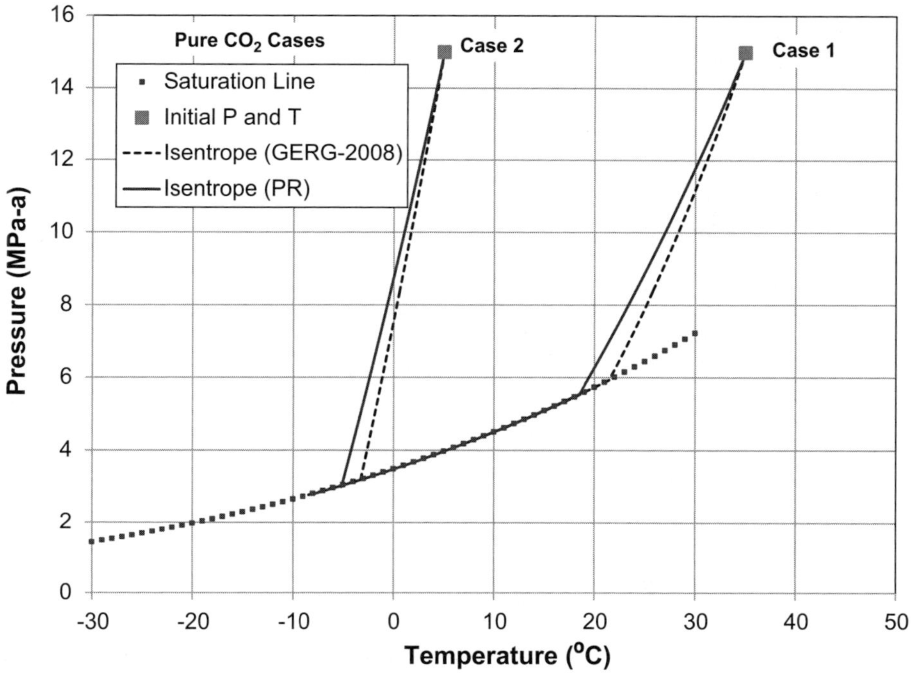

Figure 6-11. Pressure–temperature isentropes based on GERG-2008 and PR EOS of pure CO_2 (two cases)

6.5 DECOMPRESSION WAVE SPEED IN CO_2+N_2 BINARIES

Two CO_2+N_2 binary cases (Case 3 and Case 4) from different initial temperatures, as shown in Table 6-1, are next analyzed. These two cases have the same initial pressures. Figure 6-12 shows comparison of the predictions of the decompression wave speed based on GERG-2008 and PR EOS for Case 3. Notice the trend in the upper part of the decompression wave speed of GERG-2008 being slower than its PR counterpart. This is a reflection of the slope of the pressure–density profile along the corresponding isentropes, as shown in Figure 6-13, where the PR pressure–density profile is steeper than GERG-2008. Note also that the phase crossing and hence the plateau pressure prediction by PR is below that by GERG-2008, as was observed in the pure CO_2 cases described in the previous section. The lower part of the decompression wave speed curve (below the plateau) predicted by GERG-2008 also shows slower decompression wave speed at the same pressure. This may become critical in cases where the fracture propagation speed curve will be tangent to the decompression wave speed for the determination of the required arrest material toughness below the plateau, as will be discussed later in this chapter. Finally, the extent of the plateau is shorter in the case of GERG-2008 EOS due to the lower speed of sound approaching the phase envelope. The isentrope paths are also shown on a pressure–temperature diagram in Figure 6-14.

Figure 6-15 shows the decompression wave speed profiles vs. pressure for Case 4 from an initial pressure of 15 MPa-a and much lower temperature, 5°C. The speed of sound corresponding to the initial pressure and temperature predicted by PR EOS is noted to be lower than that predicted by GERG-2008 EOS, unlike the previous case (Case 3), and this is also consistent with the pure CO_2 cases. This indicates that, as en-

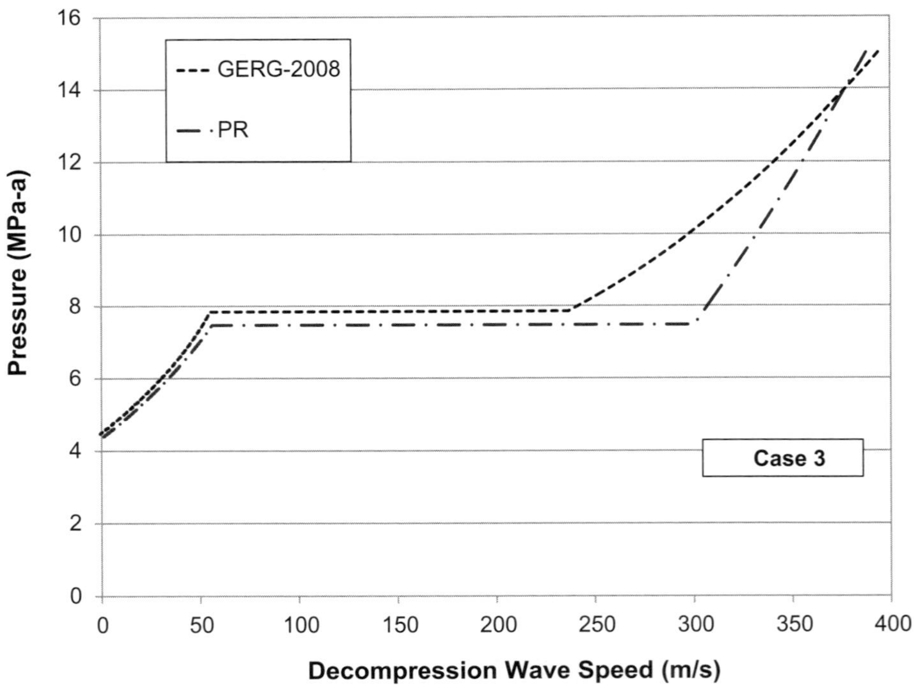

Figure 6-12. Decompression wave speed predictions based on GERG-2008 and PR EOS for CO_2+N_2 binary (Case 3)

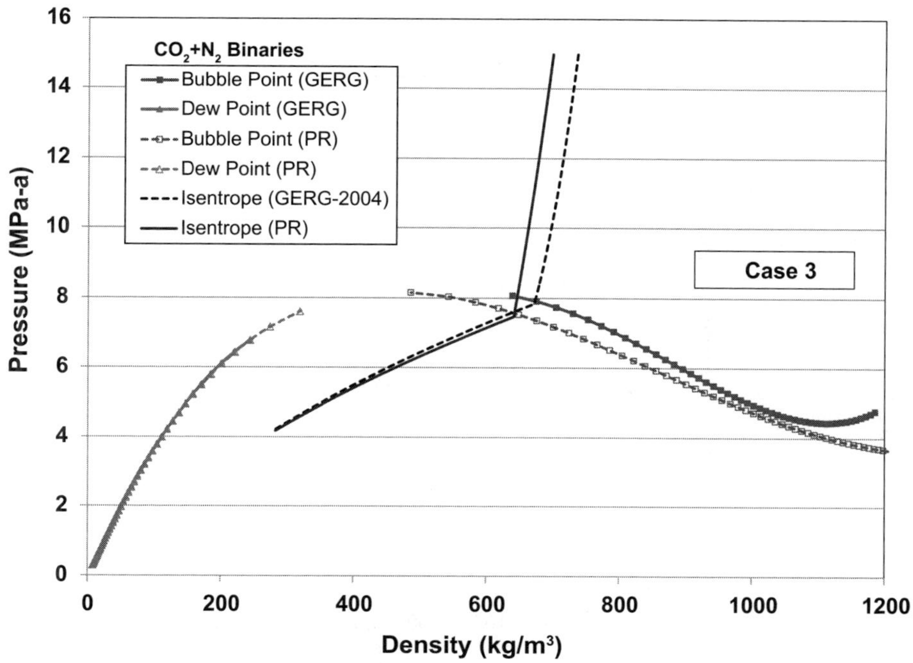

Figure 6-13. Pressure–density isentropes based on GERG-2008 and PR EOS for CO_2+N_2 binary (Case 3)

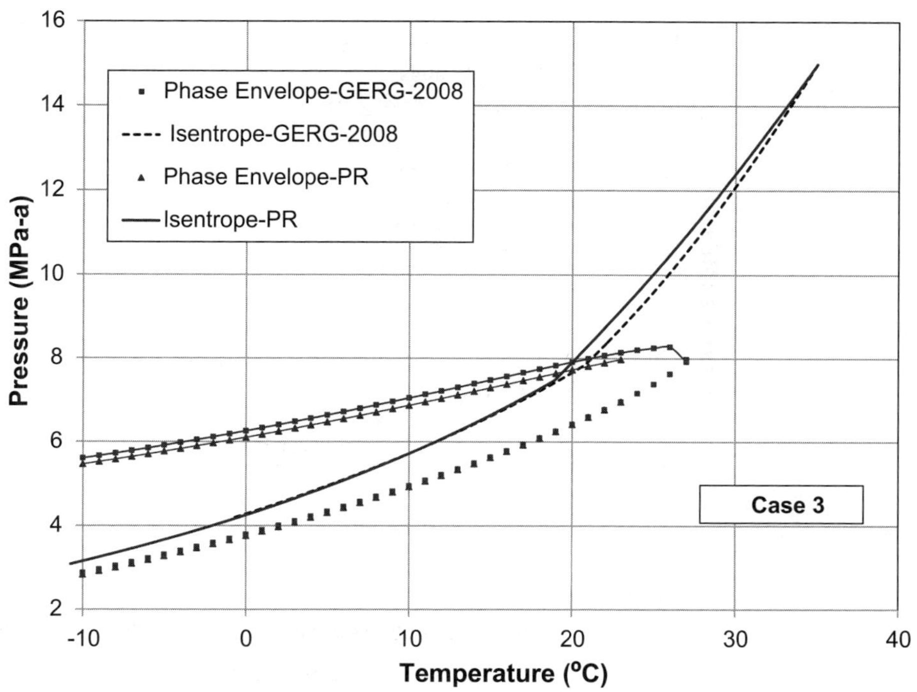

Figure 6-14. Pressure–temperature isentropes based on GERG-2008 and PR EOS for CO_2+N_2 binary (Case 3)

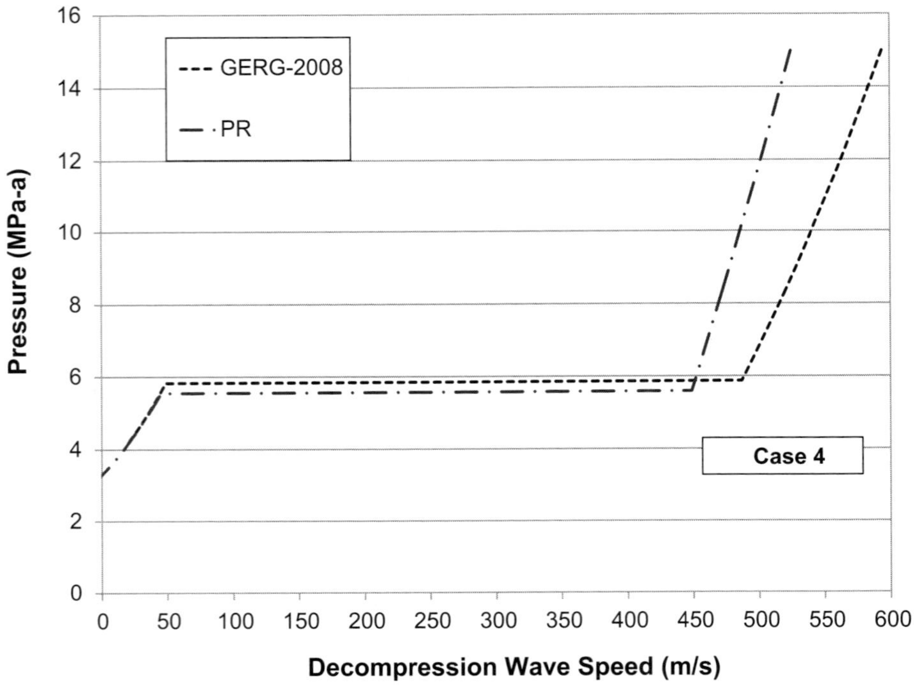

Figure 6-15. Decompression wave speed predictions based on GERG-2008 and PR EOS for CO_2+N_2 binary (Case 4)

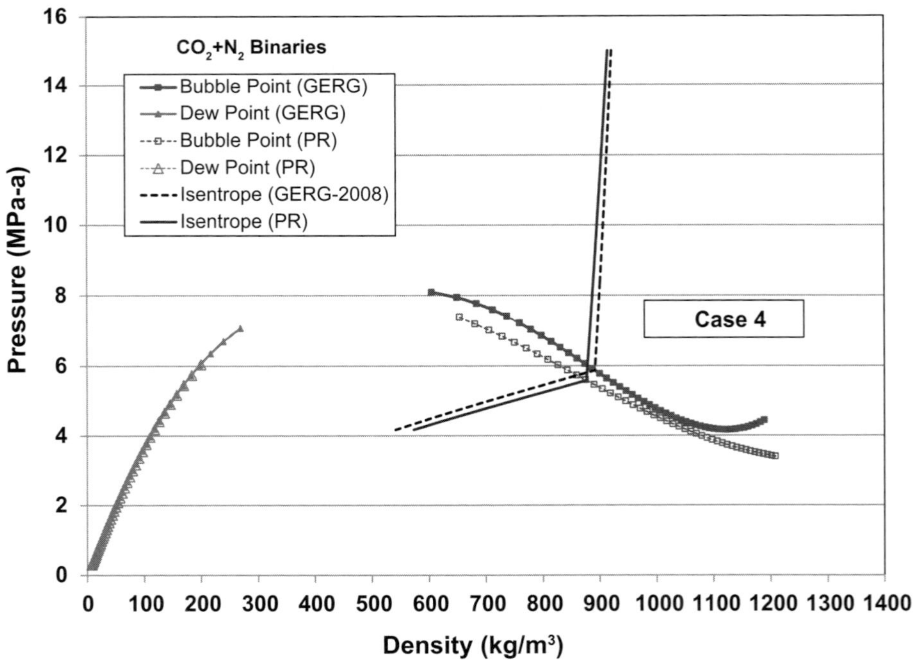

Figure 6-16. Pressure–density isentropes based on GERG-2008 and PR EOS for CO_2+N_2 binary (Case 4)

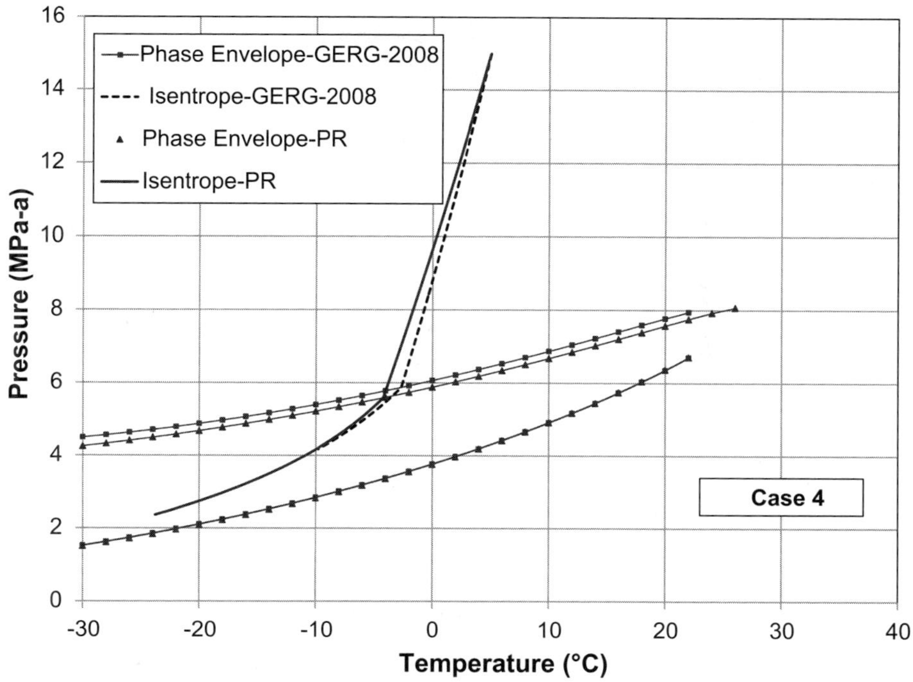

Figure 6-17. Pressure–temperature isentropes based on GERG-2008 and PR EOS for CO_2+N_2 binary (Case 4)

tropy increases, which is going towards the supercritical phase from the dense phase, the prediction of the speed of sound, and hence the decompression wave speed, by PR EOS tends to go from being lower to higher than that by GERG-2008. Again, it is noted that the plateau pressure prediction by GERG-2008 is higher than that predicted by PR, which is consistent with previously discussed cases. This is confirmed by the isentrope intersection with the corresponding phase envelope for each EOS, as shown in Figures 6-16 and 6-17 on the pressure–density and pressure–temperature diagrams, respectively. Note also that the bubble point curve crossings by the respective isentropes are far from the critical point and well into the low-temperature dense phase. This results in a larger drop in the decompression wave speed, as a result of the corresponding larger drop in the local speed of sound at phase crossing. The outcome is the longer extent of the plateau, as manifested in Figure 6-15.

6.6 DECOMPRESSION WAVE SPEED IN CO_2+O_2 BINARIES

Two CO_2+O_2 binary cases (Case 5 and Case 6) are next analyzed. Results of the decompression wave speed for Case 5, involving 3% O_2, are shown in Figure 6-18. In this case, the initial pressure was 15 MPa-a and the temperature was 35°C. It is shown that the deviation in prediction by PR EOS relative to GERG-2008 EOS, in the two aspects discussed earlier, is evident in this case as well, namely, the deviation in the part of the decompression wave speed curve above the plateau and the plateau pressure itself. Figures 6-19 and 6-20 show the isentropes, as predicted by GERG-2008 and PR, on pressure–density and pressure–temperature diagrams, showing the phase crossing which again correspond to the predicted plateau pressure. The GERG-2008 EOS prediction of

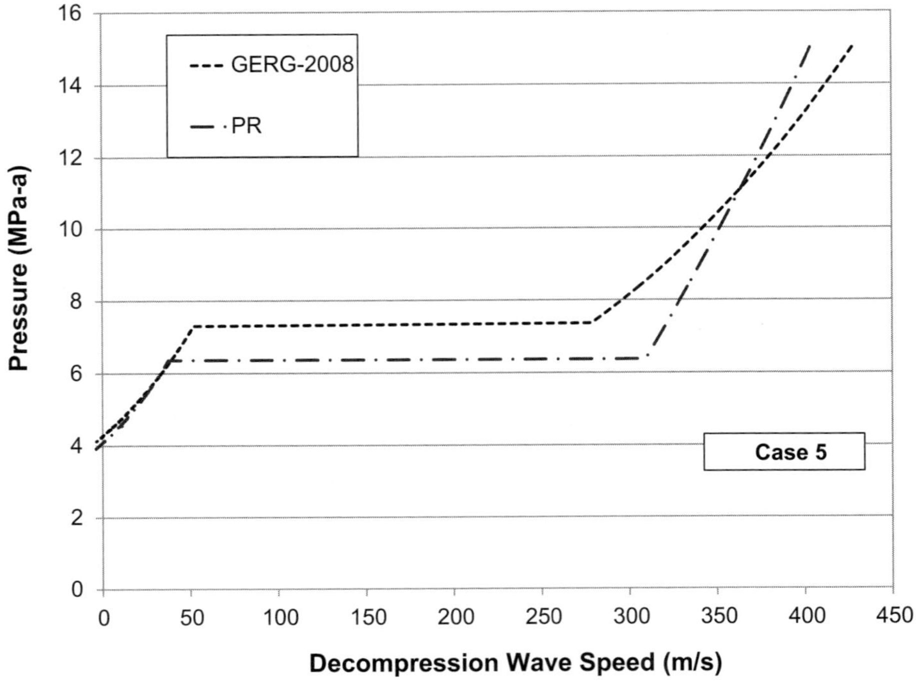

Figure 6-18. Decompression wave speed predictions based on GERG-2008 and PR EOS for CO_2+O_2 binary (Case 5)

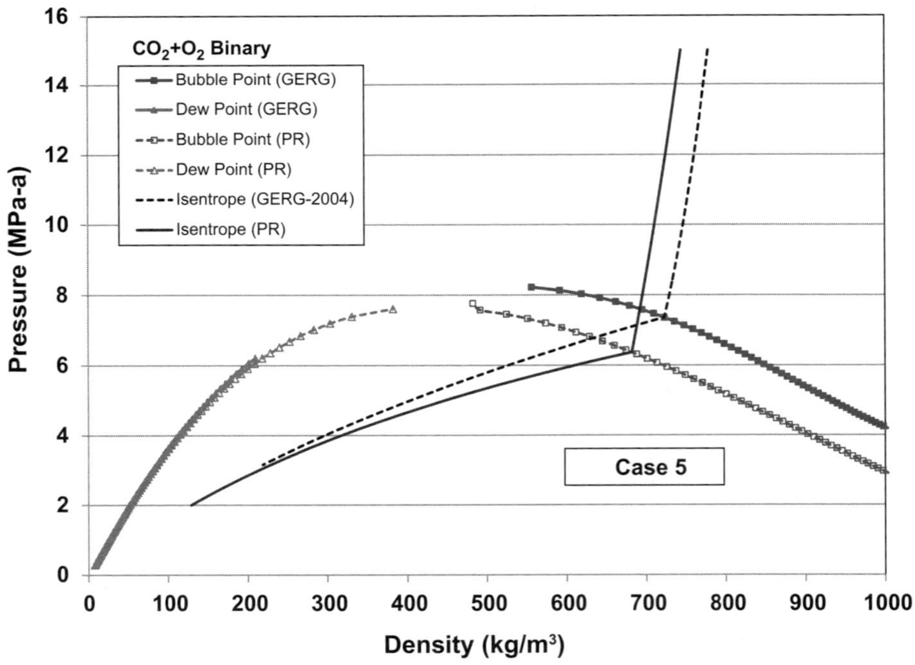

Figure 6-19. Pressure–density isentropes based on GERG-2008 and PR EOS for CO_2+O_2 binary (Case 5)

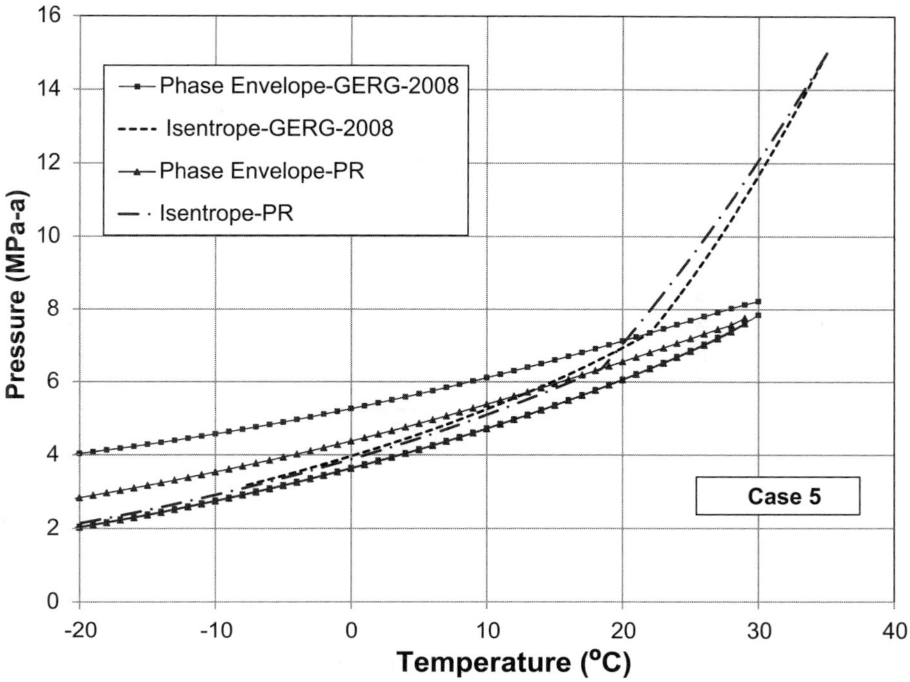

Figure 6-20. Pressure–temperature isentropes based on GERG-2008 and PR EOS for CO_2+O_2 binary (Case 5)

the plateau pressure is much higher than that predicted by PR EOS, which would result in a higher required Charpy energy for these conditions. PR EOS prediction would clearly result in a less conservative estimate of the required Charpy energy. The increase in the deviation between the predicted plateau pressures by both EOS is brought about by the considerable difference is the phase envelopes, as shown in Figure 6-20, though there is only a relatively small percentage of O_2 as impurity (3%). This is of some concern, as it will have significant implications for the accurate determination of the material toughness required. Therefore, experimental validation of the accuracy of either EOS for CO_2+O_2 binaries with different oxygen concentrations and for different initial conditions is essential.

Case 6 involves the same binary mixture composition and initial pressure. The only difference is the much lower initial temperature, 5°C (see Table 6-1). Again, because of the low temperature, the initial state of the mixture is in the dense phase, and hence a much faster pressure drop to the plateau is expected. Figure 6-21 shows the decompression wave speeds predicted by GERG-2008 and PR EOS for comparison. The PR EOS once more predicts a much lower initial speed of sound, which in turn affects the upper part of the decompression wave speed as shown. The respective isentropes based on both GERG-2008 and PR EOS predictions are shown in Figures 6-22 and 6-23. Note the large difference in the prediction of the bubble point curve of the respective phase envelopes as predicted by the two EOS, which explains the reason for the large difference in the predicted plateau pressures of the respective decompression waves.

The primary observation from the analysis of the above two cases of CO_2+O_2 binaries is that experimental work is needed to validate these EOS and that refinement of one or both EOS would be necessary for this binary.

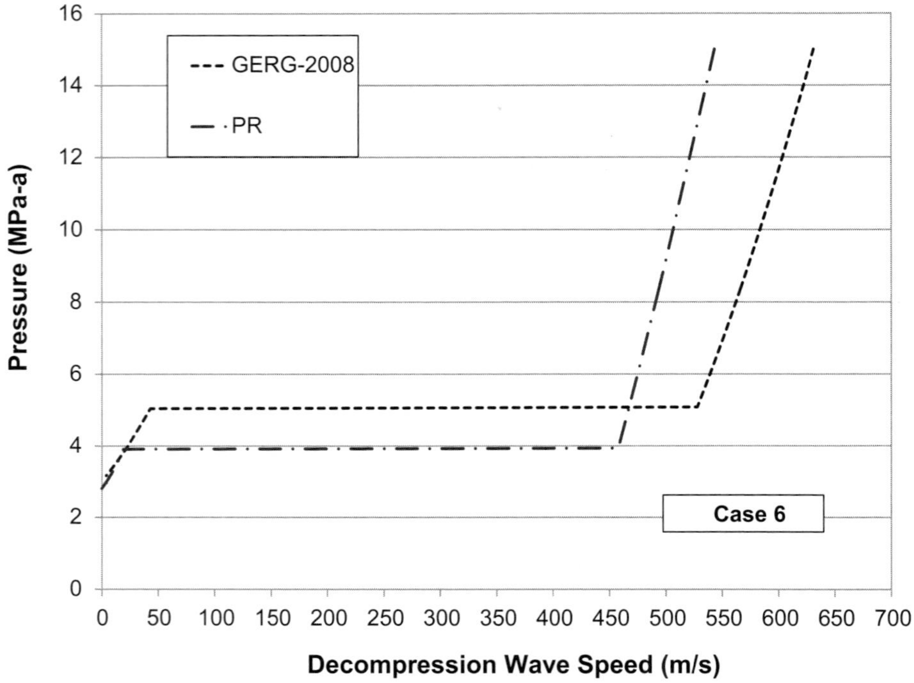

Figure 6-21. Decompression wave speed predictions based on GERG-2008 and PR EOS for CO_2+O_2 binary (Case 6)

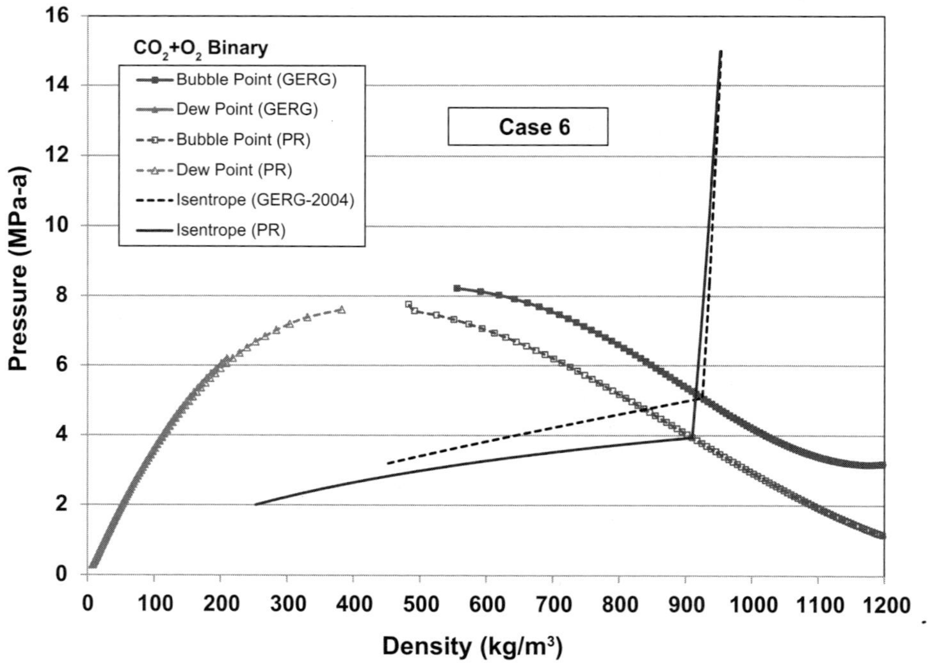

Figure 6-22. Pressure–density isentropes based on GERG-2008 and PR EOS for CO_2+O_2 binary (Case 6)

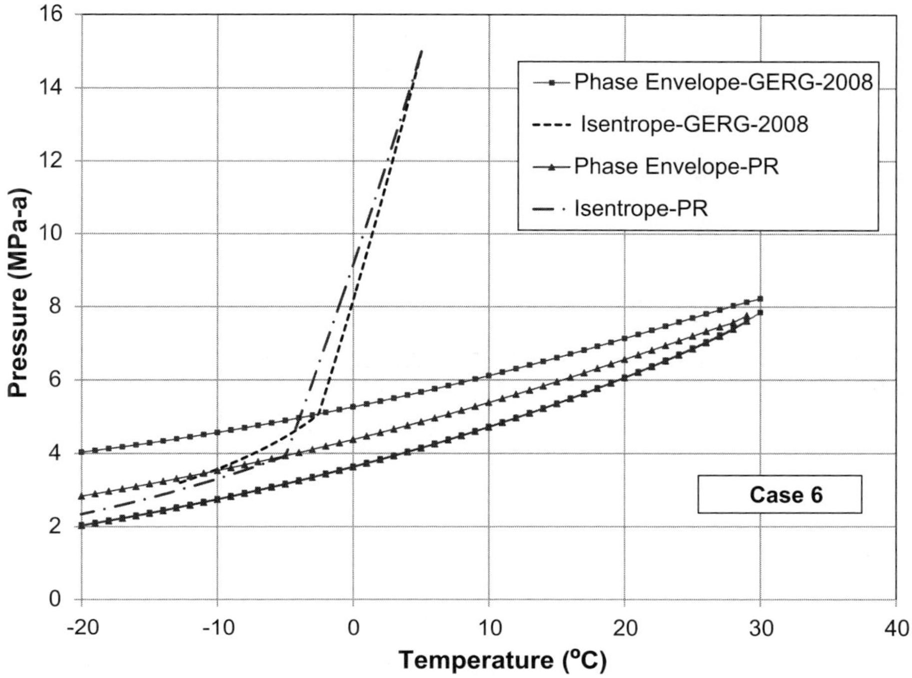

Figure 6-23. Pressure–temperature isentropes based on GERG-2008 and PR EOS for CO_2+O_2 binary (Case 6)

6.7 DECOMPRESSION WAVE SPEED IN CO_2+CH_4 BINARIES

Two CO_2+CH_4 binary cases (Case 7 and Case 8) are next analyzed. For both cases, the initial pressure is 15 MPa-a, but the temperatures are different (35°C for Case 7 and 5°C for Case 8). Results of the decompression wave speed for Case 7, involving 3% CH_4, is shown in Figure 6-24. The deviation in prediction by PR EOS relative to GERG-2008 EOS in the part of the decompression wave above the plateau is evident. The speed of sound as predicted by PR EOS is again lower than that by GERG-2008 EOS. The plateau pressure predictions by both EOS are relatively close to each other, although the fact that the plateau pressure predicted by PR EOS is always lower than that predicted by GERG-2008 EOS is evident in this case as well. This is primarily due to different isentrope paths from the initial condition to the point of intersection with the respective phase envelopes. Despite the fact that prediction of the phase envelopes by both EOS are close to each other, the different isentrope paths bring about the apparent deviation in the plateau pressure, as shown in Figures 6-25 and 6-26. These phase envelopes are narrower than those of the CO_2+O_2 binary for the same percentage of impurity.

Case 8 involves the same binary mixture composition and initial pressure, but much lower temperature (5°C). Again, because of the low temperature, the initial state of the mixture is in the dense phase, and hence a much faster pressure drop to the plateau is expected. Figure 6-27 shows the decompression wave speeds predicted by GERG-2008 and PR EOS for comparison. The PR EOS once more predicts a much lower initial speed of sound, which in turn affects the upper part of the decompression wave speed as shown. The respective isentropes based on both GERG-2008 and PR EOS predictions are shown in Figures 6-28 and 6-29. Note that the upper parts of the decompression wave speed curves,

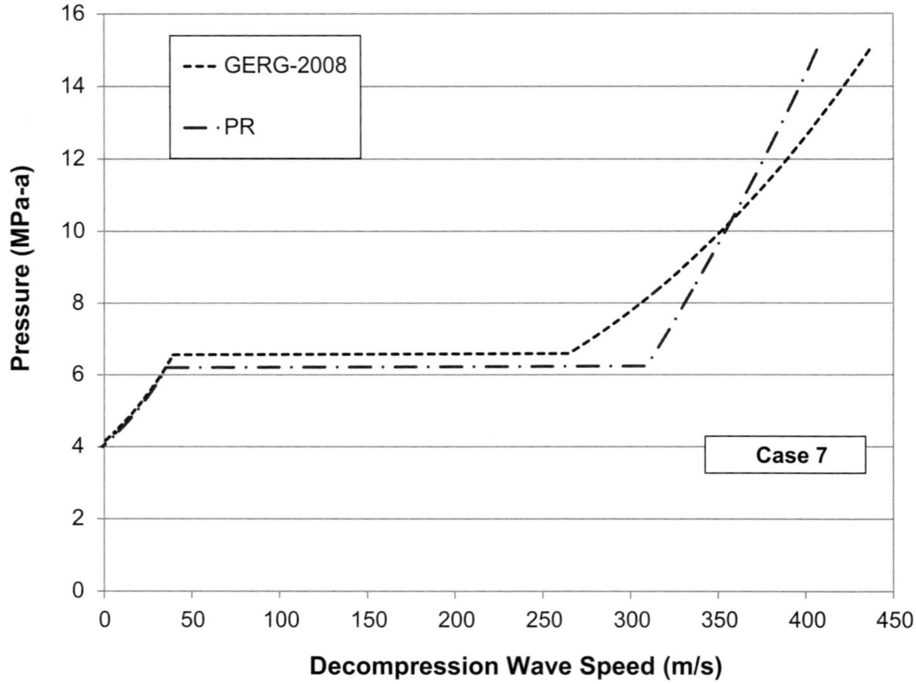

Figure 6-24. Decompression wave speed predictions based on GERG-2008 and PR EOS for CO_2+CH_4 binary (Case 7)

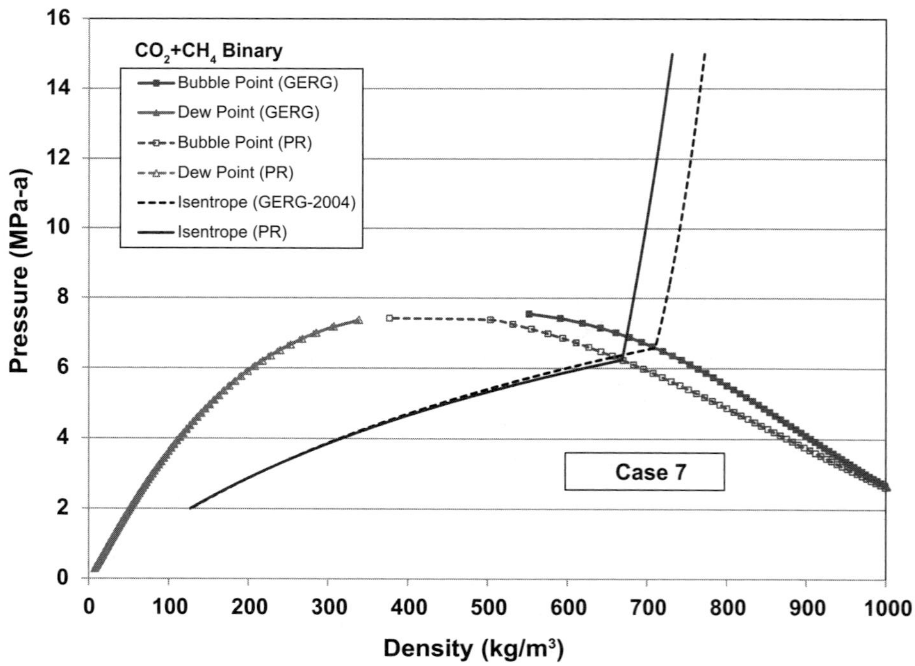

Figure 6-25. Pressure–density isentropes based on GERG-2008 and PR EOS for CO_2+CH_4 binary (Case 7)

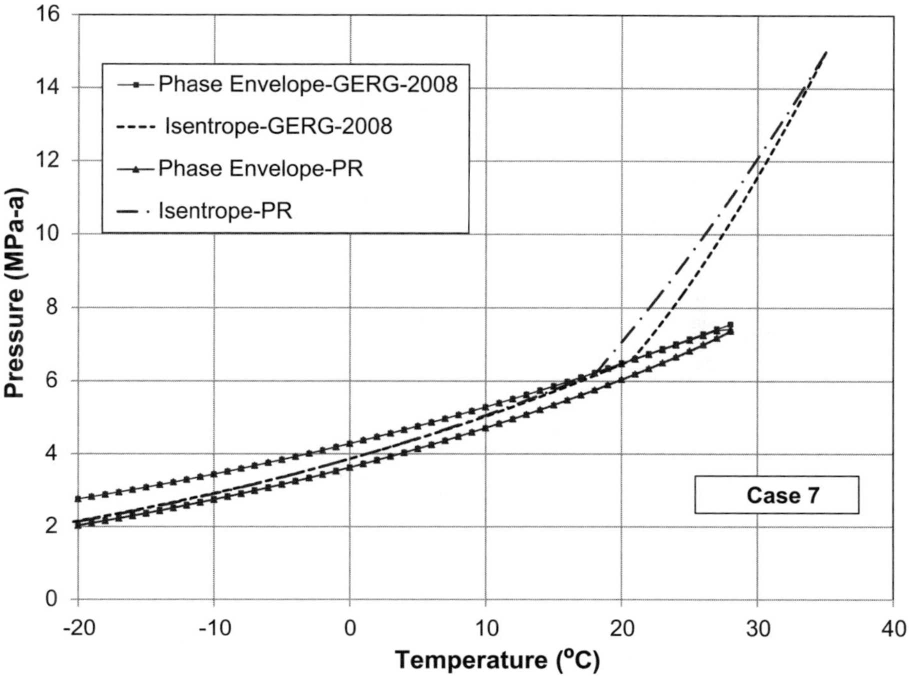

Figure 6-26. Pressure–temperature isentropes based on GERG-2008 and PR EOS for CO_2+CH_4 binary (Case 7)

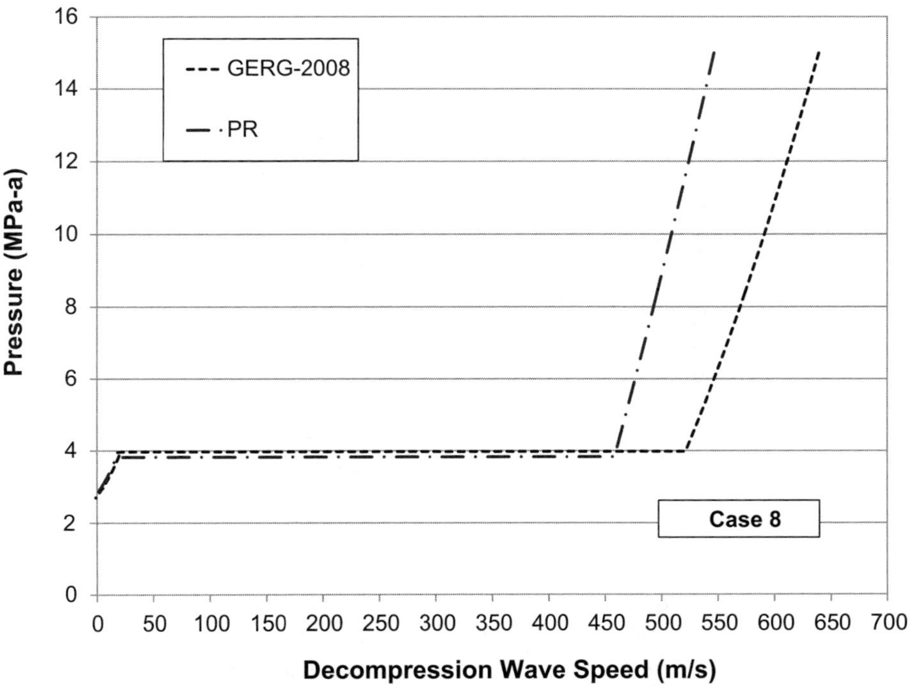

Figure 6-27. Decompression wave speed predictions based on GERG-2008 and PR EOS for CO_2+CH_4 binary (Case 8)

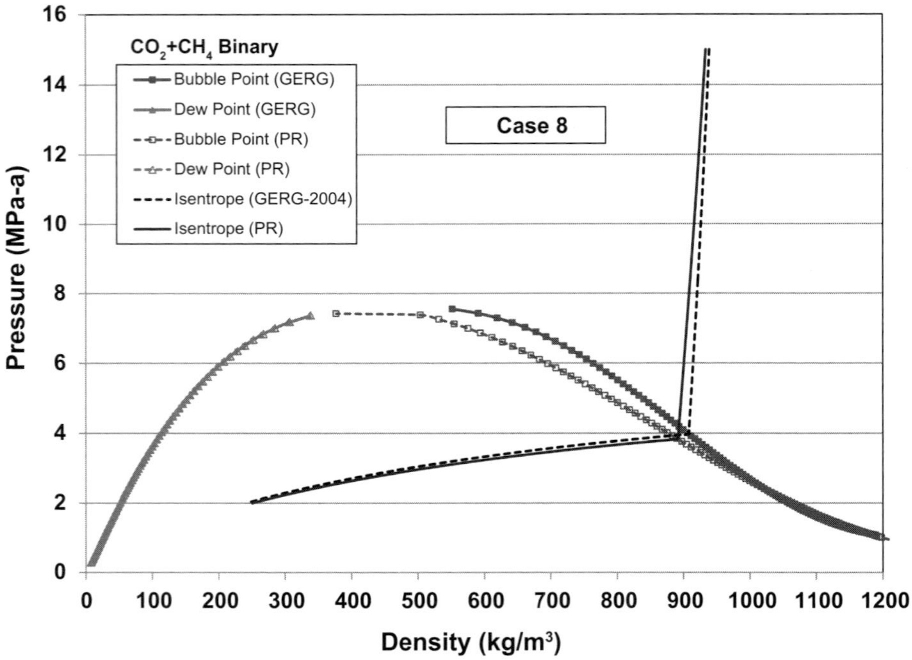

Figure 6-28. Pressure–density isentropes based on GERG-2008 and PR EOS for CO_2+CH_4 binary (Case 8)

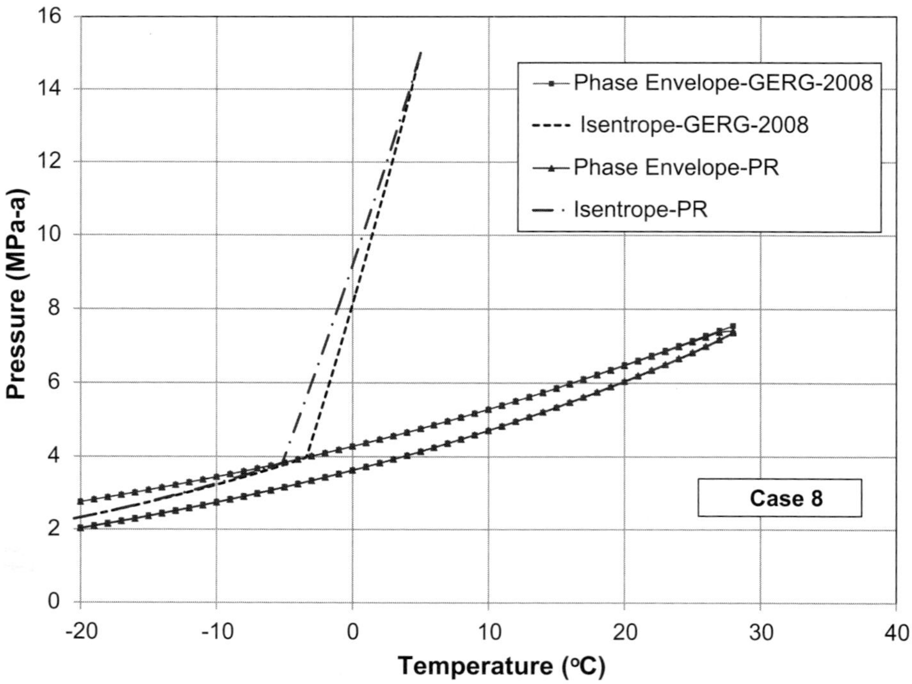

Figure 6-29. Pressure–temperature isentropes based on GERG-2008 and PR EOS for CO_2+CH_4 binary (Case 8)

as well as the isentropic paths, are very different. Nevertheless, because of offsetting differences in the isentropic paths and the phase boundaries, the predictions of the phase crossing (and hence the plateau pressures) are close. The extent of plateau is very long, indicating that the drop in the speed of sound at phase crossing is large, which is always observed when the phase crossing is far to the left of the critical point of the mixture.

6.8 DECOMPRESSION WAVE SPEED IN CO_2+H_2 BINARIES

Next, two cases with CO_2+H_2 binary mixtures are analyzed; Case 9 with a high initial temperature (35°C) and Case 10 with a lower initial temperature (5°C). Binary compositions and initial pressures are shown in Table 6-1. Figure 6-30 shows the decompression wave speeds for Case 9 as predicted by the GERG-2008 and PR EOS from the indicated initial pressure and temperature. Note that the plateau pressure as predicted by GERG-2008 EOS is now slightly below that predicted by PR EOS. This is different from all the other CO_2 binaries considered thus far. This is also manifested in Figures 6-31 and 6-32, which show that the bubble point curve of the phase envelope as predicted by GERG-2008 EOS is at a lower pressure than that predicted by PR EOS. Figures 6-31 and 6-32 also show the isentrope paths as predicted by GERG-2008 and PR on pressure–density and pressure–temperature diagrams. Note the intersections of the isentropes with the bubble point curves, showing the phase boundary crossing pressures, and the correspondence to the respective plateaus in Figure 6-30. As a result, the GERG-2008 prediction would clearly result in a lower estimation of the required Charpy energy for fracture arrest.

The second case with a CO_2+H_2 binary is Case 10, which has same initial pressure to the previous case but much lower temperature (5°C). This case exhibits a much faster drop

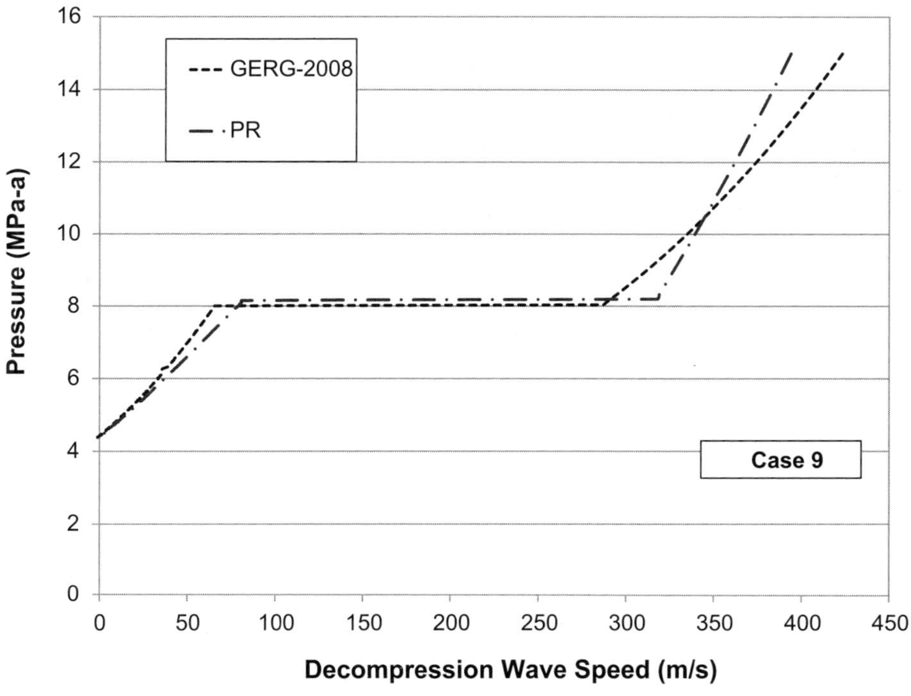

Figure 6-30. Decompression wave speed predictions based on GERG-2008 and PR EOS for CO_2+H_2 binary (Case 9)

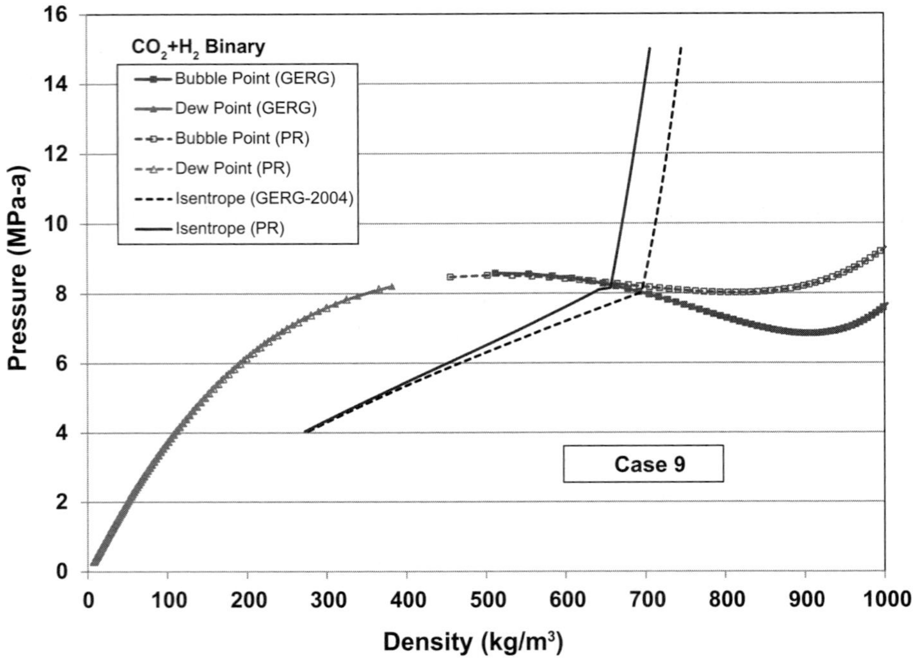

Figure 6-31. Pressure–density isentropes based on GERG-2008 and PR EOS for CO_2+H_2 binary (Case 9)

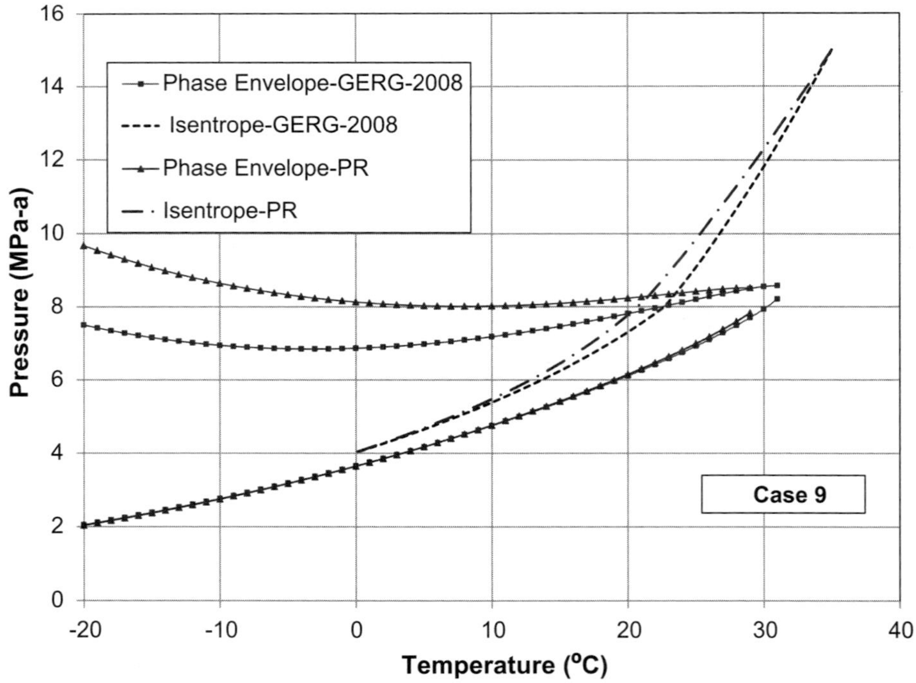

Figure 6-32. Pressure–temperature isentropes based on GERG-2008 and PR EOS for CO_2+H_2 binary (Case 9)

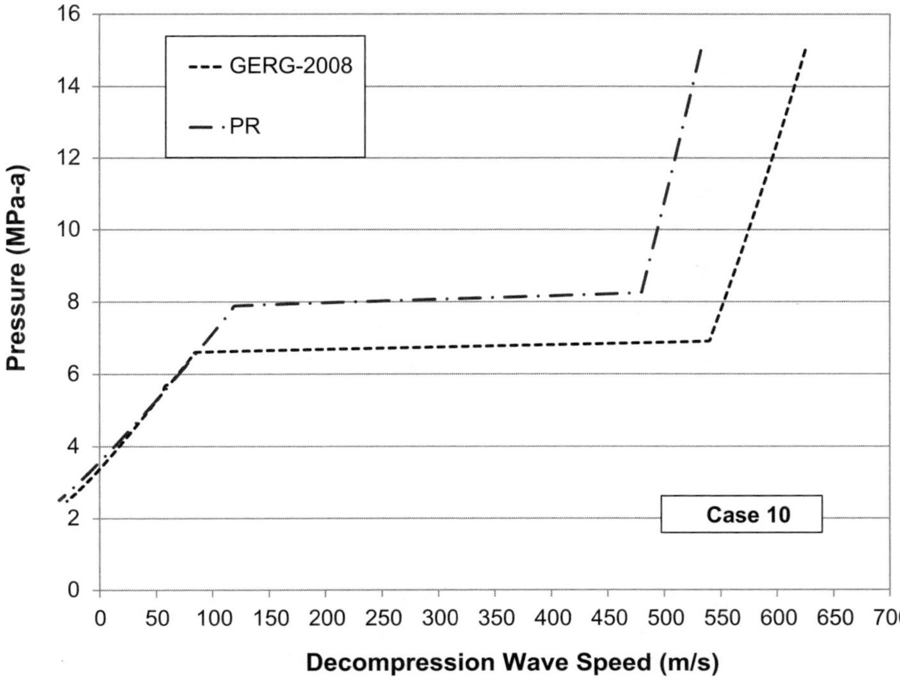

Figure 6-33. Decompression wave speed predictions based on GERG-2008 and PR EOS for CO_2+H_2 binary (Case 10)

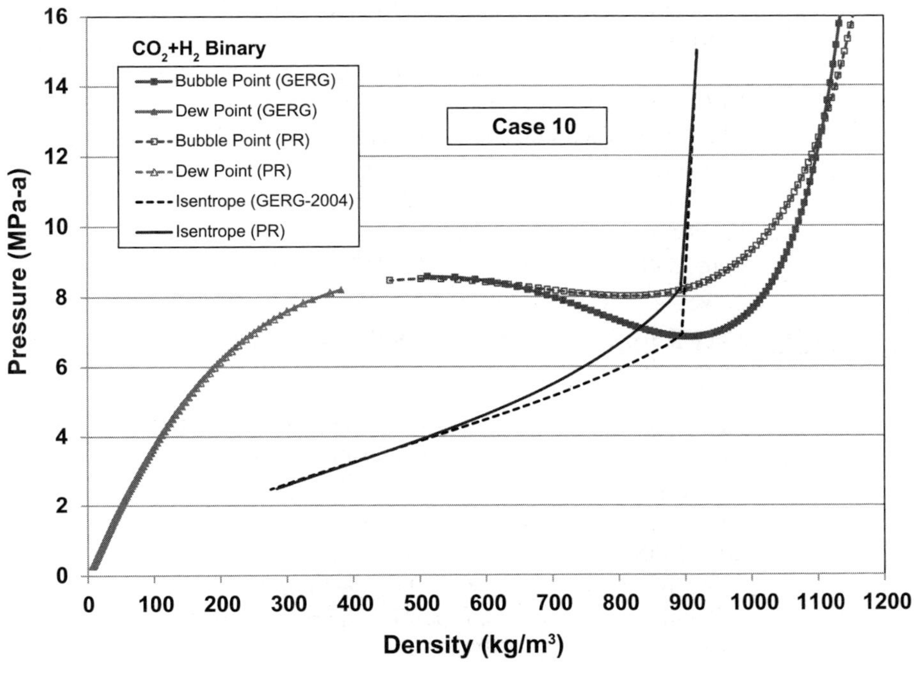

Figure 6-34. Pressure–density isentrope based on GERG-2008 EOS for CO_2+H_2 binary (Case 10)

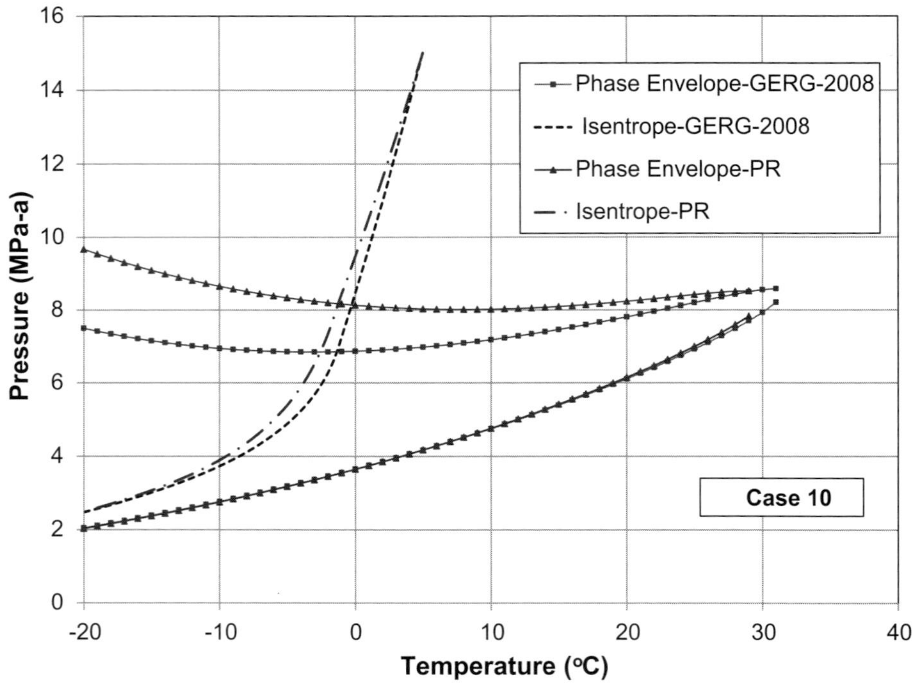

Figure 6-35. Pressure–temperature isentropes based on GERG-2008 and PR EOS for CO_2+H_2 binary (Case 10)

in pressure to the plateau, as shown in Figure 6-33. This is because the initial condition of the binary mixture is in the dense phase (see Figures 6-34 and 6-35), whereas, in the previous case, the initial mixture condition was in the supercritical phase (see Figure 6-32). Note that the PR and GERG-2008 EOS predictions of the upper part of the curve are quite far apart. Once again, as in the previous case, the prediction of the plateau pressure by PR EOS is higher than that by GERG-2008 EOS; in this case, the difference is considerable, i.e., 8.1 MPa-a for PR EOS as compared with 6.6 MPa-a for GERG-2008 EOS (see Figure 6-33). This reflects the considerable increase in the difference between the two bubble-point curves as the temperature decreases.

Based on the significant differences in the predictions of the different EOS relative to the decompression behavior of this CO_2+H_2 binary, it appears that further developments of both EOS are needed both in the $P\rho T$ and *VLE* aspects for CO_2-H_2 mixtures.

6.9 DECOMPRESSION WAVE SPEED IN CO_2+CO BINARIES

Next, two cases involving a CO_2+CO binary are analyzed. These are Case 11 and Case 12 shown in Table 6-1 and involve 3% CO. In both cases, the initial pressure is again 15 MPa-a and the temperatures are 35°C and 5°C, respectively. Figure 6-36 shows the decompression wave speeds profiles vs. pressure, as predicted by the GERG-2008 and PR EOS. GERG-2008 EOS prediction of the plateau pressure has now returned to the consistent trend of being higher than that predicted by PR EOS. Recall that this has been the case for all binaries with the exception of CO_2+H_2. Figures 6-37 and 6-38 show the isentropes as predicted by GERG-2008 and PR EOS on pressure–density and pressure–temperature diagrams, showing the phase boundary crossing, and hence the plateau pressures. The GERG-2008 EOS prediction would clearly result in a higher estimation of the required Charpy energy.

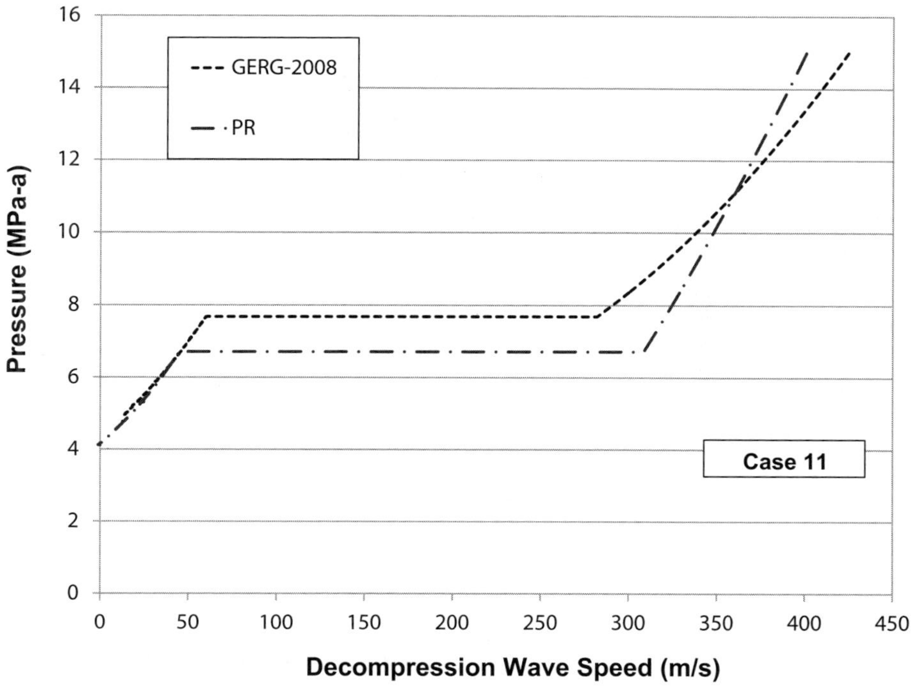

Figure 6-36. Decompression wave speed predictions based on GERG-2008 and PR EOS for CO_2+CO binary (Case 11)

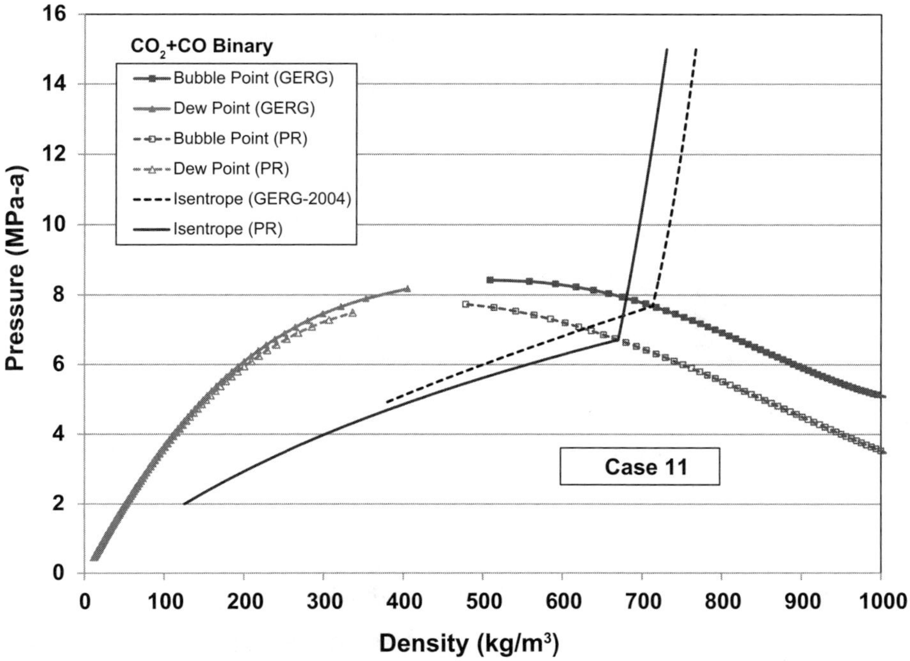

Figure 6-37. Pressure–density isentropes based on GERG-2008 and PR EOS for CO_2+CO binary (Case 11)

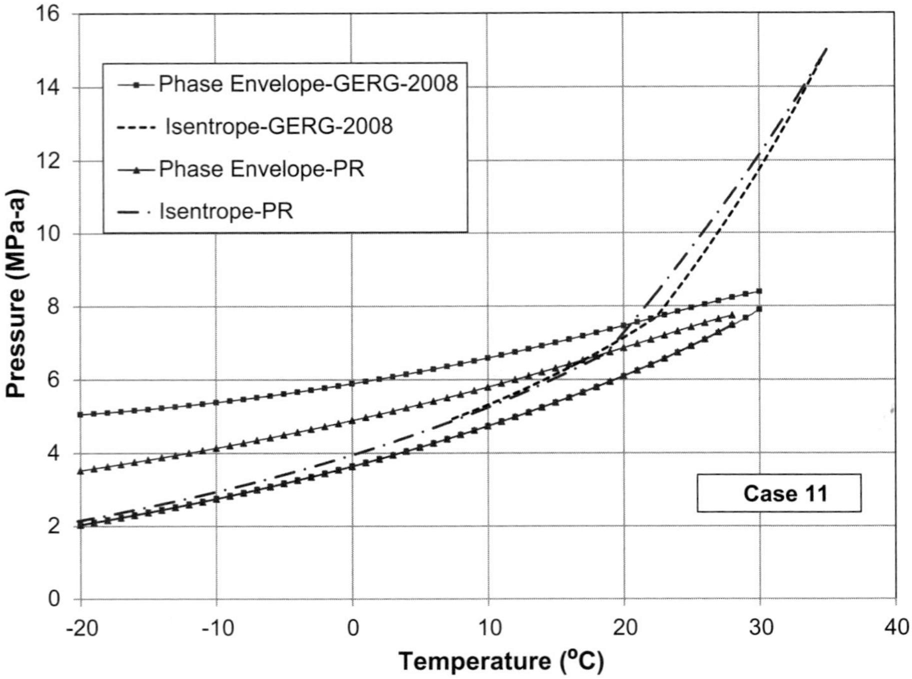

Figure 6-38. Pressure–temperature isentropes based on GERG-2008 and PR EOS for CO_2+CO binary (Case 11)

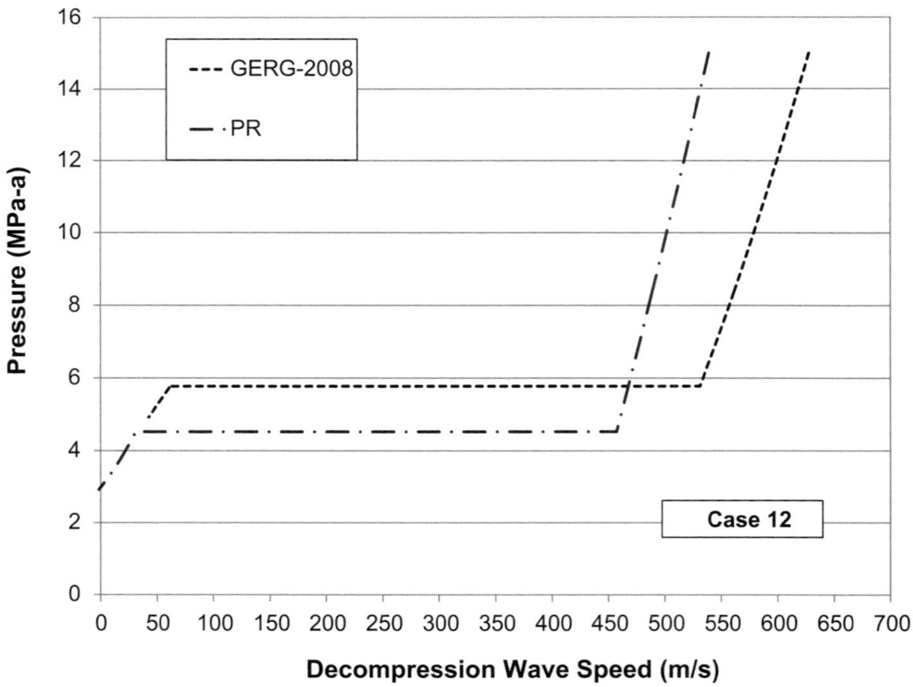

Figure 6-39. Decompression wave speed predictions based on GERG-2008 and PR EOS for CO_2+CO binary (Case 12)

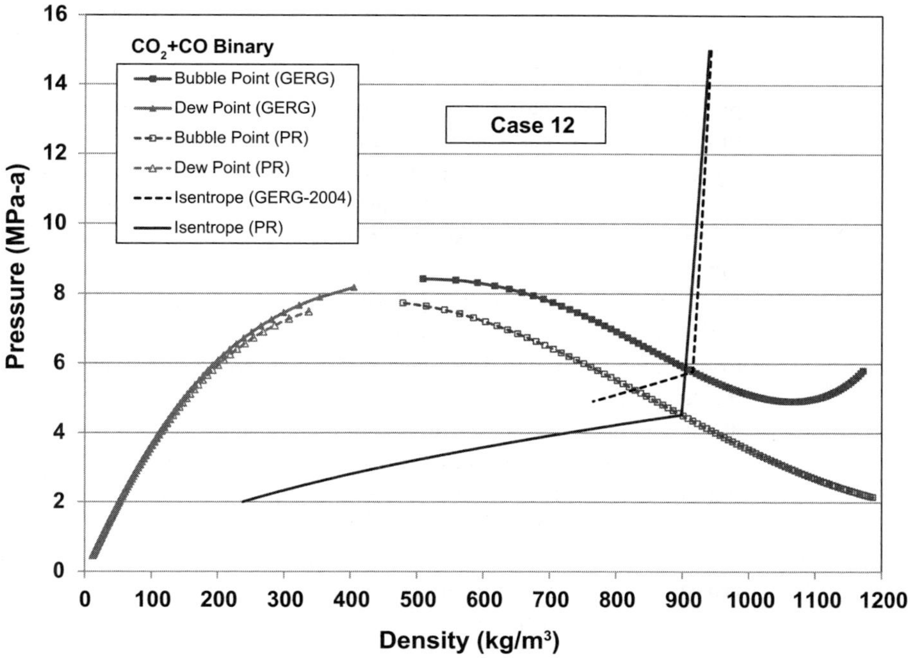

Figure 6-40. Pressure–density isentropes based on GERG-2008 and PR EOS for CO_2+CO binary (Case 12)

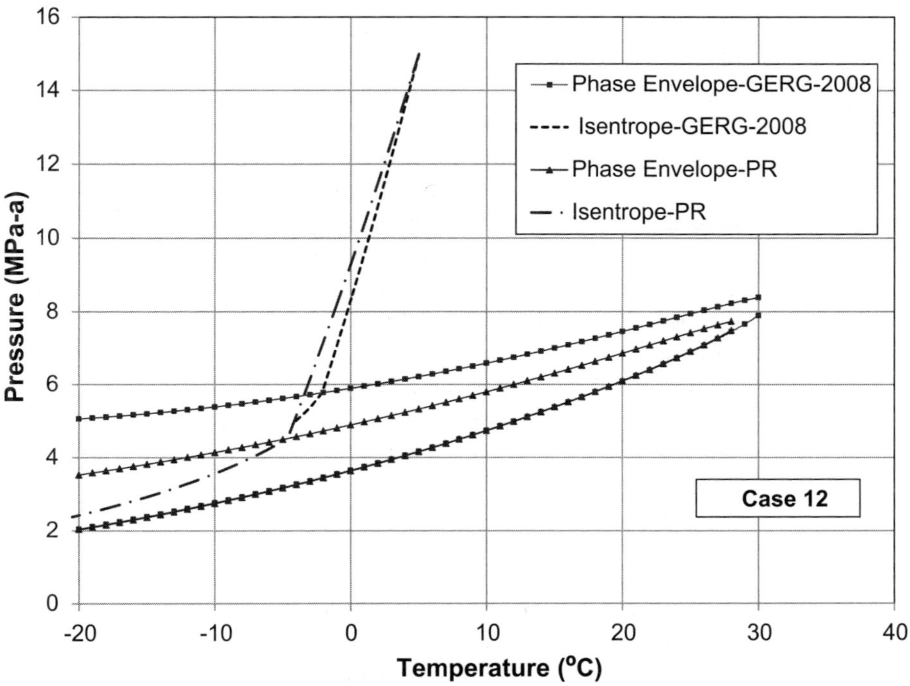

Figure 6-41. Pressure–temperature isentropes based on GERG-2008 and PR EOS for CO_2+CO binary (Case 12)

As shown in Figure 6-36, the difference in the predictions of the decompression wave speed by GERG-2008 and PR is large for the CO_2+CO binary considered. Finally, it could be argued that GERG-2008 has the advantage of being more conservative with respect to plateau pressure prediction and hence the material toughness required to arrest ductile fracture. However, excessive conservatism could lead to increased cost and even to an erroneous conclusion that a design based on pipe body arrest is not practical (see Chapter 7).

Case 12 involves the same binary mixture composition and initial pressure, but much lower temperature (5°C). Again, because of the low temperature, the initial state of the mixture is in the dense phase, and hence a much faster pressure drop to the plateau is expected. Figure 6-39 shows the decompression wave speeds predicted by GERG-2008 and PR EOS for comparison. The PR EOS again predicts a much lower initial speed of sound, which in turn affects the upper part of the decompression wave speed as shown. The respective isentropes based on both GERG-2008 and PR EOS predictions are shown in Figures 6-40 and 6-41. Note that the upper parts of the decompression wave speed curves, as well as the isentropic paths, are different. Furthermore, the difference between the bubble-point curves in the region of the phase boundary crossing is much greater than in the previous case (Figures 6-38 and 6-41), which accounts for the greater difference in predicted plateau pressure. The extent of the plateau is very long, indicating that the drop in the speed of sound at phase crossing is large, which is always observed when the phase crossing is far to the left of the critical point of the mixture.

6.10 DECOMPRESSION WAVE SPEED IN CO_2+Ar BINARIES

Two cases (Case 13 and Case 14) with CO_2+Ar binary mixtures from the same initial pressure but different temperatures (Table 6-1) are next analyzed. The Ar concentration is 3% in both cases. The continuing trend of the plateau pressure predicted by PR EOS being lower than that predicted by GERG-2008 EOS is observed in Figure 6-42. Again, recall that this was shown to be the case for all binaries with the exception of CO_2+H_2. Similar to previous binaries, PR EOS predicts a lower speed of sound at the initial pressure and temperature than GERG-2008 EOS. Figures 6-43 and 6-44 show the isentropes as predicted by GERG-2008 and PR EOS on pressure–density and pressure–temperature diagrams, showing the phase boundary crossing, and hence the plateau pressures.

The second case (Case 14) involves the same binary mixture composition and initial pressure, but lower initial temperature at 5°C (see Table 6-1), such that the initial state of the binary mixture is in the dense phase. Hence, a much faster drop in the pressure to the plateau is expected, as seen in Figure 6-45. The predictions of the decompression wave speeds by both GERG-2008 and PR EOS in Figure 6-45 show that PR EOS exhibits a similar performance to that in previous cases where initial conditions are in the dense phase, in that the initial speed of sound prediction is considerably lower than for GERG-2008 EOS. This affects the entire upper branch of the decompression wave speed curve, as shown. The isentropes based on GERG-2008 and PR EOS predictions are shown in Figure 6-46 on a pressure–density diagram and in Figure 6-47 on a pressure–temperature diagram. Note that the bubble point curve crossings by the respective isentropes are far from the critical point and well into the low-temperature dense phase. This results in a larger drop in the decompression wave speed, as a result of the corresponding larger drop in the local speed of sound at phase crossing. The outcome is the longer extent of the plateau, as manifested in Figure 6-45. Similar to the discussion on the CO_2+H_2 and CO_2+CO binaries, the CO_2+Ar binary also needs experimental investigation to validate or enhance the prediction of one or both EOS. It is evident (as was the case with CO_2+H_2 and CO_2+CO binaries) that the difference in the predictions of the decompression wave speed by GERG-2008 and PR is large for the CO_2+Ar binary as well.

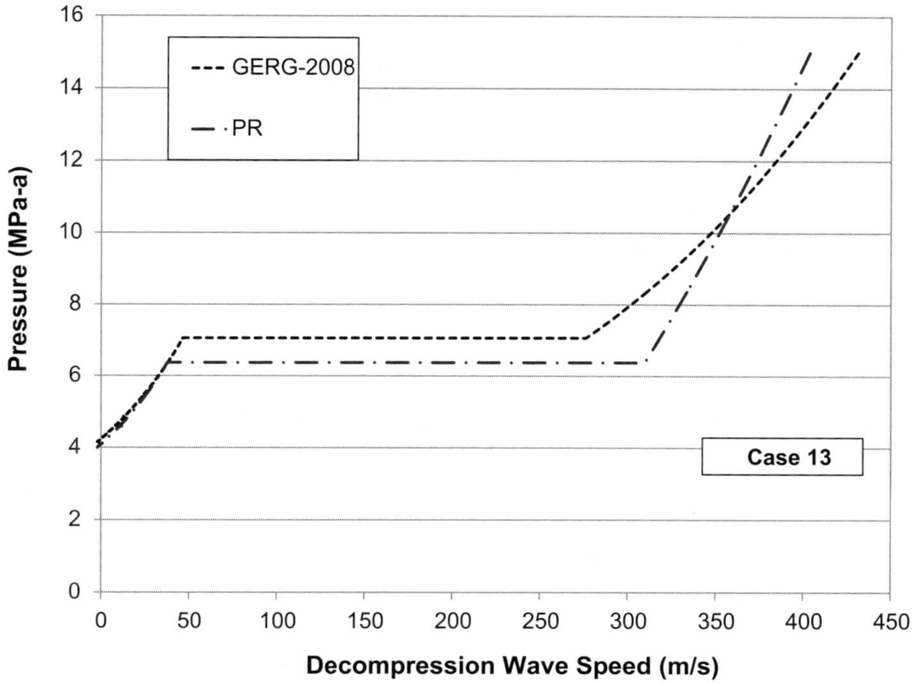

Figure 6-42. Decompression wave speed predictions based on GERG-2008 and PR EOS for CO_2+Ar binary (Case 13)

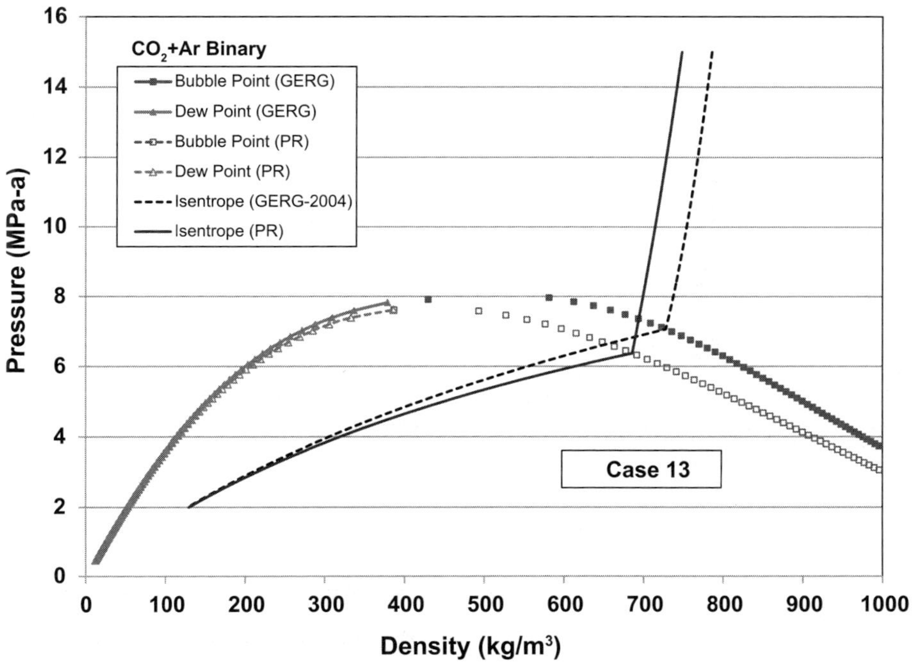

Figure 6-43. Pressure–density isentropes based on GERG-2008 and PR EOS for CO_2+Ar binary (Case 13)

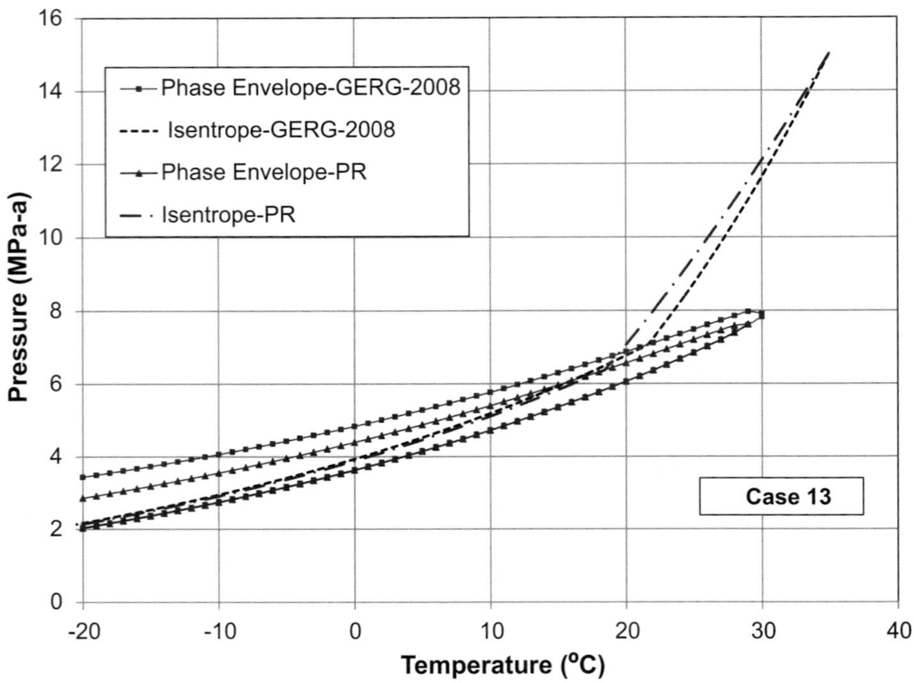

Figure 6-44. Pressure–temperature isentropes based on GERG-2008 and PR EOS for CO_2+Ar binary (Case 13)

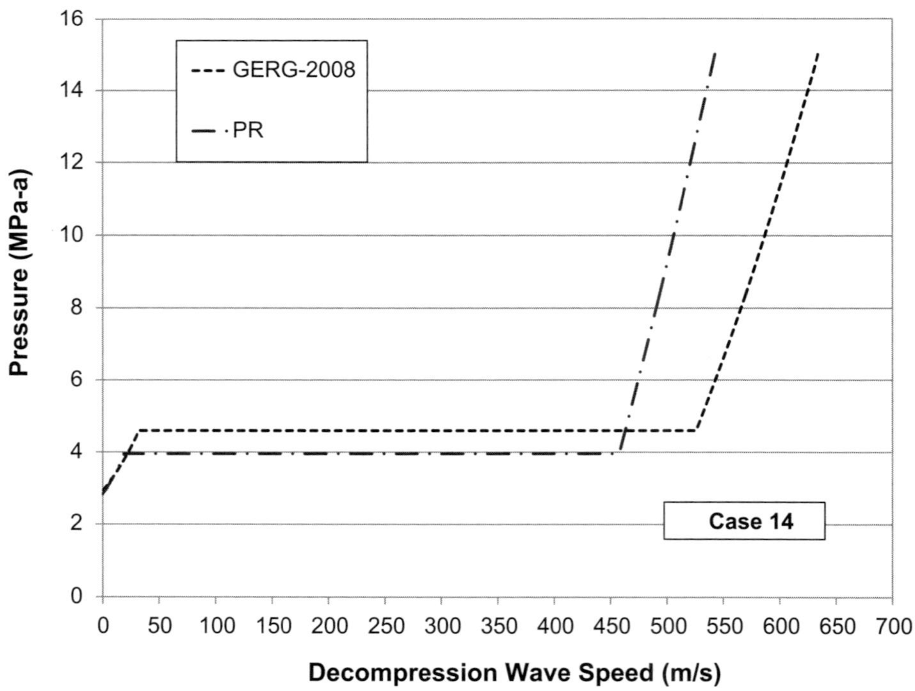

Figure 6-45. Decompression wave speed predictions based on GERG-2008 and PR EOS for CO_2+Ar binary (Case 14)

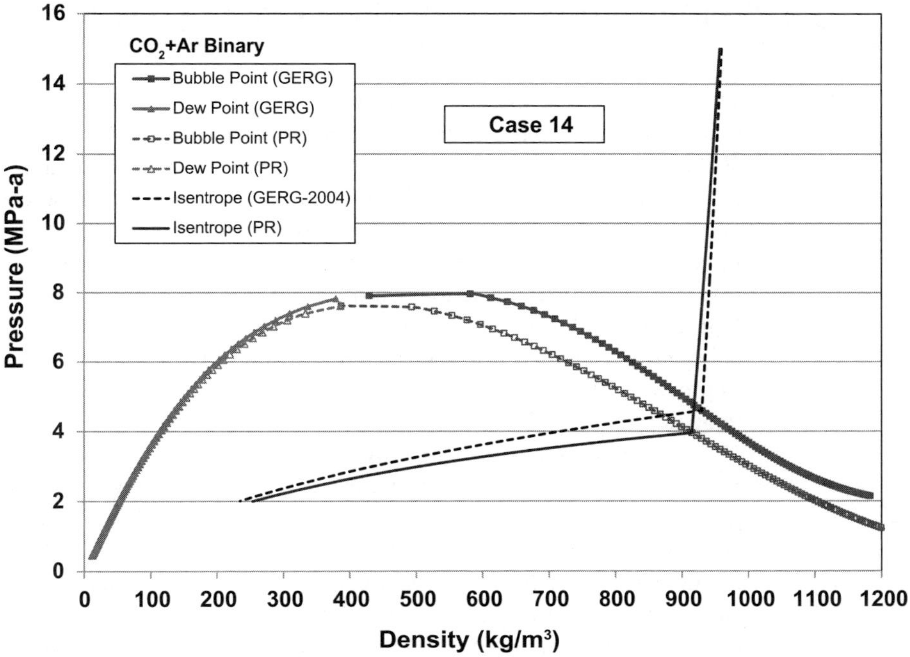

Figure 6-46. Pressure–density isentropes based on GERG-2008 and PR EOS for CO_2+Ar binary (Case 14)

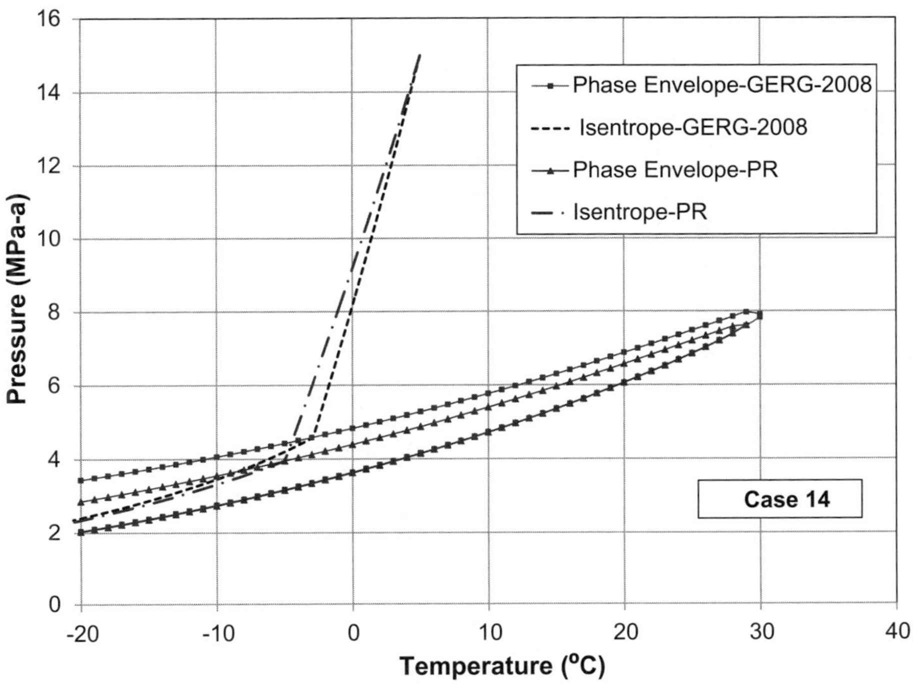

Figure 6-47. Pressure–temperature isentropes based on GERG-2008 and PR EOS for CO_2+Ar binary (Case 14)

6.11 DECOMPRESSION WAVE SPEED IN CO_2+H_2S BINARIES

The last two cases analyzed of binary mixtures are Case 15 and Case 16, with CO_2+H_2S binary mixtures from the same initial pressure but different temperatures (Table 6-1). The H_2S concentration is also 3% in both cases. Figures 6-48, 6-49, and 6-50 show the results of the analysis based on GERG-2008 and PR. Note that the phase envelope is vanishingly narrow for this mixture and lies very close to and slightly below that of pure CO_2. It was found that even for a binary mixture up to 10% H_2S, the bubble point curve and the dew point curve are very close to each other, i.e. the phase envelope is very narrow (see also Figures 4-4 and 4-5). The continuing trend of the plateau pressure predicted by PR EOS being lower than that predicted by GERG-2008 EOS is observed in Figure 6-42). Similar to previous binaries, PR EOS predicts a lower speed of sound at the initial pressure and temperature than GERG-2008 EOS.

The second case (Case 16) involves the same binary mixture composition and initial pressure, but lower initial temperature at 5°C (see Table 6-1). A much faster drop in the pressure to the plateau is seen in Figure 6-51, due to the fact the initial condition lies in the dense phase. The predictions of the decompression wave speeds by both GERG-2008 and PR EOS in Figure 6-51 show that PR EOS exhibits a similar behavior to that in previous cases where initial conditions are in the dense phase, in that the initial speed of sound prediction is considerably lower than for GERG-2008 EOS. The isentropes based on GERG-2008 and PR EOS predictions are shown in Figure 6-52 on a pressure–density diagram and in Figure 6-53 on a pressure–temperature diagram. Due to the fact that the bubble point curve crossings by the respective isentropes are far from the critical point and well into the low-temperature dense phase, a larger drop in the decompression wave speed is noticed, and hence a longer extent of the plateau.

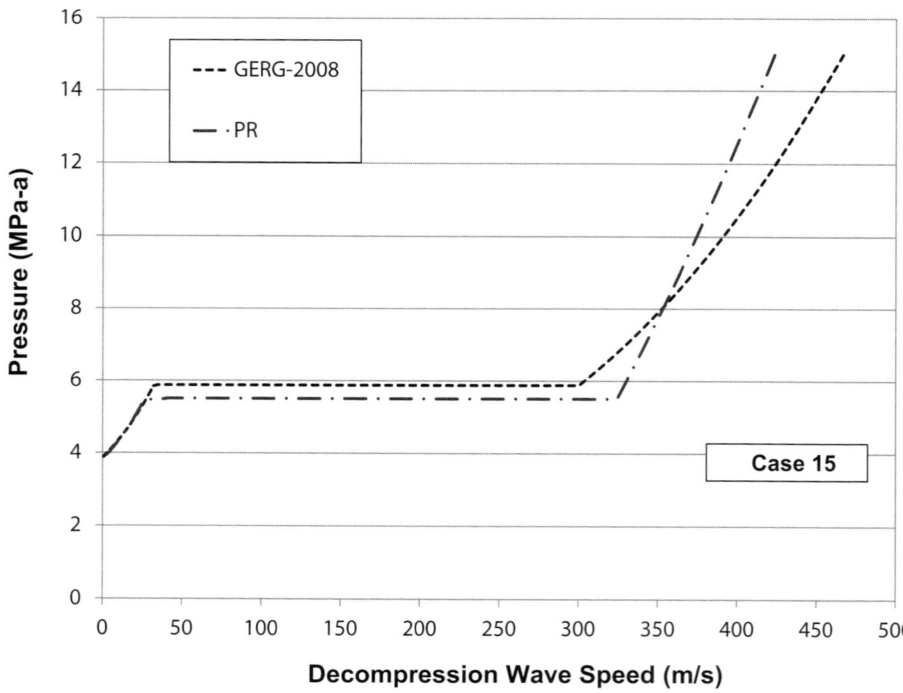

Figure 6-48. Decompression wave speed predictions based on GERG-2008 and PR EOS for CO_2+H_2S binary (Case 15)

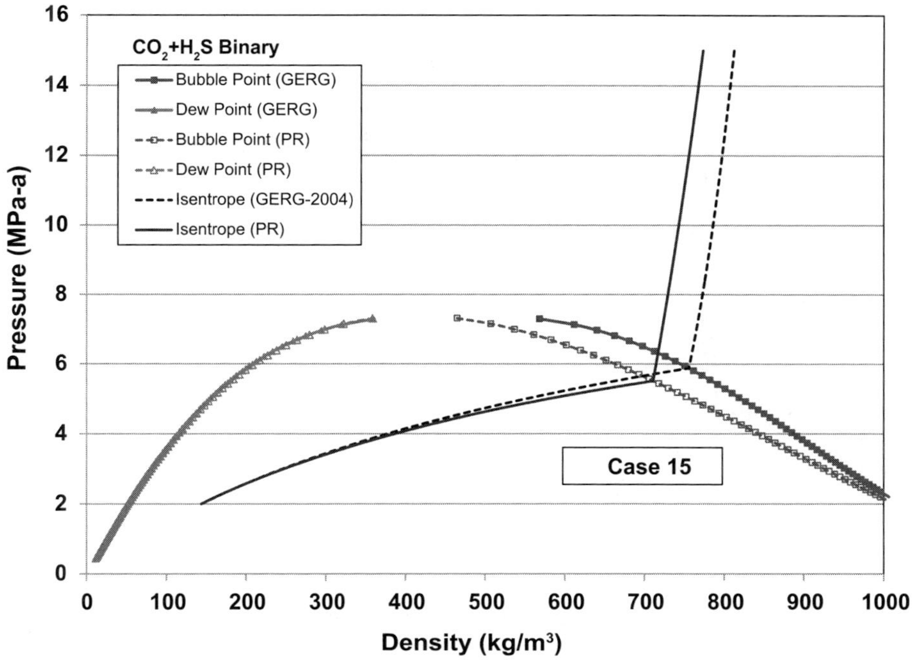

Figure 6-49. Pressure–density isentropes based on GERG-2008 and PR EOS for CO_2+H_2S binary (Case 15)

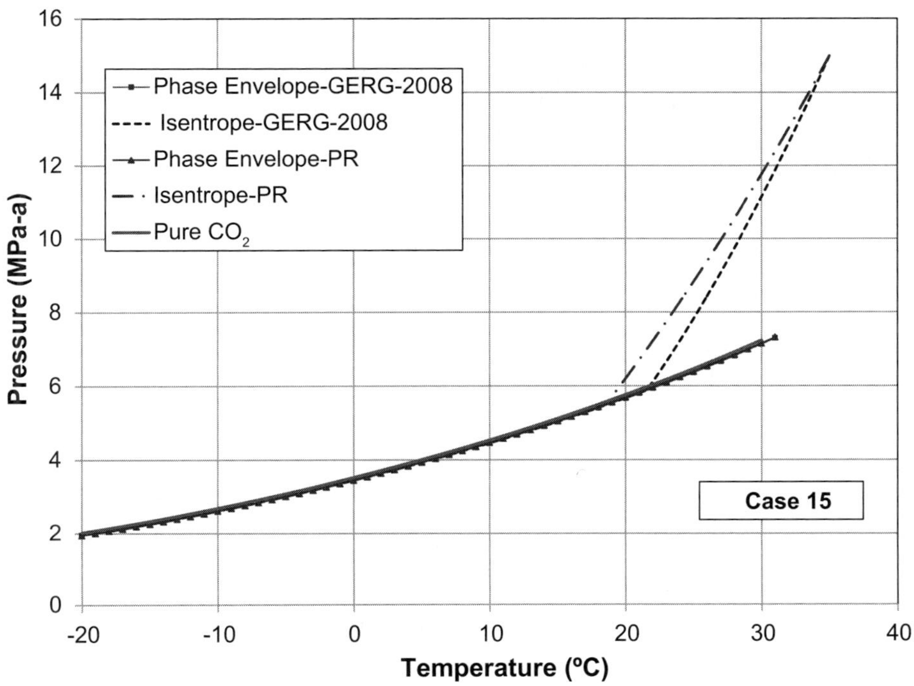

Figure 6-50. Pressure–temperature isentropes based on GERG-2008 and PR EOS for CO_2+H_2S binary (Case 15)

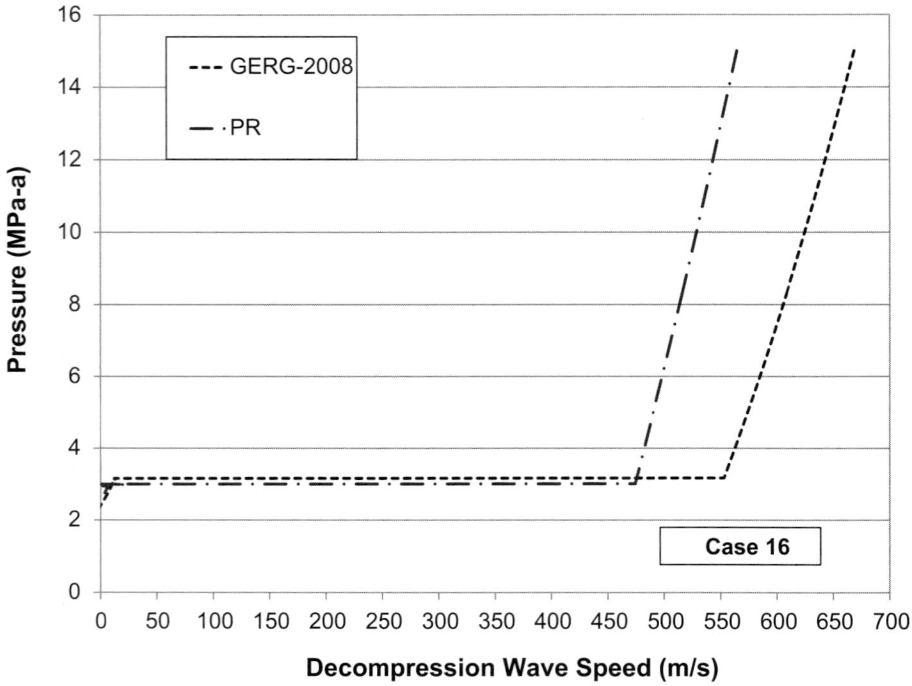

Figure 6-51. Decompression wave speed predictions based on GERG-2008 and PR EOS for CO_2+H_2S binary (Case 16)

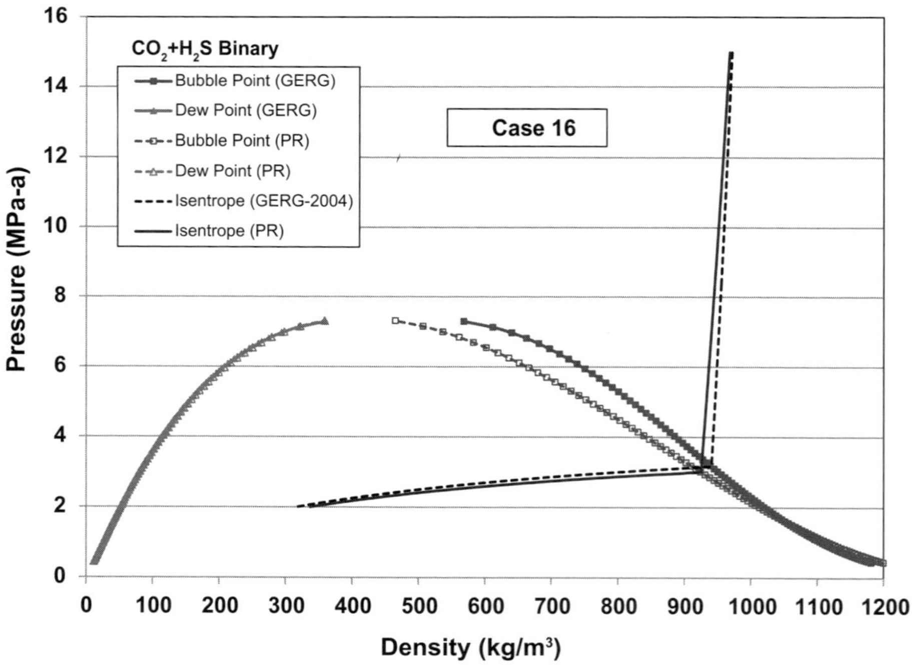

Figure 6-52. Pressure–density isentropes based on GERG-2008 and PR EOS for CO_2+H_2S Binary (Case 16)

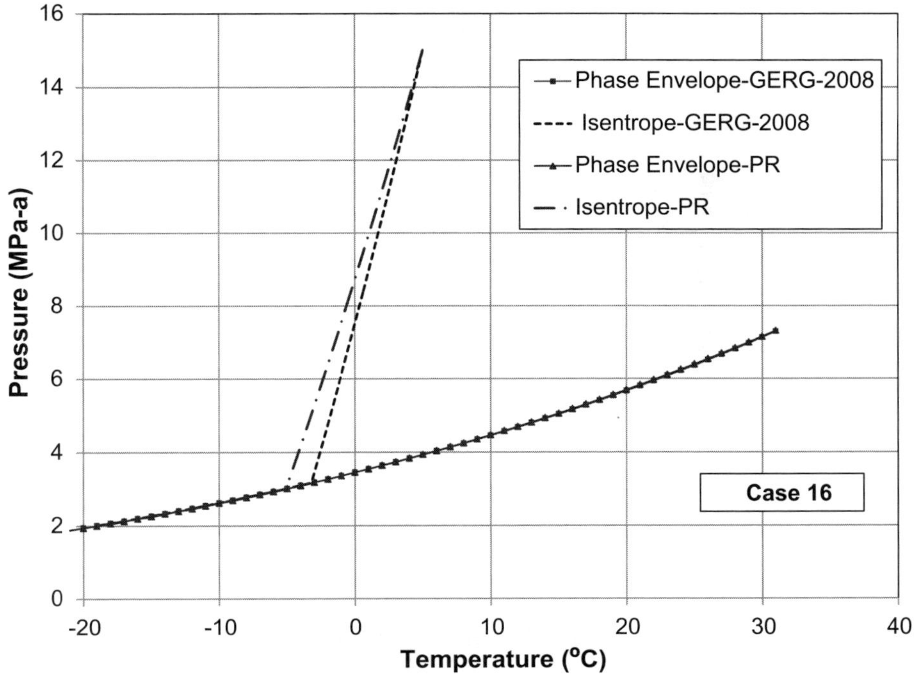

Figure 6-53. Pressure–temperature isentropes based on GERG-2008 and PR EOS for CO_2+H_2S binary (Case 16)

6.12 IMPLICATIONS RELATIVE TO FRACTURE TOUGHNESS REQUIREMENTS FOR CO_2 PIPELINES

The results presented in the preceding sections have shown that there can be significant differences between the decompression behavior calculated using the GERG-2008 and PR EOS. It is important to understand whether the implications of these differences are significant in terms of fracture behavior, and particularly whether they have a practically important effect on the pipe toughness required for the arrest of a propagating ductile fracture.

Arrest toughness is typically calculated using the Battelle two-curve method (TCM), which is implemented iteratively by varying fracture resistance (expressed as Charpy shelf energy) to establish tangency between curves expressing the variation of fracture speed and decompression wave speed as a function of pressure [11]. The background, history and derivation of this method are discussed extensively in Chapter 7.

As is apparent in the results already described, the decompression curve for CO_2-based mixtures typically shows a very long, constant pressure plateau. Figure 6-54 shows an example of the TCM for pure CO_2. In this case, it can be seen that tangency occurs on the pressure plateau; in fact, there will a limited number of cases for which practically achievable arrest conditions can be determined below the plateau. As a result, the procedure for determining arrest toughness for most CO_2 pipelines can be simplified; it is necessary to ensure that the arrest pressure (the intercept of the fracture velocity curve on the pressure axis) is higher than the saturation pressure (corresponding to the plateau pressure) for all conditions within the pipeline operating envelope.

The arrest pressure can be determined, for specific combinations of strength, diameter, wall thickness. and toughness, using the same equation for arrest stress as is used in the two-curve approach:

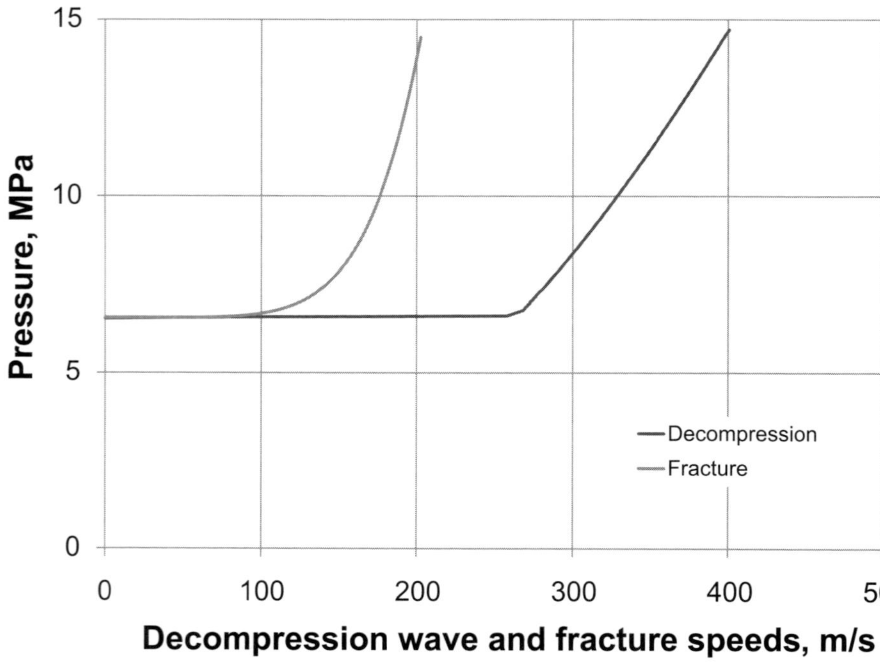

Figure 6-54. Example of determination of arrest toughness for pure CO_2 using the two curve method

$$\sigma_a = \frac{2\bar{\sigma}}{3.33\pi} \arccos \exp\left[\frac{-12.5\pi.CV.E}{24\bar{\sigma}^2\sqrt{Rt}}\right] \qquad (6-4)$$

where

- σ_a is the arrest stress, MPa;
- $\bar{\sigma}$ is the flow stress, conventionally set equal to SMYS + 69 MPa;
- CV is the Charpy V-notch energy measured on a full-size specimen;
- E is Young's modulus, MPa;
- R is one-half the pipe OD, mm; and
- t is the pipe wall thickness, mm.

The form of this equation is such that the arrest pressure increases with increasing Charpy energy, but is ultimately limited by a flow-stress controlled value that depends only on pipe geometry and strength. As an example, Figure 6-55 shows the arrest pressure curve for a 406.4 mm OD, 12.7 mm wall thickness, API Grade 415 (×60) pipeline operating at 14.5 MPa.

The curve in Figure 6-55 can be used to illustrate the effect of variations in saturation pressure on arrest toughness. If the arrest pressure is compared with the results for Case 5, it can be seen that an increase in saturation pressure from 6.36 MPa, as predicted by the PR EOS, to 7.3 MPa, as predicted by the GERG-2008 EOS, leads to an increase in minimum required toughness from approximately 28 J to 42 J. This is clearly not a trivial difference, in terms of pipe specifications. Note also the asymptotic behavior of the arrest pressure curve. When the saturation pressure is higher, arrest can only be achieved by increasing the wall thickness. This example illustrates the importance of predicting the saturation pressure with good accuracy, and emphasizes the need for further progress in refining equations

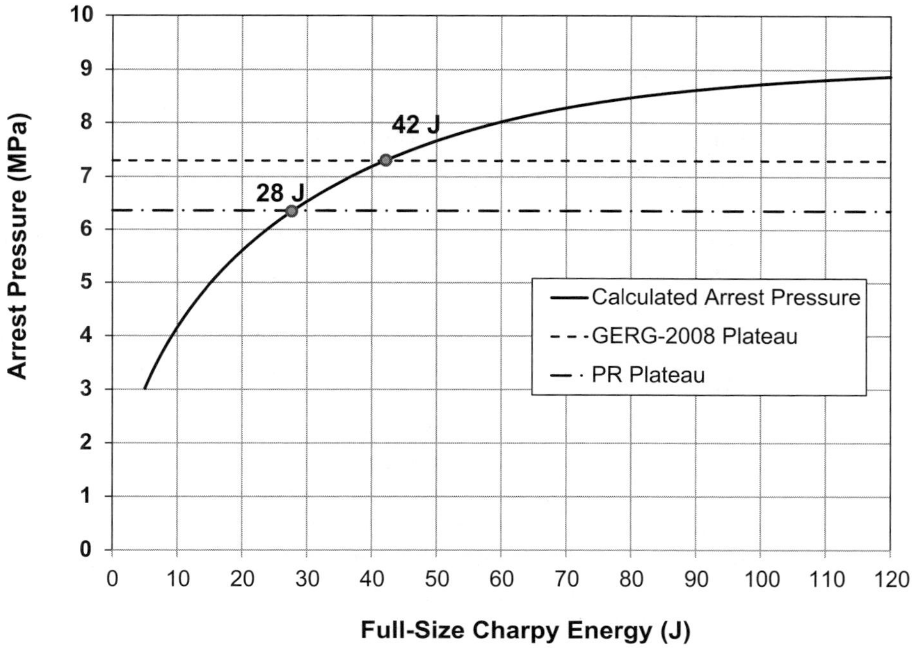

Figure 6-55. Effect of Charpy energy on arrest pressure for 14.5 MPa, 406.4 mm OD, Grade 415 (×60)

of state for CO_2-based mixtures, particularly those containing impurities with low critical temperatures.

As mentioned earlier, in Section 6.4, the deviations that have been seen, in some cases, in the initial speed of sound and in the upper branch of the decompression curve may appear to be of limited significance for the present application, since the saturation pressure alone controls the fracture behavior. However, in the method that has been applied in the work reported here, the initial speed of sound and the decompression path determine the point at which the decompression isentrope intersects the phase boundary, and they are thus of primary importance. As has been mentioned, it is also possible to determine the saturation pressure, for a given mixture and initial conditions, simply by calculating the initial entropy, and comparing it with the phase boundary expressed in terms of pressure and entropy [4,5]. Unless the software used to implement the EOS allows for the direct computation of phase diagrams in terms of entropy, however, this process may be more laborious than the calculation of the full decompression curve.

6.13 EXPERIMENTAL DETERMINATION OF DECOMPRESSION WAVE SPEED

As has been emphasized, given the sometimes large discrepancies between the predictions of different EOS and the potentially serious implications for specification of material toughness, it is of vital practical importance to obtain experimental validation, as well as indications of areas where improvement in thermodynamic calculation methods is required. Extensive experimentation has been carried out using a shock tube facility to conduct such research on natural gas mixtures. Several years of experience, involving a wide range of compositions and initial conditions, have indicated that shock tube testing is an effective and economical means of carrying out such investigations [14–21]. In particular, decompression wave characteristics can be accurately determined for a fraction of the cost involved in full-scale fracture testing. Similar test methods could easily be applied to CO_2 mixtures.

The facility has been constructed at TransCanada Gas Dynamic Test Facility in Didsbury, Alberta, Canada, for the specific purpose of studying decompression behavior from typical pipeline conditions. It has been in operation since 2002 [14–21]. The tube consists of four spool pieces of NPS 2 stainless-steel pipe making up a total length of 42 m, as shown in Figure 6-56. They are made of NPS 2 × 11.1 mm WT, SCH XX, ASTM A312, 316 SS seamless tube (I.D. = 38.1 mm). All individual spools were designed for 41.370 MPa pressure, design factor = 0.8 and location factor = 0.625 (according to the latest edition of the CSA Standard Z662). ANSI 2500 RTJ flanges (PN 420 Standards) ASTM A182-316 SS were used. All spools were shop-tested to a minimum hydrostatic pressure of 62.055 MPa and a maximum hydrostatic pressure of 62.755 MPa for a period of one hour. Spool #4 is equipped with an automated blowdown and a pressure relief valve.

All spool pieces were internally honed to a roughness R_Z less than 25 μ-in. (0.635 μm). This surface roughness was achieved by a combination of grades and grit sizes of abrasive stones and use of a double spindle hone, and simultaneous rotation and reciprocation of the hone within the bore of the tube. Kerosene is used as a coolant and lubricant, and the final finish is achieved by a polishing/buffing head. Roughness measurements according to DIN (4777) and ISO (4287:1997) were conducted using a Hommel Profile Stylus Model T500 at various locations inside the tube from both ends up to the maximum distance permitted by the Stylus reach.

A total of 16 Endevco dynamic pressure transducers are mounted along the length of the shock tube, 13 of which are mounted on the front spool where the rupture disc is located. These transducers are mounted flush to the internal surface of the shock tube to a tolerance of 0.0 mm to –0.2 mm using a specially-designed fitting. Table 6-2 shows the locations of these pressure transducers measured from the raised face of the front flange, which is the location where the flow will choke following disc rupture. The exact locations of the pressure transducers on the front spool of the shock tube were used in the determination of the speed of sound and the propagation speed of the decompression wave. There is also a static pressure transducer (by Yokogawa) mounted on spool #1. Two fast response temperature transducers (by Watlow) are located at indicated locations of spools #1 and #2, and a static temperature probe by Yokogawa on spool #3 (see locations in Table 6-2).

A rupture disc is placed at the front end of the tube, which, upon rupturing, causes a decompression wave to propagate up the 42 m pressurized shock tube. The rupture discs used are commercial Fike SCRD BT FSR type, along with Fike's SRX rupture disc-holder. The discs are *x*-scored, which provides non-fragmenting rupture when ruptured

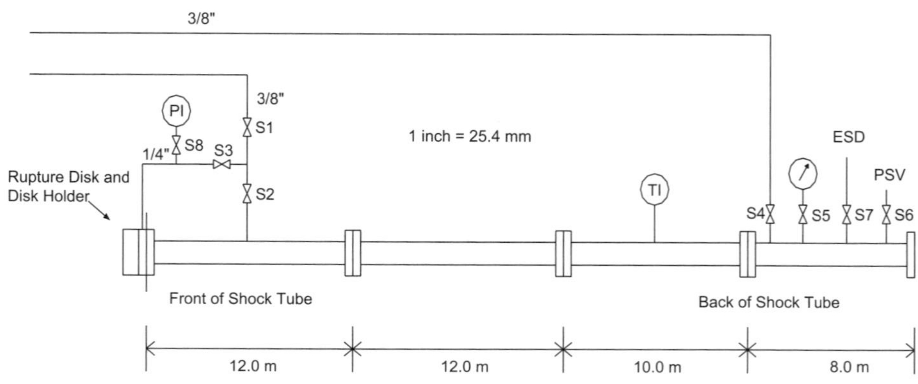

Figure 6-56. Schematic of the Shock Tube Setup

TABLE 6-2. Location of the 13 Endevco pressure transducers mounted on spool #1 and temperature probes on spools #1, #2, and #3

	Location	Distance(X) from Choked Location (mm)	Normalized Distance (X/D)
1st Spool	(Flange RF)	0	0.00
	PT1A	92.4	0.77
	PT1B	102.8	2.70
	PT2	200	5.25
	PT3	350	9.19
	PT4	500	13.12
	PT5	700	18.37
	PT6	900	23.62
	PT7	1100	28.87
	PT8	3100	81.36
	PT9	5100	133.86
	PT10	7100	186.35
	PT11	9100	2388.85
	Temp (1)	4000	104.99
2nd Spool	Temp (2)	18000	472.44
3rd Spool	Temp (3)	23000	603.67

as shown in Figure 6-57. The disc holder is made out of carbon steel and is designed to achieve a tight metal-to-metal seal, while maintaining tight rupture tolerances. Once the disc is ruptured, choked flow conditions will be established at the front surface of the front flange on the shock tube. The open membrane area matches the I.D. of the disc holder which is 63.5 mm in diameter, while the tube I.D. is 38.1 mm. This means that choking is guaranteed at the face of the aforementioned tube flange and not at the rupture disc. Additionally, the triangular pieces of the opened ruptured disc adhere to the rupture disc holder (I.D. = 63.5 mm).

In order to control the temperature of the gas mixture in the shock tube and to achieve a desired temperature before rupture, the shock tube is heat traced and insulated with 50 mm thick insulation (thermal conductivity = 0.036 W/m.K). The heat tracer is rated for 300 W/m^2, which translates to 56.8 W/m length of the shock tube based on one run along the length of the tube. Figure 6-58 shows photos of the shock tube setup.

TABLE 6-3. Uncertainty analysis in determining the decompression wave speed

Parameter	Value	Units
P_i	15,000	(kPa)
T_i	35	(°C)
t_o	2	(ms)
L_o	320.5	(mm)
dL	2	(mm)
d_{t1}(resolution in Data Acquisition)	19.53	(μs)
d_{t2} (due to filtering/smoothing)	9.77	(μs)
dL/L_o	0.624025	(%)
d_{t1}/t_o	0.976500	(%)
d_{t2}/t_o	0.488250	(%)
(1) Uncertainty in decompression wave speed due to geometry and time	1.258	(%)
(2) Uncertainty in decompression wave speed due to 0.267% uncertainty in pressure	0.050	(%)
(3) Uncertainty in decompression wave speed due to 0.5°C uncertainty in temperature	0.081	(%)
(4) Uncertainty in decompression wave speed due to mixture composition	0.183	(%)
Total Uncertainty (1 through 4 above)	1.274	(%)

Figure 6-57. Rupture Discs (before and after rupture) [16]

The shock tube is connected to an auxiliary system inside a heated building close to the tube. The auxiliary system includes a vacuum pump, a charge compressor, as well as different gas cylinders as required to achieve the compositions required for testing. The piping and instrumentation diagram for this auxiliary system is shown in Figure 6-59. The vacuum pump is an Edwards Model # E2M30, which consists of a direct drive, rotary vane, double stage vacuum pump, with a displacement of 22.3 cfm (631 L/min), pumping speed 19.5 cfm (552 L/min), and ultimate vacuum total pressure of 7.5×10^{-4} torr (without gas ballast). The charge compressor is a Haskel Gas booster model AGT-30/75, which consists of a large area reciprocating air driven piston directly coupled by a connecting rod to a small area gas piston. The compressor is capable of a discharge pressure up to 110 MPa (pressure ratio of 60:1).

Figure 6-58. Photos of the Shock Tube Facility at TransCanada Gas Dynamic Test facility, Didsbury, Alberta, Canada [16]

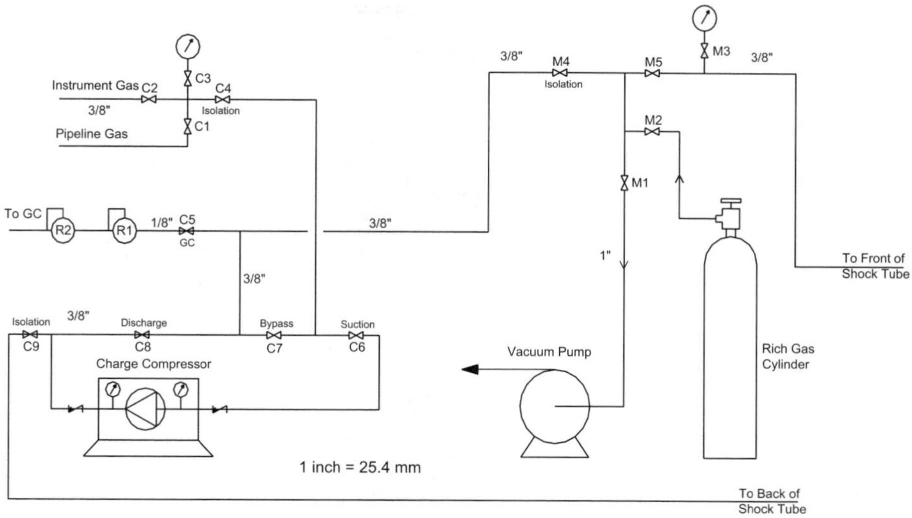

Figure 6-59. Schematic of the Shock Tube Auxiliary System [16]

A desired gas mixture in the test section is attained by mixing gas from the supply cylinders, starting with the heavier components and topped off with the predominant component (methane for natural gas mixtures). A simple mass balance calculation is performed to determine the amount and composition of the supplementary gas required, based on the main loop gas composition, the desired final composition, the volume of the test section together with the connecting system, and the final pressure. This calculation is also confirmed by actual measurement of the final gas composition by gas chromatography (GC) of a gas sample taken after sufficient mixing time within the shock tube loop.

The fast response pressure transducers are Endevco Model 8511A-5K piezoresistive pressure transducers, which are rated for safe operation up to 138 MPa, and for temperatures down to –73°C. They are supplied with certificates of traceable calibration curves with an overall uncertainty of 40 kPa and a frequency response up to 20 kHz. They feature an active four-arm strain gage bridge diffused into a sculptured silicon diaphragm for maximum sensitivity and wideband frequency response. Endevco transducers also feature excellent linearity and high rupture resistance (138 MPa). Calibrations of these transducers are performed before and after rupture of each test using a Yokogawa pressure transmitter (Model EJA530A/HAC).

The temperature probes are Watlow fast response monolithic thermocouples together with Analog Devices AD594/AD595 amplifiers with cold junction compensation. The tip of the probe has been specially tapered from 1.65 mm down to 0.51 mm. The pressure isolation is provided by a standard 1/8 NPT "Swagelok" plug that has been drilled through, and the 1.65 mm probe has been welded into it. The response time is around 20 ms, and the uncertainty is 1°C from –40°C to +100°C. They are calibrated against a static Yokogawa temperature transmitter (Model YTA110). Figure 6-60 shows photos of the mixing station and the compression setup.

The data acquisition system is a HP E1431A and E1432A High Speed DAS, capable of $\pm$40 kPa of measurement uncertainty at 1 kHz, and the corresponding cross channel phase matching is $\pm$0.125 deg. Data is collected in a block of 16,384 data points; hence, for a window of 320 ms, the data resolution is exactly 19.53 μs.

The test procedure involves six main steps starting from evacuating the main decompression tube, associated header and tubing, up to the final step of rupturing the disc to atmosphere. These are: (1) purging with N_2 gas, (2) evacuation, (3) gas charging from

Figure 6-60. Photos of the Mixing Station and the Compression Setup [16]

heaviest to lightest, (4) topping off with natural gas and charging with the Haskel compressor up to 90% to 92% of the burst pressure, (5) recycling the loop to ensure good mixing, and (6) disc rupture by means of increasing pressure in the tube. Temperature control is achieved by the heat tracer with a set point at the desired gas temperature of the particular test. Figure 6-61 shows a photo taken following a disc rupture.

Figure 6-62 shows an example of the pressure–time traces as measured by the 13 pressure transducers along the first spool of the rupture tube. The experimentally determined decompression wave speed is obtained from the slope of the distance-time linear fit of the time of arrival of the decompression wave at a given pressure at all the front 4 or 5 locations of the pressure transducers. The slope of the linear fit to this data represents the decompression wave speed at this pressure. The procedure is repeated for the full pressure range, from

Figure 6-61. Photo taken during a disc rupture in one of the shock tube tests [16]

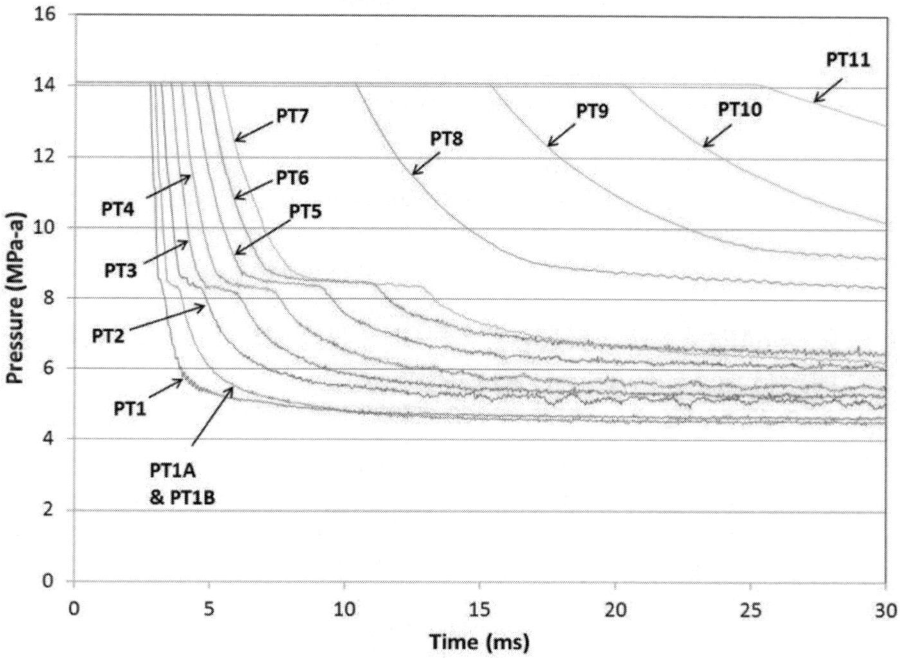

Figure 6-62. Example of typical pressure–time traces obtained from a Shock Tube Test on rich natural gas [19]

the initial pressure to the minimum pressure attained by the farthest transducer of the set used in the linear fit. Note the apparent kinks in the pressure–time profiles around 8.15 MPa in Figure 6-62. This is a manifestation of the decompression wave isentrope crossing the phase envelope into the two-phase region. It is known that at a phase crossing, a sharp drop in the speed of sound occurs due to the dramatic drop in the thermodynamic derivative.

A similar, but larger diameter, shock tube installation (DN150) is described by Cosham et al. [13].

The uncertainty in the determination of the decompression wave speed from the present measurements is introduced by the following four uncertainties:

(1) The uncertainty in the measurements of distances and time according to the following definition:

$$\frac{\delta W}{W} = \sqrt{\left(\frac{\delta L}{L_o}\right)^2 + \left(\frac{\delta L}{t_o}\right)^2 + \left(\frac{\delta t_2}{t_o}\right)^2} \qquad (6-5)$$

where

- L_o is the distance used in the determination of the decompression wave speed.
- t_o is the time between signals from the transducers a distance (L_o) apart.

The uncertainty δt has two components, namely,

- δt_1 is the uncertainty due to resolution of the data acquisition system and hence uncertainty in the determination of the time of arrival of the frontal wave;
- δt_2 is the uncertainty due to filtering/smoothing of the pressure–time profile.

(2) The uncertainty in the pressure.
(3) The uncertainty in the temperature.
(4) The uncertainty in mixture composition, which is assumed to be 2% of the mole % value of all components in the mixture.

Table 6-3 gives estimated values for the above individual uncertainties based on the linear fit of the four or five *X-t* points. It is shown that the uncertainty in the decompression wave speed is estimated at 1.27%, and the main contributor to it is δt_1 (uncertainty due to sample rate of the data acquisition procedure).

6.14 EXAMPLE OF SHOCK TUBE TEST RESULTS FOR A RICH NATURAL GAS MIXTURE

Results of one shock tube test will be presented here to demonstrate the usefulness of conducting shock tube tests to validate decompression wave speeds in a more cost effective way, rather than conducting full-scale rupture tests. The test conditions for this example are given in Table 6-4, and the corresponding phase envelope, as predicted by PR EOS, is shown in Figure 6-63.

Figure 6-62 shows the pressure–time traces as measured by the 13 pressure transducers on the first spool of the rupture tube in this test. Note the apparent kinks in the pressure–time profiles around 8.2 MPa. This is a manifestation of the decompression wave isentrope crossing the phase envelope into the two-phase region. It is known that at a phase crossing, a sharp drop in the speed of sound occurs due to the dramatic drop in the thermodynamic derivative. As mentioned earlier, the experimentally determined decompression wave speed is obtained from the slope of the linear fit of the time of arrival of the decompression wave at a given pressure against distance from the origin, at all the front 5 locations of the pressure transducers (i.e., PT1 through PT3). The slope of the linear fit to this data represents the decompression wave speed at this pressure. The procedure is repeated for the full pressure range, from the initial pressure to the minimum pressure reached by the farthest transducer of the set used in the linear fit. The results are shown in Figure 6-64, along with prediction by GASDECOM. Three parts are shown in the decompression wave speed curves, the part above the plateau, the plateau, and the part below the plateau. A good agreement is shown

TABLE 6-4. Conditions and mixture composition for the example shock tube test [19]

Pi (kPa-a)	14,182
Ti(°C)	12.20
	Mole %
C1	80.4083
C2	7.3598
C3	7.3078
iC4	0.8173
nC4	1.0930
iC5	0.0000
nC5	0.0000
C6+	0.0013
N_2	0.4619
CO_2	2.5507
Sum	100.00

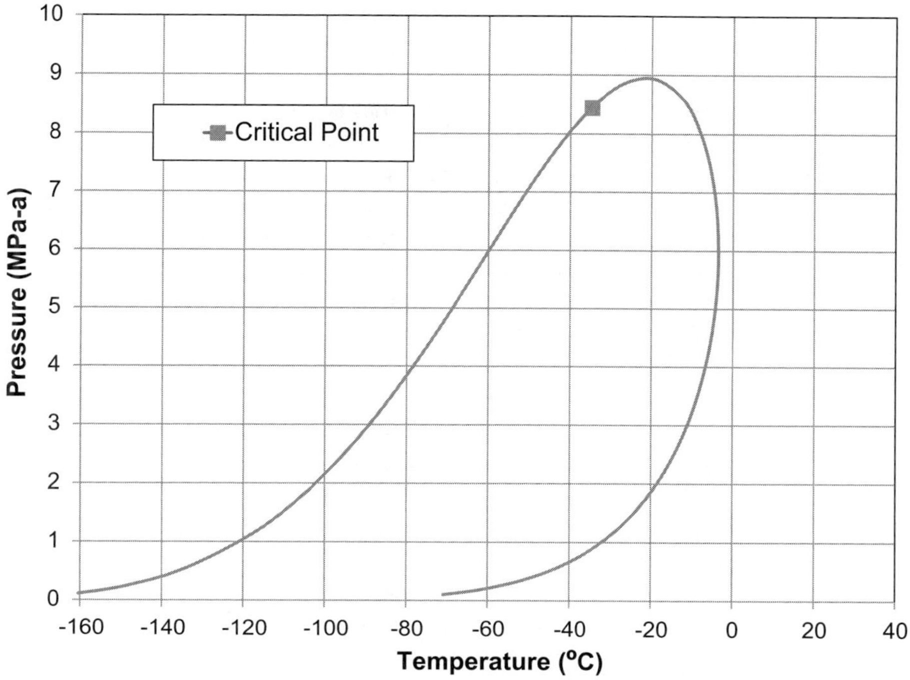

Figure 6-63. Phase envelope of the gas mixture used in the example shock tube test

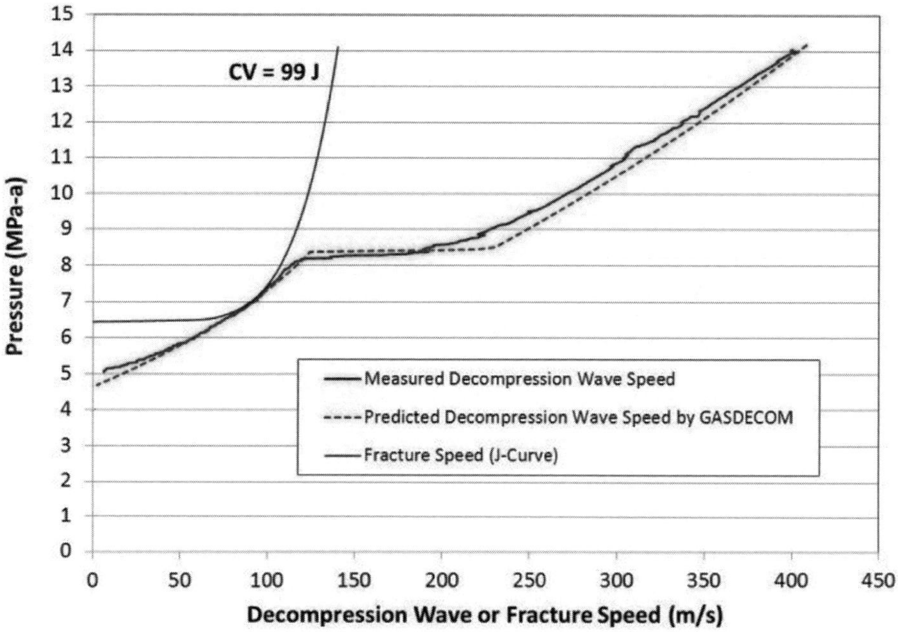

Figure 6-64. Measured vs. GASDECOM prediction of the decompression wave speed for the example shock tube test [19]

between measured and predicted speeds on both upper and lower branches, and also between the measured and calculated plateau pressure.

In order to evaluate the potential effect of such results on the required arrest toughness, an example of pipe design conditions is considered. The example is for pipe of OD = 406.4 mm, steel grade = 483 MPa, design factor = 0.72, and design pressure corresponding to the initial test pressure specific the shock tube test presented here. For the example shock tube test results, the required material toughness to arrest ductile fracture was found to be 99 J. The tangency condition is also shown in Figure 6-64. Note that, for this case, the tangent point occurs on the lower branch of the decompression wave speed curves; the measured and predicted wave speeds are in very good agreement in the tangency region, so that the toughness value of 99 J fits both curves.

The above findings were also applicable when the measured decompression curve and those predicted according to the GERG-2008 and PR EOS were compared, as shown in Figures 6-65 and 6-66, respectively. This is despite the fact that the predictions of the plateau pressures by both EOS are higher than that measured and predicted by GASDECOM (see Figure 6-64). Nevertheless, because tangency occurs in a region where all three methods are in agreement, GASDECOM, GERG-2008, and PR EOS lead to the same arrest toughness.

Figure 6-67 shows a comparison between decompression wave speeds (normalized w.r.t. initial speed of sound) obtained from a full-scale rupture test [27] and that obtained from a shock tube test conducted for the same natural gas mixture composition and from initial pressure and temperature given in Table 6-5. Very good agreement is demonstrated between the two set of test results, again indicating that shock tube testing is an effective and economical means of carrying out such investigations.

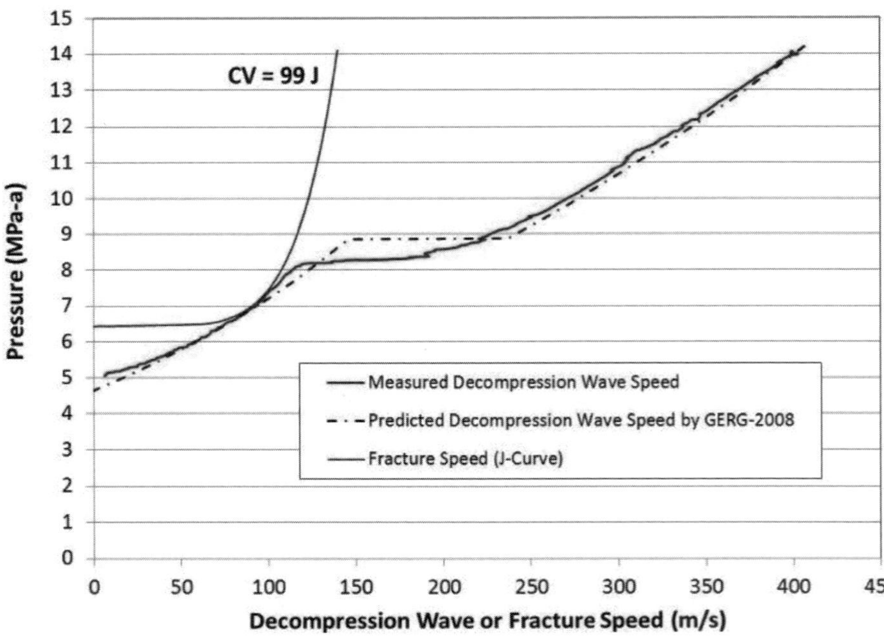

Figure 6-65. Measured vs. GERG-2008 prediction of the decompression wave speed for the example shock tube test

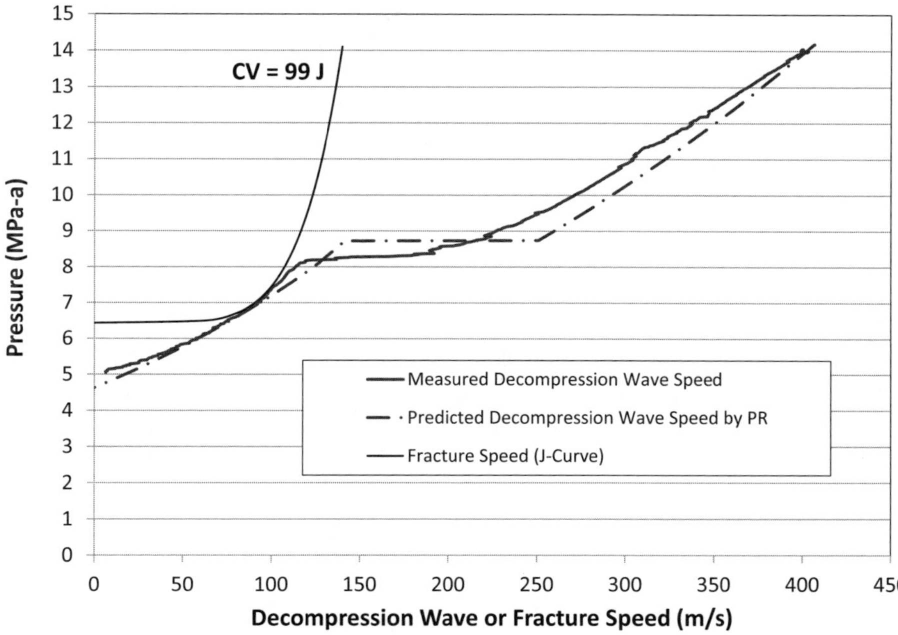

Figure 6-66. Measured vs. PR prediction of the decompression wave speed for the example shock tube test

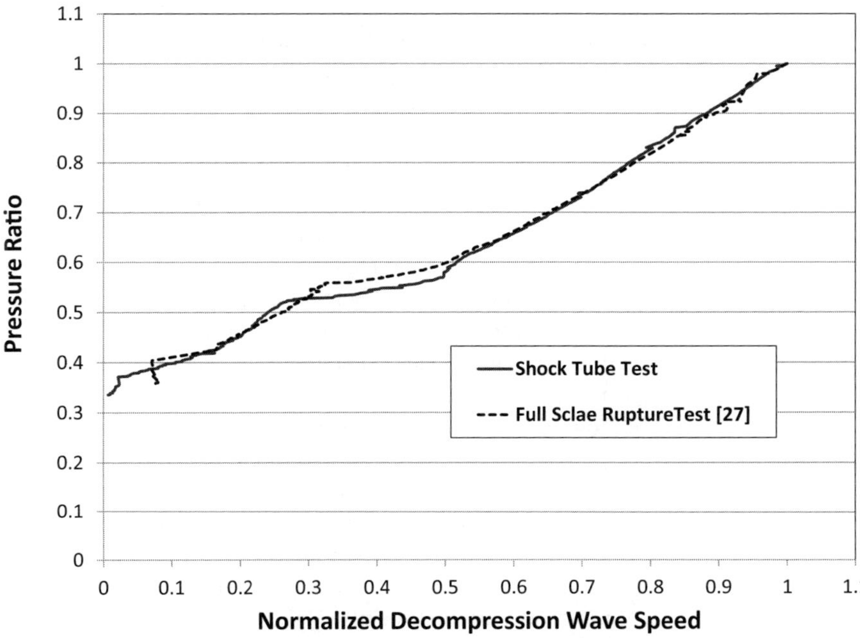

Figure 6-67. Comparison between results of a full-scale rupture test and a shock tube test on the same natural gas mixture composition and initial conditions

TABLE 6-5. Conditions and mixture composition for the example full-scale rupture test and the shock tube test shown in Figure 6-67 [27]

Pi (kPa-a)	12,105
Ti(°C)	16.50
	Mole %
C1	80.09
C2	15.22
C3	3.29
iC4	0.272
nC4	0.608
iC5	0.024
nC5	0.0000
C6+	0.001
N_2	0.045
CO_2	0.442
Sum	100.0060

6.15 SUGGESTIONS FOR FUTURE WORK

1. It appears that GERG-2008 EOS differs from PR EOS in overall prediction of decompression wave speed of mixtures of CO_2 with impurities. In the development of the GERG-2008 EOS, it was noted that limited data was used to fit three of the CO_2 binaries with impurities, namely H_2, CO and Ar. For the other CO_2 binaries, extensive data was available to allow better regression fit within GERG-2008 EOS. The following are suggestions for future work to enhance the prediction capability of GERG-2008 and/or PR EOS, in particular for impurities containing H_2, CO and Ar:
   a) Thorough literature search to find additional *PρT* and *VLE* data for the aforementioned three binaries that lacked data (and others, as deemed necessary to support analysis of anthropogenic mixtures that were not included among the data set used in the initial development of GERG-2008 and PR). For example, the same regression procedure could be performed to determine a new set of coefficients inherent in GERG-2008 and test the newly developed version with experimental data of all binaries and anthropogenic mixtures, particularly any speed of sound data that are published post GERG-2004 and GERG-2008.
   b) Find a means, based on fundamental thermodynamic principles, to extract the *PρT* data along isentropes generated from experimental shock tube tests. The challenge here is that the decompression wave speed contains both speed of sound and outflow velocity, both of which are *derivatives* from *PρT* relationships. The phase boundary crossing manifested in the plateau pressure will also give information about the *VLE* behavior of each binary and anthropogenic mixture tested. Again, a method based on equilibrium principles is needed to extract the information required to better fit the binary coefficients used in GERG-2008 and/or PR EOS from the data generated from shock tube tests, in an attempt to determine "adjusted" constants and coefficients. Validation would then follow to test that these "adjusted" coefficients, when implemented in GERG-2008 or PR EOS, will lead to a better fit to the experimental decompression wave speed and plateau pressure. This will constitute a significant contribution and advance to this EOS.

2. Once an improved version of GERG-2008 or PR EOS is obtained, development of a decompression model based on the newly developed EOS will be very straightforward, easy to implement and will be a useful tool for use with CO_2 containing impurities.

6.16 CLOSING REMARKS

Various design and operation aspects of CO_2 pipeline transportation, including, in particular, the aspect of decompression wave speed and implications for the correct determination of the material toughness to arrest ductile fracture, has been recognized since early 1980s [28,29]. Many authors have since paid attention to multiple issues related to the effects of impurities on decompression wave speed and this has resulted in significant contributions by several studies [e.g., 3–6,13,30–33]. These recent efforts have been driven predominantly by the heightened attention to CCS as a potential method to deal with CO_2 emission and concerns surrounding climate change.

REFERENCES

[1] IPCC Special Report on Carbon Dioxide Capture and Storage, B. Metz O. Davidson H. de Coninck, M. Loos and L. Meyer, Eds., 2005, Intergovernmental Panel on Climate Change, Cambridge University Press, Cambridge.

[2] Gale, J. and Davidson, J., 2004, "Transmission of CO_2-safety and economic considerations," J. Energy Progress 6(4), p. 219.

[3] Nahas, G. and Mohitpour, M., 2010, "Engineering Pipelines for Transportation of CO_2 With Impurities," Paper IPC 2010-31651, Proceedings of the 8th International Pipeline Conference, ASME, Calgary, Alberta, Canada.

[4] Cosham, A., Eiber, R. J. and Clark, E. B., 2010, "GASDECOM: Carbon Dioxide and Other Components," Paper IPC 2010-31572. Proceedings 8th International Pipeline Conference, ASME, Calgary, Alberta, Canada.

[5] Cosham, A. and Eiber, R. J., 2008, "Fracture Control in Carbon Dioxide Pipelines-the Effect of Impurities," Paper IPC 2008-64346. Proceedings 7th International Pipeline Conference, ASME, Calgary, Alberta, Canada.

[6] Seevam, P. N., Race, J. M., Downie, M. J. and Hopkins, P., 2008, "Transporting the Next Generation of CO_2 for Carbon Capture and Storage: The Impact of Impurities on Supercritical CO_2 Pipelines," Paper IPC2008-64063, Proceedings of the 7th International Pipeline Conference, Calgary, Alberta, Canada.

[7] Eiber, R. J., Bubenik, T. A. and Maxey, W. A., 1993, "Fracture Control for Natural Gas Pipelines." PRCI report, Catalogue No. L51691.

[8] Eiber, R. J., Carlson, L. and Leis, B., 2004, "Fracture Control Requirements for Gas Transmission Pipelines." Proc. 4th Int. Conf. on Pipeline Technology, Ostend, Scientific Surveys, Beaconsfield, UK (2004), p. 437.

[9] Eiber, R. J. and Maxey, W. A., 1977, "Full-Scale Experimental Investigation of Ductile Fracture Behaviour in Simulated Arctic Pipeline," Materials Engineering in the Arctic, ASM, Metals Park, OH, p. 306.

[10] Maxey, W. A., Kiefner, J. C., Eiber, R. J. and Duffy, A. R., 1972, "Ductile Fracture Initiation, Propagation and Arrest in Cylindrical Vessels", Fracture Toughness, ASTM STP 514, Philadelphia, PA, p. 70.

[11] Maxey,W. A., 1974, "Fracture Initiation, Propagation and Arrest", 5th Symposium on Line Pipe Research, PRCI Catalogue No. L30174, p. J-1.

[12] Rothwell, A. B., 1997, "Fracture Propagation Control Measures for Gas Pipelines," Proc. Int. Seminar on Fracture Control in Gas Pipelines, ISBN 0 909539-73-1, WTIA, Sydney, Australia, part. 6-1.

[13] Cosham, A., Jones, D., Armstrong, K., Allason, D. and Barnett, J., 2011, "The decompression Behaviour of CO2 in the Gas Phase," International Forum on the Transportation of CO_2 by Pipeline Hilton Gateshead-Newcastle Hotel, Gateshead, UK, 22-23 June, 2011.

[14] Botros, K. K., Studzinski, W., Geerligs, J. and Glover, A., 2004, "Determination of Decompression Wave Speed in Rich Gas Mixtures," Can. J. Chem. Eng., vol. 82, number 5, pp. 880–891, October 2004.

[15] Botros, K. K., Geerligs, J., Zhou, J. and Glover, A., 2007, "Measurements of Flow Parameters and Decompression Wave Speed Following Rupture of Rich Gas Pipelines, and Comparison with GASDECOM," Int. Journal of Pressure Vessel and Piping, vol. 84, pp. 358–367.

[16] Botros, K. K., Geerligs, J. and Eiber, R. J., 2010, "Measurement of Decompression Wave Speeds in Rich Gas Mixtures at High pressures (370 bar) Using Specialized Rupture Tube," ASME Journal of Pressure Vessel Technology, vol. 132, p. 051303-1, October 2010.

[17] Botros, K. K., Geerligs, J. and Eiber, R. J., 2010, "Decompression Wave Speed in Rich Gas Mixtures at High Pressures and Implications on Fracture Control Toughness Requirements in Pipeline Design," ASME 8th International Pipeline Conference & Exhibition, TELUS Convention Centre, Calgary, Alberta, Canada, September 27 through October 1, 2010.

[18] Botros, K. K., Geerligs, J., Rothwell, B., Carlson, L., Fletcher, L. and Venton, P., 2010, "Effects of Pipe Internal Surface Roughness on Decompression Wave Speeds in Natural Gas Mixtures," ASME 8th International Pipeline Conference & Exhibition, TELUS Convention Centre, Calgary, Alberta, Canada, September 27 through October 1, 2010.

[19] Botros, K. K., Geerligs, J., Rothwell, B., Carlson, L., Fletcher, L. and Venton, P., 2010, "Transferability of Decompression Wave Speed Measured by a Small Diameter Shock Tube to Full Size Pipelines and Implication for Determining Required Fracture Propagation Resistance," Int. Journal of Pressure Vessel and Piping, vol. 87, pp. 681–695.

[20] Botros, K. K., 2010, "Measurements of Speed of Sound in Lean and Rich Natural Gas Mixtures at Pressures up to 37 MPa using a Specialized Rupture Tube," Int. J. of Thermophysics, Vol. 31, pp. 2086–2102.

[21] Botros, K. K., 2011, "Advances in Decompression Wave Speed Measurements in Natural Gas Mixtures Using a Specialized Shock Tube," Proceedings of Rio Pipeline Conference & Exposition 2011, Paper #1474_11, Rio de Janeiro, Brazil, September, 20–22.

[22] GASDECOM, computer code for the calculation of gas decompression speed-included in "Fracture Control Technology for Natural Gas Pipelines," by R. Eiber, T. Bubenik, and W.A. Maxey, NG-18 Report 208, A.G.A. Catalogue No. L51691 (1993).

[23] Kunz, O., Klimeck, R., Wagner, W. and Jaeschke, M., 2007, "The GERG-2004 Wide-Range Equation of State for Natural Gases and Other Mixtures," Groupe Européen de Recherches Gazières (GERG), Technical Monograph, GERG TM15.

[24] AspenTech, Aspen Plus® V7.1 Simulation Engine, 200 Wheeler Road, Burlington, Massachusetts 01803, [http://www.aspentech.com/].

[25] NIST Standard Reference Database 23, NIST Reference Fluid Thermodynamic and Transport Properties Database (REFPROP): Version 9.0, www.nist.gov/ts/msd/srd/nist23.cfm.

[26] Peng D. Y. and Robinson, D.B. 1976. Ind. Eng. Chem. Fundamentals, vol. 15.

[27] Eiber, R. and Carlson, L., 2000, "Fracture Control for the Alliance Pipeline," ASME Int. Pipeline Conference, Volume 1, pp. 267–277.

[28] King, G. G., 1986, "Design Considerations for CO_2 Pipe Lines," Pipe Line Industry, Nov. 1981, pp. 125–132.

[29] Maxey, W. A.: "Long Shear Fractures in CO_2 Lines Controlled by Regulating Saturation, Arrest Pressures," Oil and Gas Journal, p. 44–46.

[30] Mohitpour, M., Jenkins, A. and Nahas, G., 2008, "A generalized overview of requirements for the design, construction, and operation of new pipelines for CO_2 sequestration, Journal of Pipeline Engineering, Vol. 7, No. 4, p. 237–251.

[31] Oosterkamp, A. and Ramsen, J., 2008, "State-of-the-Art Overview of CO2 Pipeline Transport with relevance to offshore pipelines, Report No. POL-O-2007-138-A, Polytec, Norway.

[32] King, G., Zarzour, O., Afachtal, T., Kumar, S. and Balasubramanian, M., 2009, "Design of dense phase carbon dioxide pipelines," 3rd International CO_2 Forum, SOGAT Conference, Abu Dhabi, UAE, 2009 April 2.

[33] King, G. G. and Kumar, S., 2010, "Designing CO_2 Transmission Pipelines without Crack Arrestors," ASME 8th International Pipeline Conference & Exhibition, TELUS Convention Centre, Calgary, Alberta, Canada, IPC2010-31678, September 27 through October 1, 2010.

Chapter 7

PIPE MATERIAL SELECTION AND FRACTURE CONTROL

7.1 GENERAL INTRODUCTION

Despite the very different properties of CO_2, relative to natural gas, conventional pipeline materials and construction practices can be used provided certain basic precautions are observed. Clearly, the most important of these is to ensure that no free water is present in the system (see Chapter 3) since the formation of carbonic acid can lead to extremely aggressive corrosion wherever the steel internal surface is directly exposed. This requires limits on water content that are appropriate to prevent formation of free water under the most stringent operating conditions; potential upset conditions also need to be considered. In addition to very close control, during operation, of the pipeline receipt composition, it is also necessary to pay strict attention to dewatering and drying after hydrostatic testing, to ensure that water cannot remain in low points, dead legs, crevices or appurtenances. It may be necessary to delay the installation of certain types of valve, whose internal design might make complete drying difficult and which do not have corrosion-resistant trim, until after hydrostatic testing. If these precautions are adhered to, then the use of corrosion-resistant alloys or liners will not be necessary. Caution should be exercised in the use of internal coatings intended for flow optimization since they may become disbonded as a result of CO_2 permeation or, for some coating formulations, solvent action. If severe, such disbondment and detachment could have significant consequences for the operability of downstream facilities.

Essentially all CO_2 pipelines in operation today have used conventional, carbon steel piping materials. Obviously, where the impure CO_2 stream contains H_2S in excess of the limiting concentration specified by the governing standards and regulations, additional requirements for sour service come into force. Typically, these involve provisions for resistance to environmental cracking and limited maximum hardness in the pipe body and weld. Similarly, conventional field welding processes and consumables have been successfully applied in the construction of CO_2 pipelines, though more stringent limitations on internal flaws may be advisable, to minimize the potential for crevices that could trap hydrostatic test water. Again, for sour service, additional limitations typically apply, primarily related to procedure qualification, wall thickness transitions, maximum weld hardness, and non-destructive inspection acceptance standards (see, for example, Canadian Standards Association [1] Z662 2011 Clause 16.7).

The most important aspect of material selection for CO_2 pipelines relates to fracture control. The implications of the specific thermodynamic characteristics of CO_2 relative to the requirements for ductile fracture propagation control can be extremely important and can significantly influence the choice of pipe strength and wall thickness for a specific design pressure. In many cases, it will no longer be possible to select the thinnest pipe that is consistent with the requirements of design standards for pressure containment and minimum allowable wall thickness or maximum allowable D/t ratio. For these reasons, fracture control design is discussed in considerable detail in the following section.

Some caution needs to be exercised relative to the use of non-metallic materials. With the exception of gaseous phase service, glass fiber reinforced plastic pipe is not suitable for CO_2 pipelines. Polymers and other organic materials that are used in such applications as seals and gaskets need to be specified to resist chemical degradation in CO_2 service and, where applicable, to be suitable for the low temperatures that can occur at locations where rapid pressure reduction can occur. Some greases and other lubricants that are used in contact with the service fluid can be dried out by contact with CO_2 and again should be selected accordingly. In equipment which may experience a rapid drop in pressure, elastomeric seals may need particular attention since CO_2 can diffuse into elastomers and subsequently cause explosive decompression and blistering when the external pressure is released [2, 3].

7.2 FRACTURE CONTROL

7.2.1 Background

As was stated in the previous section, the most important single aspect of pipe materials selection and specification is the requirement for fracture control. While CO_2 does not present the fire or explosion hazards that hydrocarbon fluids do, it is an asphyxiating gas that can cause significant safety consequences in the event of a major leak or rupture, and in some instances, toxic impurities may be present. Because of this, pipeline standards and regulations require that materials for CO_2 pipelines be adequately specified for the prevention of brittle fracture, resistance to fracture initiation, and control of ductile fracture propagation. For example, CSA Standard Z662 clause 5.2.2.2 requires the use of Category II pipe (which has specified toughness properties) for all CO_2 pipelines with outside diameter 114.3 mm or more and wall thickness 6.0 mm or more [1]. Clause 5.2.2.3 also requires consideration of measures for positive control of fracture propagation. These requirements parallel those for pipelines transporting natural gas and HVP liquids. Similarly, ASME B31.4 paragraph 402.5 requires measures to limit the occurrence and length of fractures, with specific consideration of brittle fracture prevention and control of ductile fracture propagation [4]. DNV-RP-J202 Section 5.5 requires CO_2 pipelines to have adequate resistance to propagating fracture unless the hoop stress utilization factor is below 40% [5]. Overall, then, it can be seen that a detailed fracture control plan needs to be established for CO_2 transportation pipelines.

Despite the different nature of the hazard presented by the occurrence of fracture in a CO_2 pipeline, relative to a gas, the elements of a fracture control plan are the same. This is because CO_2 is highly compressible, and, unlike liquids, is capable of supplying large quantities of energy for the continuation of the fracture process once it is initiated. The three specific elements to be considered are as follows:

- Prevention of brittle fracture propagation,
- Optimization of resistance to fracture initiation, and
- Limitation of the length of ductile fracture propagation.

These subjects will be discussed in the appropriate detail in the following sections.

7.2.2 Prevention of Brittle Fracture

The prevention of brittle fracture propagation is the corner-stone of any fracture control plan, since brittle fracture can propagate for long distances driven only by the elastic energy in the pipe wall. Arrest only occurs as a result of major discontinuities in material properties or operating stress levels. The fundamental requirement for the avoidance of brittle fracture is that pipe operates above its fracture appearance transition temperature for propagation (FPTT). It was established in the 1960s that the drop-weight tear test (DWTT) was an effective way of determining the full-scale fracture behavior of pipe; a shear area over 85% in the DWTT at the minimum service temperature will ensure that brittle fracture will not occur in a pipeline [6]. As a result, a requirement for a high percentage of shear area in the DWTT has been a part of line pipe specifications for over 40 years.

For CO_2 pipelines with an OD 508 mm, a requirement for 85% shear area in the DWTT at the minimum service temperature should always be invoked. For smaller diameters, the same requirement is to be preferred, where it is feasible to carry out the test. It has been shown that the DWTT can be carried out on pipe down to 323.9 mm OD [7, 8], though there may be practical difficulties for the heavier wall thicknesses often required for CO_2 pipelines. If DWT testing is not possible, then a similarly high shear area in the Charpy V-notch test should be specified, and if required, the test temperature should be adjusted, relative to the minimum service temperature, to account for any transition temperature shift associated with the difference between the thickness of the Charpy specimen and the wall thickness of the pipe [9, 10].

In determining the minimum service temperature for the specification of toughness properties, consideration should be given to the lower temperatures that may be experienced during depressurization events. This is not generally a concern relative to fracture propagation control since the process is so rapid that no significant local cooling of the pipe, close to the fracture front, can occur. However, the ductile–brittle transitional behavior of piping elements that may be particularly affected by operational depressurizations (for example blowdown assemblies and bypass piping) needs to be considered, and low-temperature materials may need to be specified for some items.

7.2.3 Fracture Initiation Resistance

The process of optimizing fracture initiation resistance for CO_2 pipelines is not different from that used for pipelines transporting other fluids. Even though supercritical CO_2 pipelines operate at higher pressures than are usual for gas and oil pipelines, there is no intrinsic effect on the fracture initiation process, which, for axial flaws, is primarily driven by the hoop stress. Nevertheless, the background is described in full here for the sake of completeness.

The approach to fracture initiation in general use in the pipeline industry is to specify appropriate minimum levels of Charpy V-notch toughness, with the aim of demonstrating that each pipe and component can withstand the presence of a large flaw without a failure that would lead to a major release incident. In general terms, fracture mechanics principles that relate critical flaw size to applied stress and material toughness parameters can be used for this purpose. In order to translate these principles into material property values that can be applied in specifications, industry-specific formulations and practices have been developed that have been used successfully for many years, and these are explained below. While many alternative fracture mechanics-based approaches could be applied, they are generally less transparent, more complex, and offer no significant benefits in terms of developing material specifications.

In principle, an axial flaw caused by any mechanism can cause immediate through-wall failure, or can remain as a stable surface flaw that may grow in operation until failure of the remaining ligament occurs, resulting in penetration of the pipe wall. Depending on its length at penetration, the resulting through-wall flaw may remain stable, causing a leak, or extend unstably, resulting in a rupture. Because the consequences of a rupture are potentially much more serious than those of a leak, the approach to the specification of initiation toughness in general use has concentrated on achieving a high resistance to the initiation of an axial rupture, by ensuring a long critical through-wall flaw length. This will inherently also tend to maximize resistance to through-wall failure of surface flaws. The framework for this analysis was developed by Battelle in the late 1960s and early 1970s, on the basis of existing fracture mechanics formulations adapted for tough materials and cylindrical geometries [11, 12]. The resulting equation for the hoop stress at failure, for a sharp, axial, through-wall flaw, is as follows:

$$\sigma_T = \frac{2\bar{\sigma}}{M_T \pi} \arccos \exp \left[\frac{-\pi K_c^{\,2}}{8c\bar{\sigma}^2} \right] \qquad (7-1)$$

where
σ_T is the hoop stress at failure, MPa
$\bar{\sigma}$ is the flow stress, MPa, originally set equal to SMYS + 69 MPa
K_c is the critical (plane stress) stress intensity factor, $N/mm^{3/2}$
c is the half-length of a through-wall crack, mm
M_T is the Folias factor for a through-wall flaw, $M_T = \left[1 + 1.255\left(\frac{c^2}{Rt}\right) - 0.0135\left(\frac{c^4}{R^2 t^2}\right) \right]^{0.5}$
R and t are the nominal radius and wall thickness of the pipe, respectively, mm.

The Folias factor accounts for stress concentration due to bulging of the pipe surrounding the flaw.

For line pipe materials, which are in general highly ductile under static conditions, Kiefner et al. established that K_c could be expressed in terms of Charpy shelf energy as follows:

$$K_c = [1000\ CV.E]^{0.5} \qquad (7-2)$$

where
CV is the Charpy shelf energy per unit area, J/mm^2
E is the elastic modulus, MPa.

The resulting formulation was originally validated against burst tests on 92 pipe vessels containing axial, through-wall flaws, and can be shown to represent the results of these tests extremely accurately (Figure 7-1). For this reason, it has continued in use to the present day, though minor modifications have been proposed for high-strength materials having high toughness.

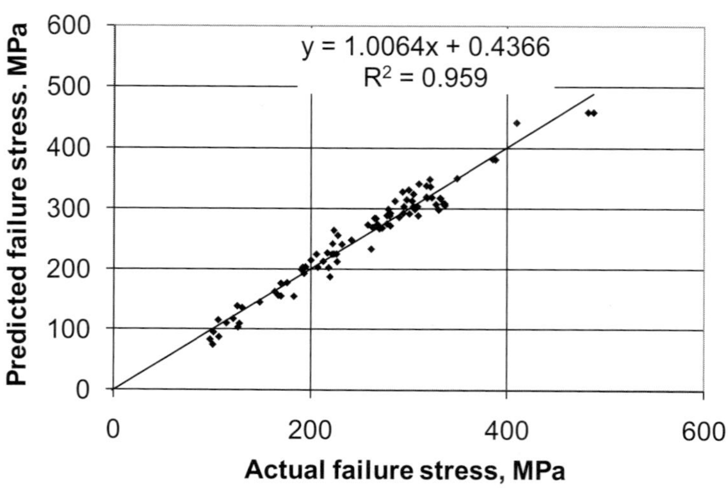

Figure 7-1. Validation of Battelle critical through-wall flaw size equation

The form of Eq. (7-1) is such that, as the toughness increases, the exponential term tends to zero, and the failure stress becomes a function of the flow stress and the Folias factor only; failure occurs by tensile overload. As a result, the critical defect length, for a specific hoop stress, increases with increasing Charpy energy, but is ultimately limited by the flow-stress dependent (FSD) length, which it approaches asymptotically. Figure 7-2 illustrates this behavior. For any given combination of pipe characteristics and applied stress, there will be a point beyond which very large increases in toughness yield insignificant improvements in critical flaw size. Because of this, it has become common practice to specify a minimum pipe body Charpy energy corresponding to a critical defect length that is a high proportion of the maximum achievable, FSD length, typically 80% or 90%.

The minimum (any heat) Charpy energy requirements in pipe specifications and standards such as API 5L (for PSL2 pipe) and CSA Z245.1 (for Category II pipe) will usually provide adequate resistance to fracture initiation, but it is good practice to verify that, for the specific design conditions considered, a long critical TW defect size, as defined above, is reached. It should also be noted that the absolute critical defect size is limited by pipe geometry, and for smaller pipes, it may in some instances be advisable to use a heavier wall thickness than the minimum permissible, in order to improve fracture resistance. For example, for a 273.1 mm OD X70 pipe with a wall thickness of 5.2 mm operating at 14.8 MPa, the 90% FSD critical TW flaw length is only 44 mm. An X60 pipe with a wall thickness of 8.1 mm, under the same conditions, has a 90% FSD TW flaw length of 90 mm. The issue of the intrinsic limitations in fracture resistance of smaller, thinner pipes will be discussed further in relation to fracture propagation control.

Since the possibility exists of fracture initiation from seam weld flaws, weld zone and, for submerged-arc welded (SAW) pipe, heat-affected zone (HAZ) Charpy testing should be specified, and, if industry specifications and standards such as API 5L/ISO 3183 or CSA Z245.1 are used, the adequacy of the required values for the specific design under consideration should be reviewed in the light of the preceding considerations. Since the dimensions of flaws that will escape the mill inspection process are expected to be small, the values specified in API 5L are likely to be adequate for most of the pipe sizes and strength grades used for CO_2 pipelines.

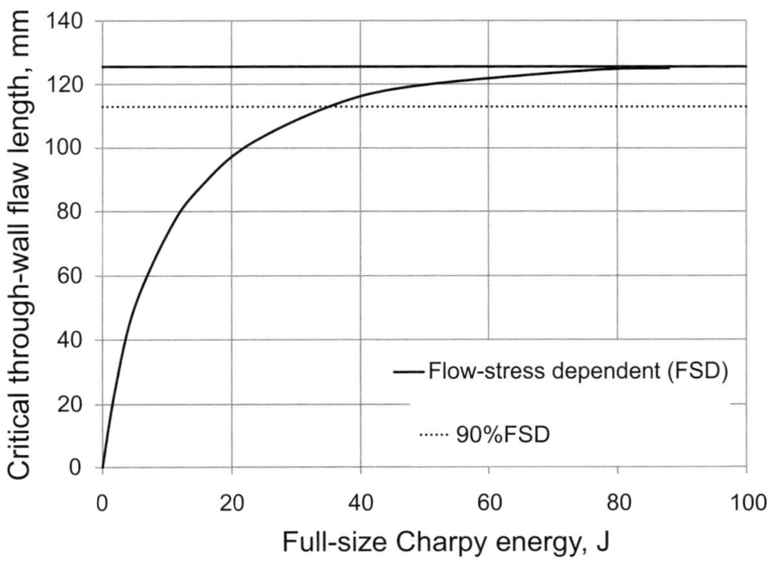

Figure 7-2. Variation of critical through-wall defect length with Charpy energy for a 508 mm OD, 11.8 mm WT, X65 pipe operating at 14.8 MPa g

7.2.4 Fracture Propagation Control

7.2.4.1 Background

It was first realized in the 1960s that ductile (shear) fractures could, like brittle fractures, propagate for long distances [13]. This results from the fact that a compressible fluid does not immediately expand to atmospheric pressure when a breach is formed in the pressure envelope as a result of rupture; the resulting pressure at the moving crack tip may be sufficient to drive continuing ductile fracture propagation. In order to understand ductile fracture propagation and arrest, the detailed decompression behavior of the contained fluid thus needs to be considered.

For a static, quasi-instantaneous opening (as in shock-tube experiments), a constant-pressure choke is formed at the opening, and a pressure wave begins to travel down the pipe, away from the outlet. The pressure falls at a progressively slower rate as the distance from the opening increases. This behavior has been discussed in detail in Chapter 6, and is illustrated schematically in Figure 7-3.

As was noted in Chapter 6, the distance-time relationship for any given pressure level is linear, and thus each pressure level within the expansion wave travels away from the origin at a constant speed. If this behavior is now transferred to the case of a moving fracture front, the ductile fracture propagation phenomenon can thus be understood as an interactive one, in which the fracture front moves at a speed which, at any moment in time, is determined by the pressure in the plane of the crack tip. This, in turn, is determined by the propagation speed of different pressure levels within the expansion wave, relative to the fracture speed.

A full transient analysis of this problem is complex, not least because, for a real rupture in a pipeline, a full-bore opening is not formed instantaneously, but rather as the result of the first few diameter lengths of fracture propagation. As a result, the initialization of the decompression calculation can be problematic [14], and is likely to be influenced by a number of factors such as the presence and consistency of backfill and the bending resistance of the pipe wall. A general solution of the transient analysis problem has been sought for some 20 years (see, for example, Sugie et al [15], Fonzo et al [16], and Misawa et al [17]) and some success has been achieved. However, none of the methods proposed for transient analysis has advanced to such a stage that it is widely used for the determination of required fracture resistance for specification purposes. Instead, a much earlier model remains the de facto standard in the pipeline industry for this purpose. This approach was developed by Maxey and his co-workers at Battelle in the late 1960s and early 1970s, and is generally referred to as the Battelle two-curve method (TCM) [6]. In order to understand clearly the application of this method to "difficult" fluids such as CO_2-based mixtures, it is helpful to consider its origins and background.

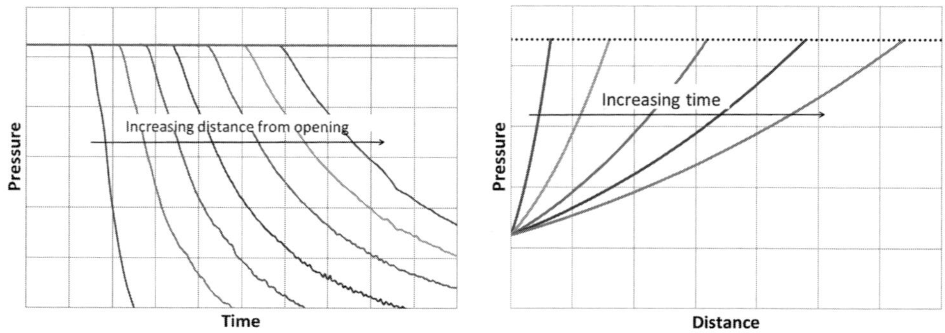

Figure 7-3. (a) Schematic illustration of variation of pressure with time after rupture (note false pressure zero). (b) Variation of pressure with distance from choked opening

In framing the two-curve approach, Maxey circumvented issues associated with near-origin transients by addressing steady-state conditions. In other words, he addressed the problem in terms of the possibility of continued ductile fracture propagation, once steady state conditions had been reached at the fracture front; he did not concern himself with the time or distance needed to reach a steady state. As a result, the interaction between the gas decompression process and continued fracture propagation can be visualized in terms of comparisons between curves expressing the variation of fracture speed and of decompression speed with pressure as in Figure 7-4. Note that this figure and later figures of the same type in this chapter are presented with pressure on the abscissa (i.e., treated as the independent variable), to facilitate discussion of the effect of pressure on the decompression wave and fracture speeds.

For all practical pipeline conditions and ductile materials, the initial gas decompression speed exceeds the fracture speed (as in Figure 7-4), so that, once a full-bore opening is established, the crack tip pressure will initially fall. This will result in a decrease in fracture speed along the corresponding curve, and a further drop in crack tip pressure will occur, as more of the decompression curve runs ahead. This process can be envisaged as recurring sequentially and continuing unless and until there is an intersection between the gas decompression curve and the fracture speed curve representative of the specific pipe material (upper curve in Figure 7-4). In this case, the pressure level within the decompression wave and the fracture are propagating at the same speed. Thus, no further reduction in crack tip pressure can occur, and fracture propagation will continue indefinitely at the speed corresponding to that pressure, until pipe geometry or properties change in such a way as to increase the propagation resistance. If, on the other hand, there is no intersection between decompression and fracture speed curves (lower curve in Figure 7-4), all pressure levels will outrun the fracture front, and fracture arrest will quickly occur. The limiting condition for fracture arrest, for a given pipe geometry and strength, is thus indicated by tangency between the gas decompression and fracture speed curves (middle curve in Figure 7-4). The problem of establishing the required fracture resistance or "arrest toughness" thus requires the determination of a decompression curve, for a specific gas composition and initial temperature and pressure, and of fracture speed curves corresponding to different toughness levels. The determination of decompression curves for CO_2-based mixtures has been discussed in detail in Chapter 6. However, some aspects of the historical development

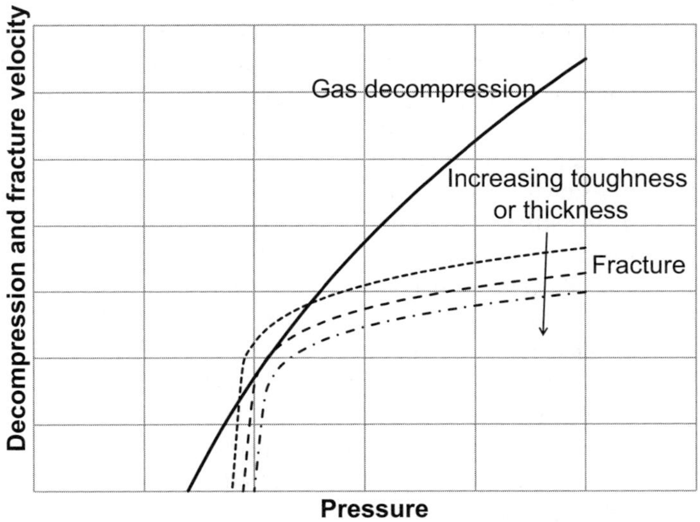

Figure 7-4. Schematic representation of the Battelle TCM

of this part of the TCM will be briefly recalled in the next section, as an aid to understanding the progressive modifications that have needed to be applied to the original analytical methods, in order to accommodate a wider range of fluids and conditions.

7.2.4.2 Development of Understanding of Decompression Wave Speed

The gas decompression curve expresses the speed of each pressure level within the expansion wave that travels into the pipe, away from the rupture origin. It is determined by the initial pressure, temperature, and gas composition, and, for conditions such that ideal gas behavior is a reasonable assumption, can be described by the following simple, closed-form solution.

$$P_d = P_i \left[\frac{2}{\gamma+1} + \left(\frac{\gamma-1}{\gamma+1} \right) \frac{V}{V_a} \right]^{\frac{2\gamma}{\gamma-1}} \qquad (7-3)$$

where

P_d is the decompressed pressure within the expansion wave;
P_i is the original pressure;
γ is the specific heat ratio of the gas;
V is the propagation speed of the pressure level P_d; and
V_a is the speed of sound under the original conditions of pressure and temperature.

The TCM was originally developed for lean natural gas transmission pipelines operating at pressures of 7 MPa and below. It was found that this equation provided a good approximation to the actual decompression behavior observed in full-scale burst tests conducted under such conditions. However, as richer gas compositions, higher pressures, and lower temperatures began to be explored, the assumption of ideal gas behavior was no longer valid, and decompression curves had to be calculated using a one-dimensional thermodynamic/fluid dynamics model that calculates thermodynamic quantities at each step of an isentropic decompression path and uses them to calculate speed of sound and outflow velocity, as described in Chapter 6. The propagation velocity of each pressure level within the expansion wave can then be calculated as the algebraic sum of outflow velocity and the local speed of sound in the gas. The most commonly used commercial model for this purpose is GASDECOM [9], which has been widely used in the pipeline industry, and has performed remarkably well for conditions considerably outside the range for which it was originally validated. At the present time, it is also relatively simple to calculate decompression wave speed using widely available thermodynamic packages and spreadsheet calculations as was shown in Chapter 6.

The importance of accurate decompression wave speed calculation methods that do not depend on ideal gas assumptions relates to the differences in phase behavior between lean natural gas and fluids such as rich gas and CO_2-based mixtures. For lean natural gas, the phase boundary lies well below the temperature range over which the isentropic decompression process takes place. However, for gas containing significant proportions of higher hydrocarbons, the phase boundary moves toward higher temperatures, and the decompression isentrope may intersect it, as shown in Figure 7-5, leading to liquid formation.

When this occurs, there is a discontinuity in the acoustic velocity of the gas mixture, leading to a step decrease in the decompression velocity, as shown in Figure 7-6. This leads to higher arrest toughness values since the fracture velocity curve needs to be moved further to the right and downwards to achieve tangency.

For supercritical CO_2 pipelines, this discontinuous decompression behavior is usually even more pronounced since the isentropic decompression path will usually intersect the phase boundary on the bubble-point side. Figure 7-7 shows an example for a pure CO_2 pipeline. It can be seen that the decompression velocity curve shows a constant pressure region that extends to very low velocities; this behavior was confirmed experimentally almost 30 years ago [18].

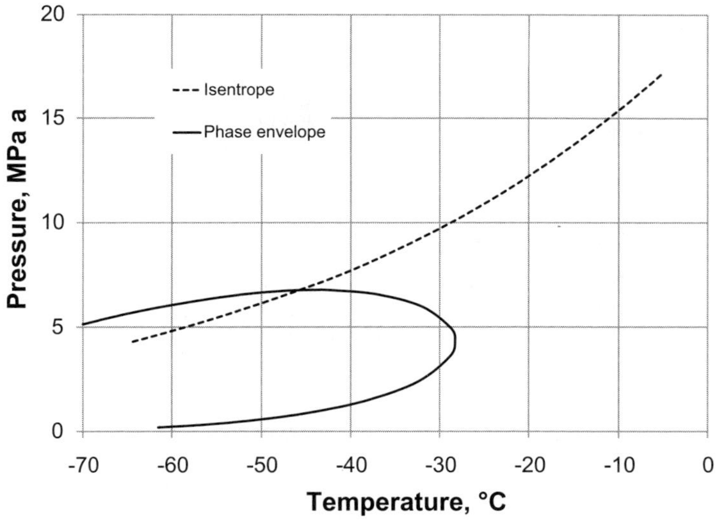

Figure 7-5. Example of phase boundary and decompression isentrope for rich natural gas

As was noted in Chapter 6, it is often difficult to calculate decompression curves for CO_2 mixtures using commercially available software packages, such as GASDECOM, which were originally developed for hydrocarbon mixtures, since they can show major computational instabilities under certain conditions. In addition, they often do not accommodate some of the impurities that may be present in CO_2 mixtures. However, it was shown in Chapter 6 that available thermodynamic software packages do behave stably for many relevant CO_2-based mixtures, and decompression curves can thus be calculated. In addition, the extensive constant pressure region in Figure 7-7, which is typical of most CO_2-based mixtures expanding from an initially supercritical or dense-phase state, indicates that determination of a full decompression velocity curve may often not be necessary in order to determine arrest toughness. It can be seen in Figure 7-7 that tangency between the fracture velocity curve and the decompression curve occurs on the pressure plateau. For this reason, the arrest toughness is determined solely by the need to ensure that the arrest pressure (the intercept of the fracture velocity curve on the pressure axis) is higher than the saturation

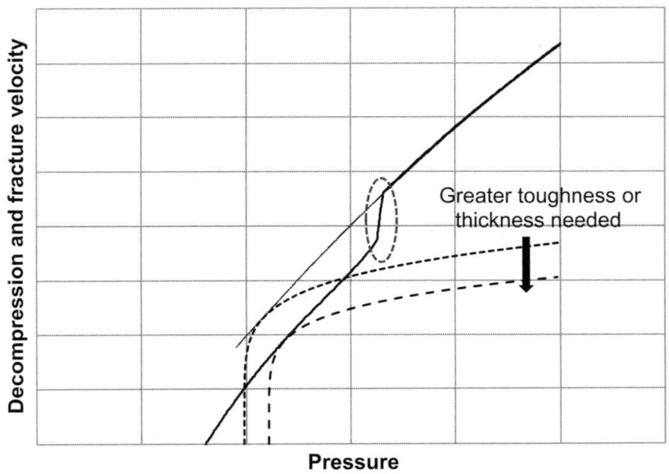

Figure 7-6. Effect of rich gas decompression on arrest toughness

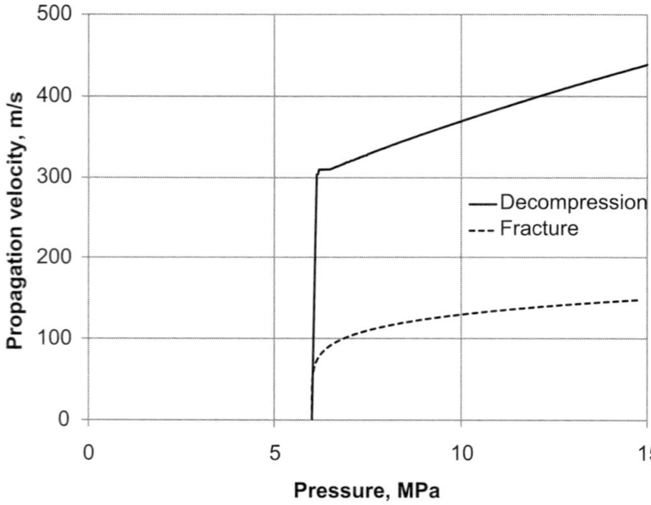

Figure 7-7. Determination of arrest toughness for a pure CO_2 pipeline

pressure (the pressure at which the decompression isentrope intercepts the phase boundary, corresponding with the pressure plateau on the wave speed curve). The saturation pressure can be determined by the direct comparison of decompression isentropes with the phase boundary, in pressure–temperature, pressure–density, or pressure–entropy space, while the arrest pressure can be calculated as a function of pipe diameter, wall thickness, flow stress, and toughness, as will be described in the following section.

As has been discussed extensively in earlier chapters, the addition of impurities to the CO_2 stream can modify the phase boundary greatly, in general opening the space between dew-point and bubble-point curves and moving the cricondenbar to higher pressures. For given initial conditions, this will usually increase the saturation pressure, leading to more difficult fracture arrest. Figure 7-8 shows an example, in this case, for a simple binary mixture of CO_2–7.5% N_2. The phase boundary for pure CO_2 is also shown for comparison.

Figure 7-9 shows an example of the calculation of arrest toughness using the Battelle TCM for the same CO_2 mixture composition. The fracture velocity curve is for a 273.9 mm

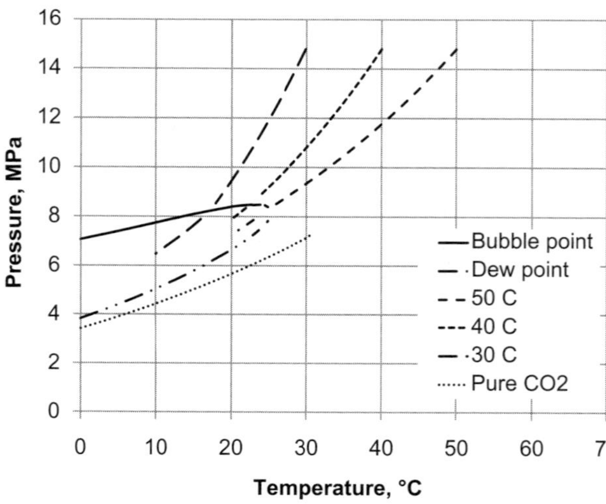

Figure 7-8. Phase boundary and isentropes from 14.8 MPa and a range of temperatures for a CO_2–7.5% N_2 mixture

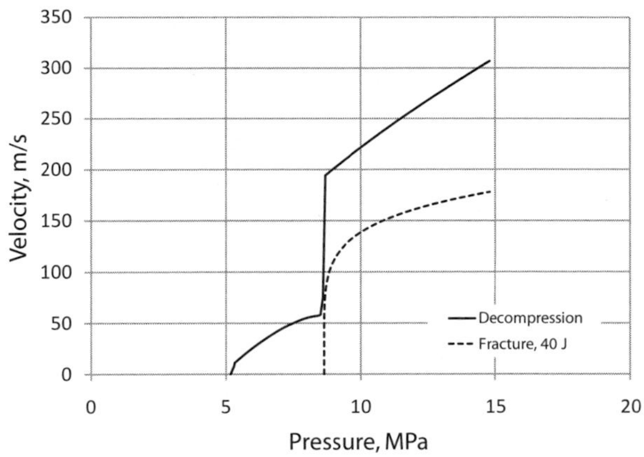

Figure 7-9. Determination of arrest toughness for impure CO_2 (7.5% N_2)

OD X60 pipeline with a wall thickness of 9.3 mm. The initial pressure and temperature were 14.8 MPa and 40°C, respectively. Although the constant pressure region is somewhat less extensive than for pure CO_2, it is still not possible, for any practical design, to achieve tangency between decompression and fracture velocity curves on the lower branch of the decompression curve, so that required arrest toughness is still controlled by the plateau pressure. Since the plateau pressure has moved up to over 8.5 MPa, this will require a significant increase in toughness, relative to a corresponding design for pure CO_2.

7.2.4.3 Calculation of the Fracture Velocity Curve

The calculation of the curve expressing fracture velocity as a function of pressure in the TCM is a two-step process. The first step, which in many cases is the only one that will be required for CO_2 pipelines, is the determination of the arrest pressure. The second step is the calculation of the fracture velocity for pressures exceeding the arrest pressure.

The calculation method for the arrest stress (which is obviously related to the arrest pressure through the pipe *D*/*t* ratio) was initially developed by Maxey and his colleagues on the basis of the critical through-wall flaw size equation for initiation, previously presented here as Eqs. 7–1 and 7–2 [13]. They reasoned that the arrest event itself was quasi-static since the fracture velocity was declining rapidly towards zero, and thus arrest was the mirror image of initiation. As a result, the arrest stress could be calculated using the initiation equation in conjunction with an equivalent half-length for the propagating crack. Clearly, this does not correspond to half the total length of the fracture since there is a great deal of outward movement of the fractured pipe from the original circumferential surface. By empirical calibration against full-scale experimental results, they found that the effective half crack length could be expressed as $3\sqrt{(Rt)}$; thus the corresponding Folias factor M_T was 3.33. If these values are inserted into Eq. 7–1, and K_c is substituted from Eq. 7–2 as before, the following equation for arrest stress results.

$$\sigma_a = \frac{2\bar{\sigma}}{3.33\pi} \arccos \exp \left[\frac{-1000\pi . CV.E}{24\bar{\sigma}^2 \sqrt{Rt}} \right] \qquad (7-4)$$

where

σ_a is the arrest stress, MPa;
$\bar{\sigma}$ is the flow stress, conventionally set equal to SMYS + 69 MPa;
CV is the Charpy V-notch energy per unit area, J/mm^2;
E is Young's modulus, MPa;
R is one-half the pipe OD, mm; and
t is the pipe wall thickness, mm.

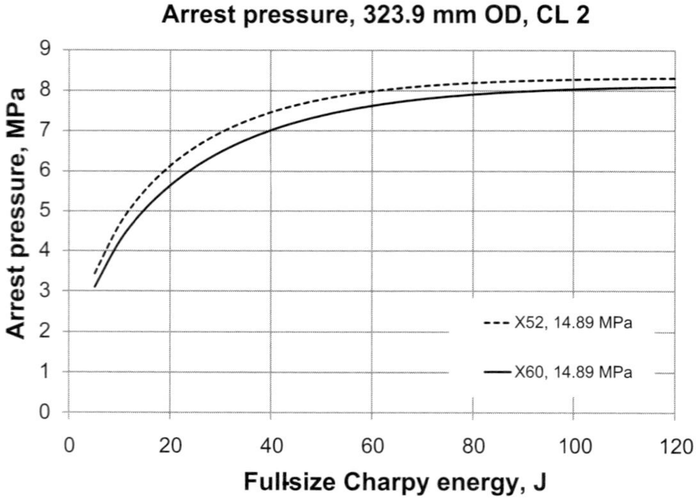

Figure 7-10. Example of the variation of arrest pressure with toughness and strength grade

Clearly, as is the case for fracture initiation, the form of this expression is such that the arrest pressure increases with increasing Charpy energy, but is ultimately limited by a flow-stress-dependent value that depends only on the strength and geometry of the pipe. An example of the variation of arrest pressure with Charpy energy is shown in Figure 7-10; the effect of strength grade, at constant design factor, is also illustrated in this figure. The thicker, lower-strength material gives slightly higher arrest pressure values.

An important consequence of the shape of such curves is that it may be impossible, for certain combinations of design conditions, to achieve an arrest pressure that exceeds the saturation pressure. In the example shown, for saturation pressures higher than about 8.3 MPa, no level of Charpy energy would be sufficient to achieve arrest, and the only solution would be to increase the wall thickness further or, if this is not economically acceptable, resort to the use of crack arrestors.

The remaining step in the calculation of the fracture velocity curve is to determine the variation of propagation velocity with pressure for pressures higher than the arrest pressure. The basis of the fracture velocity equation was derived from consideration of the propagation rate of a plastic strain field [6]. The equation was calibrated against existing data for steady-state ductile fracture propagation to give the following form:

$$V_f = \text{const}\frac{\bar{\sigma}}{\sqrt{CV}}\left[\frac{P_d}{P_a}-1\right]^{1/6} \qquad (7-5)$$

where P_d is the decompressed pressure (MPa gauge) at the crack tip, and P_a is the arrest pressure corresponding to the arrest stress calculated using Eq. 7-4. The constant includes several physical quantities and also varies according to whether the pipe is backfilled or not. For velocity in m/s, flow stress in MPa and Charpy energy in J/mm^2, the constant is 0.275 for backfilled pipe and 0.379 for unbackfilled pipe. A constant has also been proposed for underwater pipelines (0.234), but this is not well validated and in fact, for deep water, the arrest mechanism can be intrinsically different, with the external pressure wave playing a dominant role.

7.2.4.4 Limitations on the Use of Charpy Energy for Arrest From the Two-Curve Method

Research into the conditions governing fracture propagation and arrest conducted over the last three decades has indicated that there are limitations in the applicability of the Charpy energy, calculated using the TCM, as a measure of arrest toughness [9, 20, 42]. In general, caution should be exercised in using such values for high strength steels (X70 or higher) and when the calculated arrest toughness exceeds 90-100 J (full size Charpy energy) since the measured Charpy energy may overestimate the actual fracture propagation resistance. The cited references provide useful guidance regarding the concerns that may arise and methods that may be used for correcting the "raw" calculated arrest toughness. There is also some evidence that the fracture velocity equation presented above may become inaccurate for pipe diameters smaller than about 457 mm [41]. In addition, it has been shown that observed decompression velocities may become progressively slower than predicted by models such as GASDECOM as the ratio of surface roughness to pipe diameter increases [21]. All of these potential sources of inaccuracy (particularly the diameter and surface roughness effects) will be of little significance when the required toughness is calculated purely on the basis of the arrest pressure. For supercritical CO_2 pipelines, this will be the case for most practical designs, and in case of uncertainty, this approach will present a conservative alternative to designs and toughness specifications based on achieving tangency between the decompression curve and the variable portion of the fracture velocity curve.

7.2.4.5 Historical Development of Practical Methods for Determining Arrest Toughness for CO_2 Pipelines

The challenge, relative to fracture propagation control, presented by supercritical CO_2 pipelines, was pointed out by King and by Maxey in the early 1980s [22, 23, 18], and Maxey showed that the very long region of constant pressure predicted by theory was in fact observed in experiments. Maxey also proposed that the most practical way to control ductile fracture for supercritical pipelines was to ensure that the arrest pressure, for a proposed design, was higher than the highest saturation pressure anticipated following an operational rupture [24]. Following this work, a number of additional studies on the determination of arrest toughness were published (see, for example, Rothwell [25] and Wadsley and Rothwell [26]), but, until relatively recently, all were focused on the behavior of pure CO_2. This was understandable since the CO_2 that was then being transported from geological sources for EOR applications was relatively pure, and the potential effects of impurities on the phase boundary and on decompression behavior were not widely understood. More recently, the issues associated with the impurities present in CO_2 captured from anthropogenic sources, using different technologies, became apparent, and the challenges associated with determining the decompression behavior of such fluids, and of providing adequate fracture resistance, have been appreciated (see, for example, Cosham and Eiber [27], Cosham et al. [28] and King and Kumar [29]).

7.2.4.6 Practical Determination of Necessary Charpy Energy for Ductile Fracture Arrest in CO_2 Pipelines

Having reviewed the background, origins and calculation methods used in the application of the two-curve method, it is now possible to consider how it can be applied in practice to determine the necessary toughness for ductile fracture arrest. This Charpy energy is typically included in pipe specifications as an all-heat average (AHA) value, though if there are specific reasons for limiting any rupture to the lowest possible length, it may also be specified as a minimum (any heat) value.

Generally speaking, determination of the required toughness is an iterative process that must interact with the pressure-containment design to arrive at an optimum solution for the transportation of the required volumes. This process is well described in King and Kumar [29]. The required operating pressure and pipe size are determined by optimizing unit transportation costs for the required throughput (see Chapter 8). Steel strength grade is typically chosen to give the minimum possible wall thickness for pressure containment, consistent with the regulatory requirements for the location class, practical *D/t* ratios and limitations due to secondary stresses. However, because of the intrinsic limitation on ductile fracture resistance discussed in Section 7.2.4.3, it may not always be possible to use the thinnest pipe that meets these criteria, and additional optimization may be necessary. The initial lowest cost design must be reviewed to determine whether an arrest pressure that exceeds the maximum possible saturation pressure can be reached at a practically achievable level of toughness. Clearly, what is "practically achievable" will depend on the pipe production route (e.g., electric welded pipe from strip or submerged-arc welded pipe from plate), among other factors.

The way in which fracture control can influence the choice of pipe grade, and the corresponding wall thickness, can best be appreciated by reference to a worked example, which is as follows.

Problem: recommend a suitable strength grade, wall thickness and Charpy energy for the following design case;

Pipe OD: 406.4 mm
MAOP: 15.3 MPa (gauge)
Minimum delivery pressure: 8 MPa
Operating temperature range: 20°C to 40°C
Fluid composition (simplified): 96.5 mole% CO_2, 3.5 mole % O_2

Step 1. Review potential choices of strength grade and nominal wall thickness.

Table 7-1 gives several potential designs that would meet regulatory requirements in Class Location 1 (and in some cases in higher class locations) in many jurisdictions. Case 1 probably represents the lowest practical wall thickness, since grades above X70 are not widely available in HFW pipe. Cases 2 and 3 involve lower design factors that will result in greater fracture propagation resistance. Cases 4 to 6 duplicate cases 1 to 3 using X60 pipe.

Step 2. Review the operating envelope and the phase boundary to determine the initial conditions that lead to the highest saturation pressure on isentropic decompression.

This process is illustrated by reviewing the phase boundary and the decompression isentropes starting from the MAOP and temperatures of 20, 30, and 40°C. In the present case, the GERG 2004 equation of state was used, as implemented in REFPROP [30, 31]. The entropy at each of the combinations of initial temperature and pressure was first calculated. This was then used to generate isoproperty tables, showing pressure and temperature at constant entropy throughout the decompression process. The tables were calculated using a constant temperature decrement from the initial conditions since the calculations were found to be slightly more stable than when using a pressure decrement. Bubble point and dew point curves were also calculated, and the final results are shown in Figure 7-11.

The isentropes show the characteristic discontinuous change of slope as the phase boundary is encountered; this is associated with the step change in the speed of sound that occurs at the transition from single-phase to two-phase conditions (see Table 7-2). As expected, the highest initial temperature gives the highest saturation pressure, 7.9 MPa a or 7.8 MPa gauge. In general, it is good practice to review the entire operating envelope

TABLE 7-1. Possible combinations of strength grade and wall thickness for worked example

Case	Grade	Design Factor*	Wall Thickness, mm
1	485 (X70)	0.80	8.0
2	485 (X70)	0.72	8.9
3	485 (X70)	0.60	10.7
4	415 (X60)	0.80	9.4
5	415 (X60)	0.72	10.4
6	415 (X60)	0.60	12.5

*Product of design factor and location factor in the terminology of Canadian Standard Z662.

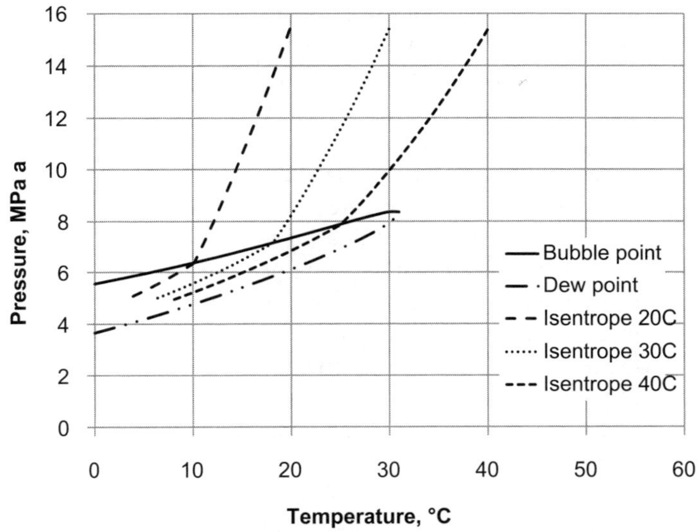

Figure 7-11. Phase boundary and decompression isentropes for worked example

in pressure–temperature coordinates to ensure that there are no other conditions that lead to higher saturation pressures. In the present case, it is assumed for simplicity that the temperature will always decrease sufficiently with falling pressure to ensure that the corresponding isentropes will lie to the left of the isentrope starting from MAOP and 40°C. In cases of doubt, the highest possible saturation pressure can be taken to correspond with the cricondenbar, but this will often be excessively conservative.

If software is available that directly determines phase boundaries in terms of entropy and pressure (rather than requiring external calculations), a simpler method presents itself for determining the range of saturation pressures that could be encountered. This was mentioned by Wadsley and Rothwell, in connection with pure CO_2, and described in detail by Cosham and Eiber [26, 27]. If the phase boundary is developed in pressure–entropy space, the decompression isentrope is now a straight line, originating at the entropy corresponding

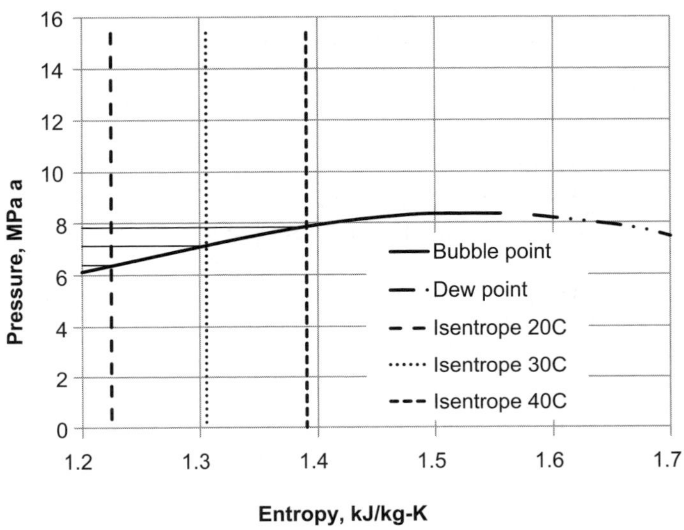

Figure 7-12. Phase boundary and isentropes in pressure–entropy coordinates

to the initial conditions of pressure and temperature under consideration. It thus involves none of the complications and occasional computational instabilities involved in calculating isentropes in pressure–temperature coordinates. This approach can be illustrated for the case considered in the worked example. Figure 7-12 shows the phase boundary, determined using GERG 2004 in REFPROP. The same application was used to determine the entropy under the three sets of initial conditions considered in developing Figure 7-11; the decompression isentropes in Figure 7-12 are represented as vertical lines projected downwards from the corresponding values of entropy and pressure.

It can be seen that the point of intersection of the decompression isentropes with the phase boundary (the saturation pressure) can be determined with reasonable accuracy by inspection, and the accuracy can be refined (though this will not usually be necessary for the purpose of determining toughness requirements) by curve-fitting the relevant portion of the phase boundary. This procedure is simpler than using the phase boundary in pressure–temperature coordinates, together with isentropes determined from calculated isoproperty tables.

At this stage, two approaches are possible in determining fracture arrest conditions. The first, which requires more complex calculations, involves the determination of the full decompression curve, from the initial conditions that will give the highest saturation pressure, as well as full fracture velocity curves for a range of Charpy energies. This is often advisable for unfamiliar compositions, for which the extent of the constant pressure plateau on the decompression curve may not be easy to estimate *a priori*.

Approach (a)

Step 3. Determine the full decompression curve for conditions that lead to the highest saturation pressure.

This process was again carried out using the GERG 04 EOS in REFPROP, using the methodology described in Chapter 6. Again, an isoproperty table was calculated based on constant entropy starting from the initial state point at 15.4 MPa a and 40°C. In this case, temperature, pressure, density, and speed of sound at each step are needed for the calculation. It is important to choose a pressure or temperature decrement that is small enough to give sufficiently accurate calculation of the necessary terms in the calculation, but not so small that the calculated gradients become unstable. In the present case, a temperature decrement of 0.03°C was used. Table 7-2 shows the first few rows of the isentropic property table, as well as the two rows on either side of the discontinuity. It should be noted that REFPROP does not calculate the speed of sound in the two-phase region, and the speed of sound below the discontinuity was calculated from

$$c = (\partial P/\partial \rho)_s^{0.5}$$

where

c is the local speed of sound

$(\partial P/\partial \rho)_s$ is the partial derivative of pressure with respect to density at constant entropy.

The outflow velocity was calculated using the relationship

$$(\partial u/\partial \rho)_s = 1/\rho.c$$

where

$(\partial u/\partial \rho)_s$ is the partial derivative of outflow velocity with respect to pressure at constant entropy.

The outflow velocity is calculated by adding the incremental u at each step to the value of u at the previous step.

TABLE 7.2 Isentropic property table showing calculation of decompression velocity

Temperature (°C)	Pressure (MPa a)	Density (kg/m³)	Enthalpy (kJ/kg)	Enthalpy (kj/kg-K)	Sound Speed (m/s)	delta u m/s	U m/s	W m/s	Pressure (MPa a)
40	15.398	741.11	293.07	1.3902	398.54	0.0643516	0.0643516	398.47565	15.398
39.97	15.379	740.99	293.04	1.3902	398.31	0.1253602	0.0610086	398.18464	15.379
39.94	15.361	740.88	293.02	1.3902	398.09	0.0644436	0.0644436	397.9002	15.361
39.91	15.342	740.76	292.99	1.3902	397.87	0.0644905	0.2542943	397.61571	15.342
39.88	15.323	740.64	292.97	1.3902	397.64	0.0611407	0.315435	397.32456	15.323
24.97	7.8917	675.1	282.52	1.3902	270.41	0.621839	31.240144	239.16986	7.8917
24.94	7.8842	674.35	282.51	1.3902	87.10	0.1124415	31.352585	55.746249	7.8842

Figure 7-13 shows the resulting decompression velocity curve; the step change in decompression velocity (from 239 m/s to 56 m/s) associated with the step change in the speed of sound (from 270 to 87 m/s) when the phase boundary is encountered, at 7.89 MPa, is apparent. Fracture velocity curves are also shown, and these will be discussed next.

Step 4. Determine fracture velocity curves for potential design options and Charpy energy levels, and identify options for which there is no intersection with the decompression velocity curve.

In order to determine conditions of tangency with the decompression curve established as shown above, fracture velocity curves for different potential designs and toughness levels need to be calculated, using the approach previously described. Normally speaking, for any given combination of strength grade and wall thickness, fracture velocity curves are calculated for a range of toughness levels intended to span the region in which tangency will occur. It should be noted that Eq. 7–5 was developed for gauge pressure, whereas thermodynamic calculations are typically carried out using absolute pressures, so that the requisite, small adjustment needs to be applied. Once the required toughness has been bracketed, additional refining calculations can be carried to determine the toughness that leads to tangency with greater precision. Some key results for some of the examples from Table 7-1 are shown in Figure 7-13. The following conclusions can be drawn:

- For case 3 (X70 with a wall thickness of 10.7 mm), a Charpy energy of 80 J is sufficient to avoid intersection with the decompression curve. The arrest pressure is

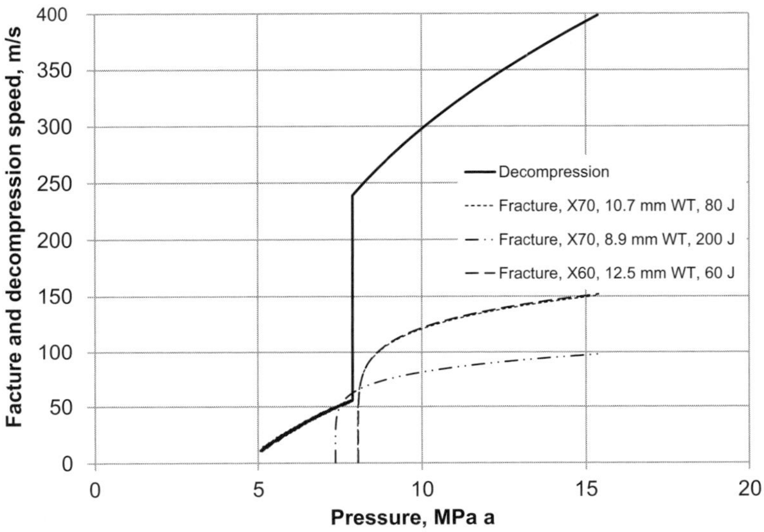

Figure 7-13. Decompression and fracture velocity curves

slightly higher than the plateau pressure. 80 J would thus be a suitable value for pipe specifications for this grade and wall thickness.

- For case 2, (X70 with a wall thickness of 8.9 mm), even a Charpy energy of 200 J is insufficient to avoid intersection with the decompression curve. Particularly in the light of the corrections that need to be applied to Charpy energy values as high as this, this combination of strength grade and wall thickness is clearly impractical from the point of view of ductile fracture control based on pipe body arrest. This case also indicates the impracticability of fitting a tangent fracture velocity curve to the lower branch of the decompression curve for this fluid and initial conditions. In addition, it follows that the case 1 option (X70 with a wall thickness of 8.0 mm) is not a viable one.
- For case 6 (X60 with a wall thickness of 12.5 mm), the fracture velocity curve corresponding to a Charpy energy of 60 J is virtually coincident with that for case 3 at 80 J. This is because, in Eq. 7–5, the changes in flow stress and the square root of Charpy energy, together with the slight change in arrest pressure, almost exactly balance each other. 60 J would thus be an appropriate specification value for this case.

Approach (b)

It has been shown that, for the present example, there is no practical possibility of finding tangency on the lower branch of the decompression curve and, as has been stated previously, this opens the way to simpler methods of determining the required conditions for fracture arrest. It is only necessary to ensure that the arrest pressure is higher than the saturation pressure. The governing value of the saturation pressure was previously established (in *Step 2*) as 7.9 MPa a or 7.8 MPa gauge.

Step 3. Calculate the arrest pressures for the potential design options and Charpy energy levels, and determine practical combinations of strength, wall thickness and Charpy energy that result in an arrest pressure higher than the design saturation pressure.

The arrest pressures, for different values of Charpy energy, can be determined from Eq. 7–4, as illustrated previously in Section 7.2.4.3. The results, for the six design cases in Table 7-1, are plotted in Figure 7-14.

It is very quickly apparent from these figures that the wall thicknesses corresponding to the two highest design factors are not suitable for this application, while the heaviest wall thickness, for both strength levels, allows an arrest pressure greater than the design saturation pressure to be achieved at reasonable levels of Charpy energy (respectively 60 and 80 J for X60 and X70).

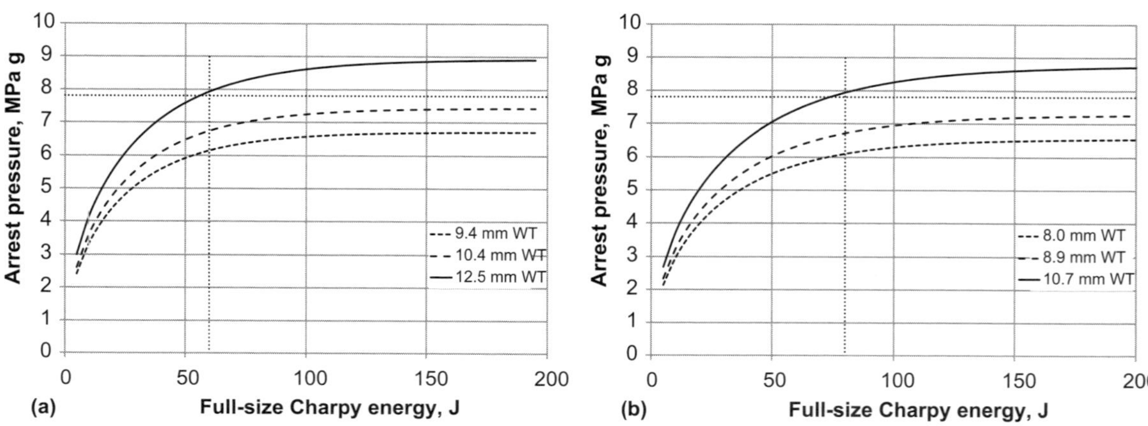

Figure 7-14. Variation of arrest pressure with Charpy energy (a) X60 cases (b) X70 cases

Recommendation

It is clear that, for this example, the highest permissible design factors and highest strength levels (cases 1, 2, 4, and 5 in Table 7-1) do not allow a fracture propagation control design based on pipe body toughness to be realized. While a design based on crack arrestors (see below) could be contemplated, it will often be preferable, where feasible, to increase the wall thickness sufficiently for pipe body arrest to be achieved. In the present case, pipe body arrest is achievable at reasonable wall thicknesses (cases 3 and 6 in Table 7-1), and this approach would generally be recommended. Of the two viable options from the six cases considered, the X70 option would usually have lower material costs since any difference in grade extra between X60 and X70 would usually be outweighed by the increased pipe tonnage for the thicker wall. Field welding costs would also be somewhat lower for the X70 option. Thus, case 3 using X70 pipe would be the primary design recommendation. However, depending on whether the arrest toughness is to be specified as all-heat average (AHA) or minimum value for any heat (which will depend on the perceived consequences of incremental fracture length), the lower toughness value required for the X60 option may have some influence. If the required toughness for X70 is difficult to achieve, or limits the potential sources of pipe, the lower strength, lower toughness approach of case 6 may be preferred.

Check on initiation resistance

If the minimum required (any heat) API 5L PSL2 Charpy energy for these pipe sizes and grades, 27 J, is applied, the resulting critical through-wall flaw lengths (calculated using Eqs. 7–1 and 7–2) are 111 mm for case 3 and 130 mm for the case 6, respectively, 80% and 85% of the FSD values. If, as is most common, the arrest toughnesses calculated for these cases are applied as AHA requirements, the minimum Charpy energy could be increased to 40 J for case 3 and 35 J for case 6, which will achieve 90% of the FSD critical flaw size, respectively, 123 mm and 136 mm. Though the increase is modest in absolute terms, the revised minimum values are consistent with the recommended AHA and thus are unlikely to result in significant incremental cost.

7.2.5 Fracture Propagation Control in Gaseous Phase CO_2 Pipelines

Although the low throughput and efficiency that can be achieved in CO_2 pipelines operating in the vapor phase region make it unlikely that new pipelines of significant length will be built to operate in this mode, the possibility exists of the conversion of existing, underutilized natural gas or liquid pipelines. The fracture resistance of such pipelines would have to be reviewed to ensure that they are suitable for the changed service.

As has been pointed out by Wadsley and Rothwell and by Cosham, the constant pressure plateau is much shorter when the decompression isentrope enters the two-phase region from the vapor side than when it enters from the liquid side [26, 32]. This is beneficial from the point of view of toughness requirements for fracture arrest, but there is usually a counteracting influence of the lower initial speed of sound. These factors will be illustrated by analysis of a vapor phase design for the same pipe diameter and CO_2 mixture considered in the preceding section. As before, the decompression isentropes were calculated using the GERG 2004 EOS, and superimposed on the phase diagram; the results can be seen in Figure 7-15. Two isentropes are shown, for an initial pressure of 6 MPa and temperatures of 30°C and 25°C; clearly, for vapor phase initial conditions, lower initial temperatures lead to higher saturation pressures (as for most natural gas pipelines) and require higher toughness for ductile fracture arrest. This is the opposite trend, of course, to that observed in CO_2 pipelines for supercritical and dense phase conditions.

Because the constant pressure plateau on the decompression curve is expected to be short, it is necessary to compare the full decompression curve with the fracture velocity curves; basing the required Charpy energy on arrest pressure alone is likely to lead to unnecessarily conservative assessments of arrest toughness.

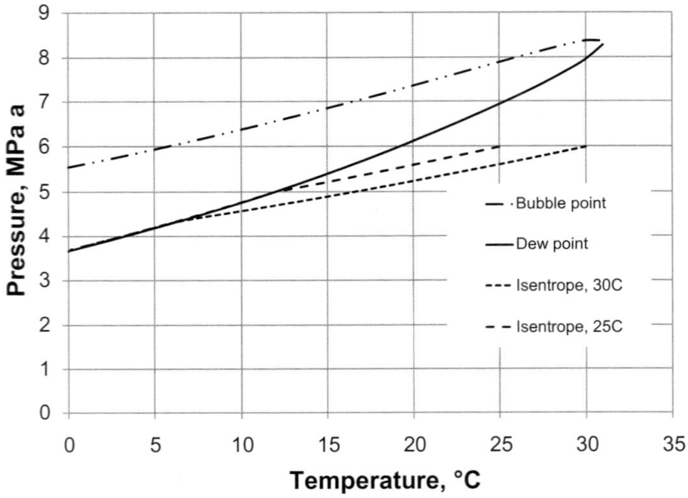

Figure 7-15. Phase boundary and isentropes for gaseous phase operating conditions

Figure 7-16 shows the decompression curve from 25°C and 6 MPa, together with fracture velocity curves for a 406.4 mm OD, Grade L415 (X60) pipeline with a wall thickness of 6 mm.

It can be seen that a Charpy energy of 43 J represents the threshold value between propagation and arrest. This relatively high value may appear surprising since the MAOP is low and the effective design factor is less than 0.5. By comparison, a lean natural gas pipeline operating under the same conditions would have an arrest toughness of less than 20 J. The main cause of this difference is the low acoustic velocity of most CO_2-based mixtures at pressures below the dew-point curve. In the present case, it can be seen that the initial acoustic velocity is only about 215 m/s compared with a value of around 400 m/s for lean natural gas under the same conditions.

This example also provides an illustration of the potential difficulty of converting pipelines (particularly older ones) from natural gas or liquid service to CO_2 service. The

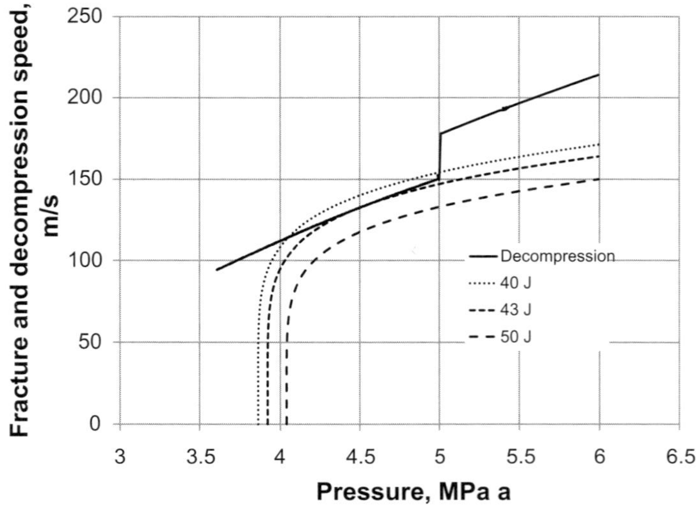

Figure 7-16. Decompression and fracture velocity curves for gaseous phase example

pipeline in the present example would have been suitable for natural gas service at up to 8.8 MPa, with a design factor of 0.72. Even after downrating to 6 MPa to ensure that the fluid remains in the gaseous phase during operation in CO_2 service, a relatively high arrest toughness is required. Particularly for older pipelines, the actual toughness distribution may not be adequate for fracture control purposes. In some cases, retrofitting of crack arrestors, at a suitable spacing, may be an acceptable solution. However, in some jurisdictions and locations, this approach may not be viewed as adequate.

7.2.6 Crack Arrestors

It is evident that there will be cases in which ductile fracture propagation control based on pipe body arrest will not be practical. In some cases, the increase in wall thickness that is necessary to reach a practical arrest toughness may compromise the economic viability of a project. In other cases, particularly for impurities with low critical temperatures, such as hydrogen and helium, the saturation pressure will be too high for the arrest toughness to be achieved for any practical combination of strength and wall thickness. In such cases, fracture control may be based on the use of crack arrestors. Cosham and Eiber [27] provides an indicative diagram for combinations of conditions for which the use of crack arrestors may be necessary, but case-specific analysis of the type described above is generally required.

Crack arrestors have been used on CO_2 pipelines in a number of instances and, in particular, have been retrofitted to pipelines that were designed and constructed in the 1970s before the implications of the decompression behavior of CO_2 on fracture control were fully realized. For natural gas service, crack arrestors have been used, or proposed in design studies, for pipelines for which the service conditions made pipe body arrest doubtful or unachievable, based on the material properties and fracture information available at the time. For example, crack arrestors were used in the original construction of the Northern Borders pipeline, which was designed to transport rich gas from Alaska from the termination of the Foothills Pipe Lines system at Monchy, SK to the US Mid-West. Early designs for the Alaska portion of the original Alaska Highway project were also based on crack arrestors. They have also been used to deal with changes of service, for example, from liquid to gas, where requirements for the control of ductile fracture propagation could not be met on the basis of pipe body arrest (see, for example, Porter and Patrick [33]).

Crack arrestor design background

Several reviews of crack arrestor types and design principles have been published, for example. Venton and Dietsch [34], Hillenbrand et al. [35], and Wilkowski et al. [20]. The two main categories are in-line arrestors (generally full joints of pipe having higher thickness, higher toughness, or both) and external arrestors (devices applied to the pipe that limit opening at and behind the fracture front and thus reduce the driving force). Both types have been shown, by full-scale testing, to be effective when properly designed. Several different types of external arrestor have been evaluated, including grouted steel sleeves, reinforced composite wrap (including Clockspring® and CRLP™), wire rope, and "toroidal" (steel rod) arrestors [20]. Both CRLP™ and grouted sleeve arrestors have been shown to be effective by large-scale testing, including tests carried out under conditions involving very high fracture driving force [36].

In-line arrestors

For new pipelines, there may be some perceived advantages in the use of in-line arrestors, in that they are not intrinsically outside normal pipeline design and construction experience, and, provided any increase in thickness is not excessive, they can be installed with minimal disruption to logistics and construction practices. All that is required is a suitable machined transition at the end of the crack arrestor pipes.

Different approaches to the design of in-line crack arrestors are possible, but the basic principle is clearly that they should be capable of arresting any propagating ductile fracture within a single joint of pipe. A suitable criterion was proposed by Venton and Dietsch [34] that has the advantage of using the same technical framework as has already been described for the evaluation of arrest toughness for line pipe. Essentially, it requires that the arrest pressure in the arrestor be higher than the calculated pressure in the fracture plane for the highest possible steady state fracture velocity (lowest toughness) in the line pipe. The highest fracture velocity corresponds with the intersection of the fracture curve for the lowest toughness pipe with the decompression curve. Suppose that, for economic motives, it had been decided to choose the design in case 1 in Table 7-1 (X70, 8.0 mm WT) and to use the minimum toughness requirement in API 5L for PSL 2 pipe of this diameter and strength grade (27 J). Figure 7-17 indicates that the intersection between fracture and decompression curves occurs at a pressure of 8.26 MPa, at a fracture velocity of 253 m/s. Thus, the highest steady-state crack-tip pressure that can occur in the pipeline is 8.26 MPa. In order to reach an arrest pressure that exceeds this value in an X70 pipe with a 10.7 mm wall thickness (case 3 in Table 7-1), a Charpy energy of 110 J is required (see Figure 7-14 (b)). In ordering pipe for crack arrestors, of course, the required Charpy energy must be specified as a minimum, not AHA. If this value is found to be too high, increasing the wall thickness to 12 mm reduces the required Charpy energy to 65 J.

Options for external arrestors

As has been mentioned, various types of external crack arrestors have been proposed, and several have been evaluated in full-scale tests. For highly ductile materials, the main driving force for fracture propagation derives from the pressure of the escaping gas on the opening "flaps" behind the crack front. External arrestors limit or prevent this opening, so that the applied crack opening angle (the driving force from a fracture mechanics viewpoint) is greatly restricted. Steel sleeve arrestors were the first to be studied; their effectiveness is strongly influenced by the annular clearance between the internal surface of the sleeve and the external surface of the pipe [20]. Filling the annulus with grout makes the sleeve perform as well as an ungrouted sleeve with very limited annular clearance, and the larger physical clearance that is then possible allows easier and more consistent installation. Gen-

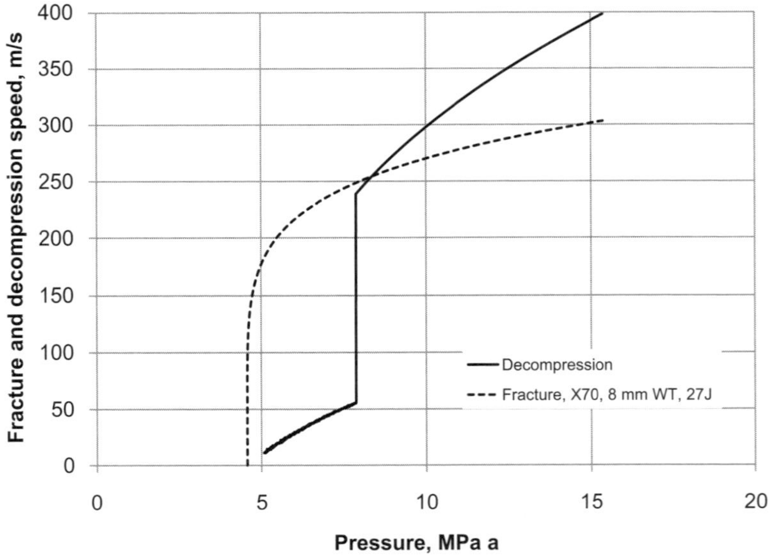

Figure 7-17. Determination of maximum steady state fracture velocity for case 1

erally speaking, the unit strength (product of tensile strength and thickness) of the sleeve is made at least equal to that of the carrier pipe; the required length depends on the incoming fracture speed and the effective annular clearance. Wilkowski et al. [20] provides empirically based criteria that can be used to determine an appropriate sleeve length, for different maximum fracture velocities, for both grouted and ungrouted sleeves. As an example, these guidelines indicate that the minimum required length of a grouted sleeve crack arrestor of the same unit strength as the pipe, for a crack speed of about 250 m/s (as in the case illustrated in Figure 7-17) would be 30% of the pipe diameter. Since, within limits, it is not anticipated that arrestor length would be a dominant cost element in overall project economics, it would probably be prudent to increase this to at least 50% of the pipe diameter.

Crack arrestors of this type have been installed on the Northern Borders natural gas pipeline system, and also on some CO_2 pipelines. To maintain their long-term effectiveness, steel sleeves need to be protected against corrosion; of course, it is also vital that their application does not damage the pipe coating, and that their presence does not interfere with the cathodic protection of the pipe itself. The arrestors on the Northern Borders pipeline have been in service since 1982 apparently without incident.

Clock Spring® crack arrestors (spirally-wound, multiple-layer, glass-reinforced polyester composite) have been evaluated and shown to be effective in large-scale tests involving high fracture driving force [37]. They have been applied to a number of CO_2 and natural gas pipelines and have the advantage that they can easily be retrofitted to existing pipelines, for example, to deal with changes in service. Clock Springs are currently available for all practical pipe sizes, though the width is limited, which can lead to a need for multiple, abutting installations for large-diameter, high-pressure pipelines.

An alternative to Clock Springs that may be more suitable for new, long distance pipelines requiring large numbers of arrestors is the direct installation of filament-wound composite arrestors to individual pipes, prior to construction. An example of this type of arrestor is CRLP™ [38]. This technology was developed primarily for the mass production of line pipe and gas transport modules having increased pressure capacity, as a result of an outer layer of glass-reinforced resin, unidirectionally wound in the hoop direction. CRLP™ has been shown to be effective in arresting propagating fracture; it is also resistant to external damage and deterioration and can be applied over epoxy and polyolefin corrosion coating with minimal surface preparation. An example of a similar type of arrestor has also been described by Hillenbrand et al. [35]. Like Clock Springs, such crack arrestors have been shown to be effective at high fracture driving force [36]. While the design of composite arrestors can be accomplished in a similar way to that of steel sleeves, by determining an "effective radial clearance" that accounts for the lower stiffness of the glass-based composite, typical designs have tended to be very conservative [20]. A more refined methodology based on finite element analysis is to be preferred for major projects, since it leads to more efficient designs with a known degree of conservatism [36, 39]. In general, since the arrestor material strength, elastic modulus, and strain to failure are constant, the design variables are reduced to incoming fracture speed and arrestor length and thickness. The latter can be optimized through a series of simulations that determine whether a fracture encountering the arrestor at a given speed penetrates beyond the arrestor or is arrested within its length. The full range of fracture speeds that could be encountered, based on the pipe characteristics and fluid composition, temperature, and pressure, needs to be considered. Designs for high driving force conditions have typically required a thickness equal to or somewhat greater than the pipe wall thickness and a length in the range one to two pipe diameters. If a high-productivity application process is used, the cost of a somewhat conservative choice of arrestor length will be insignificant, relative to total project costs. Excessive conservatism in thickness is generally undesirable since a high arrestor thickness can lead to complications in the pipeline construction process and a local change in stiffness that may have adverse effects under longitudinal bending.

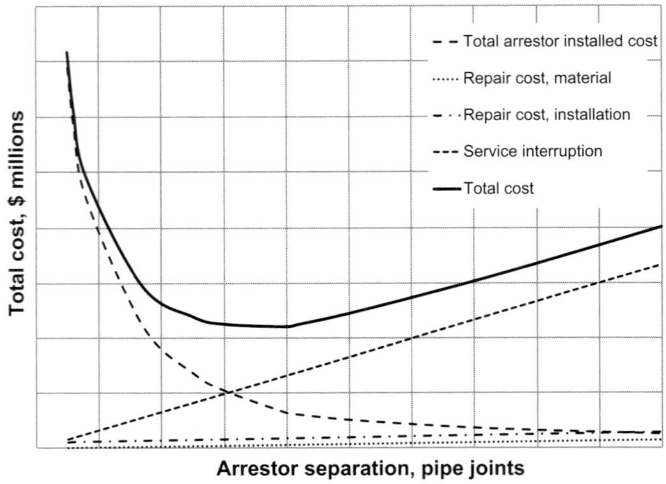

Figure 7-18. Effect of arrestor separation on cost of ruptures (schematic)

Crack arrestor spacing considerations

The appropriate spacing for crack arrestors is driven by both economic and safety-related considerations. Economic optimization involves balancing the incremental costs of fractures exceeding a minimum length (below which the major cost drivers are notionally constant) and the cost of installing additional arrestors [25, 40]. Since pipeline ruptures, particularly for new construction, are expected to be extremely rare events, such calculations are somewhat artificial, but do provide some rational basis for the choice of arrestor spacing. Figure 7-18 provides a schematic illustration of the way in which such an analysis can be carried out. It is assumed that any rupture will propagate for the entire distance between arrestors.

The installed cost of arrestors includes the cost of arrestors as installed on a pipe and any additional costs related to logistics and construction; this cost element obviously decreases with increasing arrestor separation. It is assumed that any rupture has a cost that is made up of two components:

- a constant value, corresponding to a minimum outage time, which is determined by the time to mobilize a repair crew, transport repair materials to the failure site, complete a short repair (one to five joints), and recommission the pipeline;
- a variable portion that is proportional to the incremental fracture length (and hence the incremental outage time). The incremental outage time will be determined by the additional length to be repaired and can be calculated by dividing the incremental fracture length by a notional repair rate (joints per day). Production rates for repairs are typically much slower than for original construction.

The variable component of the incremental cost of fractures contains both the repair costs and the costs associated with the service interruption. For typical high-throughput pipelines, the latter are usually dominant. They mainly relate to lost revenue associated with failure to deliver, but in certain cases may also include a penalty component; the details will be determined by the tariff structure of a specific pipeline.

Crack arrestor spacing on natural gas and CO_2 pipelines, determined on an ad hoc basis, or using the type of economic analysis just described, has ranged from 150 m to over 3 km. More typical spacing would be from about 300 to 800 m.

Safety considerations related to incremental fracture length are hard to quantify in any general way. DNV-RP-J202 recommends that crack arrestor spacing be based on safety evaluation and cost of repair but indicates that the risk associated with a propagating fracture may not necessarily be significantly worse than for a shorter rupture [5]. It is clear that the longer the fracture, the greater the potential for property damage or injury related to debris impact, but this type of consequence has historically been a very minor contributor to overall consequences of pipeline failures. Relative to what is usually considered the dominant hazard, the release to atmosphere of a large volume of CO_2 (in some cases containing toxics, such as H_2S), followed by atmospheric dispersion leading to potentially hazardous concentrations extending to significant distances, the influence of arrestor spacing *per se* is not clear. In most cases, the siting and spacing of automatic or remote-controlled block valves, which can limit the volume of CO_2 released, will be more significant. As with all releases of hazardous materials that can lead to dense cloud formation, attention needs to be paid to terrain and the position of potential receptors in the event of a rupture. The appropriate siting of both block valves and crack arrestors can help to limit the potential hazard to vulnerable populations. In general, in sensitive locations, it will be wise to conduct detailed dispersion studies to determine the potential influence of factors such as valve spacing, location, and closure times, as well as crack arrestor spacing and position. On the other hand, for many locations where population is sparse, or where the topography and meteorological conditions are such that formation of stable, hazardous concentrations of CO_2 is unlikely, crack arrestor spacing based primarily on economic considerations will generally be appropriate.

REFERENCES

[1] Canadian Standards Association, 2011, Standard Z662, "Oil and gas pipeline systems", Canadian Standards Association, Mississauga, Ontario.

[2] NACE, 2003, TM192-2003 "Evaluating Elastomeric Materials in Carbon Dioxide Decompression Environments", NACE International, Houston, TX.

[3] NACE, 2008, TM0297-2008, "Effects of High-Temperature, High-Pressure Carbon Dioxide Decompression in Elastomeric Materials", NACE International, Houston, TX.

[4] ASME, 2009, "Pipeline Transportation Systems for Liquid Hydrocarbons and other Liquids", ASME, New York, NY.

[5] DET NORSKE VERITAS, 2010, DNV-RP-J202 "Design and Operation of CO_2 Pipelines", DET NORSKE VERITAS, Høvik, Norway.

[6] Maxey, W. A., 1974, "Fracture Initiation, Propagation and Arrest", Paper J, *Proc. 5th Symposium on Line Pipe Research*, Catalogue # L30174, PRCI, Falls Church, VA.

[7] Cosham, A., Jones, D. G., Eiber, R., and Hopkins, P., 2009, "Don't drop the drop weight tear test", *Proc. 5th International Pipeline Technology Conference*, Paper No. Ostend 2009-090, Great Southern Press, Beaconsfield, UK.

[8] Barbaro, F., Fletcher, L., Dinnis, C., Piper, J., and Gray, J. M., 2011, "Design and Specification of Line Pipe and Line Pipe Steels for Weldability, Constructability, and Integrity," *Proc. 18th PRCI-APIA-EPRG Joint Technical Meeting on Pipeline Research*, San Francisco, CA.

[9] Eiber, R. J., Bubenik, T. A., and Maxey, W. A., 1993, "Fracture Control Technology for Natural Gas Pipelines", NG-18 Report No. 208, Catalogue # L30174, PRCI, Falls Church, VA.

[10] Rosenfeld, M. J., 1997, "Procedure improves line pipe Charpy test interpretation", *Oil & Gas Journal*, April, pp. 40–44.

[11] Kiefner, J. A., 1969, "Fracture Initiation", Paper G, *Proc. 4th Symposium on Line Pipe Research*, Catalogue # L30075, PRCI, Falls Church, VA.

[12] Kiefner, J. A., Maxey, W. A., Eiber, R. J., and Duffy, A. R., 1973, "Failure Stress Levels of Flaws in Pressurized Cylinders", ASTM STP 536, "Progress in Flaw Growth and Fracture Toughness Testing", p. 461, ASTM, Philadelphia, PA.

[13] Maxey, W. A., Kiefner, J. A., Eiber, R. J., and Duffy, A. R., 1972, "Ductile Fracture Initiation, Propagation and Arrest in Cylindrical Vessels", ASTM STP 518, ASTM, West Conshohocken, PA.

[14] Rothwell, A. B., and Wadsley, M. W., 1997, "Fracture Propagation Control Measures for Gas Pipelines — Methane", Proc. WTIA/APIA/CRC-MWJ International Seminar on Fracture Control in Gas Pipelines, Paper 6, WTIA, Lidcombe, NSW, Australia.

[15] Sugie, E., Matsuoka, M., Akiyama, T., Tanaka, K., and Tsukamoto, M., 1988, "Notch ductilities of rich natural gas transmission line pipes for arresting propagating shear fracture", *Trans. ASME, Journal of OMAE*, 109, pp. 2–8.

[16] Fonzo, A., Salvini, P., Di Biaggio, M., and Mannucci, G., 2002, "Full history burst test through finite element analysis", *Proc. 4th International Pipeline Conference*, Paper IPC2002-27120. ASME, New York, NY.

[17] Misawa, K., Imai, Y., and Ihara, S., 2009, "Coupled numerical model of gas-decompression and unstable ductile crack propagation in high-pressure gas pipelines", *Proc. 5th International Pipeline Technology Conference*, Paper No. Ostend 2009-009, Great Southern Press, Beaconsfield, UK.

[18] Maxey, W. A., 1983, "Gas Expansion Studies". NG-18 Report No. 133, Catalogue # L51435, PRCI, Falls Church, VA.

[19] Leis, B. L., 2000, "Predicting Fracture Arrest Behaviour based on a Relationship between Charpy V-Notch Toughness and Dynamic Crack Propagation Resistance", *Proc. 3rd International Pipeline Technology Conference*, Brugge, pp. 407–420, Elsevier, Amsterdam, Netherlands.

[20] Wilkowski, G., Rudland, D. L., and Rothwell, A. B., 2006, "How to Optimize the Design of Mechanical Crack Arrestors", *Proc. 6th International Pipeline Conference*, Paper IPC2006-10357, Proc. IPC2006, ASME, New York, NY.

[21] Botros, K., Geerligs, J., Rothwell, B., Carlson, L., Fletcher, L., and Venton, P., 2010, "Effects of Pipe Internal Surface Roughness on Decompression Wave Speed in Natural Gas Mixtures", *Proc. 8th International Pipeline Conference*, Paper IPC2010-31667. ASME, New York, NY.

[22] King, G. G., 1981, "Design Considerations for Carbon Dioxide Pipe Lines", Pipe Line Industry, pp. 125–132.

[23] King, G. G., 1982, "CO_2 Pipeline Design — Conclusions: Propagating fractures, pipe sizing probed for high-operating-pressure CO_2 pipelines", *Oil and Gas Journal*, pp. 100–102.

[24] Maxey, W. A., 1986, "Long Shear Fractures in CO_2 Lines Controlled by Regulating Saturation, Arrest Pressures", *Oil and Gas Journal*, pp. 44–46.

[25] Rothwell, A. B., 1988, "Fracture Control in Natural Gas and CO_2 Pipelines" Proc. Conf. on Microalloyed HSLA Steels, pp. 95–108. ASM International, Materials Park, OH.

[26] Wadsley, M. W., and Rothwell, A. B., 1997, "Fracture Control for Pipelines Carrying Other Gases — HVPL, CO_2 and Others", Proc. WTIA/APIA/CRC-MWJ International Seminar on Fracture Control in Gas Pipelines, Paper 9, WTIA, Lidcombe, NSW, Australia.

[27] Cosham, A., and Eiber, R. J., 2008. "Fracture Control in Carbon Dioxide Pipelines — The Effect of Impurities", *Proc. 7th International Pipeline Conference*, Paper IPC2008-84346. ASME, New York, NY.

[28] Cosham, A., Eiber, R. J., and Clark, E. B., 2010, "GASDECOM: Carbon Dioxide and Other Components", *Proc. 8th International Pipeline Conference*, Paper IPC2010-31572. ASME, New York, NY.

[29] King, G. G., and Kumar, S., 2010, "Designing CO_2 Transmission Pipelines without Crack Arrestors", *Proc. 8th International Pipeline Conference*, Paper IPC2010-31678. ASME, New York, NY.

[30] Kunz, O., Klimeck, R., Wagner, W., and Jaeschke, M., 2007, "The GERG-2004 Wide-Range Equation of State for Natural Gases and Other Mixtures", Groupe Européen de Recherches Gazières (GERG), Technical Monograph, GERG TM15.

[31] NIST, 2007, Standard Reference Database 23, NIST Reference Fluid Thermodynamic and Transport Properties Database (REFPROP): Version 8.0, www.nist.gov/ts/msd/srd/nist23.cfm.

[32] Cosham, A., 2009, "It's a gas, Jim, but not as we know it", *Proc. 5th International Pipeline Technology Conference*, Paper No. Ostend 2009-091, Great Southern Press, Beaconsfield, UK.

[33] Porter, P. C., and Patrick, A. J., 2002, "Using Composite Wrap Crack Arrestors Saves Money on Pipeline Conversion", *Pipeline and Gas Journal*, October, pp. 65–67.

[34] Venton, P. B., and Dietsch, A. E., 1997, "Design of Crack Arrestors". Proc. WTIA/APIA/CRC-MWJ International Seminar on Fracture Control in Gas Pipelines, Paper 10, WTIA, Lidcombe, NSW, Australia.

[35] Hillenbrand, H.-G., Brauer, H., and Knauf, G., 2004, "Crack Arrestors", *Proc. 4th International Pipeline Technology Conference*, Ostend, pp. 501–513, Scientific Surveys Ltd, Beaconsfield, UK.

[36] Mannucci, G., Di Biaggio, N., Demofonti, G., Fonzo, A., Salvini, P., and Edwards, A., 2004, "Crack arrestor design by finite element analysis for an X100 gas transportation system", *Proc. 4th International Pipeline Technology Conference*, Ostend, pp. 483–499, Scientific Surveys Ltd, Beaconsfield, UK.

[37] Clock Spring Company LP, 2010, "Crack Arrestors Stop Long Running Fracture. Applications for CO_2 Pipelines and Carbon Sequestration", *Clock Spring Quarterly*, June, http://www.clockspring.com/PDF/Clock%20Spring%202007%20Quarterly%202.pdf.

[38] Fawley, N. C., 1994, "Development of fibreglass composite systems for natural gas pipeline service", *Final Report to Gas Research Institute*, GRI-94/0072.

[39] Fonzo, A., Meleddu, A., Di Biaggio, M., Mannucci, G., Demofonti, G., Peterson, C. W., and Biery, N.E., 2006, "Crack Propagation Modelling and Crack Arrestor Design for X120", *Proc. 6th International Pipeline Conference*, Paper IPC2006-10319. ASME, New York, NY.

[40] Rothwell, A. B., and Fearnehough, G., 1984, "Application of Fracture Technology and Design Philosophies for Major Gas Transmission Pipelines", *Proc. CBMM—Bardin Institute Seminar on Fracture in Gas Pipelines*, Moscow, pp. 135–154, CBMM, Sao Paulo, Brazil, 1985.

[41] Rudland, D., Shim, D.-J., Xu, H., Rider, D., Mincer, P., Shoemaker, D., and Wilkowski, G., 2007, "First Major Improvements to the Two-Curve Ductile Fracture Model", PRCI Report Catalogue # L52245, PRCI, Falls Church, VA.

[42] Wilkowski, G., Rudland, D., Xu, H., and Sanderson, N., 2006, "Effect of Grade on Ductile Fracture Arrest Criteria for Gas Pipelines", *Proc. 6th International Pipeline Conference*, Paper IPC2006-10350. ASME, New York, NY.

Chapter 8

CO_2 PIPELINE SYSTEM DEVELOPMENT AND DESIGN

8.1 INTRODUCTION

The CO_2 transported by CCS infrastructure will have been captured at source using a variety of capture technologies. As mentioned earlier, the CO_2 stream from the capture process contains CO_2 and other components that will depend on the fuel source, the capture technology and the postcapture treatment employed. The composition of the CO_2 stream to be transported is usually specified by legislation/regulation, such as pipeline codes and standards, which are mostly based on health, safety, and environmental requirements and whether it is to be stored or used for enhanced oil recovery (EOR). It was also mentioned that the presence of other components in the stream influences its physical properties and has a significant effect on matters including hydraulic characteristics, compression, and transport efficiency as well as environmental impact.

The phase diagram for pure CO_2 has been shown before, and is repeated here in Figure 8-1 to aid in the discussion of compression and pumping thermodynamics. This diagram contains two distinct features: the 'triple point' at (0.52 MPa-a, –56°C), where the CO_2 can exist at all three phases (solid, liquid, or vapor); and the 'critical point' at (7.382 MPa-a, 30.97°C). The two vertical and horizontal lines passing through the critical point divide the map into four distinct and important regions where the pure CO_2 fluid takes a distinct phase in each. These are: vapor phase, supercritical phase, dense phase and liquid phase in this order, if one goes around the critical point in a counterclockwise direction as shown in Figure 8-1.

Two phase flow (vapor and liquid) in a pipeline can occur if the local pressure and temperature of the pure CO_2 or the mixture of CO_2 with impurities is such that the state of the fluid lies within the phase envelope corresponding to the mixture. Clearly, it is not desirable to transport the CO_2 stream in such a two phase state due to several operational and economic reasons as well as potential equipment damage unless the equipment is specifically designed for multiphase flow. Typically, it is desirable to maintain the state of the CO_2 mixture at pressures above the cricondentherm of the mixture, or above the critical pressure for pure CO_2. This means that the phase of the fluid will be either supercritical or dense phase according to the map of Figure 8-1. It is also possible to transport CO_2 mixtures in the vapor phase but that would not be economical as it is associated with large pressure drops. For this reason, it is necessary to know the location of the critical point on the phase diagram and the boundaries of the envelope. However, the exact composition of the CO_2 stream may vary, and may not be known exactly. In the case of pure CO_2, the phase envelope is actually the vapor–liquid equilibrium saturation line (shown in Figure 8-1), and the desired operating range for a pipeline transporting this fluid is shown in Figure 8-2. The upper pressure level of the operating range envelope is associated with ANSI 2500 (~35 MPa) rating, while the lower pressure is dictated by the saturation line. Typically, CO_2 pipelines operate around 15 MPa which corresponds to ANSI 900 rating, or 25 MPa, which corresponds to ANSI 1500.

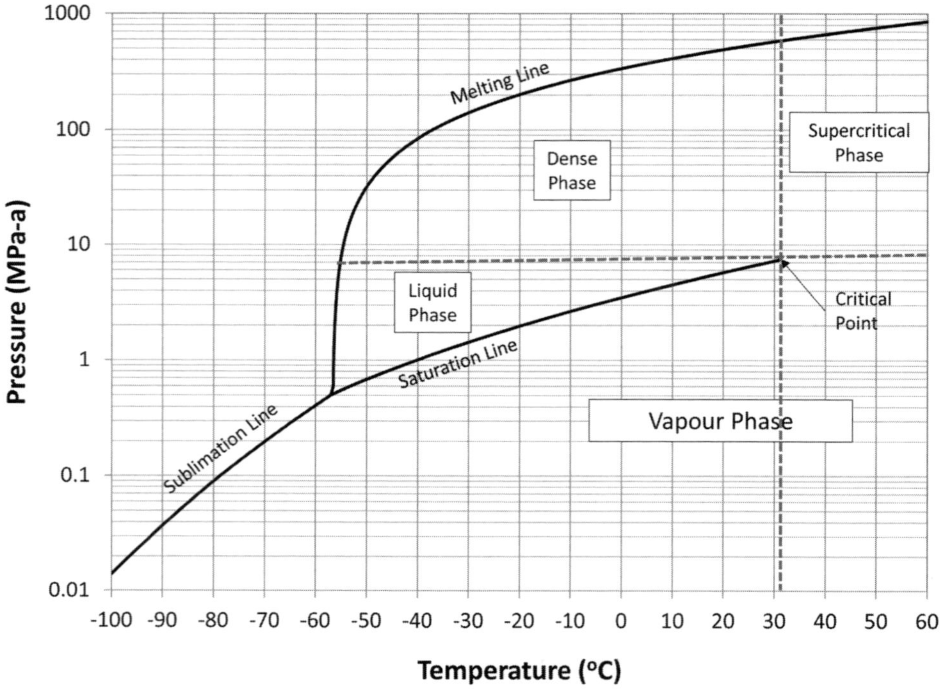

Figure 8-1. Phase diagram of pure CO_2

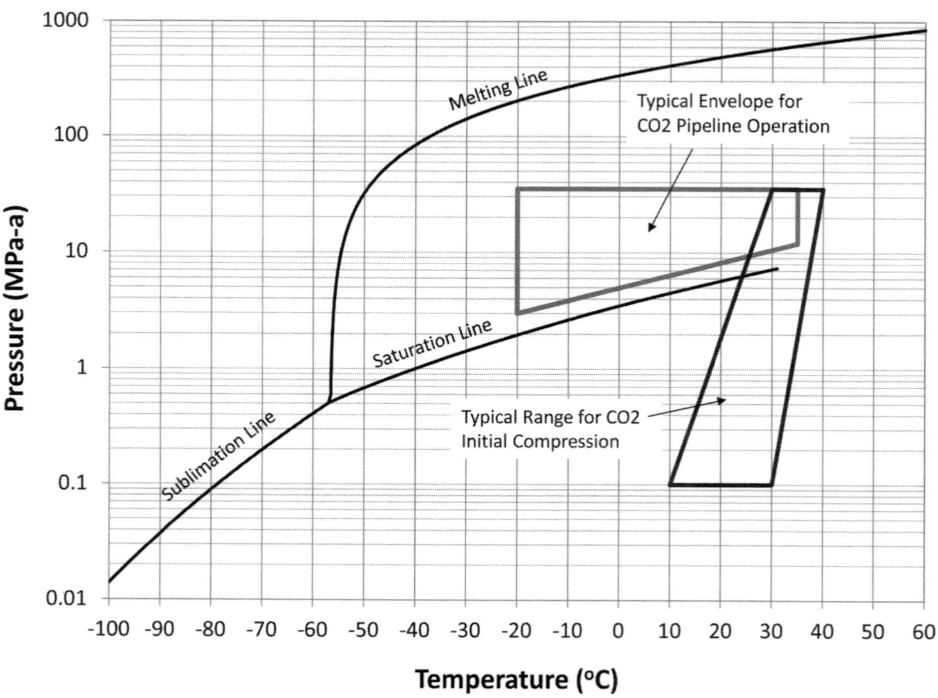

Figure 8-2. Typical CO_2 pipeline operating envelope and initial compression range

Current CO_2 pipelines operate over an internal temperature range of around 4°C to 38°C so as to transport the CO_2 stream in dense phase. The pipeline inlet pressure is likely to have to be in excess of 10 MPa and the pressure drop over the pipeline length sufficiently small to avoid the two-phase envelope. Before the CO_2 is introduced into the transmission pipeline, it requires initial compression from the capture pressure, which is typically atmospheric, to the operating pressure of the pipeline, say 15 MPa. This is much too high of a pressure ratio to be performed by one stage compression. Typically, it takes up to five stages of compression utilizing normally high compression ratio compressors (e.g., reciprocating) with intercooling to achieve such high pressure ratios. Figure 8-2 shows the typical operating range of pressures and temperatures through the initial compression process.

8.2 CODES AND STANDARDS

Although CO_2 transport by pipeline could be considered to be mature (greater than 7000 km of pipeline already exists or is planned worldwide) and that extensive experience has been gained throughout these years, guidelines and standards for the design and operation of these pipelines are still considered inadequate. The main reason is that most of the existing CO_2 pipelines are located in remote areas (in particular in Texas, New Mexico, etc.), where the population density is very low and the probability of external interference, which is one of the greatest causes of failure in a buried pipeline, is low. The statistics in the last five years for these pipelines showed that about 0.2 leak events occurred for every 1000 km of line per year, as indicated in the 7th EGIG report, 2008 [1]. Figure 8-3 shows statistics of CO_2 pipeline incidents between 1986 and 2008.

Moreover, a significant amount of this experience has been developed on the transportation of pure CO_2 mostly from natural sources. This experience cannot be automatically applied to anthropogenic CO_2 mixtures. As mentioned several times throughout the book, the presence of impurities (H_2S, SO_2, N_2, O_2, Ar, CO, H_2, CH_4, etc.) and in particular H_2O, can also have a significant impact on the physical properties of the mixture, which creates the following issues: transportation (recompression distances), operation and maintenance (corrosion and stress corrosion control), and design (fracture control and corrosion/stress

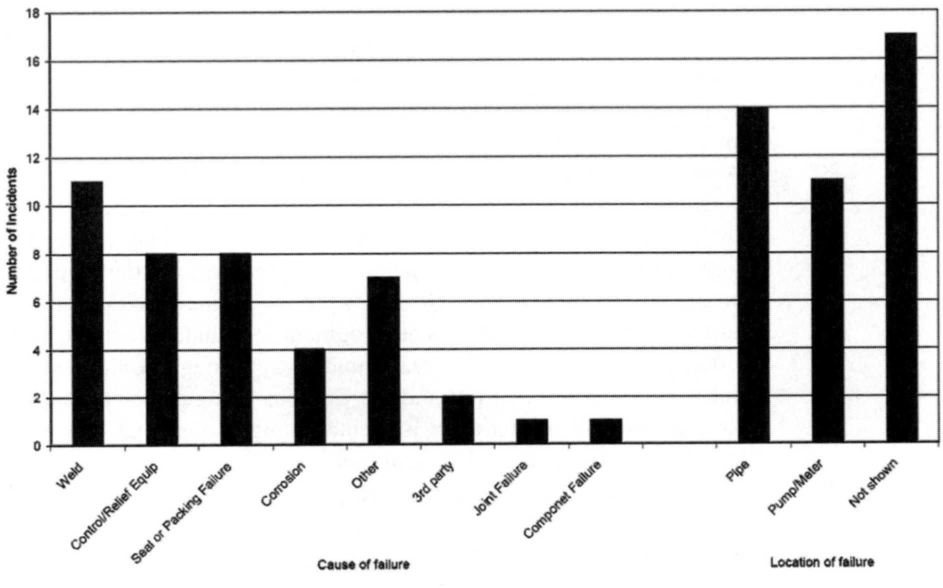

Figure 8-3. CO_2 pipeline incidents from 1986 to 2008 [1]

corrosion prevention). Furthermore, networks connecting multiple anthropogenic carbon dioxide capture and sequestration areas need proper specification. In particular, the major challenges for these networks can be summarized in the following four key issues:

- Regulatory, financing, legal issues, codes/standards.
- Engineering design of pipelines from the long-term integrity point of view.
- Supply and demand balance.
- Overall cycle cost and capacity.

Regarding the first point, review of existing codes and standards devoted to the transport of fluids in pipelines has revealed that suitable guidelines for anthropogenic CO_2 transportation are not available. Codes such as ASME B31.8, ASME B31.4, IP6, BS EN 14161, BS PD 8010, and DNV OS-F101 may be applicable to pipelines transporting CO_2, but they do not explicitly include transportation of anthropogenic CO_2 in supercritical/dense/liquid phase. Up to now, a specific design or fitness-for-service procedure for CO_2 pipelines within the international regulatory frameworks does not exist. DNV recently proposed (April, 2010) a new Recommended Practice (DNVRPJ 202 "Design and Operation of CO_2 Pipelines") [2] as conclusion of the first phase of a JIP ("CO_2PipeTrans Phase 1") aimed at assessing the possibilities of updating the DNV-OS-F101 code to include offshore transportation of CO_2.

All of these key issues are going to impact the development of anthropogenic CO_2 capture and storage technologies and their further industrial "in field" application worldwide. The lack of CCS operational know-how joined with the uncertain long-term financial and regulatory drivers and acceptability of selected locations, as well as storage capacities evaluation, affects the commercial development of large-scale CCS pipeline infrastructures. To reduce these barriers, efforts to increase the understanding of integrity of anthropogenic carbon dioxide pipelines and their management through credible codes, standards, and guidelines, will significantly contribute towards reducing costs and risks, which in turn encourage significant investment in CO_2 pipeline networks.

The aforementioned existing international pipeline standards were thoroughly reviewed in a IEA Greenhouse Gas R&D program, Rules and standards for the transmission and storage of CO_2, [3]. Although for the most part these standards are very helpful, it was concluded that some reinforcement should be considered regarding design, operation, health, safety, and environmental aspects. The one issue that stands out is the block valve locations in the existing codes. The reason for this is that small leaks from pipelines were identified as the most significant hazard from a CO_2 pipeline based on a dispersion modeling study. The existing codes could be improved by using industry standard Quantitative Risk Assessment and goal-setting procedures to set block valve location requirements. The adoption of such practices should help develop confidence that industry best practices are being employed to reduce the risk of leaks in the event of pipeline failures, which will become important when large pipeline infrastructures are developed and in particular in more densely populated areas than those in which they are currently deployed [3]. With regard to international standards development, it is understood that unification of the various standards on pipelines is already underway through bodies such as the International Standards Organisation (ISO) and the European Committee for Standardization (CEN). Also, it is understood that there is an increasing widespread acceptance of one standard (ISO 13623, Petroleum and natural gas industries — Pipeline Transportation systems) in the petroleum industry and with regulatory bodies [4]. It is encouraging to see that although existing standards and codes vary between countries, there is gradual unification of these documents. Table 8-1 highlights gap areas where codes and standards can be further developed to cover various aspects of the design and operation of anthropogenic CO_2 pipelines.

TABLE 8-1. Gaps and issues where codes and standards can further be developed

- Process conditions
  - Properties
  - Operating conditions at entry and exit
  - Flow calculation method
  - Transient (surge) modelling
  - Flow characteristics
  - Typical carbon dioxide compositions
- Piping design
  - Fracture propagation
  - Blow down assembly design
  - Blow down rate basis and calculation
  - Line break controls
  - Pig traps
  - Depth of cover
  - Routing topography
- Safety and environmental
  - Ambient/ground temperature
  - Blow down rate basis and calculation
  - Dispersion pattern
  - Frequency and position of block valves
  - Leak detection systems
  - Line inventory
- Material selection
  - Pipeline materials
  - Carbon equivalent
  - Hardness value
  - Fracture strength
  - Valve, fitting and trim types
  - Seal, packing materials of construction
  - Valve actuators
- Cleaning and strength testing
  - Cleaning
  - Hydrostatic testing/drying/dewatering
- Construction techniques
  - Corrosion monitoring
  - External corrosion
  - Fracture propagation
  - Special construction and welding
  - Stress relief
- Pipeline operation
  - Refrigeration effects during start-up/blow down
  - Start-up/shutdown methodology
  - Line pressuring
  - Requirement for blow down noise control
  - Environmental considerations
  - Operational problems
  - Operational safety
- Measurement
  - Custody transfer methods
  - Moisture analysis

Finally, public concern could highlight the issue of CO_2 releases from transportation systems, either through rupture or minor leaks. It is possible that standards may be changed in the future to address specific public concerns. Odorants are often added to domestic low-pressure gas distribution systems, but not to gas in long-distance pipelines; they could, in principle, be added to CO_2 in pipelines. Mercaptans, which are naturally present in the Weyburn pipeline system, are the most effective odorants but are not generally suitable for this application because they are degraded by O_2, even at very low concentrations [5]. Disulphides, thioethers and ring compounds containing sulphur are alternatives. The value and impact of odorization could be established by a quantitative risk assessment.

8.3 DESIGN REQUIREMENTS

The physical, environmental, and social factors that determine the design of CO_2 pipelines are typically summarized in a design basis, which then forms the input for the conceptual design. This includes a system definition for the preliminary route and design aspects for cost-estimating and concept-definition purposes. It is also necessary to consider the process data defining the physical characteristics of the mixture transported, the optimal sizing and pressures for the pipeline, and the mechanical design, such as operating, valves, pumps, compressors, seals, etc. The topography of the pipeline right-of-way must be examined. Topography may include mountains, deserts, river and stream crossings, and for offshore pipelines, the differing challenges of very deep or shallow water, and uneven seabed. It is also important to include geotechnical considerations. For example, is this pipeline to be constructed on thin soil overlaying granite? The local environmental data need to be included, as well as the annual variation in temperature during operation and during construction, potentially unstable slopes, frost heave, and seismic activity. Also included are water depth, sea currents, permafrost, ice gouging in Arctic seas, biological growth, aquifers, and other environmental considerations such as protected habitats. The next set of challenges is how the pipeline will accommodate existing and future infrastructure — road, rail, pipeline crossings, military/governmental restrictions, and the possible impact of other activities — as well as shipping lanes, rural, or urban settings, fishing restrictions, and conflicting uses, such as dredging. Finally, this integrated study will serve as the basis for a safety review.

8.3.1 Conceptual Design

The conceptual design includes the following components:

- Mechanical design: follows standard procedures, e.g., those described in detail in [6].
- Stability design: standard methods and software are used to perform stability calculations, offshore or onshore, though the offshore methods have been questioned. New guidelines for stability published by Det Norske Veritas (DNV) [2] gives a good basis for the design.
- Protection against corrosion: a well-understood subject of which the application to CO_2 pipelines is based on the specific mixture composition and the pipeline environment. This is governed by the corrosion rate, which can be established by a number of predictive tools. For piping applications with high concentrations of carbonic acid, particularly in turbulent flow areas (such as downstream of control valves, or near pumps), corrosion-resistant stainless steels provided with erosion protection may have to be used.
- Trenching and backfilling: onshore lines are usually buried to depth of 1 m. Offshore lines are almost always buried in shallow water. In deeper water, pipelines narrower than 400 mm are trenched and sometimes buried to protect them against damage by fishing gear.
- CO_2 pipelines may be more subject to longitudinal running fracture than hydrocarbon gas pipelines. Fracture arresters are installed at intervals of about 500 m.

As an example, the conceptual design parameters of the SACROC CO_2 pipeline are described here [7]. The initial transportation options examined were the tank truck and rail options which cost more than twice as much as a pipeline. The refrigerated pipeline was rejected because of cost and technical difficulties with liquefaction. The dense phase (Option II) was found to be 20% cheaper than a low-pressure CO_2 gas pipeline (Option I). The intermediate 4.8 to 9.6 MPa pressure range was avoided so that two-phase flow would not occur. An added advantage of dense-phase transport was that high delivery pressures were required for CO_2 injection. The final design conforms to the ANSI B31.8 code for gas pipelines and to the DOT regulations applicable at the time. The main 290 km section is 406.4 mm (16 in.) outside diameter and 9.53 mm wall thickness made from grade X65 pipe (specified minimum yield stress of 448 MPa). A shorter 60 km section is 323.85 mm (12.75 in.) outside diameter, 8.74 mm wall thickness, grade X65. Tests showed that dry CO_2 would not corrode the pipeline steel; 304 L corrosion-resistant alloy was used for short sections upstream of the glycol dehydrator. The line is buried to a minimum of 0.9 m, and any point on the line is within 16 km of a block valve. There are six compressor stations, totalling 60 MW, including a station at the SACROC delivery point. The compressor stations are not equally spaced, and the longest distance between the two stations is about

160 km. This is consistent with general practice, but some long pipelines have 400 km or more between compressor stations. Significant nitrogen and oxygen components in CO_2 would shift the boundary of the two-phase region towards higher pressures and would require a higher operating pressure to avoid two-phase flow.

8.3.2 Detailed Design

A transmission pipeline can experience a wide range of ambient temperatures, so maintaining stability of a supercritical or dense single phase is important in order to avoid considerations of a two-phase flow that could result in pressure surges. Typical pipeline detailed engineering considerations for CO_2 include:

- operating pressure;
- operating temperature;
- gas mixture composition;
- corrosivity;
- ambient temperatures; and
- pipeline control (SCADA).

Examples of additional specific design considerations in place for CO_2 pipelines are the following:

- effect of cooling from pressure changes;
- requirement for dehydration of CO_2;
- routing topography;
- dispersion pattern;
- valve materials;
- compressor, seal, and auxiliary materials;
- requirement to minimize flow transients; and
- risk assessment focused on CO_2 and impurities toxicity.

Table 8-2 provides a list of the various parameters and issues to be addressed in the detailed design of CO_2 pipeline. Material selection is a key design issue in several aspects. One is the control of ductile fracture, which has been dealt with in more detail in Chapters 6 and 7, in addition to consideration of the risk of brittle fracture due to low temperatures during blow down or even small leak. The other is the design factor and population proximities. In the USA, where CO_2 is in supercritical or dense phase, ASME B31.4 uses a design factor of 0.72 and provides no specific guidance on proximities. Therefore, where indicated by service or location, users of this code may elect to use a design factor less than 0.72. In Europe, PD 8010 Part 1 proposes design factors ranging from 0.72 to 0.3 based on location classes and population densities. The population densities are calculated using individual risk contours or using a design formula that takes diameter, pressure, and substance into account. However, the applicability of this formula for use on high-pressure CO_2 pipelines needs to be confirmed. In class 1 locations, for pipelines intended to convey category C, category D, and category E substances, the design factor should under normal circumstances not exceed 0.72. In class 2 locations, the design factor should normally not exceed 0.3, but this may be raised to a maximum of 0.72 provided that it can be justified to a statutory authority by a risk analysis carried out as part of a safety evaluation for the pipeline. Pipelines designed to convey category D and category E substances in class 2 locations should either have a nominal wall thickness of at least 9.52 mm or be provided with impact protection to reduce the likelihood of penetration from mechanical interferences. Routing of pipelines conveying categories C (hazardous), D, and E substances through class 3 locations should be avoided.

TABLE 8-2. List of the various parameters and issues to be addressed in the design of a CO_2 pipeline

CO_2 Pipeline Design Elements	
System Design/ Flow Assurance	Properties of CO_2+impurities EOS Flow calculation and system hydraulics Routing topography Ambient and ground temperature effects
Design	Ductile fracture control (shock tube) Ductile fracture control (full scale) CO_2DECOM (based on best EOS) Fracture arrest Brittle fracture Blowdown assembly design Pig traps Depth of cover Spacing of block valves Meter station design Pump and compressor station design Large releases (rupture) Blowdown releases Small leak releases Fatigue
Operation	Transient flow modeling Blowdown rate calculation CO_2 solid sublimation during blowdown Leak detection Line inventory, packing, drafting phenomena SCADA SCC Hydrogen induced cracking External corrosion Start-up/shutdown methodology Pressurizing and depressurizing Blowdown noise Hydrate formation
Integrity Monitoring Program	Integrity management system Pipeline integrity during blowdown Leak detection methodologies pigging and ILI
Health/Safety/ Environment	Large releases (rupture) Blowdown releases Small leak releases Dispersion modeling Failure Hazard Analysis
Material Selection	Line pipe Effects of cooling during release Valves, fitting and trims Elastomers Lubricants Packing and sealants Gaskets Valve actuators Mixture effects and specifications.
Construction	Welding practices Cleaning Hydrostatic testing Dewatering/drying
Codes/Standards/ Guidelines/ Recommend Practice	Design factors Location classes Population densities and proximities Hazardous substance categories ASME B31.8 and ASME B31.4 (US) PD 8010 Part 1 (Europe) DNV-RP-J202 CSA Z662 (Canada) AS 2885 (Australia)

For pipelines conveying category C substances and having a design factor not exceeding 0.3, the final route may be established by allowing a minimum distance of 3 km between the pipeline and occupied buildings. The minimum distance for category D substances is the same as category C provided the wall thickness is greater than or equal to 11.91 mm. DNV-RP-J202 recommends offshore utilization factors from 0.77 to 0.67 and onshore utilization factors from 0.83 to 0.45 depending on location classes defined in ISO 13623.

8.3.3 Hydrate Formation

The presence of free water (as opposed to dissolved water) in the CO_2 stream may lead to hydrate formation in the pipeline. The issue with hydrate formation in pipelines is that hydrates are solid compounds with similar properties to ice; consequently, they can block the pipeline and plug or foul other equipment such as flow meters and heat exchangers. Under CO_2 pipeline operating pressures, and with certain water contents, it would be possible for hydrates to form at around 10°C to 11°C [8, 39], and therefore there is a requirement to control water levels. There is a consensus that this water must be removed prior to pipeline transportation, and there exists a number of effective drying technologies and processes, however there is no general consensus on what the maximum final level of water should be. Both the Dynamis and Ecofys projects recommended a water content of 500 ppm to ensure that no free water is present in the pipeline. The Dynamis project did conclude, however, that the 500 ppm limit should be reviewed for offshore pipelines which could be operating at temperatures around 4°C. At this temperature, the solubility of water is below 500 ppm for pressures below 4 MPa. Although the Dynamis report considered the cross-effects of H_2S and CH_4 on water solubility in CO_2, the effects of other impurities in the CO_2 stream (e.g., O_2) had not been investigated. A conservative limit for water content might therefore be 200 ppm, i.e., 60% of the minimum solubility at 4°C [35].

Therefore, the specification of an acceptable level of water in the pipeline is dependent on the solubility of water in the fluid at the operating temperature and pressure. The amount of water must be significantly less than the maximum that can be dissolved in the CO_2. An indicative limit could be taken as 60% of the minimum solubility achieved by pure CO_2 at the design temperature, e.g., 200 ppm at 4°C, although opinion may differ on what constitutes an acceptable margin of safety. As mentioned above, impurities in the CO_2 stream may affect when hydrates form and the nature of the hydrates that form. In some locations for hydrocarbon pipelines, small amounts of inhibitors are added to reduce the risk of hydrate formation, although this may be subject to regulatory approval. There is very limited understanding of the appropriate margin of safety required to eliminate the risk of hydrates. There is only a limited amount of data available, and more laboratory experiments are required to determine the solubility of water and potential for hydrate formation in CO_2 containing the types of impurities to be expected from capture over a range of operating temperatures and pressures. Until this data becomes available, there is a risk that developers and regulators will make inappropriate drying specifications. Clearly, insufficient drying could lead to mechanical damage from hydrates. In contrast, a conservative approach may raise drying costs unnecessarily. In turn, this could raise overall CCS system costs and reduce the likelihood that sources will connect [35].

8.4 ROUTE SELECTION

The source and storage points affect the overall pipeline system design. The locations of the sources and storage points determine the pipeline route and the locations of facilities and control points [6]. Following the identification of CO_2 sources and storage locations, and as a prelude to pipeline design, a preliminary route selection is undertaken. Determining the pipeline route will influence design and construction in that it affects requirements for line size (length and diameter), as well as compressor or pumping facilities and their location [6].

Urban areas

Pipeline construction in urban areas is very complex from a planning, legal, safety, and technical perspective. Strict regulations are applied to urban construction, and urban construction considerably adds to the costs because of accessibility to construction and additional safety measures required. In addition, planning procedures tend to be more time consuming. In this way, urban areas are avoided, and the pipeline trajectory should go around such areas [7].

Existing cables and pipeline corridors
In the planning of onshore pipelines, it is recommended to follow existing pipeline trajectories because this will reduce costs and limit delays in planning procedures. The pipeline trajectory should follow existing cable and pipeline corridors where possible [7].

Land cover
Not all land covers are equally suitable for land pipelines to be placed. For example, steep slopes and unstable peaty soils add significantly to the costs.

Connection to the mainland
For the offshore storage of CO_2, the pipeline must make a crossing from the onshore to offshore area via a CO_2 export terminal. The most consistent sites for such CO_2 export terminals would be at or close to existing landfall reception terminals for gas and oil pipelines arriving at the coast as facilities and infrastructures can be shared and connection to the existing offshore pipeline network goes relatively easy. The use of existing pipeline terminals and landfalls depends on the distance between the capture location and existing landfalls, and whether the pipeline can be channelled through it [7–10].

Sensitive areas
The identification of areas that are of special interest when planning a pipeline route is indispensable to identify because of nature protection, biodiversity, or other environmental constraints. In some cases, it might be forbidden to construct pipelines through sensitive areas and the magnitude of impact of pipeline construction and operation on sensitive areas defines whether it is necessary to relocate the pipeline and circumvent these areas [7–10].

Obstacles
Linear features such as roads, railway tracks, streams, and rivers are major obstacles in a pipeline route. The construction costs may increase due to these obstacles, however, there are techniques to let pipelines cross such obstacles. In order to rank the routes on obstacles, the number, complexity, and location of obstacles along alternative pipeline trajectories need to be identified [7–10].

8.5 PIPELINE CAPACITY AND FLOW ASSURANCE

The key inputs for the design of CO_2 pipelines are flow rate, length, initial pressure, pressure drop, and storage pressure in order to calculate the diameter. When designing pipelines, physical properties of the fluid and design parameters may be used in specific calculations. Key physical properties and key design parameters are the pipe diameter, pipe length, specific gravity, density, compressibility, viscosity, and thermal capacity of the CO_2 and impurities, temperature and pressure, elevation profile, and flow friction factor. A variety of friction factors are used in pipeline design equations. They are determined empirically and are related to the roughness of the inside pipe wall. One of the first items of information required for design is the amount of fluid that must flow through the pipeline. Estimates of pipeline input and delivery volumes must be made based on the data on production of CO_2 and expected storage capacity, among other data. With projected volumes and the origin and destination of the pipeline known, the basic steps involved in a simplified preliminary design of a single pipeline are as follows:

- A required delivery pressure is determined at the pipeline's destination.
- Pressures losses due to friction and the pressure required to overcome changes in elevation are added to the delivery pressure to determine the inlet pressure.

A trial-and-error procedure may be involved because it is necessary to choose a tentative pipe size in order to calculate pressure losses. If pressure loss is too high, the resulting inlet pressure may exceed the pressure rating of the pipe or an excessive amount of pumping or compression horsepower may be required. In this case, a larger pipeline is selected and the calculations are repeated. The goal is to select a pipe size that can operate at the pressure required. With the line size and operating pressure determined, the pumping or compression power needed to deliver the desired volume of the fluid at the specified delivery pressure can be accurately calculated. If more than one pump or compression station is required, the location and the size of additional stations is set by calculating pressure loss along the line and determining how much pump and compressor horsepower is needed to maintain the minimum operating pressure. A pump/compressor can also be installed at the injection point. Economic calculations are usually performed to compare design with other combinations of line size, operating pressure, and power in order to choose the best system.

In this simplified outline for the design of a single pipeline, no branch connections are considered, neither are alternative routes nor significant changes in the throughput during the lifetime of the pipeline. Few pipelines systems are this simple, and because of this, most pipelines are designed by sophisticated computer programs, built on basic flow equations used to design a simple pipeline manually, but the computer can perform repeated calculations on a larger number of alternative solutions quickly. Different simulations are generally run for different pipe diameters in order to calculate the most economical and efficient pipe size that can be operated at a pressure permitted by regulations. Many system variables are interdependent. For example, operating pressure depends, in part, on pressure drop in the line. Pressure drop, in turn, depends on flow rate, and maximum flow rate is dictated by allowable pressure.

8.5.1 Flow Capacity

The pipeline flow capacity refers to the amount of fluid through the pipeline per unit of time, and it can be expressed as a volume or mass flow rate. The flow capacity is a function of a number of parameters, some of which are related to the customer's requirements (e.g., the volumes to be delivered and required delivery pressure) and others depend on the technical solutions used for construction, route selection, and on the physical properties of the gas (e.g., pipe diameter, changes in elevation along the pipeline path, pressure losses, viscosity and properties of CO_2 and impurities, etc.). The estimation of the mass flow rate to be handled throughout the life of the pipeline, or flow capacity, is one key element for a successful pipeline project and will influence route planning. Flow capacities are commonly calculated as steady-state, isothermal flow, in which it is assumed that the volume flow and composition of the gas transported remain constant with time (steady-state flow). A general equation for the steady-state isothermal flow of compressible fluids in pipelines will be derived below.

The general steady, one-dimensional, compressible pipe flow equation is derived from the conservation of mass and momentum along the length of the pipe, together with an equation of state describing the $P\rho T$ or PVT relation for the fluid. The following assumptions are made in the derivation of the general flow equation:

- The flow is one-dimensional, i.e., all flow parameters can vary only with distance (x) along the pipe.
- The flow is isothermal, or the temperature drop is small along a short distance of the pipe such that the fluid properties along this short distance can be averaged over the length of the pipe.
- The change in the fluid kinetic energy over this short distance is neglected.
- Frictional losses are represented by the Darcy–Weisbach friction factor (f), which is

assumed to be constant along this short distance of the pipeline. The explicit expression of this friction factor can be obtained from the Colebrook–White equation as follows [33]:

$$f = \frac{0.25}{\left[\log\left(\frac{\varepsilon}{3.7D} + \frac{5.74}{\mathrm{Re}^{0.9}}\right)\right]^2} \quad (8-1)$$

where:

D - internal pipe diameter
Re - Reynolds number based on internal pipe diameter and mean flow velocity
ε - internal surface roughness expressed as Nikuradse equivalent sand grain size.

- The real gas behavior is accounted for by the compressibility factor, Z, which is represented by a constant average value at the average pressure and temperature along the pipeline.

The governing conservation equations for steady, one-dimensional, and isothermal flow can then be written as:

Conservation of mass:

$$\frac{d}{dx}(\rho u) = 0 \quad (8-2)$$

Conservation of momentum:

$$\frac{d}{dx}\left(\rho u^2 + P\right) + \frac{f \rho u^2}{2D} + \rho g \sin\theta = 0 \quad (8-3)$$

Equation of state:

$$\rho = \frac{P}{ZRT} \quad (8-4)$$

where:

g - acceleration due to gravity
m - mass flux, ($m = \rho u$)
u - mean fluid velocity
M - mass flow rate, $M = m\left(\frac{\pi}{4} D^2\right)$
P - static pressure
T - fluid temperature
R - gas constant
ρ - fluid density
θ - pipe inclination (positive when upward)

Substituting the mass flux, m, to eliminate the fluid velocity, u, from the continuity Eq. (8-2), we get:

$$\frac{d}{dx}(m) = 0 \quad (8-5)$$

And also in the momentum Eq. (8-3),

$$\frac{d}{dx}\left(\frac{m^2}{\rho} + P\right) + \frac{f \rho m^2}{2\rho D} + \rho g \sin\theta = 0 \quad (8-6)$$

Applying the chain-rule to expand the derivatives in the above equation:

$$\frac{2m}{\rho}\frac{dm}{dx}-\frac{m^2}{\rho^2}\frac{d\rho}{dx}+\frac{dP}{dx}+\frac{f\,\rho m^2}{2\,\rho D}+\rho g\sin\,\theta=0 \qquad (8-7)$$

Substituting the continuity Eq. (8-5) into the above equation yields:

$$-\frac{m^2}{\rho^2}\frac{d\rho}{dx}+\frac{dP}{dx}+\frac{f\,\rho m^2}{2\rho D}+\rho g\sin\theta=0 \qquad (8-8)$$

Dropping the first term in the above equation since the kinetic energy is neglected, we get:

$$\frac{dP}{dx}+\frac{f\rho m^2}{2\rho D}+\rho g\sin\theta=0 \qquad (8-9)$$

Substituting the equation of state to eliminate the density and replacing the mass flux (m) with the mass flow (M):

$$\frac{dP}{dx}+\frac{8fM^2ZRT}{\pi^2D^5P}+\frac{Pg\sin\theta}{ZRT}=0 \qquad (8-10)$$

Integrating the above equation along a segment of the pipeline of length (L):

$$\int_{P_1}^{P_2}\frac{\pi^2D^5ZRT\,dP}{8fM^2Z^2R^2T^2+\pi^2D^5Pg\sin\theta}+\int_0^L dx=0$$

which results in:

$$P_1^2-P_2^2\exp\left[\frac{2gL\sin\theta}{Z_{avg}RT_{avg}}\right]-\frac{8fLZ_{avg}^2R^2T_{avg}^2M^2ZRT\left\{\exp\left(\frac{2gL\sin\theta}{Z_{avg}RT_{avg}}\right)-1\right\}}{\pi^2D^5gL\sin\theta}+\frac{Pg\sin\theta}{ZRT}=0 \qquad (8-11)$$

and re-arranging in terms of the mass flow (M):

$$M=\frac{\pi}{4}\sqrt{1/f}\;D^{5/2}\sqrt{\frac{\left[P_1^2-\exp(s)P_2^2\right]s}{Z_{avg}RT_{avg}L(\exp(s)-1)}} \qquad (8-12)$$

where:

- L - length of the pipe segment
- P_1 - pipeline upstream pressure
- P_2 - pipeline downstream
- P_{avg} - average pressure, $P_{avg}=2/3\left[P_1+P_2-\left(\frac{P_1P_2}{P_1+P_2}\right)\right]$
- T_{avg} - average temperature
- s - elevation term, $s=\frac{2g\Delta y}{Z_{avg}RT_{avg}}$
- Z_{avg} - compressibility factor at average pressure and temperature

When elevation changes in the pipeline are neglected, Eq. (8-12) becomes:

$$M=\frac{\pi}{4}\sqrt{1/f}\;D^{5/2}\sqrt{\frac{P_1^2-P_2^2}{Z_{avg}RT_{avg}L}} \qquad (8-13)$$

Finally, the heat transfer with the surrounding medium can be accounted for by the balance of energy in the following equation:

$$\int_{L_1}^{L_2} U\pi D\, dx = \int_{h_1}^{h_2} M\, dh \qquad (8-14)$$

where:

U - overall heat transfer coefficient between flowing fluid inside the pipe and ground surface based on internal pipe diameter (D)

h - specific enthalpy of the fluid which can be determined from the appropriate EOS for any given pressure and temperature

8.5.2 Effects of Impurities on Pipeline Flow Capacity

The composition of the transported CO_2 stream will depend on its source. The majority of anthropogenic CO_2 pipelines transport CO_2 with impurities, such as hydrocarbons, water, hydrogen sulphide and nitrogen. It is difficult to determine the exact levels of impurities that may be present in the captured CO_2 as this will be affected not only by the capture technology but also by economics (i.e., the increased capture cost associated with the removal of impurities to low levels), and also legislative and regulatory requirements, specifications and safety considerations.

As mentioned earlier in the book, there are three main process routes for capturing anthropogenic CO_2:

- Post-combustion (capture of CO_2 from flue gas);
- Pre-combustion (capture of CO_2 before combustion) or IGCC (Integrated Gasification Combined Cycle) and
- Oxyfuel combustion.

CO_2 with impurities such as SO_x, NO_x, and argon has not been transported before by pipeline and therefore their effects on pipeline design parameters are not fully understood. Variations in the impurity levels in the captured CO_2 were noted in the literature, e.g., [12], due to differences in processes within the same capture technology. It is highlighted that water is not considered in this specification as it is assumed that the CO_2 will be dehydrated due to stringent specifications and only trace amounts will be present [12]. However, it is recognized that the solubility of water in CO_2 is dependent on the operating pressure, temperature, and composition. Further detailed analysis is necessary to determine the solubility limits of water in anthropogenic CO_2 since it influences corrosion and hydrate formation. Pipeline steel has low corrosion resistance to wet CO_2 and therefore if wet CO_2 has to be transported (i.e., the CO_2 cannot be dried) then alternative pipeline materials will have to be considered.

The Maximum Allowable Operating Pressure (MAOP), minimum operating pressure, pipeline length, and operating temperatures are required in the initial design stages of a pipeline:

- The maximum pressure is usually set by the design codes, regulations, and economic considerations;
- The minimum pressure should be set to avoid two-phase flow. This pressure depends on the type, combination, and quantity of impurities present.
- The pipeline length is governed by the route chosen between the CO_2 source and the CO_2 sink.
- Operating temperatures (other than the compressor discharge temperature) are usually determined by the ambient temperature. This is an important operating variable

as it markedly affects the transport properties of CO_2 in terms of density, compressibility and static headlosses (i.e., increasing discharge temperature lowers static head loss) [15].

These properties and parameters change with the presence of impurities. The relationship between the flow of the fluid in the pipeline and the properties of both the pipe and the fluid at a given pressure and temperature is important in determining the pipeline capacity. Pipeline capacity, or flow rate, is dependent on critical properties such as pressure, temperature, compressibility, density, elevation, and friction factor, most of which depend on the composition of the transported fluid. Due to CO_2's unique properties and operating parameters, careful consideration needs to be given to the interrelationship of these variables, in particular how they interact with each other and consequently how they change along the pipe length for both pure CO_2 and CO_2 with impurities. Using Eq. (8-12) above, the relationship between the internal diameter of a pipeline and its capacity for a given pressure drop can be determined. Figure 8-4 shows an example of the relationship between pipeline capacity and pipe diameter for a given pressure drop of 20 kPa/km [16]. Figure 8-4 also shows the capacity of the potential CO_2 streams in the pipeline from the respective capture technologies mentioned above as well as other binary mixtures. There are several important conclusions that can be drawn from this data:

- For a given pressure drop, the presence of impurities reduces the pipeline capacity. The reduction in capacity compared to pure CO_2, at the same adjusted pressure drop is more significant at larger diameters.
- The capacity is strongly affected by the presence of impurities. As illustrated by Figure 8-4, for a given diameter of 29 in., the oxyfuel technology shows the highest reduction in capacity of 25% followed by the 95% CO_2–5%H_2 binary composition which results in a reduction of 11.5% when compared to pure CO_2. The pre-combustion,

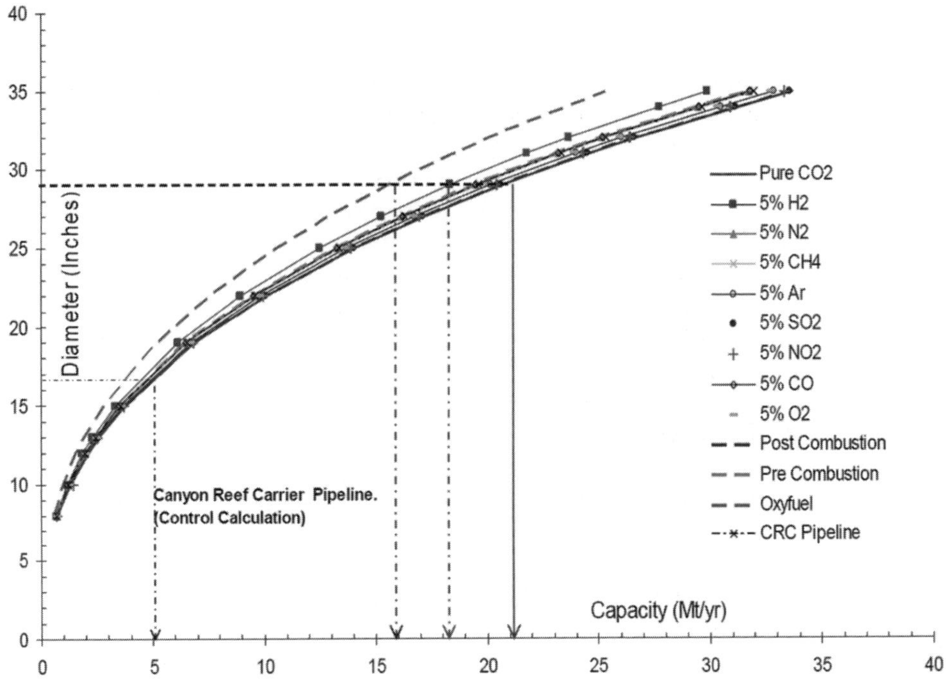

Figure 8-4. Comparison of pipeline capacity (operating at a pressure drop of 0.0002 bar/m) for different combination of impurities and capture technologies [16]

95% CO_2–5%N_2 composition and the 95% CO_2–5%CO, have a reduction of approximately 6%.

- For a given diameter, binary combination with 5%impurities other than H_2 and CO, do not show much. Deviation in capacity from pure CO_2. However, in a given combination, changes in pipeline capacity are possible due to the interactions that occur. This is clearly shown by the combination of impurities present in the oxyfuel and pre-combustion technology.
- For the three different technologies analysed here, oxyfuel shows the highest reduction in capacity followed by pre-combustion. The post-combustion technology does not show any significant deviation.

The Canyon Reef Carrier (CRC) 16 in. pipeline [18, 19] was plotted as an example validation of the calculation. The actual size of this pipeline matches that sized by the calculation in Figure 8-4. The calculation was based on a capacity of 5 Mt/year with MAOP of 14 MPa, and the same pipe size of 16 in. It should be emphasized that if the flow rate is expressed as the volume flow rate rather than the mass flow rate, then larger differences in capacity are obtained among the given cases due to the effect of changing density for different combinations of impurities [20]. The presence of impurities reduces pipeline capacity and also affects other design and operational aspects of supercritical CO_2 transport. These effects can be further understood by modeling the physical and transport properties of CO_2.

For example, the density of CO_2 changes with temperature and pressure and also with the level of impurity. Temperature, in particular, can have a dramatic effect on the density of CO_2. Figure 8-5 illustrates this point, where experimental data obtained from ref. [21] shows that at a temperature of 21.1°C and a pressure of 11 MPa-a, the density of pure CO_2 is approximately 860 kg/m^3; however, for the same pressure at a temperature of 4.4°C, the density is 957 kg/m^3, a resultant change of 97 kg/m^3. Conversely a change in pressure from 11 to 13 MPa-a at 4.4°C only results in a density difference of 12 kg/m^3. It can therefore be concluded that the temperature has a more significant effect on density than pressure for both pure CO_2 and CO_2 containing impurities. Increasing the temperature of the fluid decreases the density, and this would have a significant effect on pipeline operation. CO_2 fluids with high densities are desired to reduce the pipeline size and improve throughput. This reduces the investment cost but increases the running cost and vice versa if low density CO_2 is used. The other feature to note from Figure 8-5 is that there is a steep non-linearity in the density in the vicinity of the critical point for pure CO_2. This is observed from the experimental data in Figure 8-5, as the pressure decreases from 12 MPa-a to 7.6 MPa-a for temperatures increasing from 21.1°C to 43.3°C. Figure 8-5 also illustrates the non-linear relationship between pressure and density for CO_2 containing 5% of the impurities H_2, NO_2 and H_2S at a temperature of 20°C and 5°C. It can be seen that there is a discontinuity in the relationship around 50 bar for pure CO_2 because at these temperatures and pressures the fluid is approaching the vapor liquid line. At this point, a small change in either temperature or pressure can have a large effect on the density. Below this pressure, it is possible that high–density liquid and low-density gas could exist in the pipeline which would result in liquid slugs and potential damage to compressors. The addition of impurities shifts this discontinuity to a higher pressure (due to changes in the phase envelope, critical pressure, and critical temperature) as illustrated in Figure 8-5. In particular, the discontinuity in density is around 11 to 12 MPa-a with the addition of 5% H_2, which is typically around CO_2 pipeline operating pressures.

The flow capacity and pipeline hydraulics are expressed by three main parameters, namely, CO_2 mixture density, pressure, and temperature profiles along the pipeline. In terms of the pipeline, the effect of these parameters translates into pumping intervals and pump/compressor power requirements, which would have both capital and operating cost impli-

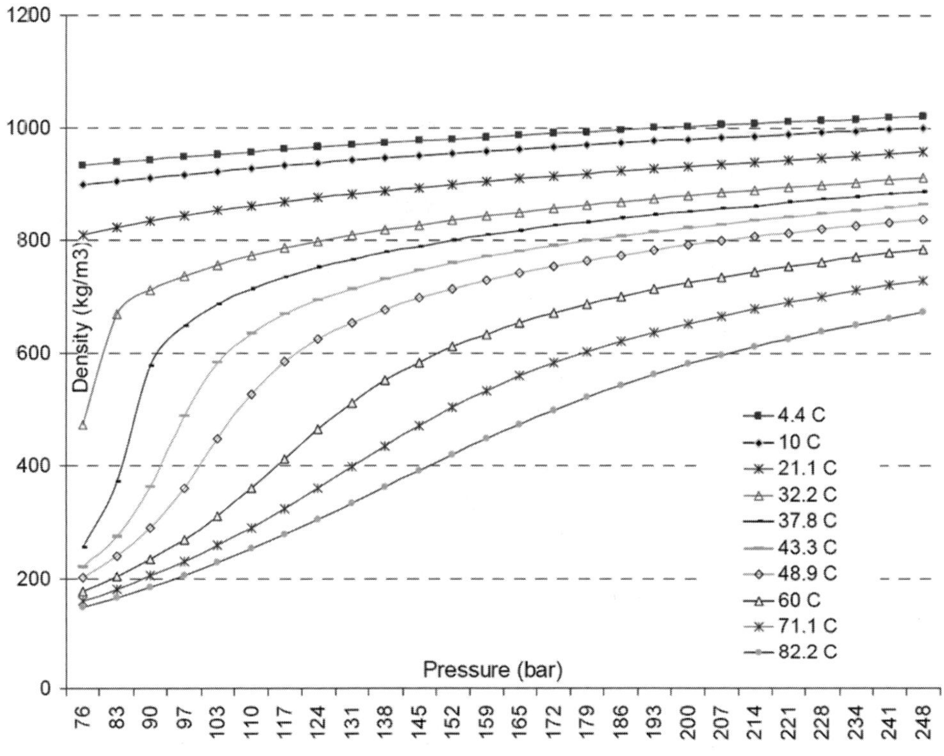

Figure 8-5. Densities for pure CO_2 at various temperatures and pressures based on the experiments conducted by Kinder Morgan [21]

cations. Figures 8-6 and 8-7 show the pressure and temperature profile for a 500-km pipeline conveying pure CO_2, binary combinations of CO_2 with 5% of different impurities and combinations of impurities that make up the potential CO_2 stream from the aforementioned three capture technologies. The impurity concentrations for each of the capture technologies presented are shown in Table 8-3 and are typical of potential ranges. In these Figures, the pipeline modeled was 406.4 mm outside diameter with a wall thickness of 12.7 mm and a total length of 500 km with no elevation changes. The inlet pressure was raised to 110 bar to ensure that the CO_2-impurity combination was in the dense phase. The outlet temperature was set at 50°C with an ambient temperature of 5°C. The average steady-state flow rate was taken to be 72 kg/s, based on figures for pre-combustion capture of CO_2 from an 860 MW Combined Cycle Power Plant [22]. A roughness of 0.0457 mm, typical for commercial pipe [23] was used.

The binary combination of CO_2 with 5% N_2, H_2, and CO, respectively, shows the largest pressure and temperature drops along the pipeline. In terms of temperature, there seems to be a significant decrease resulting in temperatures below the 5°C ambient temperature. This is also a major concern in the event of a leak or pipeline rupture where the sudden expansion causes a large temperature drop which might decrease the pipeline material's toughness.

The binary combinations of CO_2 with 5% SO_2, NO_2, and H_2S, respectively have a less significant pressure and temperature drop relative to pure CO_2 and also when compared to H_2 and N_2. The pressure profile for these binary impurity combinations is actually raised above that of pure CO_2 and can be considered as a positive effect in terms of pipeline hydraulics. However, 5% of H_2S, NO_2, and SO_2, respectively, in the CO_2 stream would be highly improbable due to regulations on health and safety. For example, in the United King-

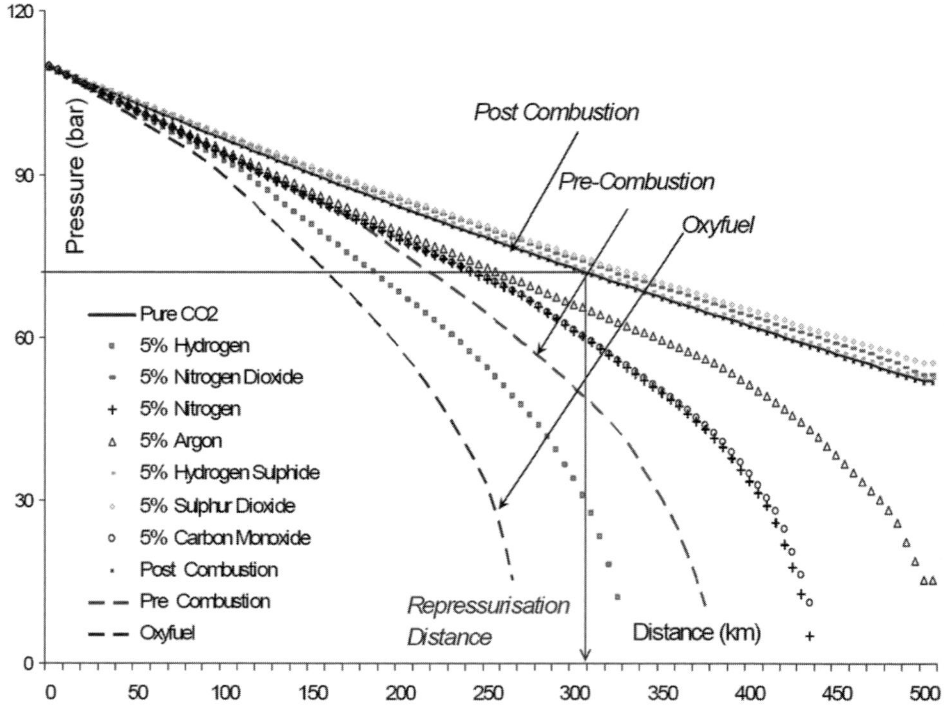

Figure 8-6. Simulated pressure profile for the three different capture streams and the binary CO_2 with 5% impurity combinations [16]

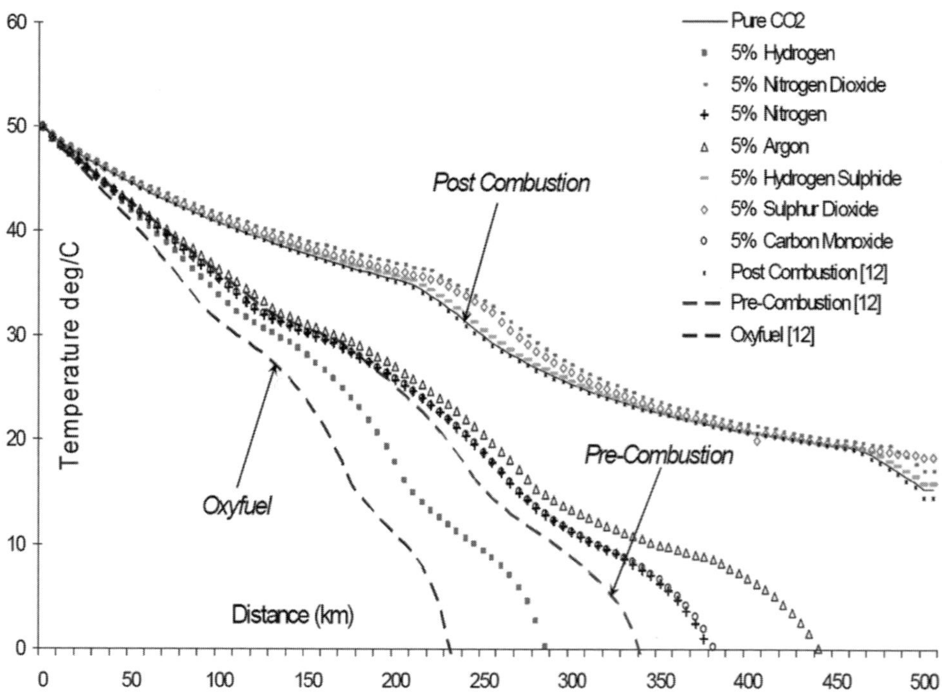

Figure 8-7. Simulated temperature profile for the various combinations of impurities, and for each of the capture streams [16]

TABLE 8-3. Typical values of impurities for three capture technologies

Comp	Post-combustion (%vol)	Pre-combustion (%vol)	Oxyfuel (%vol)
CH_4	0.01008	0.02057	–
N_2	0.17170	0.34759	3.53945
H_2S	–	1.00005	–
C_2+	0.01008	0.00588	–
CO	0.00101	0.23509	–
O_2	0.01008	–	1.80290
NO_X	0.00504	–	0.15025
SO_X	0.00101	–	1.49950
H_2	–	2.76144	
Ar	–	0.0939	3.00490

dom, the current Large Combustion Plant directive (LCPD) [24] which imposes restrictions on SO_x and NO_x emissions would also have an effect on the quantity of these impurities in the capture stream. These restrictions are also expected to tighten in the future although it should be noted that such directives or regulations may differ worldwide.

The following can be concluded in terms of transport properties for the different capture technologies and their corresponding impurities in CO_2:

- POST-COMBUSTION: As expected, the most pure CO_2 stream (99.9%) from the post-combustion capture process shows a similar temperature and pressure profile to that of pure CO_2.
- PRE-COMBUSTION: Pre-combustion shows a pressure and temperature profile closer to the CO_2–5% H_2 combination. This is expected as the pre-combustion capture stream consists of approximately 3% H_2 which is believed to dominate the phase properties. It also has about 1% of H_2S which could potentially hydraulically "improved" the profiles since it behaves similar to pure CO_2.
- OXYFUEL: The oxyfuel capture technology, with the least pure stream of CO_2 (90%), shows the largest drop in pressure and temperature profile. The presence of N_2, O_2, and Ar, all of which have a high pressure drop profile when compared to pure CO_2, would have contributed to this large drop in pressure and temperature along the pipeline.

8.5.3 Pumping Station Intervals

Re-pressurization is necessary to ensure CO_2 is in the single phase (i.e., outside the phase envelope) and remains in the dense or supercritical phase. In order to demonstrate the effect of impurities on the re-pressurization distance, the CO_2 product stream was re-pressurized when the pressure in the pipeline was 1 MPa (buffer) above the critical pressure of the transported fluid. The re-pressurization distance for the binary impurity combination is shown in Figure 8-8 and for different capture stream mixture compositions. It is shown that for the 500-km case study pipeline, pure CO_2 would require re-pressurization at 300 km in order to avoid two-phase flow. A decrease in re-pressurization distance would mean a huge additional cost of pumping stations for this pipeline. Hydrogen impurity has the greatest effect on re-pressurization distance in that addition of 3% H_2 halves the re-pressurization distance to 150 km. Of the impurities shown, H_2S has the least effect on the re-pressurization distance. Re-pressurization distance calculated for the oxyfuel stream is the shortest and would therefore require more pumps for the same 500 km pipeline modeled in this analysis, whereas the post-combustion technology has a similar re-pressurization distance to CO_2. Pre-combustion, although having approximately 3% H_2, has a

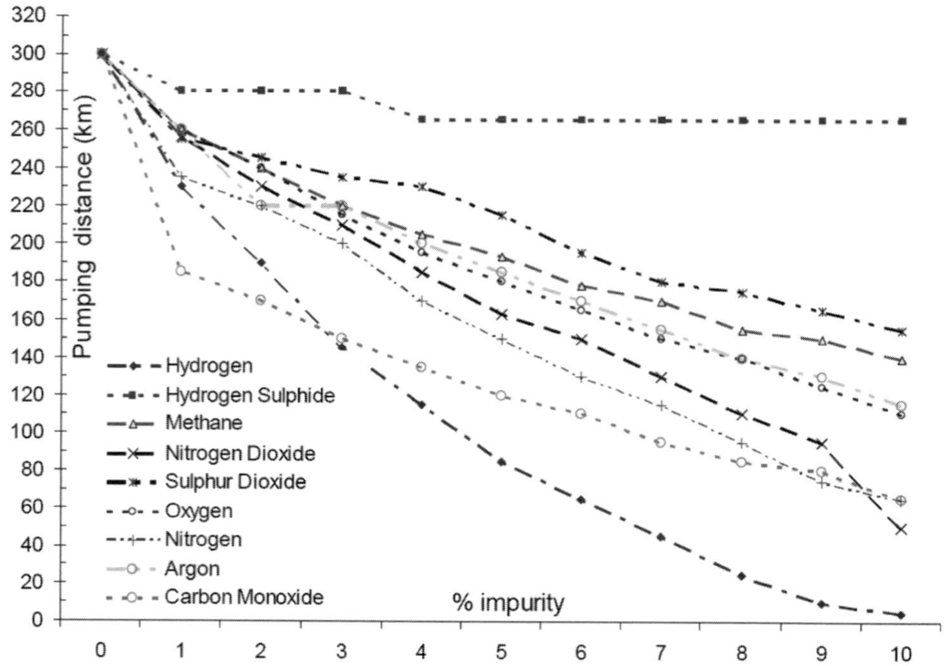

Figure 8-8. Calculated pumping intervals for the CO_2–5% impurity binary combination [16]

re-pressurization distance of 105 km which is lower than the binary combination of CO_2 and H_2 of 150 km. This is because the interaction of impurities with each other also plays a part in determining the physical and transport properties of CO_2. This impurity interaction effect is further demonstrated in the oxyfuel re-pressurization distance of 35 km, which does not correspond to the re-pressurization distance of a binary combination of CO_2 with 4% N_2, 3%Ar, and 2% O_2 in Figure 8-8, all of which are present together with the CO_2 in the oxyfuel stream as shown in Table 8-3.

8.5.4 Economic Analysis

Determination of the flow capacity of a given pipeline size and CO_2 mixture composition is only one task within an overall system design and optimization. Economic analysis must be done to evaluate various pipe sizes and associated pumping and compressor station costs. This analysis requires estimates of facility costs and other economic parameters specific to the corporation, such as rate of return on capital investment, debt to equity ratio, interest rate, inflation rate, corporate tax rate, and other regulatory parameters. Economic analysis typically incorporates a life cycle cost (LCC) tool that takes all of these parameters into consideration and determines the cumulative present value cost of service (CPVCOS) associated with the application of the various options for transporting a given volume of the CO_2 mixture under consideration. A key element in all of this is the capital cost of facilities, pipeline and pumping stations, as well as the operating cost (including energy cost).

- The cost of pipelines can be categorized into three items
    - Construction costs
    - Material/equipment costs (pipe, pipe coating, cathodic protection, telecommunication equipment; pumping stations)
    - Installation costs (labor)

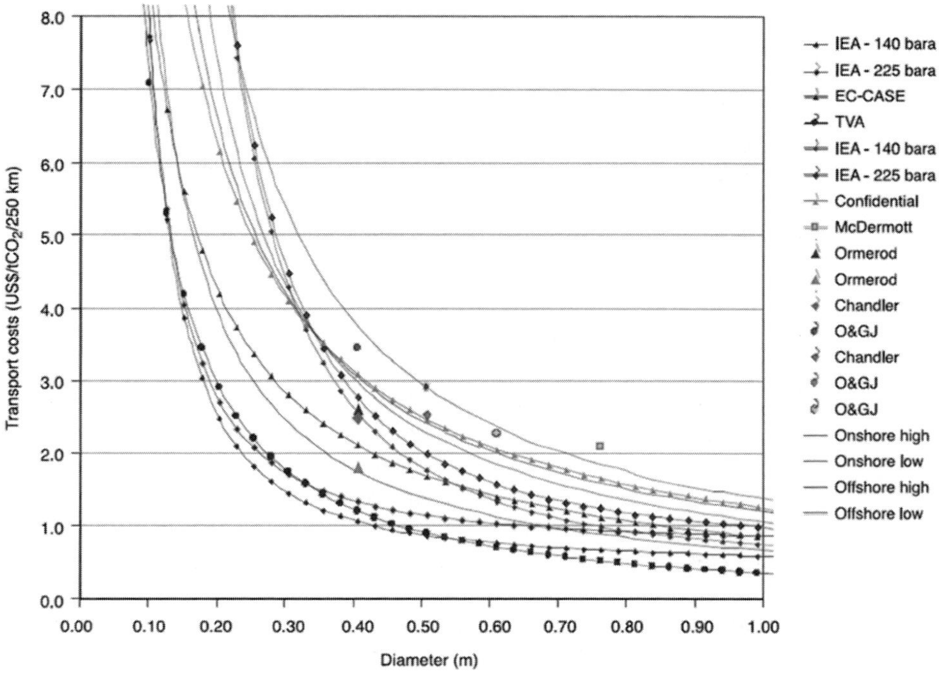

Figure 8-9. Transport costs derived from various information sources for offshore and onshore pipelines (costs exclude possible booster stations, applying a capital charge rate of 15% and a load factor of 100%) [12]

- Operation and maintenance costs
  - o Monitoring costs
  - o Maintenance costs
  - o Energy costs
- Other costs (design, project management, regulatory filing fees, insurance costs, right-of-way costs, contingency allowances)

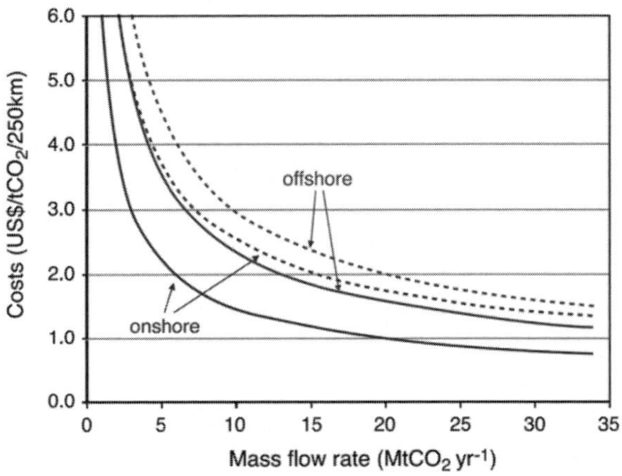

Figure 8-10. Transport costs for onshore and offshore pipelines per 250 km. High (broken lines) and low range (continuous lines) are indicated [12]

The pipeline material costs depend on the length of the pipeline, the diameter, the amount of CO_2 to be transported and the quality of the carbon dioxide. For costs, it is assumed that CO_2 is delivered from the capture system at 10 MPa. Clearly, investments are higher when compressor station(s) are required to compensate for pressure loss along the pipeline, or for longer pipelines or for hilly terrain. Compressor stations may be avoided by increasing the pipeline diameter and reducing the flow velocity. Reported transport velocity varies from 1 to 5 m/s. The actual design will be optimized with regard to pipeline diameter, pressure loss (required compressor stations and power), and pipeline wall thickness. Costs also depend on the terrain. Onshore pipeline costs may increase by 50% to 100% or more when the pipeline route is congested and heavily populated [12–14]. Costs also increase in mountains, in nature reserve areas, in areas with obstacles, such as rivers and freeways, and in heavily urbanized areas because of accessibility to construction and additional required safety measures. Offshore pipelines generally operate at higher pressures and lower temperatures than onshore pipelines, and are often, but not always, 40% to 70% more expensive. It is cheaper to collect CO_2 from several sources into a single pipeline than to transport smaller amounts separately. Early and smaller projects will face relatively high transport costs, and therefore be sensitive to transport distance, whereas an evolution towards higher capacities (large and wide-spread application) may result in a decrease in transport costs [12–14]. Implementation of a "backbone" transport structure may facilitate access to large remote storage reservoirs, but infrastructure of this kind will require large initial upfront investment decisions. Further study is required to determine the possible advantages of such a pipeline system. Figure 8-9 presents onshore and offshore transport costs versus pipeline diameter, where costs are based on investment cost information from various sources. Figure 8-10 gives a cost window for specific transport as function of the flow.

8.6 PIPELINE TRANSIENT CONSIDERATIONS

Many of the transient analyses conducted during the design phase of a CO_2 pipeline and its pump stations or for troubleshooting problems are invariably involved in the start-up and shutdown scenarios, flow controls, and in the opening and closing of valves during operation, blowdowns, and packing and drafting of the pipeline. Interaction between system hydraulic dynamics and associated control protocols could also trigger severe dynamics and catastrophic failure or system overall reliability issues. Expansion of pump stations to cope with either capacity increases or changes in conditions or products always present challenges related to associated transient phenomena. The addition of various units in series or parallel with different pump head-flow characteristics, and drivers' type and inertias require attention and thorough analysis.

In order to understand the dynamic behavior of a CO_2 pipeline and its pumping system, the governing equation describing the unsteady flow through a segment of pipe section needs to be solved in both temporal and spatial domains. A constant area pipe is described by the continuity and momentum equations. Therefore, the fundamental governing equations in one-dimension are described first and are solved using the method of characteristics. Since the pump driver exerts a torque on the rotating shaft, which in turn, conveys energy to the impeller to develop a total dynamic head across the pump, the dynamic behavior of this system is also described and combined with the unsteady flow of the connecting pipeline system. The transient-state pressures and temperatures at some locations in the system may be reduced to conditions below the bubble point curve and into the two-phase region of the CO_2 mixture. This will produce vapor cavities in the flow and may cause severe transients. Rejoining of the separated columns or collapse of the cavities results in a large pressure rise, which may damage the piping system. This section deals with these dynamic phenomena. The fundamental governing equations will be first described followed by examples of transients that could occur in CO_2 pipeline systems.

8.6.1 Governing Equations for Transient Flows in Pipe

The unsteady flow through a constant area pipe is described by the continuity and momentum equations. The following assumptions are made in the derivation of these equations.

1. The flow in the pipe is one-dimensional and the velocity distribution over the cross-section of the pipe is uniform.
2. The pipe wall and the fluid are considered linearly elastic, i.e., the respective stress is proportional to the strain.
3. The unsteady frictional losses are assumed to be the same as that pertaining to steady flow frictional losses having the same instantaneous flow and fluid properties.

The one-dimensional continuity equation can be written in the form:

$$\frac{\partial \rho}{\partial t} + U\frac{\partial \rho}{\partial x} + \rho\frac{\partial U}{\partial x} = 0 \qquad (8-15)$$

where:

ρ - density of the fluid
U - mean flow velocity $U(x,t)$
x - axial distance along the pipe
t - time

From the definition of the speed of sound in the fluid:

$$c_o = \sqrt{\frac{\partial P}{\partial \rho}} \qquad (8-16)$$

Eq. (8-15) can be written in terms of the static pressure (P) as follows:

$$\frac{\partial p}{\partial t} + U\frac{\partial p}{\partial x} + \rho c_0^2\frac{\partial U}{\partial x} = 0 \qquad (8-17)$$

TABLE 8-4. Example of bulk modulus and density for pure CO_2

Temperature (°C)	Pressure (MPa-a)	Density (kg/m³)	Bulk Modulus GPa	Bulk Sound Speed(m/s)
10	10	920.72	0.292	562.82
20	10	856.34	0.195	477.79
30	10	771.45	0.113	382.13
10	15	954.48	0.380	631.14
20	15	904.21	0.287	563.66
30	15	847.04	0.208	495.46
10	20	980.38	0.460	685.02
20	20	937.31	0.368	626.28
30	20	890.7	0.288	568.47

In a circular pipe the speed of sound (c_o) is a function of the bulk modulus of the liquid (K) and its density (ρ), as well as the bulk modulus of elasticity of the pipe material (E) and its diameter to thickness ratio (D/t), by the following relation [25]:

$$c_o = \sqrt{\frac{1/\rho}{\frac{1}{K} + \frac{1}{E}\left(D/t\right)}} \qquad (8-18)$$

Table 8-4 gives example of bulk modulus and density for pure CO_2 at different conditions based on GERG-2008 equation of state [26]. The modulus of elasticity of common steel pipe is around 200 to 220 GPa.

The one-dimensional momentum equation of the flow in a constant area pipe can be written as:

$$\frac{\partial U}{\partial t} + U\frac{\partial U}{\partial x} = -g\sin\left(\alpha\right)\frac{fU\,U}{2D} \qquad (8-19)$$

where:

f - is Darcy–Weisbach friction coefficient which is a function of the local Reynolds number and internal wall surface roughness. It can be can be determined from a derivative of the Colebrook–White equation in explicit form [27].

D - pipe internal diameter.

α - pipe inclination (positive in the positive direction of (x).

The static pressure can be replaced by the piezometric head (H), with the elevation of the hydraulic gradient above an arbitrary datum, as follows:

$$P = \rho g(H - z) \qquad (8-20)$$

where:

g - is the gravitational acceleration.

z - elevation of the pipe center-line above the arbitrary datum.

and:

$$\frac{\partial P}{\partial x} = \rho g\left[\frac{\partial H}{\partial x} - \sin(\alpha)\right] \qquad (8-21)$$

The continuity Eq. (8-17) and the momentum Eq. (8-19) can then be written in terms of the piezometric head (H), in the form:

$$\frac{\partial H}{\partial t} + U\frac{\partial H}{\partial x} + \frac{c_o^2}{g}\frac{\partial U}{\partial x} = 0 \qquad (8-22)$$

$$\frac{\partial U}{\partial t} + U\frac{\partial U}{\partial x} + g\frac{\partial H}{\partial x} + \frac{fU\,U}{2D} = 0 \qquad (8-23)$$

8.6.2 Solution Techniques

Eqs. (8-22) and (8-23) form a pair of partial differential equations in terms of the two dependent variables (U) and (H) and the two dependent variables (x) and (t). These two equations can be transformed into two ordinary differential equations along two characteristics paths defined by two slopes C^+ and C^- [28] in the form:

$$\frac{g}{c_o}\frac{dH}{dt} + \frac{\partial U}{\partial t} + \frac{fU\,U}{2D} = 0 \text{ along the path } C^+ : \frac{dx}{dt} = +c_o \qquad (8-24)$$

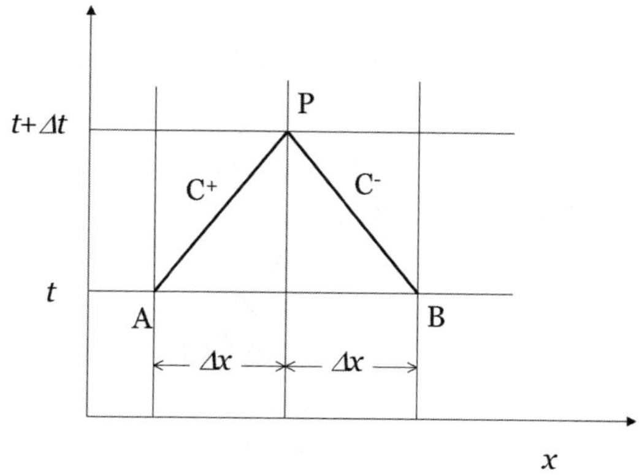

Figure 8-11. Characteristics lines in the *x–t* plane

and;

$$-\frac{g}{c_o}\frac{dH}{dt}+\frac{\partial U}{\partial t}+\frac{fU|U|}{2D}=0 \text{ along the path } C^-: \frac{dx}{dt}=-c_o \quad (8-25)$$

The above two paths, C^+ and C^-, are represented by two straight lines on the *x–t* plane, having slopes of $1/c_o$ as shown in Figure 8-11. These are called the "characteristics lines." If a pipe is divided into equal increments (Δx), and a time step is defined by

$$\Delta t=\frac{\Delta x}{c_o} \quad (8-26)$$

then the slope, C^+, will be satisfied by a positively sloped diagonal, AP as shown in Figure 8-9. If the dependent variable U and H are known at location A at time t, then Eq. (8–24) which is valid along C^+ can be integrated between A and P to reveal:

$$H_P-H_A+\frac{c_c}{gA}(Q_P-Q_A)+\frac{f\Delta x}{2gDA^2}Q_A|Q_A|=0 \quad (8-27)$$

Likewise, the slope, C^-, will be satisfied by a negatively sloped diagonal, BP as shown in Figure 8-9. If the dependent variable U and H are known at location B at time t, then Eq. (8-25), which is valid along C^- can, be integrated between B and P to reveal:

$$H_P-H_B+\frac{c_c}{gA}(Q_P-Q_B)+\frac{f\Delta x}{2gDA^2}Q_B|Q_B|=0 \quad (8-28)$$

Eqs. (8-27) and (8-28) are two algebraic equations that describe the temporal variation of the two independent variables H_P and Q_P based on these variables at time t upstream and downstream of the point P. These two equations can thus be solved algebraically simultaneously to reveal the values of H_P and Q_P at time $t+\Delta t$. If this calculation is repeated for all points (nodes) along the discretized pipe section, and progressed in time, the transient behavior of the flow in terms of the variables (Q and H) can be determined.

Boundary Conditions:
At the extreme end of piping elements, only one of the above two characteristics lines and the corresponding algebraic relations is available. For the upstream end, Eq. (8-28) holds along the C^-characteristic. Likewise for the downstream end, Eq. (8-27) holds along the C^+ characteristic. Each of these two equations is a relationship between the two variables H_P and Q_P at the boundary. Additional information (i.e., relationship) between these two

variables is needed at either end to close the system of equations. These are given by known information at the boundary, called the "boundary conditions". Examples of boundary conditions are as follows:

1. The piezometric head is a known function of time, i.e., $H_P = H_p(t)$.
2. The flow is a known function of time, i.e., $Q_P = Q_p(t)$.
3. Closed end, i.e., $Q_P = 0$.

8.6.3 Examples of CO_2 Pipeline Transients

Various transient examples are given courtesy of SPT Group. The pipeline example is a 12 in., 230 km long section of CO_2 pipeline from Karsto (on the shoreline of Norway) to the Utsira formation in the North Sea. The pipeline system is shown schematically in Figure 8-12. This line is operating at 9.5 MPa and average temperature of 5°C. Simulations of steady-state operation were first conducted to establish the initial condition of the flow field in the pipeline before transients are initiated. The steady-state pressure, temperature and liquid volume fraction profiles along the pipeline are determined and shown in Figure 8-13 [29]. The fluid properties calculations were based Span and Wagner equation of state [30] and CSP for the transport properties. Notice that the temperature profile reaches the surrounding temperature of 5°C in approximately 25 km from the pipeline inlet condition of 45°C. The static pressure peak at around 15 km is due to the significant drop in the elevation profile.

The first transient simulation is related to a shut-in/cool down scenario, where the pipeline was assumed to be shut-in at both ends and allowed to cool at the surrounding temperature of 5°C. The results of this simulation after 8 days of shut-in are shown in Figure 8-14 [29]. Notice the significant drop in pressure from the initial steady-state condition primarily due to the cooling effect. The peak pressure dropped from roughly 12 MPa to under 10 MPa.

The next simulated transient scenario is a blowdown operation, again from the steady-state condition of Figure 8-13. The results are shown in Figure 8-15 in terms of the liquid, vapor and total mass release rates from the blowdown stack as a function of time [29]. Notice that the blowdown lasts about 12 days, and a total of 14,000 tonnes of CO_2 were released. Figure 8-16 shows the various profiles along the pipeline from the start of the blowdown (Figure 8-16a), after 6 hours (Figure 8-16b), after 6 days (Figure 8-16c), and after 14 days (Figure 8-16d). Figure 8-17 shows the same results of conditions directly upstream of the blowdown valve on a pressure temperature diagram along with the pure CO_2 saturation line. Notice that the pressure drops very rapidly to the saturation line, and after about 12 hours, the lowest temperature of –20°C is reached.

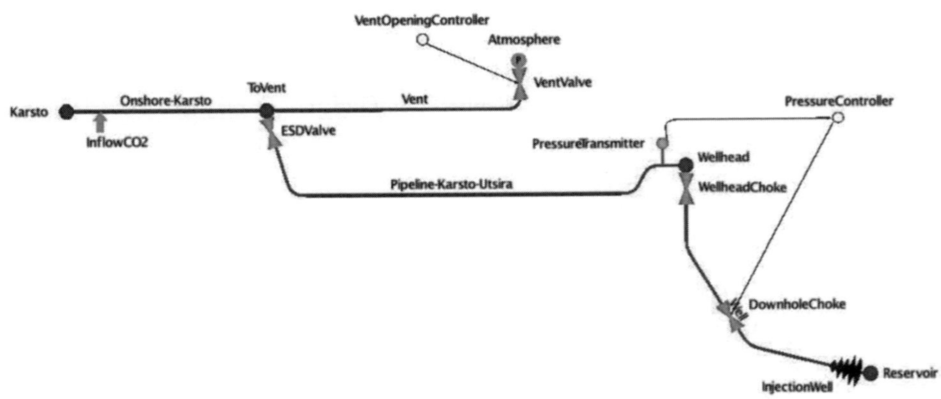

Figure 8-12. Example CO_2 pipeline from Karsto to Utsira in the North Sea [29]

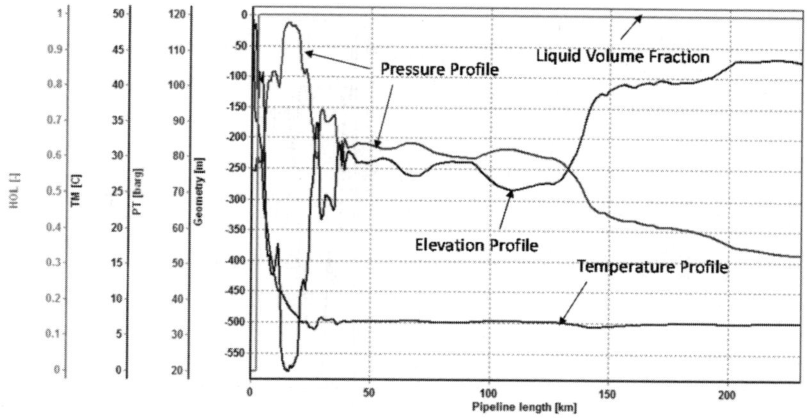

Figure 8-13. Steady-state pressure and temperature profiles along the CO_2 pipeline from Karsto to Utsira in the North Sea [29]

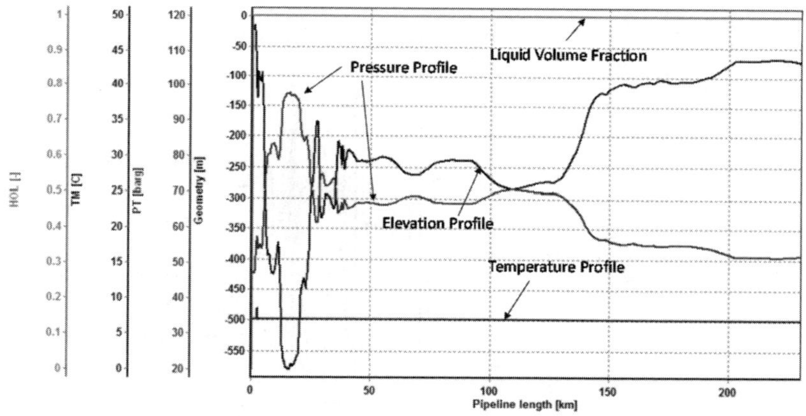

Figure 8-14. Transient simulation results of a shut-in/cool-down scenario after 6 days of the CO_2 pipeline from Karsto to Utsira in the North Sea [29]

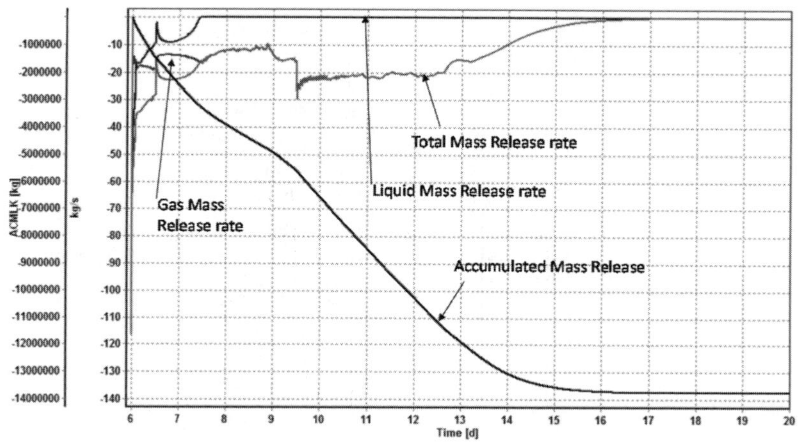

Figure 8-15. Transient simulation results of a blowdown scenario of the CO_2 pipeline from Karsto to Utsira in the North Sea [29]

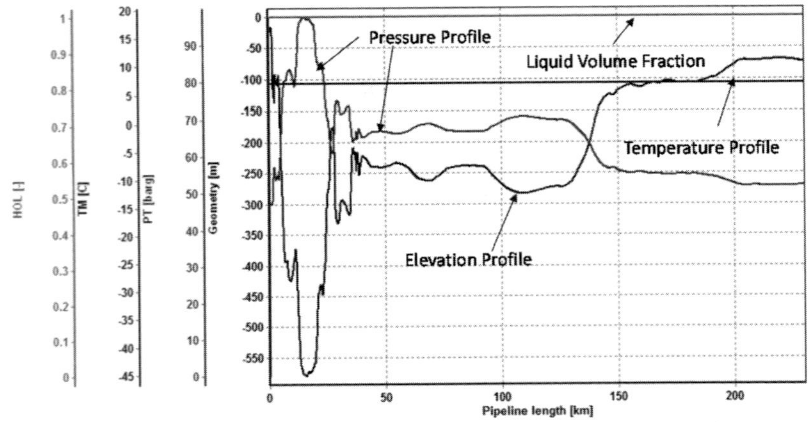

Figure 8-16a. Profiles at the start of the blowdown of the CO_2 pipeline from Karsto to Utsira in the North Sea [29]

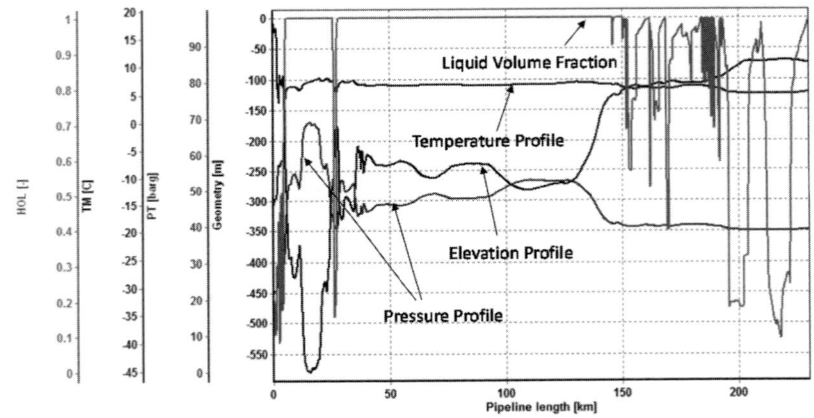

Figure 8-16b. Profiles after 12 hours of the blowdown of the CO_2 pipeline from Karsto to Utsira in the North Sea [29]

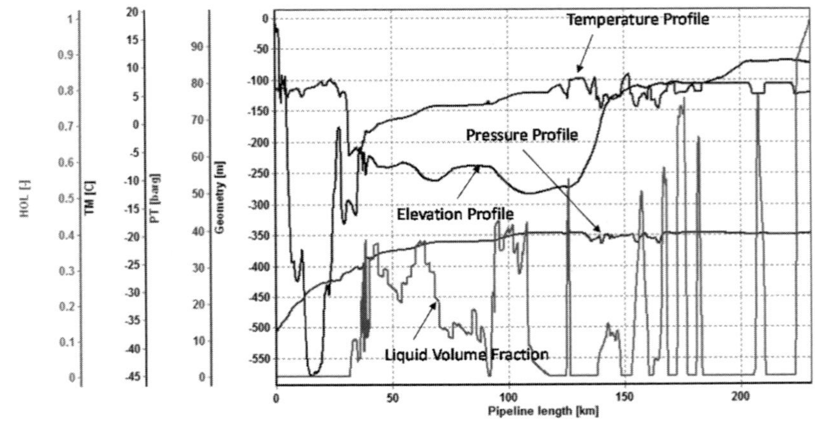

Figure 8-16c. Profiles after 6 days of the blowdown of the CO_2 pipeline from Karsto to Utsira in the North Sea [29]

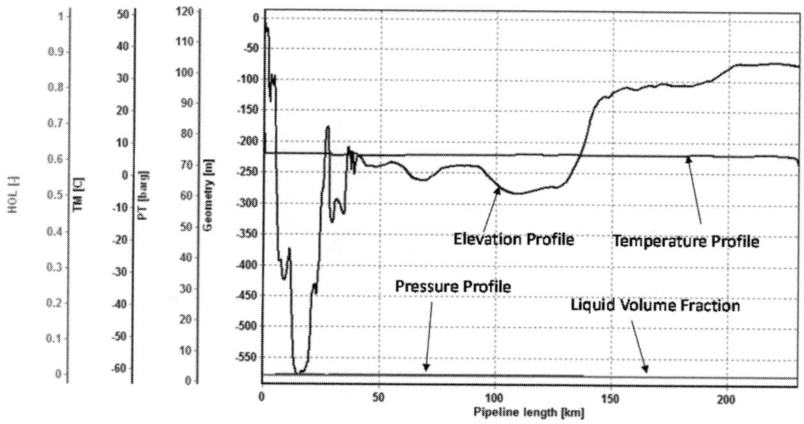

Figure 8-16d. Profiles after 14 days of the blowdown of the CO_2 pipeline from Karsto to Utsira in the North Sea [29]

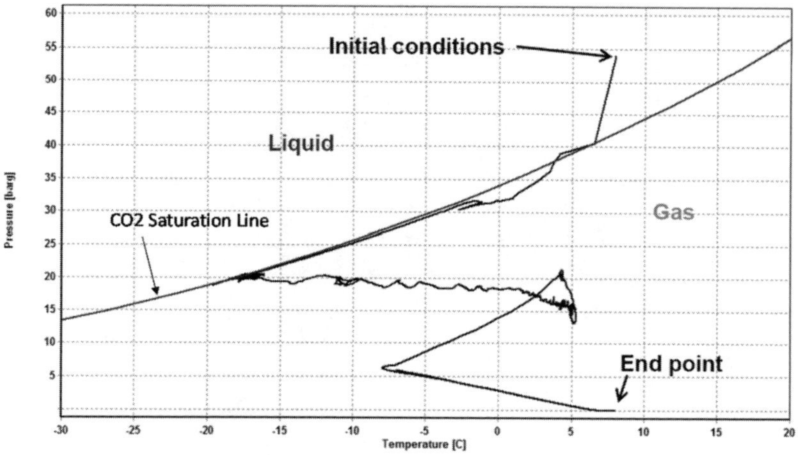

Figure 8-17. Pressure–temperature trajectory of the condition immediately upstream of the blowdown valve following blowdown of the CO_2 pipeline from Karsto to Utsira in the North Sea [29]

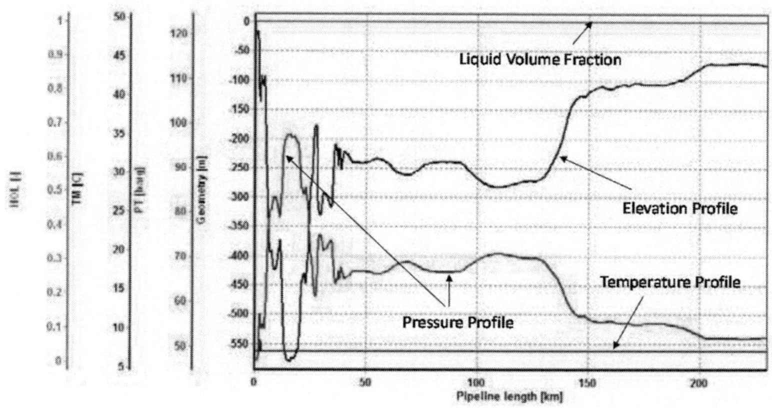

Figure 8-18a. Initial condition before start-up after shut-in/cool down scenario of the CO_2 pipeline from Karsto to Utsira in the North Sea [29]

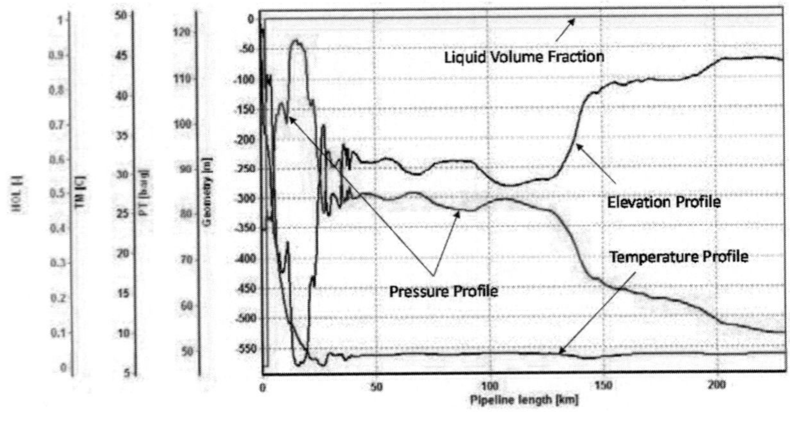

Figure 8-18b. Conditions after 12 hours of start-up after shut-in/cool-down scenario of the CO_2 pipeline from Karsto to Utsira in the North Sea [29]

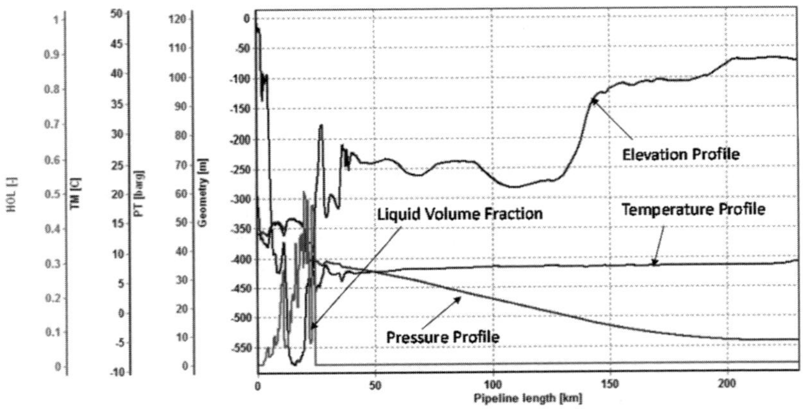

Figure 8-19a. Conditions of the pipeline after 8 hours of continuous filling of the CO_2 pipeline from Karsto to Utsira in the North Sea from ambient pressure [29]

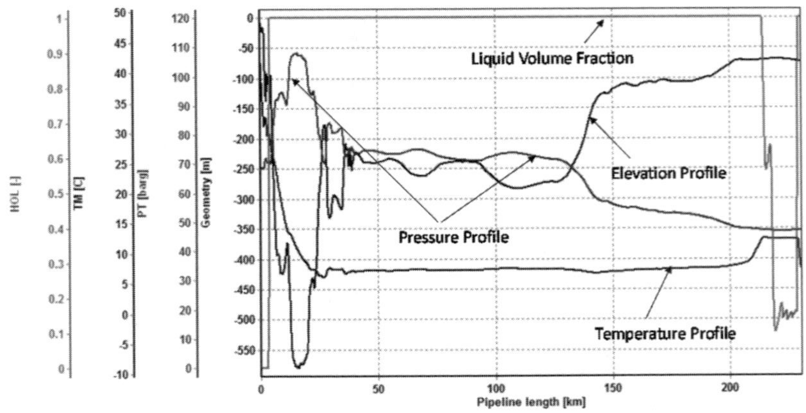

Figure 8-19b. Conditions of the pipeline after 4 days of continuous filling of the CO_2 pipeline from Karsto to Utsira in the North Sea from ambient pressure [29]

A start-up scenario after a shut-in/cool down scenario was also simulated, and results are shown in Figure 8-18. Figure 8-18a shows the initial condition before the start-up, while Figure 8-18b shows the pipeline condition after 12 hours from start-up. Finally, a filling-in transient scenario of an empty pipeline was simulated and results are shown in Figs. 8-19a (after 8 hours of filling) and in Figure 18-19b (after 4 days of filling). Notice the CO_2 volume fraction profiles showing that the CO_2 phase is primarily vapor initially, and then becomes all liquid (or dense phase) after 4 days of continuous filling.

8.7 SCADA AND SIMULATION TOOLS

Control and SCADA systems are required to monitor the CO_2 as it travels through the pipeline network. Centralized control stations collect and manage data received from monitoring and compressor stations all along the pipe. Supervisory Control and Data Acquisition (SCADA) systems provide most of the data that is received by a control station. These systems take measurements and collect data along the pipeline, usually in metering or compressor stations and valves, and transmit them to the centralized control station. Readings on the flow rate through the pipeline, operational status, pressure, and temperature may all be used to assess the status of the pipeline at any one time. These systems work in real-time, and in this way, there is little lag time between the measurements taken along the pipeline and their transmission to the control station. This allows quick reactions to equipment malfunctions, leaks, or any other unusual activity along the pipeline. Some SCADA systems are able to operate certain equipment along the pipeline remotely, such as compressor stations, allowing engineers in a centralized control center to immediately and easily adjust flow rates in the pipeline.

Most of the simulation tools for CO_2 pipelines are those which were developed for natural gas applications. Users are usually faced with the challenge of selecting the appropriate equation of state for their specific mixture compositions. By and large, the most common commercial software used in CO_2 pipeline simulation are PIPE-FLO [31], OLGA (by SPT Group) [32] and Aspen Plus® [33].

REFERENCES

[1] European Gas Pipeline Incident Data Group, Report EGIG 08, TV-B.0502, 2008.

[2] DNV Recommended Practice April, 2010.

[3] IEA Greenhouse Gas R&D Programme. 2003. Rules and standards for the transmission and storage of CO_2, Report No. PH4/23, August, 2003.

[4] International Standards Organisation at http://www.iso.org/iso/en/CatalogueDetailPage.

[5] Kunz, O., Klimeck, R., Wagner, W. and Jaeschke, M., 2007, "The GERG-2004 Wide-Range Equation of State for Natural Gases and Other Mixtures," GroupeEuropéen de RecherchesGaziéres (GERG), Technical Monograph, GERG TM15.

[6] Mohitpour, M., Golshan, H. and Murray, A., 2003, Pipeline design & construction: a practical approach. 2nd Ed. The American Society of Mechanical Engineers, Three Park Av., New York, NY 10016.

[7] Downie, M. J., Seevam, P. N. and Race, J. M., 2007, "Transport of CO_2 for Carbon Capture and Storage in the UK," In: *Proceedings of Offshore Europe* 2007, Aberdeen, UK.

[8] Fradet, A., Saysset, S., Odru, P., Broutin, P., Ruer, J. and Bonnissel, M., 2007, "Technical and Economic Assessment of CO_2 Transportation for CCS Purposes," *Journal of Pipeline Engineering*, **6**(3), pp.173–180.

[9] Decker, L. C., Checak, J. R., Heath, W.H. and Warren, R.N., 1985, "CO_2 Pipeline Design, Construction and Operation" in Facilities, Pipelines, and Measurements: A Workbook for Engineers, presented at 41st Petroleum Mechanical Engineering Workshop and Conference, Kansas City, MO, USA.

[10] Hein, M. A., 1985, "Rigorous and Approximate Methods for CO_2 Pipeline Analysis in Facilities, Pipelines and Measurements: A Workbook for Engineers," 41st Petroleum Mechanical Engineering Workshop and Conference, Kansas City, MO, USA.

[11] West, J. M., 1974, "Design and operation of a supercritical CO_2 pipeline-compression system sacroc unit, scurry county, Texas," *SPE Permian Basin Oil Recovery Conference*, Midland, Texas, 11–12 March.

[12] IPCC, 2005, "IPCC Special Report on Carbon Dioxide Capture and Storage," Working Group III of the Intergovernmental Panel on Climate Change, 2005, Eds. Metz, B., Davidson, O., de IEA GHG.

[13] McCoy, S. T. and Rubin, E. S., 2007, "An engineering-economic model of pipeline transport of CO_2 with application to carbon capture and storage," *International Journal of Greenhouse Gas Control*, pp. 219–229.

[14] McCoy, S. T., 2008, "The economics of CO_2 transport by pipeline and storage in saline aquifers and oil reservoirs," PhD thesis, Carnegie Mellon University.

[15] Farris, C., 1993, "Unusual Design Factors for Supercritical CO_2 pipeline" 3, September 1983, *Energy Progress*, **3**, pp. 150–158.

[16] Seevam, P., Race, J. M., Downie, M. J. and Hopkins, P., 2008, "Transporting the Next Generation of CO_2 for Carbon, Capture and Storage: The Impact of Impurities on Supercritical CO_2 Pipelines," *ASME Proceedings of the 7th International Pipeline Conference*, Paper #IPC2008-64063, September 29-October 3, 2008, Calgary, Alberta, Canada.

[17] Gale, J. and Davidson, J., 2004, "Transmission of CO_2-safety and economic consideration," *Journal of Energy Progress*, **6**(4), p. 219.

[18] Marsili, L., 1990, "Reducing the Risk of Ductile Fracture on the Canyon Reef Carriers CO_2 Pipeline," SPE Paper No. 20646, 65th Annual Technical Conference and Exhibition of the Society of Petroleum Engineers, New Orleans LA.

[19] Kinder Morgan Energy Partners L.P., 2002, "KMP Annual Report number10-K405SEC."

[20] Seevam P. N., Race, J. M. and Downie, M. J., 2007, "Carbon Dioxide Pipelines for Sequestration in the UK – Engineering GapAnalysis" Transmission of CO_2, H_2, and Biogas: Exploring New Uses for Natural Gas Pipelines Conference, Clarion, Amsterdam, pp. 141–162.

[21] McCollum D. L. and Ogden J. M., 2006, "Techno-Economic Model for Carbon Dioxide Compression, Transport and Storage & Correlations for Estimating Carbon Dioxide Density and Viscosity." Report No. UCD-ITS-RR-06-14, Institute of Transportation Studies, University of California.

[22] Hanne M. and Kvamsdal, T. M., 2005, "Tjeldbergodden Power/Methanol- CO_2 Reduction Efforts. SP2: CO_2 Capture and Transport," SINTEF Energy Research, Trondheim., p. 9.

[23] Brill, J. P. and Mukherjee, H., 1999, "Multiphase Flow in Wells" Society of Petroleum Engineers Monograph, Richardson, Texas, pp. 12–27.

[24] Defra, 2001, "Directive 2001/80/EC of the European Parliament and of the Council of 23 October 2001 on the limitation of emissions of certain pollutants into the air from large combustion plants." http://www.defra.gov.uk/environment/airquality/eu-int/eudirectives/lcpd/index.htm.

[25] Wylie, E. B. and Streeter, V. L., 1982, Fluid Transients, FEB Press, Ann Arbor, Michigan.

[26] Kunz, O., Klimeck, R., Wagner, W. and M. Jaeschke, 2007, "The GERG-2004 Wide-Range Equation of State for Natural Gases and Other Mixtures," Groupe Européen de Recherches Gaziéres (GERG), Technical Monograph, GERG TM15.

[27] Miller, D. S., 1990, Internal Flow Systems, 2nd Edition, Chapter 8, Gulf Publishing Company.

[28] Collatz, L., 1966, The Numerical Treatment of Differential Equations", 3rd Edition, Springer-Verlag, New York.

[29] Håvelsrud, M., 2010, "The use of OLGA software for simulation of CO_2 transport: results from a recent project", International Forum on Transport of CO_2 for Carbon Capture and Storage", Newcastle, UK. 1-2 July.

[30] Span, R. and Wanger, W., 2003, "Equations of State for Technical Applications. I. Simultaneously Optimized Functional Forms for Nonpolar and Polar Fluids", Int. J. Thermophys, 2003, **24**(1), pp. 1–39.

[31] http://www.eng-software.com/products/pipeflo/.

[32] http://www.sptgroup.com/en/Products/olga/Flow-Assurance/.

[33] AspenTech, Aspen Plus® V7.1 Simulation Engine, 200 Wheeler Road, Burlington, Massachusetts 01803, [http://www.aspentech.com/].

[34] Colebrook, C. F. and White, C. M., 1937, "Experiments with fluid friction in roughened pipes," *Proc. Royal Soc.* (A) **161**, pp. 367–378.

[35] IEA GHG GLOBAL CCS PIPELINE INFRASTRUCTURES –FINAL MARCH 2010.

[36] deWaard, C. and Lotz, U., 1993. Prediction of CO_2 corrosion of carbon steel. Corrosion/93 NACE

[37] International. Paper No. 69.

[38] White, R. A., 1998, Materials selection for refineries and associated facilities. NACE Appendix A.

[39] Wallace, C. B., 1985, Drying supercritical CO_2 demands care. Oil and Gas Journal, June 24: 98–104.

[40] Nguyen, D., 2003, "Carbon dioxide geological sequestration technical and economic reviews," *SPE International, Society of Petroleum Engineers* SPE 81199.

[41] Gale, J. and J. Davison, 2002, "Transmission of CO_2 — safety and economic considerations," In: Gale, J. and Y. Kaya (eds.): Proceedings of the 6th International Conference on Greenhouse Gas Control Technologies, Pergamon, pp. 517–522.

[42] Mohitpour, M., Botros, K. K. and Van Hardeveld, T., 2008, "Pipeline Pumping and Compression Systems: A Practical Approach," Chapter 10, ASME Press.

[43] WRI, 2008. CCS guidelines: guidelines for carbon dioxide capture, transport, and storage. Washington, DC. World Resources Institute.

[44] IEA GHG, 2009. CO_2 storage in depleted gas fields. Report Number 2009/01, International Energy Agency R&D Programme.

[45] International Energy Agency (IEA), 2007, "Capturing CO_2", Paris.

[46] Moore, J. J. and Nored, M. G., 2008, "Novel Concepts for the Compression of Large Volumes of Carbon Dioxide", *Proceedings of ASME Turbo Expo* 2008, Berlin.

[47] Romeo, L. M., Bolea, I., Lara Y. and Escosa J. M., 2008, "Optimization of Intercooling Compression in CO_2 Capture Systems", *Appl. Therm. Eng.*, Available Online.

[48] Rao, A., 2007, "An Assessment of CO_2 Compression Options for Near Zero Emission Power Plants", Panel Presentation, ASME Turbo Expo 2007, Montreal.

[49] Botero, C, et al., 2009, "Thermoeconomic evaluation of CO_2 compression strategies for post-combustion CO_2 capture applications", *Proceedings of ASME Turbo* Expo: Power for Land, Sea and Air, Paper # GT2009-60217, Orlando, Florida, USA, June 8-12, 2009.

[50] Caroll, J., 2006, "Compression and High Pressure Pipeline Transportation of Carbon Dioxide Rich Fluids," 1st Regional Symposium on Carbon Management — Carbon management Challenges and Opportunities for Petroleum Industry, May 22–24.

[51] Black, H. F. and Jenssen, D. N., "Effects of High Pressure Ring Seals on Pump Rotor Vibration," ASME, 71-WA-FE-36.

[52] 49CFR192, 2008, "Pipeline Safety: Pipeline Integrity Management in High Consequence Areas (Gas Transmission Pipeline)", Final Rule, Department of Transportation, Revision04/08.

[53] 49CFR195, 2008, "Pipeline Safety: Pipeline Integrity Management in High Consequence Areas (Hazardous Liquid Operators With 500 or More Miles of Pipeline)", Final Rule, Department of Transportation, Revision 06/08.

[54] API 5L, 2007, "Specification for Line Pipe," ANSI/API Specification 5L, 44th Edition.

[55] ASME B31.4 (2006) "Pipeline Transportation Systems for Liquid Hydrocarbon and Other Liquids", ASME B31.4-2006, American Society of Mechanical Engineers, New York, USA, 2006.

[56] ASME B31.8, 2007, "Gas Transmission and Distribution Piping Systems", ASME B31.8-2007, American Society of Mechanical Engineers, New York, USA.

[57] Anheden, M., Andersson, A., Bernstone, C., Eriksson, S., Yan, Liljemark, S. and Wall, C., 2005, "CO_2 Quality Requirement for a System with CO_2 Capture, Transport and Storage," *Proceedings of the 7th Greenhouse Gas Technologies Conference* GHGT-7, Vancouver, Vol 2 pp. 2559–2563.

[58] Barrie, J., Brown, K., Hatcher, P. R. and Schellhase, H. U., 2005, "Carbon Dioxide Pipelines: A Preliminary Review of Design and Risks", *Proceedings of the 7th International Conference on Greenhouse Gas Control Technologies*, Vancouver, Canada.

[59] Bentham, M., 2006, "An Assessment of Carbon Sequestration Potential in the UK – Southern North Sea Case Studies", Tyndall Centre for Climate Change Research Working Paper 85, BS8010-1(2004) "Code of practice for pipelines. Steel pipelines on land", British Standards Institute, 2004.

[60] Hendriks, C., Hagedoorn, S. and Warmenhoven, H., 2007, "Transportation of Carbon Dioxide and Organisational Issues of CCS in the Netherlands", Report prepared by Ecofys for EnergieNed, the Ministry of Economic Affairs and the Ministry of Housing, Spatial Planning and the Environment, March.

[61] IGE/TD/1(2001) "Recommendations on Transmission and Distribution Practice — Steel Pipelines for High Pressure Gas Transmission" IGE/TD1/1 Edition 4, Institution of Gas Engineers, London.

[62] Coninck, H. C., Loos, M. and Meyer, L. A., Cambridge University Press, Cambridge, United Kingdom and New York, NY, USA.

[63] Kantar, K. and Connor T. E., 1984, "Turkey's Largest Oil Field Poised for CO_2 Immiscible EOR Project", *Oil and Gas Journal*, **82**(48), p. 95.

[64] McCullough, D. E., 1986, "Central Basin Pipeline: A CO_2 System in West Texas," *Energy Progress*, **6**(4), p. 230.

[65] Najera, G., 1986, "Maintenance Techniques Proven on CO_2 Line", *Oil and Gas Journal*, **84**(46), p. 55.

[66] Oosterkamp, T. and Ramsen, J., 2008, "State of the Art Review of CO_2 Pipeline Transportation with Relevance to Offshore Pipelines," Report No. POL-0-2007-138-A, Polytec, Norway.

[67] PD 8010-1, 2004, "Code of Practice for Pipelines, Part 1 — Steel pipelines on Land", British Standards Institute.

[68] PSR, 1996, "The Pipelines Safety Regulations 1996 — Statutory Instrument 1996 No. 825", The Stationery Office Limited.

[69] Seiersten, M. and Kongshaug, K. O., 2005, "Materials Selection for Capture, Compression, Transport and Injection of CO_2", Carbon Dioxide Capture for Storage in Deep Geologic Formations. D. C. Thomas and S. M. Benson, Elsevier Ltd., **2**, pp. 937–953.

[70] Visser, E. and Hendriks, C., 2007, "Towards Hydrogen and Electricity Production with Carbon Capture and Storage — DYNAMIS Quality Recommendation," DYNAMIS, Project No.: 019672.

[71] Wallace, C. B., 1985, "Drying Supercritical CO_2 Demands Care," *Oil and Gas Journal*, **83**(25), p. 98.

[72] White, V., Allan, R. and Miller, E., 2006, "Purification of Oxyfuel Derived CO_2 for Sequestration or EOR," *Proceedings of GHGT-8*.

[73] Aspelund, A. and Jordal, K., 2007, "Gas conditioning — The Interface Between CO_2 Capture and Transport," *International Journal of Greenhouse Gas Control*, **1**(3): pp. 343–354.

[74] Bratfos, H. A., Leinum, B. H., Torbergsen, L. E. and Saugerud, O. T., 2007, "Challenges to the pipeline transportation of dense CO_2," *Journal of Pipeline Engineering*, **6**(3): pp. 161–172.

[75] Castello, P., Tzimas, E., Moretto, P. and Peteves, S. D., 2005, "Techno-economic assessment of hydrogen transmission and distribution systems in Europe in the medium and long term," EUR 21586 EN. EC-DG JRC-Institute for Energy, Petten, The Netherlands.

[76] Denbury Resources Inc., 2008. Notes to Consolidated Financial Statements, in: Defining Denbury — Annual Report. http://www.denbury.com/assets/docs/DNR_2008AR.pdf. Last consulted on 7 December 2010.

[77] Det Norske Veritas, 2007. Offshore Standard on Submarine Pipeline Systens DNV-OSF101.

[78] Det Norske Veritas, 2010. Recommended Practice on Design and Operation of CO_2 Pipelines DNV-RP-J202, JIP CO_2 PIPETRANS Phase 1.

[79] Ecofys Netherlands BV, 2009, "Assessment of technical experience and engineering practice for CO_2 pipelines," Deliverable report from DYNAMIS Project.

[80] Fradet, A., Saysset, S., Odru, P., Broutin, P., Ruer, J. and Bonissel, M., 2007, "Technical and economic assessment of CO_2 transportation for CCS purposes," *Journal of Pipeline Engineering*, **6**(3): pp. 173–180.

[81] Gale, J. and Davidson, J., 2004. "Transmission of CO_2 — Safety and economic considerations," *Energy*, **29**: pp. 1319–1328.

[82] Hendriks, C. A., Wildenborg, A. F. B. and Blok, K., 2000, "Costs of carbon dioxide removal by underground storage," In: Fifth International Conference on Greenhouse Gas Control Technologies, Cairns, August 13–16, 2000. p. 967–72.

[83] [23] IPCC, 2005, Special Report on Carbon Dioxide Capture and Storage. Prepared by Working Group III of the Intergovernmental Panel on Climate Change [Metz, B., O.Davidson, H.C. de Coninck, M. Loos, and L.A. Meyer (eds.)]. Cambridge UniversityPress, Cambridge, United Kingdom and New York, NY, USA, 442 pp.

[84] ISO 13623, 2009, Petroleum and Natural Gas Industries — Pipeline Transportation Systems.

[85] McCollum, D. and Ogden, J., 2006, "Techno-economic Models for Carbon Dioxide Compression, Transport, and Storage & Correlations for Estimating Carbon Dioxide Density and Viscosity," Institute of Transportation Studies. University of California. Public Power Institute: Muscle Shoals, AL.

Chapter 9

PUMPING, COMPRESSION AND OTHER APPURTENANCES

9.1 INTRODUCTION

If the CO_2 stream is being transported in supercritical or dense phase, it is important that the pressure is kept sufficiently high to avoid two-phase flow. As mentioned in the previous chapter, that pressure drops occur in the pipeline due to frictional effects and gravitational (static head) effects related to the pipeline elevation. Furthermore, the fact that the supercritical or dense phase CO_2 is as dense as most hydrocarbon liquids (600-900kg/m^3) means that the effect of static head on the pressure is significant. This will be important in areas where there are variations in topographic height along the pipeline route. This property of CO_2 can be advantageous if the elevation decrease in a pipeline can counteract any frictional losses. For example, in the Sheep Mountain pipeline, once the CO_2 leaves the production facility at 9.6 MPa, it is delivered to the oil fields at a pressure of 13.7 MPa with no requirement for intermediate compression. The increase in pressure is achieved purely through elevation changes along the 660 km pipeline route. Elevation changes are also being used to prevent offshore compression in the Snøvhit pipeline. In this pipeline the CO_2 is compressed to 15 MPa and the 150 km pipeline drops 300 m to the injection site. For systems that rely on changes in elevation to achieve the required pressure changes, it is important that a static analysis is conducted to take account of a shut in, when pipeline pressures could be increased.

Compressor stations in a CO_2 pipeline system can be sub-divided in two classes: the originating stations, which are positioned at the inlet to the pipeline, and the booster stations, which are located along the pipeline to compensate for the pressure decrease due to friction and elevation losses. In principle, the longer the pipeline and the elevation of the terrain crossed, the more compressor horsepower is required to achieve the required delivery pressure at destination. Under a fixed route and flow capacity, the number and size of booster stations depend on the circumstances and design as discussed in the previous chapter. Fewer stations might be easier to operate but the disadvantage is the need of for high inlet pressures, which are likely to require the more expensive use of thicker pipes. The present chapter discusses the various aspects of compression and pumping facility requirements as well as other appurtenances and equipment specifications associated with CO_2 transportation.

9.2 THERMODYNAMICS OF CO_2 COMPRESSION IN DIFFERENT PHASES

Before detailed description of the compression and pumping facilities are discussed, it is important to understand the thermodynamic concept of CO_2 compression in the different phases described above. Consider the three compression processes (cases) shown in Figure 9-1. Case (1) represents compression in the vapor region, Case (2) in the supercritical

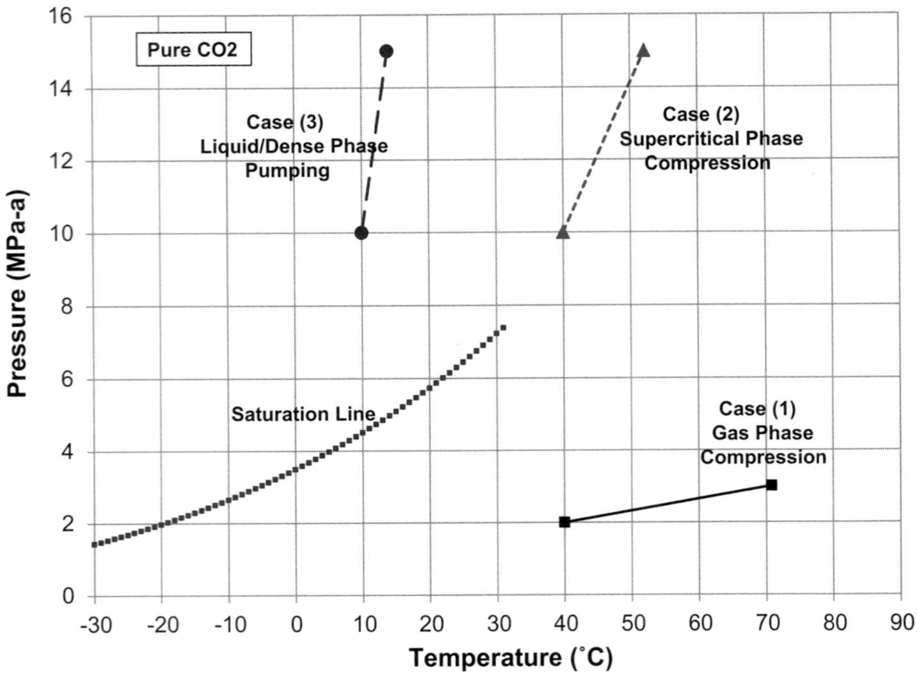

Figure 9-1. Comparison of compression/pumping on pressure–temperature diagram, and in different phase regions of CO_2

region, and Case (3) in the dense/liquid region. All three compression cases have pressure ratio = 1.5. Particulars of these three compression cases are given in Table 9-1, where isentropic compression was assumed and the properties at suction and discharge are calculated based on GERG-2008 [1] as implemented in REFPROP [2] or AspenPlus [3]. Figure 9-1 shows these isentropic compression cases on a pressure–temperature diagram along with pure CO_2 saturation line. Figure 9-2 shows the same three compression cases on a pressure–density, while Figure 9-3 show the same on a pressure–entropy diagram. Note that the saturation line is now shown as bubble and dew point curves. The isotherm (31°C) passing through the critical point is also shown which separates the supercritical phase compression (Case 2) from the dense/liquid phase compression (Case 3).

In the vapor phase, the compressors' performance characteristics are often provided in terms of polytropic head and polytropic efficiency vs. actual inlet flow. These can be converted to adiabatic head and adiabatic efficiency via the following relationships [4–6]:

TABLE 9-1. Thermodynamic properties of the three compression cases in Figure 9-3

	Temperature	Pressure	Density	Int. Energy	Enthalpy	Entropy	Comp. Factor	Isentropic Exponent	Viscosity	Sound Speed	Isentropic Enthalpy Increase (DH)
	(°C)	(MPa)	(kg/m³)	(kJ/kg)	(kJ/kg)	(kJ/kg K)	–	–	(μPa s)	(m/s)	(kJ/kg)
Case (1)	40	2	37.127	447.78	501.65	2.1766	0.91053	1.2709	15.899	261.65	22.82
	70.799	3	51.125	465.79	524.47	2.1766	0.90304	1.2643	17.508	272.38	
Case (2)	40	10	628.68	297.08	312.98	1.3561	0.26886	4.5529	47.833	269.11	7.63
	52.138	15	680.61	298.57	320.61	1.3561	0.35862	5.5667	54.237	350.26	
Case (3)	10	10	920.72	207.18	218.04	1.0395	0.20303	29.166	97.108	562.82	5.38
	13.932	15	935.4	207.39	223.42	1.0395	0.29567	22.806	101.06	604.74	

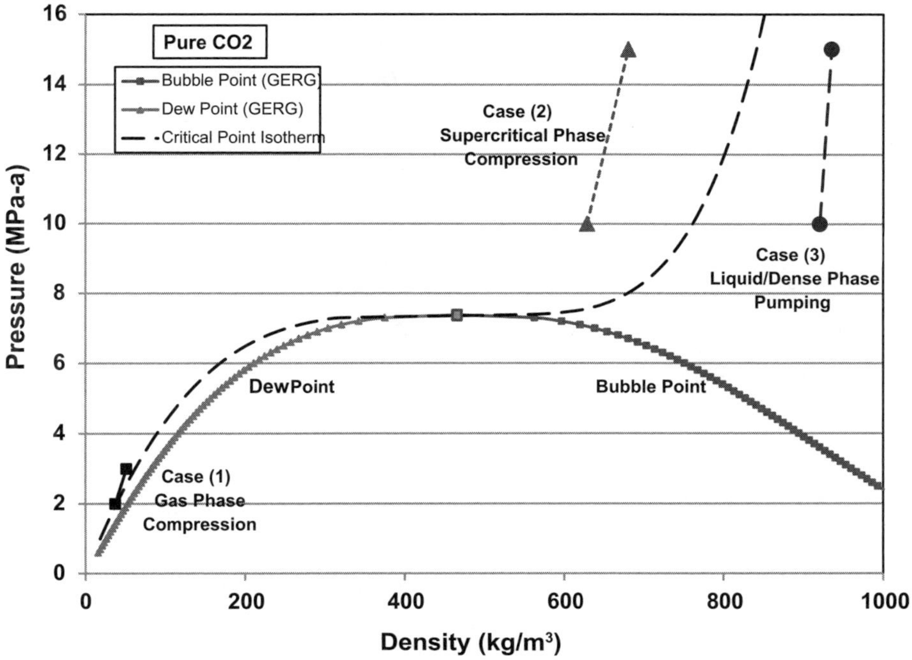

Figure 9-2. Comparison of compression/pumping on pressure–density diagram and in different phase regions of CO_2

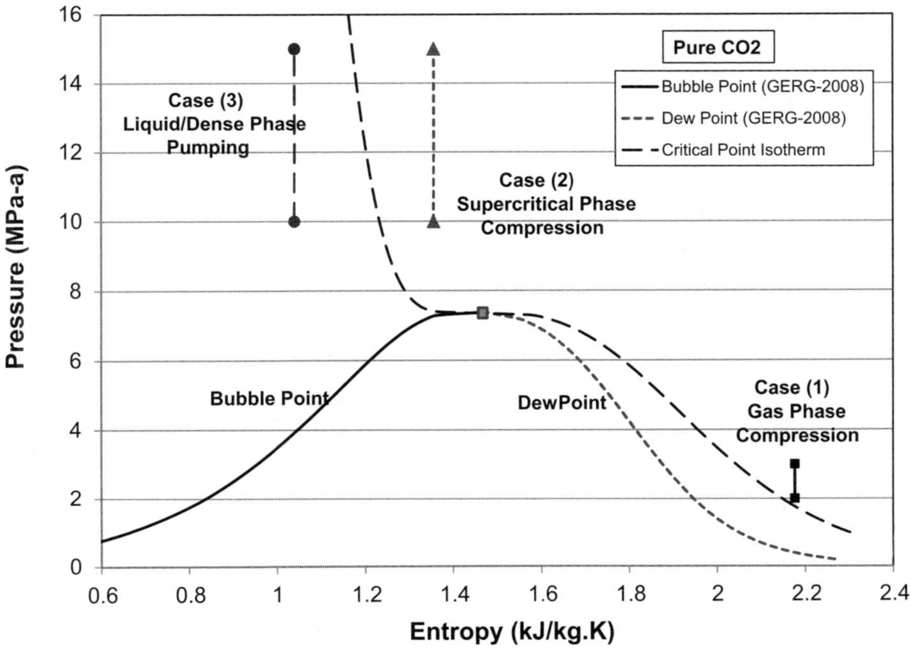

Figure 9-3. Comparison of compression/pumping on pressure–entropy diagram, and in different phase regions of CO_2

Adiabatic head:

$$H_a = \frac{Z_{av}RT_1}{\left(\frac{k-1}{k}\right)}\left[(P_r)^{\frac{k-1}{k}} - 1\right] \tag{9-1}$$

Polytropic head:

$$H_p = \frac{Z_{av}RT_1}{\left(\frac{n-1}{n}\right)}\left[(P_r)^{\frac{n-1}{n}} - 1\right] \tag{9-2}$$

Polytropic efficiency:

$$\frac{n}{n-1} = \frac{k}{k-1}\eta_p \tag{9-3}$$

Adiabatic efficiency:

$$\frac{H_a}{\eta_a} = \frac{H_p}{\eta_p} \tag{9-4}$$

where:

H_a - adiabatic head across the compressor
H_p - polytropic head across the compressor
k - average isentropic exponent of the compressed gas
n - polytropic exponent of compression
P_r - pressure ratio
R - gas constant
T_1vsuction gas temperature
Z_{av} - average compressibility factor across the compressor
η_a - Compressor adiabatic efficiency
η_p - Compressor polytropic efficiency

Given the properties of CO_2 at suction and discharge condition (Table 9-1), the adiabatic head can be calculated using Eq. (9-3) for pressure ratio = 1.5. The result of this calculation is 22.70 kJ/kg as shown in Table 9-2. Fundamentally, this adiabatic head should be equal to the enthalpy rise across the compressor and due to isentropic compression. The thermodynamic isentropic enthalpy rise can be found by calculating the enthalpy at the suction state of P = 2 MPa-a, and T = 40°C and using GERG-2008 equation of state (EOS). The

TABLE 9-2. Thermodynamic properties of the three compression cases in Figure 9-3

	Temperature	Pressure	Pressure Ratio	Average Compressibility Factor (Za)	Average Isentropic Exponent (ka)	Adiabatic Head from Eq. (9.1)	Error in Enthalpy Rise (%)	Power Required to Compress 5 Mt/year of CO2 (Ad. Eff. = 0.85)
	(°C)	(MPa)	–	–		(kJ/kg)		MW
Case (1)	40	2	1.5	0.9068	1.2676	22.70	−0.51	4.26
	70.8	3						
Case (2)	40	10	1.5	0.3137	5.0598	8.89	16.54	1.42
	52.138	15						
Case (3)	10	10	1.5	0.2494	25.9860	6.61	22.89	1.00
	13.932	15						

resulting enthalpy at suction = 501.65 kJ/kg and entropy = 2.1766 kJ/kg K. Imposing the same entropy of 2.1766 kJ/kg K at discharge and P = 4 MPa-a, the discharge temperature can be calculated (again using GERG-2008) and was found to be 70.8°C, which is higher than suction temperature of 40°C as expected. The enthalpy at the compressor discharge is also higher is equal to 524.47 kJ/kg, hence the thermodynamic enthalpy rise is 524.47-501.65 = 22.82 kJ/kg. This is very close to the adiabatic head calculated using Eq. (9-1) which was 22.70 kJ/kg. In fact the error is only –0.51%. The power requirement for this Case (1) compression can be calculated as:

The compressor power can be calculated from:

$$\dot{W}_{copm} = \frac{\dot{m}\Delta H}{\eta_a} \tag{9-5}$$

where:

ΔH - isentropic enthalpy rise across the compressor
$\dot{m}$ - CO_2 mass flow rate through the compressor
$\dot{W}_{comp}$ - compressor power requirement
η_a - Compressor isentropic (or adiabatic) efficiency

The power requirement for case (1) compression above is calculated using Eq. (9-5) for CO_2 mass flow for 5 Mt/year (158.5 kg/s) and is given in Table 9-2 as 4.26 MW based on adiabatic efficiency of 85%.

The above agreement between the adiabatic head across the compressor and the thermodynamic enthalpy rise in Case (1) compression is generally observed in vapor phase compression whether it is pure CO_2 or CO_2+ impurities or natural gas. This is not the case when the fluid phase of compression is in the supercritical region, and certainly not the case in the dense/liquid region. The simplification of Eq. (9-1) will not be applied as will be seen from the following calculations of Case (2) and Case (3) below.

Table 9-1 shows the resulting enthalpy rise across Case (2) compression from 10 MPa-a and 40°C to 15 MPa-a (i.e., the same compression ratio or 1.5). This compression case is in the supercritical region as shown in Figure 9-1. The isentropic enthalpy rise in this case is calculated as 7.63 kJ/kg. The simplified Eq. (9-1) gives adiabatic head of 8.89 kJ/kg, which is higher than the thermodynamic (actual) enthalpy rise, with an error of 16.54% (see Table 9-2). The compressor power requirement for Case (2) compression is much lower than that for Case (1), that is 1.42 MW compared to 4.26 MW (Table 9-2) for the same flow rate of 5 Mt/year of pure CO_2. This is an excellent example that shows that it is desirable to operate the pipeline in the supercritical (or dense/liquid, as will be shown below) region for this obvious economic reason.

Case (3) compression is entirely in the dense/liquid region (see Figure 9-1). Table 9-1 shows the resulting enthalpy rise across this case for compression from 10 MPa-a and 10°C to 15 MPa-1 (i.e., the same compression ratio or 1.5). Note that the suction temperature of 10°C is well below the critical point temperature of 31°C. The isentropic enthalpy rise in this case is calculated as 5.38 kJ/kg. The simplified Eq. (9-1) gives adiabatic head of 6.61 kJ/kg, which is again higher than the thermodynamic (actual) enthalpy rise, with a much higher error of 22.89% (see Table 9-3). The compressor power requirement for Case (3) compression is yet lower than that for Case (1) and Case (2), that is, 1.00 MW compared to 4.26 MW and 1.42 MW for Case (1) and Case (2), respectively, for the same flow rate of 5 Mt/year of pure CO_2 (Table 9-2). This is, again, an excellent example that shows that it is desirable to operate the pipeline in the dense/liquid region for this obvious economic reason.

TABLE 9-3. Thermodynamic properties of the different sate along compression strategy I shown in Fig. 9-5

State	Density	Enthalpy	Entropy	Temperature	Pressure	Isentropic Compression Enthalpy Rise	Isentropic Power to Compress 5 Mt/year CO_2
	(kg/m3)	(kJ/kg)	(kJ/kg-K)	(°C)	(MPa)	(kJ/kg)	(MW)
1	1.8152	501.65	2.7251	20	0.1		
2	3.4203	550.46	2.7251	76.87	0.2247	48.81	7.74
3	4.1061	500.43	2.5693	20	0.2247		
4	7.7409	548.93	2.5693	77.24	0.50491	48.5	7.69
5	9.37	497.65	2.4096	20	0.50491		
6	17.686	545.35	2.4096	77.986	1.1345	47.7	7.56
7	21.846	491.11	2.2408	20	1.1345		
8	41.316	537.03	2.2408	79.676	2.5493	45.92	7.28
9	54.15	474.52	2.0464	20	2.5493		
10	102.4	516.1	2.0464	82.963	5.7283	41.58	6.59
11	193.46	408.36	1.7079	20	5.7283		
12	427.38	453.93	1.1879	112.47	20	45.57	7.23
13	937.31	233.85	1.057	20	20		
Sum						278.08	44.09

9.3 INITIAL COMPRESSION OF CO_2 BEFORE INTRODUCTION TO PIPELINE

As mentioned above, the most efficient way to transport CO_2 is in supercritical or dense/liquid phases. CO_2 is generally transported at temperature and pressure ranges between 12°C and 44°C and 8.5 MPa and 15 MPa [7, 8]. The lower pressure limit is set by the phase behavior of CO_2 and should be sufficient to maintain supercritical condition while the upper pressure limit is mostly due to economic concerns. Regarding the temperatures, the upper temperature limit is determined by the compressor-station discharge temperature and the temperature limits of the external pipeline coating material, while the lower limit is determined by winter ground temperature [8]. An advantage of offshore pipelines for CO_2 transport is that higher design pressures can be used than onshore, potentially up to 30 MPa. This is partly due to the reduced hazard to population compared to onshore routes, which allow higher design factors to be used; and partly due to the compensatory effects of external hydrostatic pressure, particularly in deep water [9].

Before CO_2 can be transported in a pipeline, it should be compressed to ensure that single-phase (supercritical or dense/liquid). The most widely used operating pressure is between 7.4 and about 21 MPa. Clearly, a transmission pipeline can experience a wide range of ambient temperatures, so maintaining stability of this single phase is important in order to avoid considerations of two-phase flow that could result in pressure surges.

In a power plant with CO_2 capture, CO_2 is separated and compressed for further transportation to the storage or re-injection site. The energy requirement of the overall process results in power generation efficiency penalties, which depending on the type of power plant, fuel, and separation method used range between about 6 and 12 percentage points (%-pts.). While the separation of CO_2 from the usually dilute exhaust gas or fuel stream is indeed the most energy-intensive process step in CCS plants, the CO_2 initial compression process can account for up to 5%-pts efficiency loss, depending on the power plant and compression chain characteristics [10].

In light of the large contribution of the initial CO_2 compression process to the overall system performance and of the technical challenges still being faced, research interest has recently turned to the CO_2 compression process itself, see, e.g., references [11–13], whereas CO_2 captured from integrated gasification combined cycle (IGCC) power plants has been the focus of these investigations. The separation of CO_2 at multiple pressure levels and the availability of low-temperature N_2 from the air separation unit in IGCC plants constitute significant differentiators for the CO_2 compression process in such systems, when compared to other CO_2 separation routes.

In order to gain a better understanding of the possibilities and limitations of the CO_2 initial compression process for post-combustion CO_2 capture applications, different thermodynamically feasible CO_2 compression routes will be discussed and their respective power requirements quantified. This will be discussed for pure thermodynamic evaluation. Usually, full economic evaluation is based on the determination of capital cost requirements, operating cost, availability of refrigeration, and cooling systems in the plant, specific location, among other factors. The main purpose of this Section is to bring to light the main aspects of the initial compression process from fundamental thermodynamic view point.

In a state-of-the-art post-combustion capture process based on chemical absorption, CO_2 is separated from the exhaust gas stream of the power plant at close-to-ambient conditions. Despite the low CO_2 concentration of typically about 4% (v) in the exhaust of natural gas-fired plants, the large exhaust gas flows in large-scale power generation systems result in volume flows of separated CO_2. Let us consider a case where 5 Mt/year of pure CO_2 is separated and ready to be transported to an injection point by a pipeline. The captured CO_2 is at ambient pressure, and say at 20°C. This constitutes the initial state of the CO_2 gas captured. The final state before introduction into the pipeline is say 20 MPa. Thus, there is a need for a tremendous compression from ambient pressure to 20 MPa. This is illustrated on the pressure–enthalpy diagram of pure CO_2 in Figure 9-4.

As it can be seen in the figure, even though the final pressure will be above the critical value of 7.382 MPa-a, the path taken will determine the final temperature and hence whether the compressed CO_2 will exist in a subcritical (liquid) or supercritical state, i.e., below or above 31°C, which is the critical temperature of pure CO_2. Generally, the post-combustion CO_2 captured contains varying amount of water vapor (which could reach up to 304% by mole). In this case, the path taken during compression and intercooling will determine the

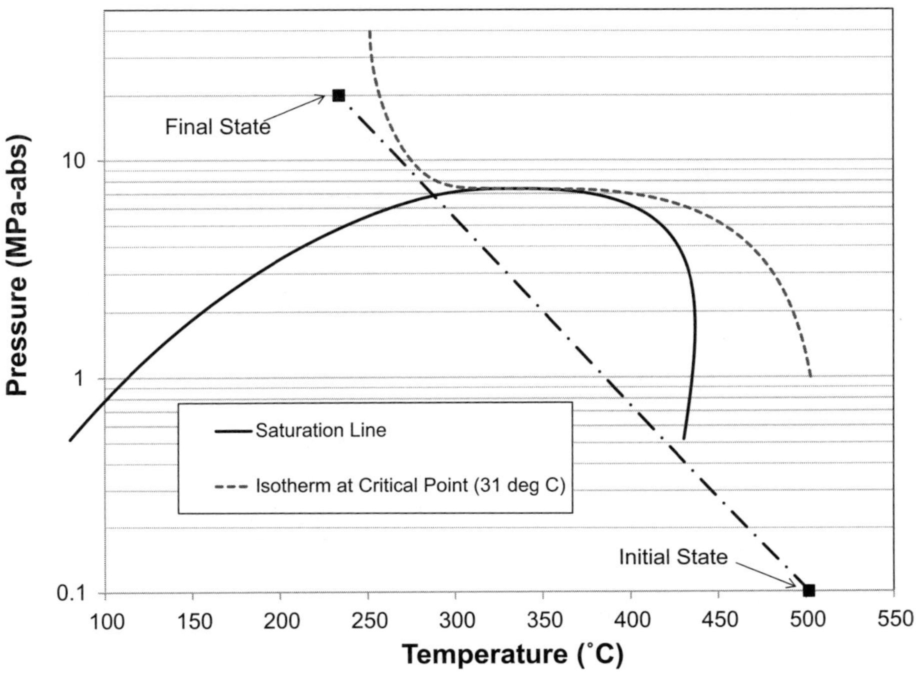

Figure 9-4. Comparison of compression/pumping in different phase regions of CO_2

purity of the CO_2 delivered for transportation since the extent to which the water vapor content of the gas condenses at the intercoolers is dependent on the compression strategy.

Let us consider three different paths (or strategies) of compression from the initial state shown in Figure 9-4 to the final state at 20 MPa-a required before introduction into the pipeline. Intercooling to 20°C using water at ambient conditions is common to all strategies. The main differentiator between these strategies is the final compression stage whether it involves a compressor or a pump. Strategy I is the base case strategy which is based on gas compression all the way to the desired pressure of 20 MPa-a, with stage intercooling as shown on the pressure–enthalpy (P-h) diagram shown in Figure 9-5. In this strategy, five stages of compressions with intercooling are involved up to pressure = 5.7283 MPa-a corresponding to saturation temperature of 20°C. Five stages of compression are assumed from the initial ambient pressure of 0.1 MPa-a to 5.7283 MPa-a at a pressure ratio of 2.247 per stage, regardless of the number of internal impellers or sub-stages in each compression section. We also assume ideal isentropic compression in each stage and intercooling to 20°C between stages for the purpose of comparison between different strategies. The final stage of compression involves a compressor as well, which compresses the saturated vapor CO_2 at 20°C and 5.7283 MPa-a (state 11) isentropically to pressure = 20 MPa-a (state 12), as shown in Figure 9-5. Final cooling will then be applied to bring the CO_2 temperature from state 12 to state 13. The pressure, temperature, enthalpy, and entropy of each state of these total of 6 stage compressions are shown in Table 9-3 based on GERG-2008 EOS. The individual ideal (isentropic) compression stage power is also calculated based on CO_2 mass flow rate of 5 Mt/year. It is shown that the total power required is 44.09 MW, and the power is approximately evenly divided among the 6 stages.

In strategy II, the CO_2 stream is brought to a pressure just above the critical pressure (8 Mpa-a) through again five compression stages intercooled with water to 20°C as before. The fifth stage cooling results in the liquefaction of the CO_2 at the compressor outlet pressure of 8 MPa-a, after which a pump is used to bring the dense phase CO_2 to the final pressure of 20 MPa-a as shown in the P-h diagram of Figure 9-6. The pressure ratio in this strategy

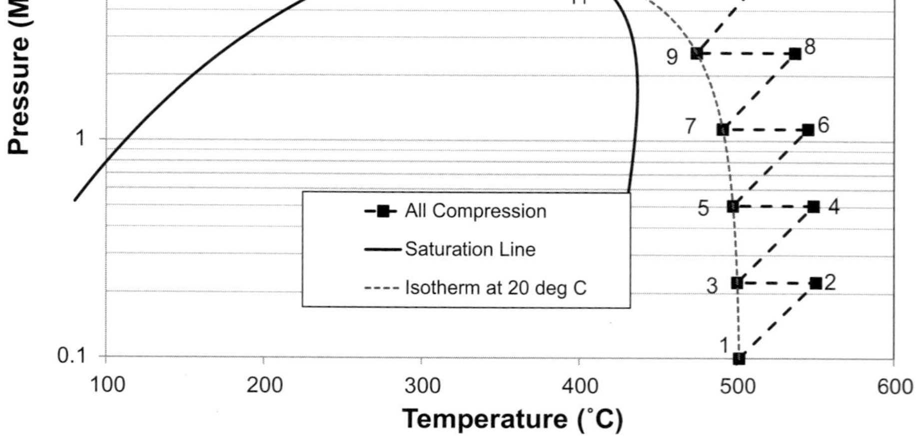

Figure 9-5. Compression Strategy I on *P-h* diagram

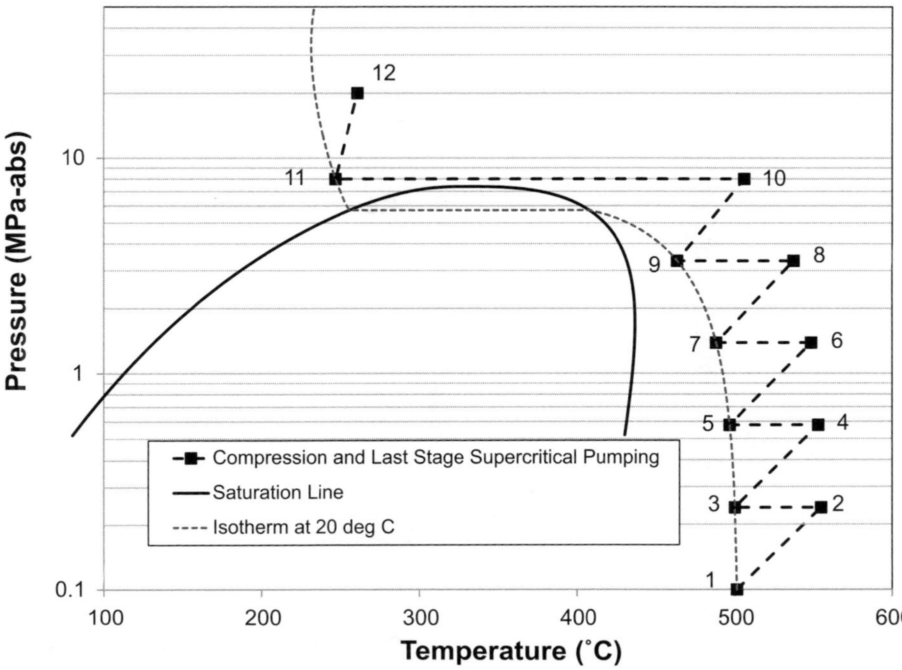

Figure 9-6. Compression Strategy II on *P-h* diagram

is slightly higher at 2.4 per stage due to the higher pressure at the discharge of the fifth stage. Ideal isentropic compression in each stage was also assumed here, with intercooling to 20°C between stages. The final stage from state 11 to state 12 in Figure 9-6 is now in the dense phase which will require a pump rather than a compressor with a much lower power requirement. The pressure, temperature, enthalpy, and entropy of each state of the total of 5 stage compressions and final stage of pumping are shown in Table 9-4 based on GERG-2008 EOS. The individual ideal (isentropic) compression or pumping stage power is also calculated based on CO_2 mass flow rate of 5 Mt/year. It is shown that the total power required is 41.80 MW which is less than strategy I power by 5.2%, mostly attributed to the lower power requirement of the final stage (compare Table 9-3 to Table 9-4).

TABLE 9-4. Thermodynamic properties of the different state along compression strategy II shown in Figure 9-6

State	Density	Enthalpy	Entropy	Temperature	Pressure	Isentropic Compression Enthalpy Rise	Isentropic Power to Compress 5 Mt/year CO_2
	(kg/m^3)	(kJ/kg)	(kJ/kg-K)	(°C)	(MPa)	(kJ/kg)	(MW)
1	1.8152	501.65	501.65	20	0.1		
2	3.6056	554.88	554.88	81.857	0.24022	53.23	8.44
3	4.3934	500.28	500.28	20	0.24022		
4	8.7326	553.11	553.11	82.311	0.57708	52.83	8.38
5	10.753	496.92	496.92	20	0.57708		
6	21.405	548.71	548.71	83.324	1.3863	51.79	8.21
7	27.119	488.36	488.36	20	1.3863		
8	54.101	537.63	537.63	85.701	3.3302	49.27	7.81
9	75.669	463.74	463.74	20	3.3302		
10	149.87	506.22	506.22	90.013	8	42.48	6.74
11	827.61	246.96	246.96	20	8		
12	877.01	261.01	261.01	32.783	20	14.05	2.23
Sum						263.65	41.80

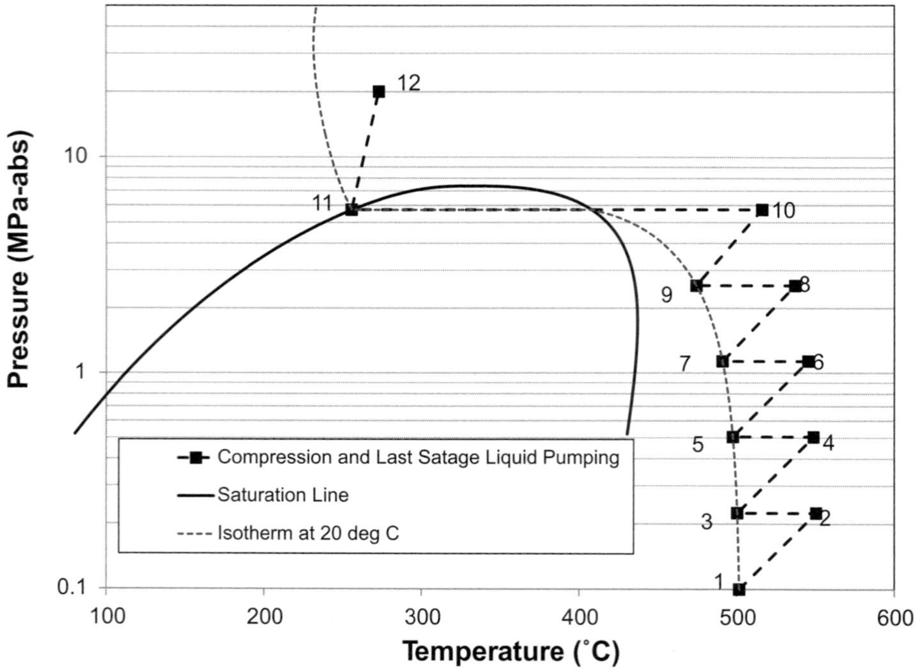

Figure 9-7. Compression strategy III on *P-h* diagram

In view of the possibility of liquefying the CO_2 at a pressure lower than that of strategy II for the cooling medium used, the third strategy (III) uses again five compression stages to bring the CO_2 to a subcritical pressure of 5.7283 MPa-a corresponding to saturation temperature of 20°C. This is similar to first five compression stages of strategy I. After this fifth stage compression, the CO_2 is liquefied to the liquid saturation state 11 in Figure 9-7. The liquid CO_2 is then pumped to the final pressure of 20 MPa-a to state 12. The pressure, temperature, enthalpy, and entropy of each state of the total of 5 stage compressions and final stage of pumping are shown in Table 9-5 based on GERG-2008 EOS. The individual ideal (isentropic) compression or pumping stage power is also calculated

TABLE 9-5. Thermodynamic properties of the different state along compression strategy III shown in Figure 9-7

State	Density	Enthalpy	Entropy	Temperature	Pressure	Isentropic Compression Enthalpy Rise	Isentropic Power to Compress 5 Mt/year CO2
	(kg/m^3)	(kJ/kg)	(kJ/kg-K)	(°C)	(MPa)	(kJ/kg)	(MW)
1	1.8152	501.65	2.7251	20	0.1		
2	3.4203	550.46	2.7251	76.87	0.2247	48.81	7.74
3	4.1061	500.43	2.5693	20	0.2247		
4	7.7409	548.93	2.5693	77.24	0.50491	48.5	7.69
5	9.37	497.65	2.4096	20	0.50491		
6	17.686	545.35	2.4096	77.986	1.1345	47.7	7.56
7	21.846	491.11	2.2408	20	1.1345		
8	41.316	537.03	2.2408	79.676	2.5493	45.92	7.28
9	54.15	474.52	2.0464	20	2.5493		
10	102.4	516.1	2.0464	82.963	5.7283	41.58	6.59
11	774.79	255.5	1.1865	20	5.7283		
12	849.33	272.98	1.1865	38.208	20	17.48	2.77
Sum						249.99	39.64

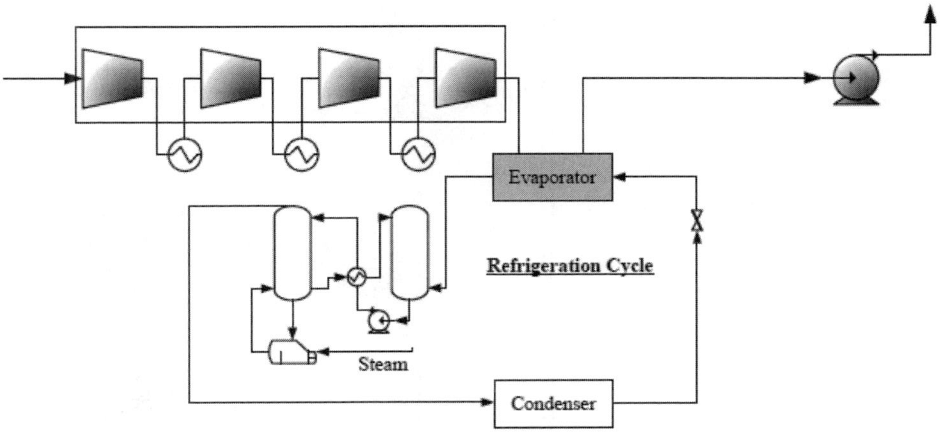

Figure 9-8. Schematic of CO_2 initial compression system involving refrigeration before last stage pump [14]

based on CO_2 mass flow rate of 5 Mt/year. It is shown that the total power required is 39.64 MW, which is even lower than the previous two strategies and is less than strategy I power by 10.12%.

The above analysis shows that the power requirement of CO_2 compression can be reduced by up to 10% if the CO_2 is liquefied during the compression process. The performance benefit arises mostly from the possibility of using a pump to bring an almost incompressible, dense fluid, to the final pressure after liquefaction. Compared to the compression-only solution (strategy I), a fraction of the overall pressure rise is allocated from the energy-intensive compressor to the pump. As indicated by the improvement of strategy II relative to strategy III, the pumping benefit can be maximized if the liquefaction is carried out at the lowest pressure possible, i.e., maximizing the fraction of the total pressure rise allocated to the pump. Minimum pressure required for liquefaction is the saturation pressure of CO_2 corresponding to its liquefaction temperature and is hence determined by the temperature of the cooling medium available for this process step. For a liquefaction temperature of 20°C, the minimum pressure required to achieve liquefaction is therefore 5.7283 MPa-a. By introducing a refrigeration cycle to lower the liquefaction temperature below that achievable at ambient conditions, the possibility of liquefying at an even lower pressure can be achieved but at the cost of a costly refrigeration system. Figure 9-8 shows a schematic of this last stage refrigeration system which was investigated by Botero et al. [14]. Clearly, a thorough thermo-economic analysis must be conducted for each plant case to determine the optimum strategy of initial CO_2 compression before introduction into the pipe line.

9.4 PIPELINE COMPRESSION AND PUMPING SYSTEMS

As mentioned in the previous section, compressors and pumps are needed to convert the captured CO_2 mixtures from atmospheric or near atmospheric pressures to the desirable pipeline pressure and state (supercritical or dense/liquid phases) before introduction into the pipeline. Compressor and pumping stations in a CO_2 pipeline system can be divided in two classes: the originating stations, which are positioned at the inlet to the pipeline, and the booster stations, which are located along the pipeline to compensate for the pressure decrease due to friction and elevation losses. In principle, the longer the pipeline and the elevation of the terrain crossed, the more compression/pumping power is required to achieve the required delivery pressure at destination. Under a fixed route and flow capacity,

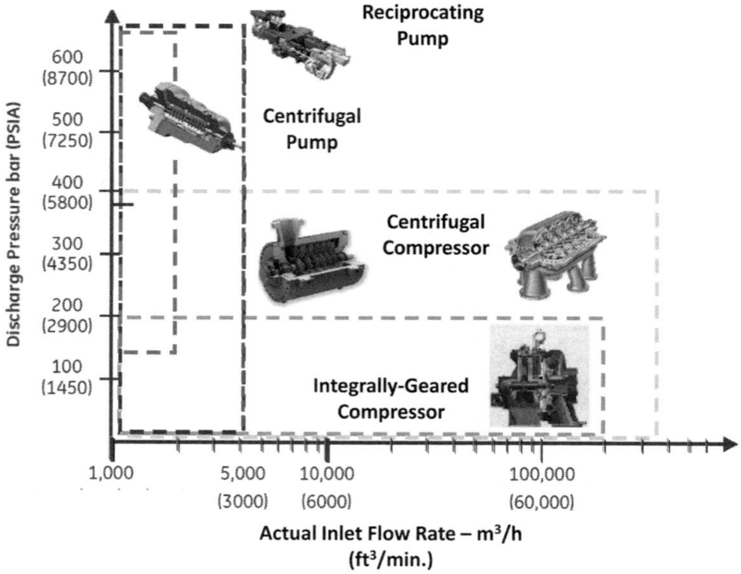

Figure 9-9. Characterization of CO_2 compression and pumping systems based on actual inlet flow rate and discharge pressure (Courtesy of GE)

the number and size of booster stations depend on the circumstances and design. Fewer stations might be easier to operate but the disadvantage is the need of for high inlet pressures, which are likely to require the more expensive use of thicker pipes. The type of compression/pumping equipment required depends primarily on the phase of the CO_2 mixtures at inlet conditions, actual inlet flow rates and desired discharge pressure. Figure 9-9 provides a good characterization map of the type of compressor/pumps required for a given actual inlet flow rate and desired discharge pressure.

9.4.1 Integrally Geared Compressors

For example, for the initial station, the suction pressure is near atmospheric, and hence the actual inlet flow rate (at suction condition) is typically very high (due to the low suction density). In this case, integrally geared compressors are used. These types of compressors involved multiple stage compressions, all driven from a single drive shaft with the use of gear systems as shown in Figure 9-10 for a four-stage compression and Figure 9-11 for an eight-stage compression. The input speed can be either 1800 or 3600 rpm. Figure 9-12 shows an example of two stages connected to one shaft whose pinion is driven from the main gear. A schematic of the integrally geared system is shown in Figure 9-13 (taken from Figure 5-20), and skid-mounted systems are depicted schematically in Figs. 9-14 and 9-15 showing the interconnected piping. Installation examples of this type of integrally geared compressors on CO_2 applications are shown in Figure 9-16 and on a test bed in Figure 9-17.

9.4.2 Centrifugal Compressors

Referring back to the characterization map of the type of compressor/pumps of Figure 9-9, in line multi-stage, multi-impeller compressors combined with centrifugal pumps are also used as an alternative to the integrally geared compressors. Figure 9-18 shows different alternative configurations of these combinations. Typically reciprocating compressors are

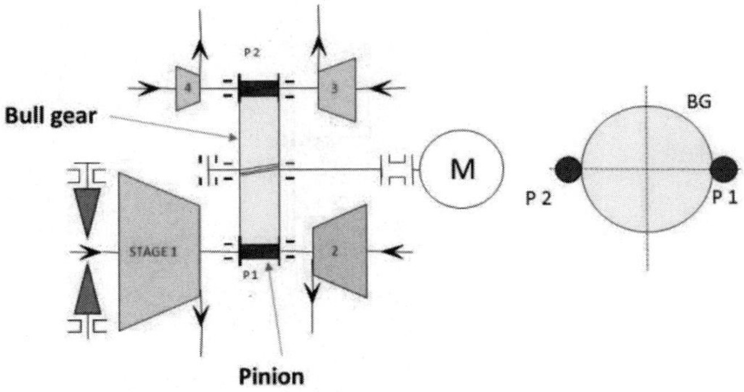

Figure 9-10. Concept of a four-stage integrally geared compressor (Courtesy of MAN Turbo)

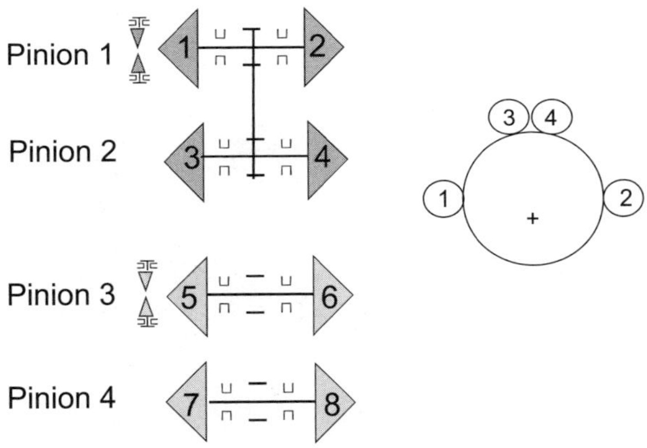

Figure 9-11. Concept of an eight-stage integrally geared compressor (Courtesy of MAN Turbo)

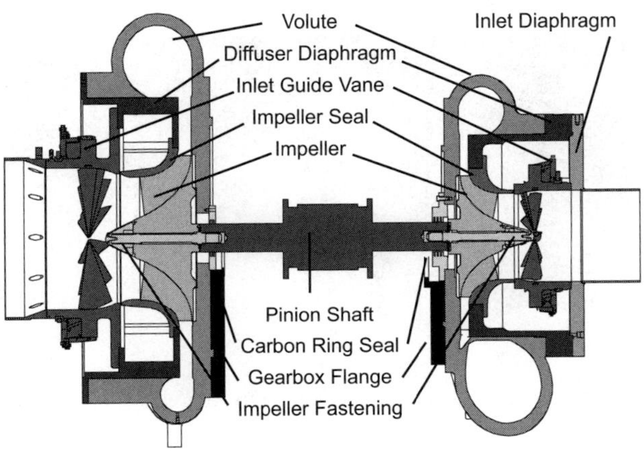

Figure 9-12. Example of two stages connected to one shaft whose pinion is driven from the main gear (Courtesy of MAN Turbo)

Figure 9-13. Schematic of the integrally geared system (Courtesy of MAN Turbo)

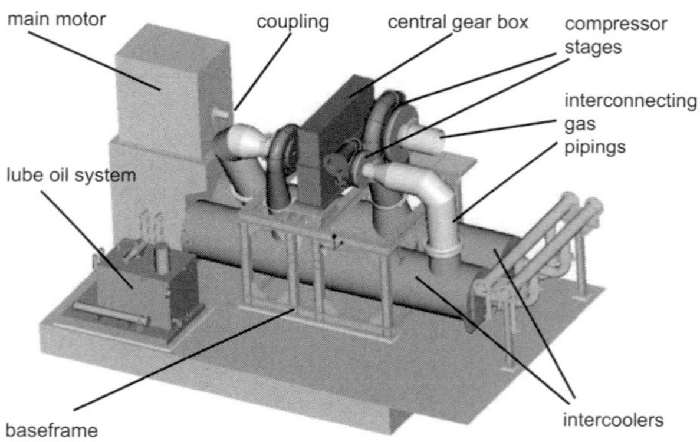

Figure 9-14. Skid mounted integrally geared system (Courtesy of MAN Turbo)

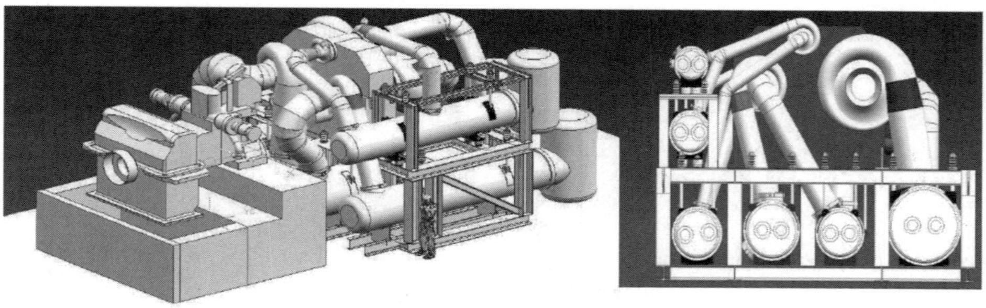

Figure 9-15. Skid mounted integrally geared system (Courtesy of GE)

Figure 9-16. Installation examples of integrally-geared compressors in CO_2 applications (Courtesy of MAN Turbo)

more cost effective for small volumes and centrifugal compressors for larger volumes although there is a wide range of recommended switching points [15]. The majority of the compressors currently used are reciprocating compressors but, while still appropriate for small volumes with potential variation in flow rate, for large constant volumes of CO_2 centrifugal compressors are more appropriate.

Figure 9-17. Integrally geared compressors on a test-bed (Courtesy of MAN Turbo)

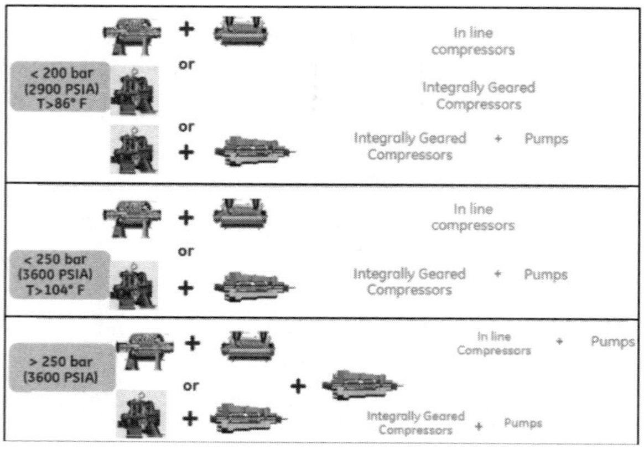

Figure 9-18. Possible CO_2 compression and pumping configuration (Courtesy of GE)

The centrifugal compressors can use either a horizontal or a radial split case. The type of case used will depend on the pressure rating of the given stage of compression with radial split casings generally being used for the higher pressure applications. Flow arrangements through the set of impellers inside the centrifugal compressor include straight through, back-to-back, double flow, and side stream configurations as shown in Figure 9-19. Figure 9-20 shows an example of a centrifugal compressor which was provided for the Weyburn CO_2 facility (suction pressure = 1.47 MPa, discharge pressure = 15 MPa, power = 11.6 MW).

The presence of water together with CO_2 creates carbonic acid which is corrosive to carbon steels. The use of stainless steel for any components in contact with wet CO_2 eliminates the problem. Additionally, special O-ring materials are required to resist explosive decompression due to entrapped CO_2. Typically, materials selected for CO_2 compressors are as follows:

- Carbon steel case with Inconel weld overlay for wetted surfaces at each wet inlet.
- SS Impellers, particularly the first impeller at each wet end.

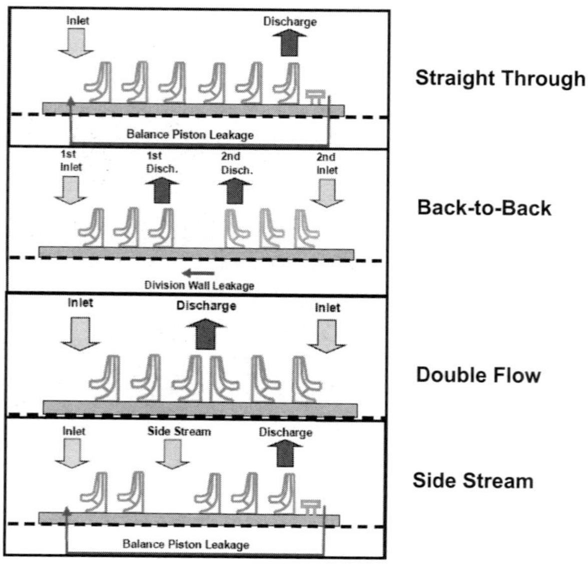

Figure 9-19. Different types of multistage, multi-impeller centrifugal compressor arrangement

Figure 9-20. Centrifugal compressor provided for the Weyburn facility (Courtesy of Dresser-Rand)

- Inconel impellers may be used depending on impurities (type and concentration).
- SS stationaries for first stage at each wet inlet.

Dry-gas seals are typically used with CO_2 compressors (see Figure 9-21). The required seal gas supply normally comes from compressor discharge or alternate supply source cab be used, particularly for tart-up. Labyrinth seals can also be used for its lower cost, but usually produce higher leakage rate. They can only be used if process gas can be vented to atmosphere and that there is no poisonous or flammable constituents in the leaked gas, otherwise, vent should be routed to a safe (unpopulated) area with adequate dispersion characteristics. Labyrinth seals still need a separation seal to keep CO_2 out of the bearing housing. In this case, the separation gas is usually air.

9.4.3 Centrifugal Pumps

Pumping rather than compression is used to maintain the pressure along CO_2 pipelines as it is more economical, efficient and reliable, and has better operating flexibility. Pumping

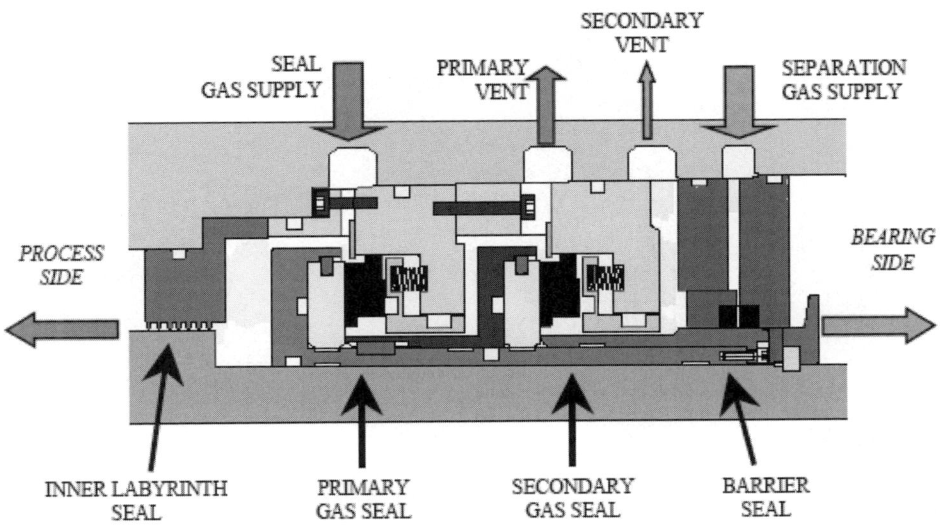

Figure 9-21. Typical dry-gas seal

is especially beneficial in pipeline boosting and final injection where suction pressures are high and temperatures are ambient. Pumps are also used in the last stage of compression at the initial station. As mentioned earlier, the state of CO_2 mixture should be in the dense or liquid phase to permit use of pumps rather than compressors, although the transition between the use of either is arguable as shown in Figure 9-22. Centrifugal pumps are typically used, and their performance prediction is complicated by the compressibility of the CO_2 mixtures; i.e., density changes throughout the pump.

Centrifugal pumps can be single stage (single impeller) or multi-stage (multi impellers). They can also be low flow-high head or high flow-low head. Figure 9-23 shows an example of a single stage pump of low head and high flow along with its performance map. Figure 9-24 shows an example of multi-stage, multi-impeller pump of low flow and high head along, again with its performance map. Typically, these pumps are of radially split case type. For a flow rate of 1,600 t/h of CO_2, and a suction pressure of 8.4 MPa, suction temperature 24°C, and discharge pressure of 14.6 MPa, three pumps of the single stage type would be required in series, or two pumps of the multi-stage type will be required in parallel.

Typical material specified for CO_2 centrifugal pumps are as follows:

• Case and Impellers	A352 GR LCB (to accommodate −90°C in case of rapid depressurization)
• Stationary Wear Parts	Graphalloy (due to low lubricity of CO_2)
• Rotating Wear Parts	A743 GR CA-15
• Shaft	A434 GR 4140 CL BC
• Bearing Housings	A216 GR WCB
• Elastomers	to be compatible with CO_2 applications

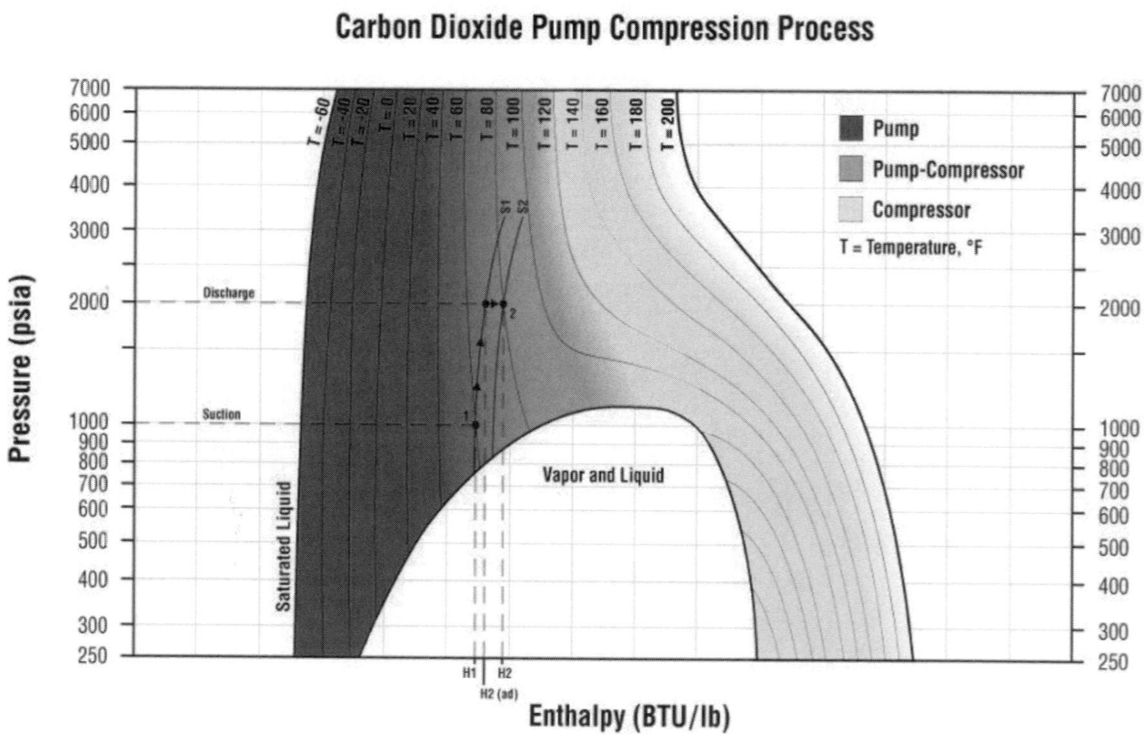

Figure 9-22. Transition zone between where pumps vs. compressors are used for CO applications

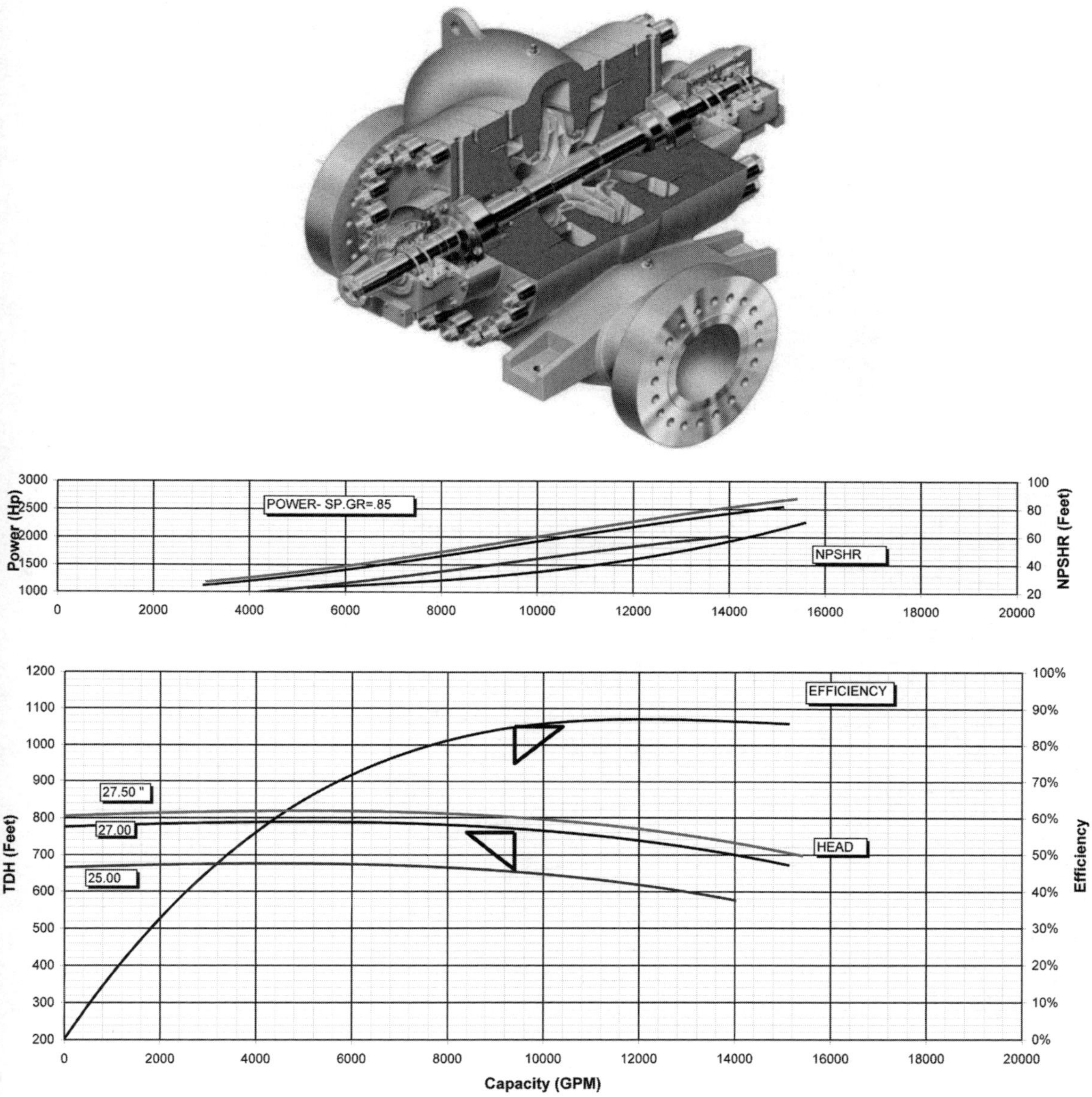

Figure 9-23. Example of a single-stage pump of low head and high flow and its performance map (Courtesy of FLOWSERVE)

The low surface tension of CO_2 and high suction pressure typical for pump application with CO_2 create challenges with the casing joint. Therefore, API axial split type pumps require crowning of casing joint and surface finish to "grip" gasket. Otherwise, radial split may be required based on ISO 13709 (API 610) if CO_2 is classified as "hazardous," especially in offshore applications.

Temperature control is also very important as the density of CO_2 is very sensitive to temperature and can therefore affect pump and compressor capacity which in turn will affect pipeline throughput. Therefore, the following instrumentations are typically specified:

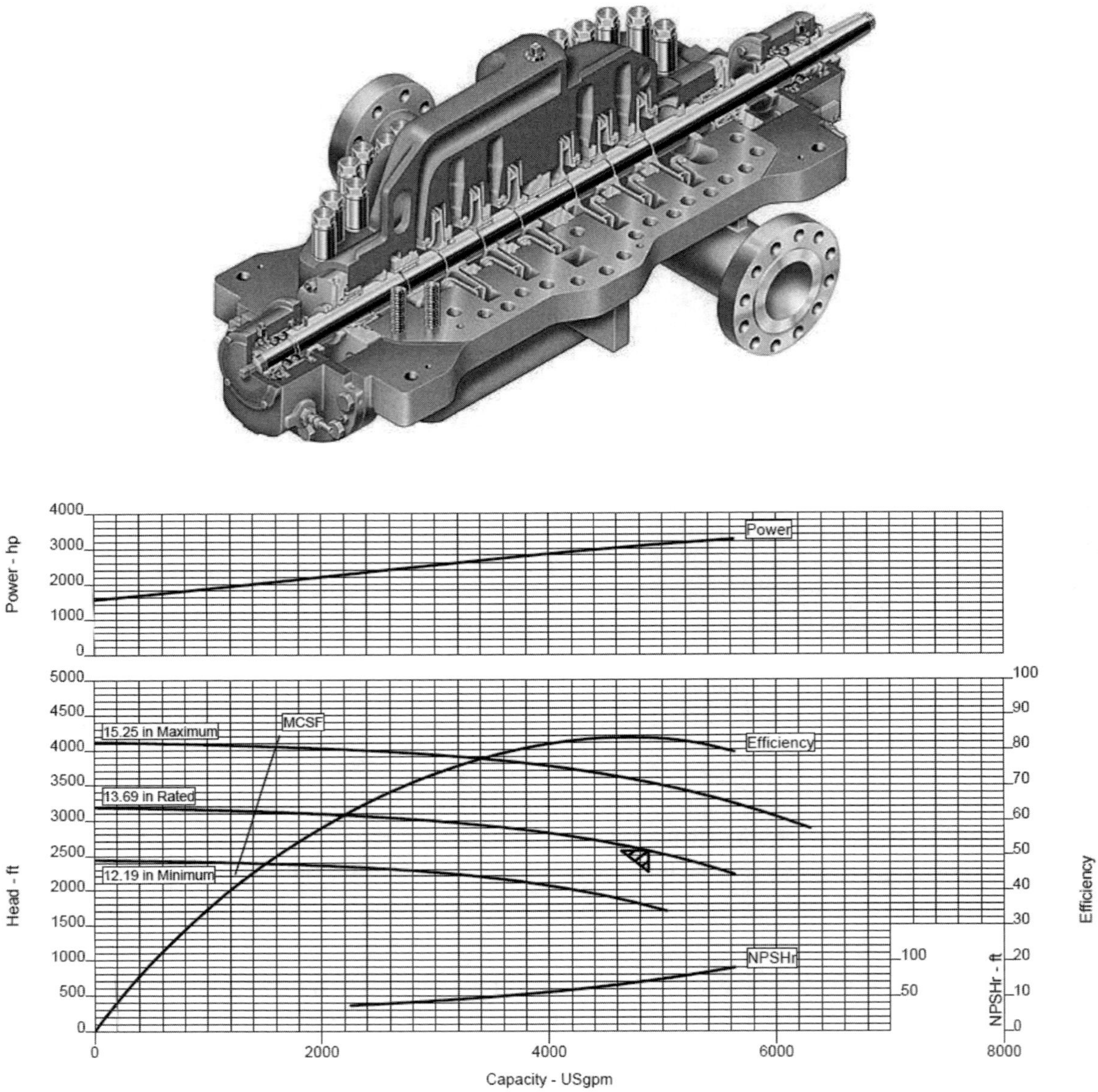

Figure 9-24. Example of a multi-stage pump of low head and high flow and its performance map (Courtesy of FLOWSERVE)

- RTD: Radial Bearings (2), Thrust Bearings (2) and Case (1);
- Accelerometer on Each Bearing Housing;
- Proximity Probes and Proximeters;
- Rigid Conduit to NEMA 4 Junction.

Three main techniques (types) of mechanical seals are used in CO_2 pump applications. These are:

Type I — Dual pressurized seal with a liquid barrier fluid (Figure 9-25): the main feature of this type is that a liquid barrier fluid is used between the inner and outer seals, hence, it is not affected nor relying on the fluid in the pipeline. As such, it is not impacted by varying operating conditions of the process fluid by any contaminants therein. However,

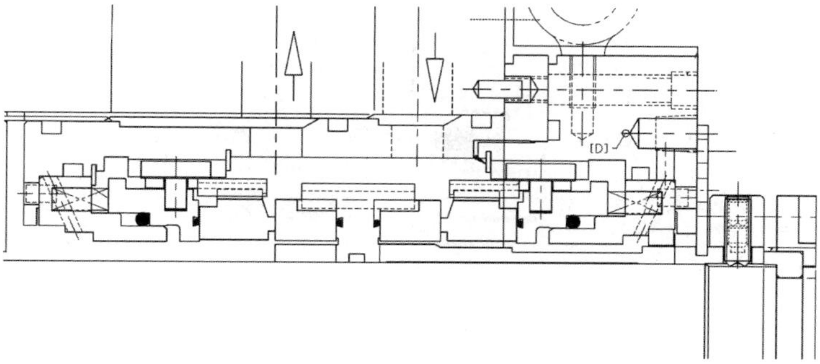

Figure 9-25. Type I mechanical seal for centrifugal pumps (Courtesy of FLOWSERVE)

it has the disadvantage of possible introduction of the barrier fluid into the process fluid, as well its requirement for extensive external support system.

Type II — Dual unpressurized seal with inner liquid seal and outer containment seal (Figure 9-26): the two seals are unpressurized in this type, where the inner one operates on the liquid phase of CO_2 and hence has poor lubrication effects. Because of this, the inner seal can create solid phase of CO_2 in the seal faces. The advantages, however, are that the seal moderately tolerant of contamination of the process fluid. The outer seal is dry running containment as a back-up. The whole seal requires a simple support system; however, it suffers inconsistent results.

Type III — Dual unpressurized seal with inner non-contacting seal and outer containment seal (Figure 9-27): this is also based on two unpressurized seal arrangement, except that the inner seal operates on vapor phase CO_2, which may require heating to vaporize liquid CO_2. The outer seal is running dry as a back-up. Contamination tolerance depends on hydrodynamic features of the inner seal. The main advantage of this type is its long life performance and its simple support system.

The materials used in all of the mechanical seals are typically specified as shown in Table 9-6.

For multi-stage, multi-impeller pumps, rotor dynamic aspects of the shaft assembly must be thoroughly assessed as CO_2 fluids has different characteristics than water for example. Figure 9-28 shows an example of critical speed curve typical for a multi-stage pump.

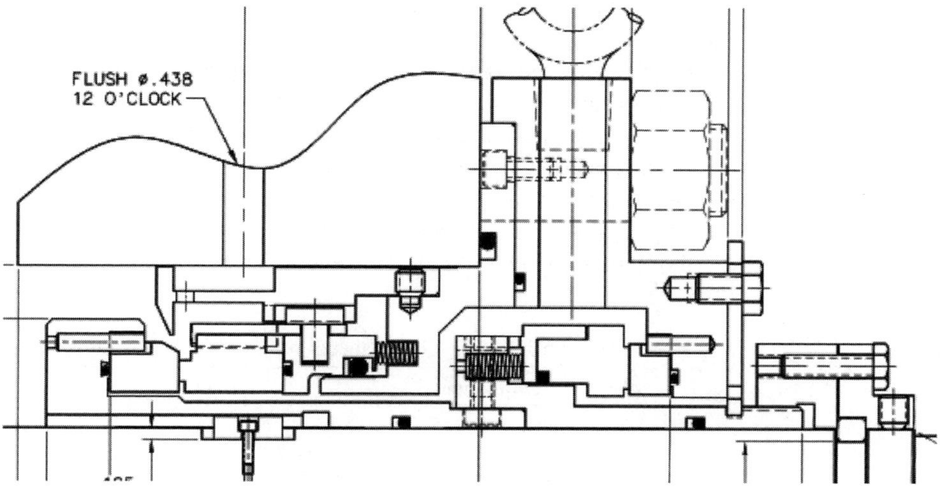

Figure 9-26. Type II mechanical seal for centrifugal pumps (Courtesy of FLOWSERVE)

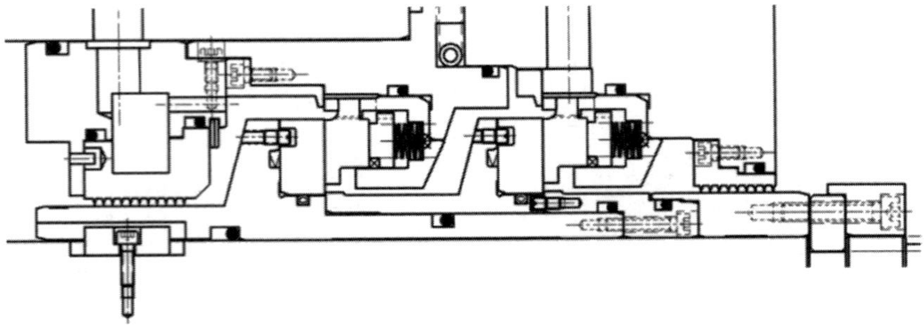

Figure 9-27. Type III mechanical seal for centrifugal pumps (courtesy of FLOWSERVE)

Note that the 1st critical in CO_2 application is 90% to 95% lower than that for water. Additionally, for high-speed VFD pump applications (e.g., speed greater than 5000 rpm), a large shaft diameter should be considered according to the typical rotor dynamics factor K defined as:

$$K = \sqrt{\frac{WL^3}{D^4}} \tag{9-6}$$

where:

W - rotor weight in Newtons
L - rotor length in mm
D - impeller diameter in mm

Figure 9-29 shows a typical design guideline limits for large and slender shafts in dry- and wet-running conditions. Clearly, for high-speed pumps, larger diameter shafts are required. This will also promote shaft stability and independence on Lomakin effects for extra bearing supports or lack of it. It will also eliminate or reduce the demand for graphalloy wear parts.

The Lomakin effects can be described briefly as follows. Pump wear rings and (in multistage pumps) inter-stage bushings and balancing devices can develop large forces in their annular-clearance space due to the circumferentially varying Bernoulli pressure drop induced as rotor eccentricity develops. This is known as the Lomakin effect and can dramatically affect pump rotor dynamics, especially in two ways:

- By changing the rotor-support stiffness and therefore the rotor natural frequencies, thereby either avoiding or inducing possible resonance between strong forcing frequencies at one and two times the running speed and one of the lower natural frequencies.
- By increasing the rotor-support cross-coupled stiffness or, as the seals wear, decreasing the rotor damping, allowing greater magnification factors during resonance, and, in the extreme, allowing destabilizing forces on the rotor to dominate those associated with stabilizing energy absorption.

TABLE 9-6. Typical seal materials for CO_2 pumps

Component	Material
Gland and sleeve	316 SS
Rotating face	Metallized Carbon, SiC, WC
Stationary face	SiC (DLC on gas seals)
Gaskets	Explosive Decompression Resistant FKM
Springs	Alloy C-276

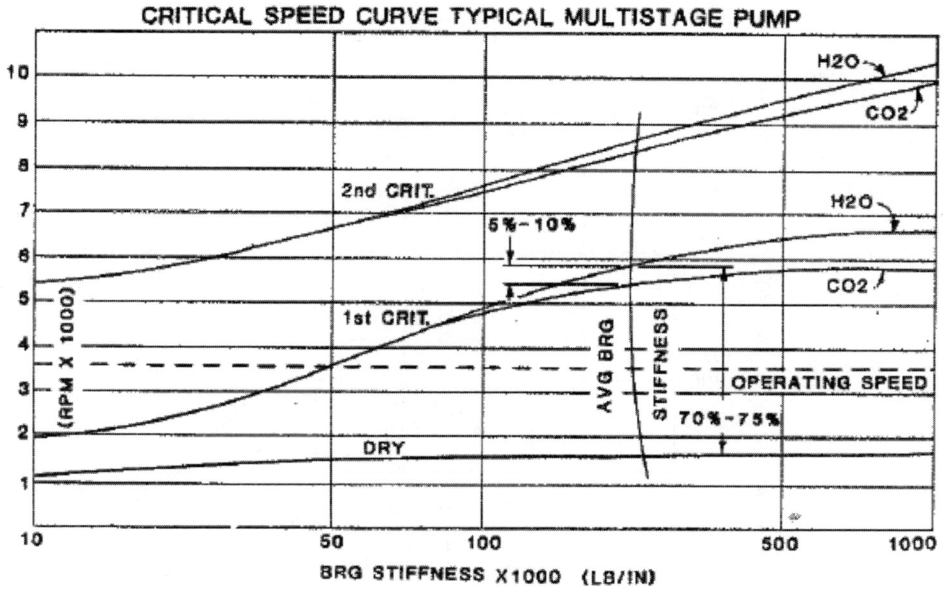

Figure 9-28. Example of critical speed curve typical for a multi-stage pump design guideline limits for large and slender centrifugal pump rotor

The Lomakin stiffness can be calculated from Black et al. [16], in terms of effective stiffness as follows:

$$K_{Lomakin} = \sqrt{\frac{RL\Delta P}{c}} K \text{ (N/mm)} \tag{9-7}$$

where:

- R - wear ring radius (mm)
- L - wear ring length (mm)
- ΔP - differential pressure across the ring (MPa)
- c - radial clearance
- K - non-dimensional stiffness coefficient [16]

These effects are particularly strong in multistage pumps because multistage rotors are relatively long and flexible. Since annular seals (such as wear rings) are primarily in the central portion of the rotor, where they exercise considerable leverage on the first bending mode of the rotor, the contribution of seal stiffness to the rotor support can be comparable to that of the rotor stiffness itself. The Lomakin effect depends directly on leakage and therefore the pressure drop across the seal, which, for constant-system flow, results in a variation of the Lomakin support stiffness with roughly the square of the running speed. However, the more usual case of roughly constant system head results in only a small variation of Lomakin effect with pump speed. The other important parameters are annular sealing cavity length, diameter, and clearance. Fluid properties are of secondary importance.

The radial-clearance effect is a strong influence, with Lomakin effect roughly proportional to its reciprocal. The physical reason for the strong influence of clearance is that it gives the opportunity for the circumferential pressure distribution, which is behind the Lomakin effect, to diminish through circumferential flow. Any seal surface which includes grooving has a similar effect as increased clearance. The primary action of the Lomakin ef-

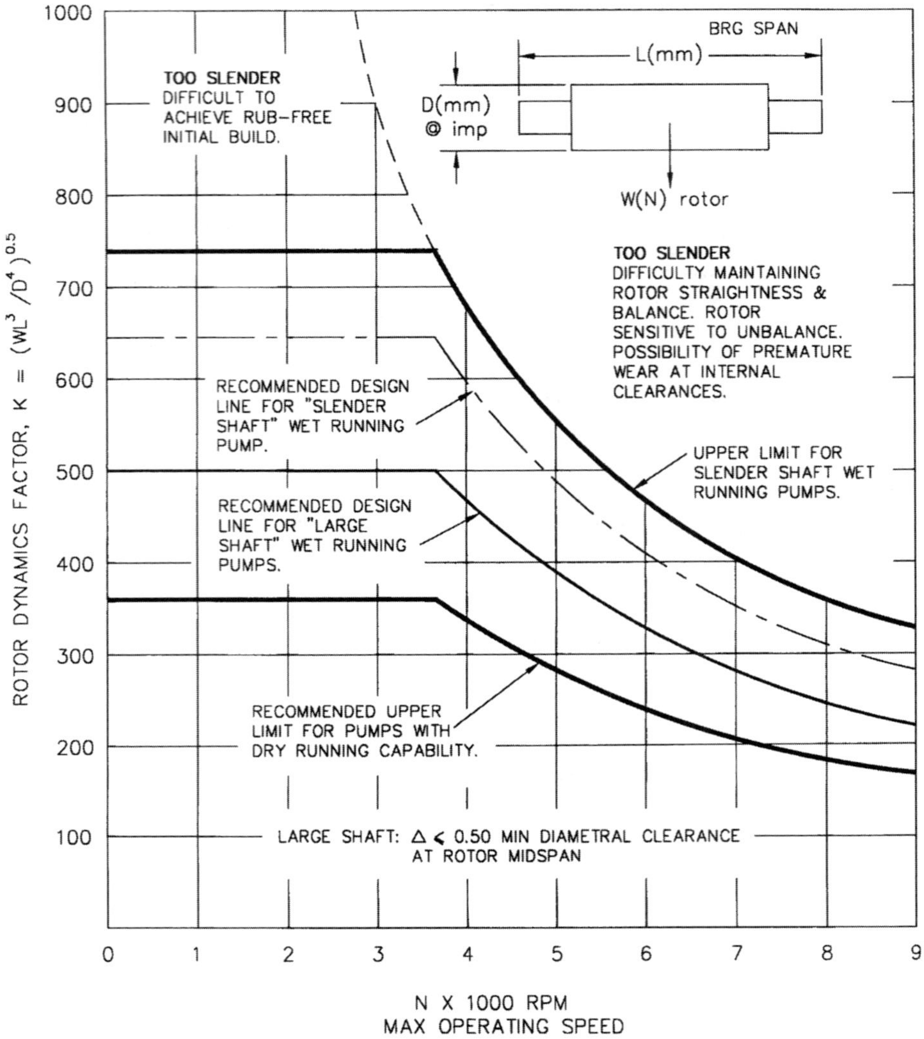

Figure 9-29. Typical design guideline limits for large and slender centrifugal pump rotor

fect is beneficial, through increased system direct stiffness and damping tending to increase the rotor natural frequency and decrease the rotor vibration response at that natural frequency. However, over-reliance on Lomakin effect can put the rotor design in the position of being too sensitive to wear of operating clearances, resulting in unexpected rotor failures due to resonance. It is important that modern rotors be designed with stiff enough shafts, so that any natural frequency that starts above running speed with new clearances remains above the running speed with clearances worn to the point that they must be replaced from a performance standpoint.

9.5 FLOW METERING

There are two purposes for flow rate metering of CO_2 mixtures. The first is for custody transfer and the second for system balance and company internal records. Clearly, the requirements for custody transfer are more stringent than that for system and internal balance. Several flow metering technologies or concepts are utilized in CO_2 pipeline industry.

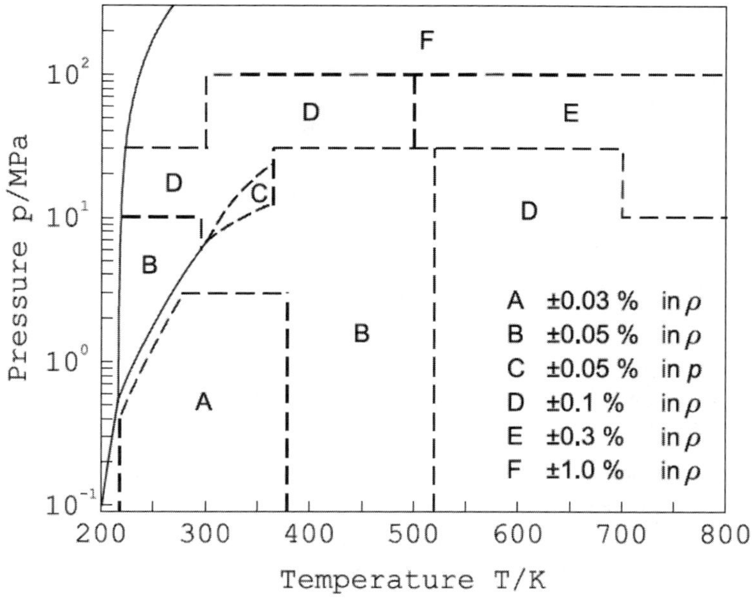

Figure 9-30. Uncertainty in density predictions by GERG-2008 EOS [1]

Those are either volumetric-based, which are then corrected for density at prevailing pressure, temperature and mixture composition, or mass-based. The correction required for the volumetric-based technologies requires utilization of agreed upon equation of state by the parties involved. The metering technology and standard represent the "best available" tools for CO_2 pipeline operators, which tend to accept resolution of EOS to predict the density to approximately 2%. However, GERG-2008 EOS, for example, is claimed to do better than 0.1% in the range of typical pipeline range of pressure and temperature as shown in Figure 9-30 [1].

Examples of volumetric-based meters used are orifice meters, turbine meters, and vortex meters. Mass flow meters are typically coriolis flow meters. The following are examples of actual CO_2 installations:

- Sell Cortez CO_2 Pipeline System (Cortez, Colorado to Denver City, Texas): orifice meters for custody transfer and vortex meters for system balance.
- Amoco Bravo Pipeline System (Bravo Dome, New Mexico to Denver City, Texas): orifice meters for custody transfer.
- Arco Sheep Mountain Pipeline System (Gardner, Colorado to Denver City, Texas): turbine meters for custody transfer and orifice meters for system balance.
- Weyburn facilities: mass flow meters for custody transfer and orifice meters for internal and system balance (in accordance to AGA-4 with pressure and temperature compensation).

Typically, electronic flow meters in accordance with API-21.1 are employed with the above devices. This consists of *primary*, *secondary* and *tertiary* devices. The *primary* device defines the basic type of the meter used, including, but not limited to, an orifice, turbine, vortex, or other devices. The *secondary* device provides such data, including, but not limited to, static pressure, temperature, differential pressure, relative density, and other variables that are appropriate for inputs into the tertiary device. Each primary device requires specific or properly configured tertiary device appropriate to the type of meter used. The *tertiary* device is an electronic computer, programmed to correctly calculate flow within

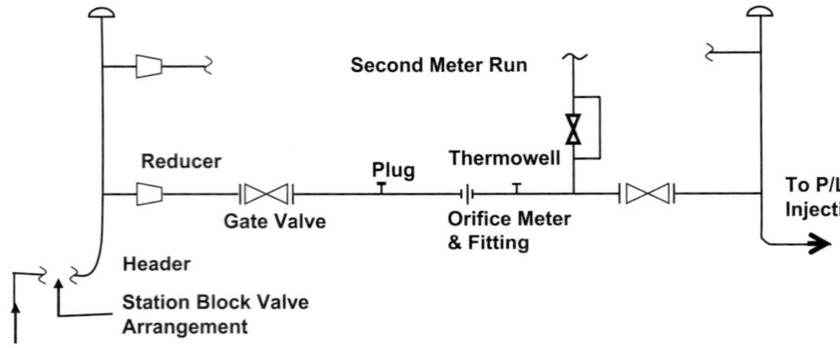

Figure 9-31. Typical orifice meter run for custody transfer

specified limits that receives information from the primary or the secondary devices. The tertiary and secondary devices, as well as the primary device, may be contained in one or more enclosures, or packaged separately.

A custody transfer meter station is generally installed at the outlet of the compressor station prior to entering the pipeline as well as at the delivery points from the pipeline. All underground piping for meter stations should be epoxy coated with a corrosion coupons installed at every station. Typical orifice meter run design is shown in Figure 9-31. Vortex meters are used for flow calibration, line integrity, and system balance. Industry recommends the use of flow computer that does not require density measurements. Density can be computed from the monitored pressure and temperature readings and use of appropriate equation of state. A typical design of a vortex meter rune is depicted in Figure 9-32.

The industry usually insulate all meter runs, as 0.5°C (1°F) temperature differential can swing pressure recordings up to 3 kPa. Orifice plates (with a 0-400 in. of water column) pressure transmitter are typically used. Meter runs are typically equipped with differential pressure and temperature recorders as well as a densitometer. A pressure relief system is also added (see Figure 9-33).

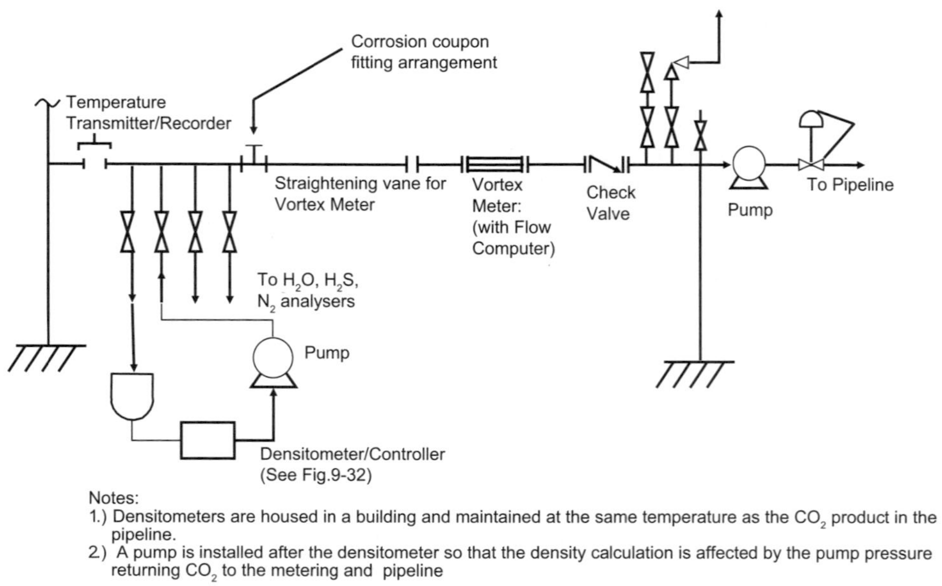

Figure 9-32. Typical vortex meter run

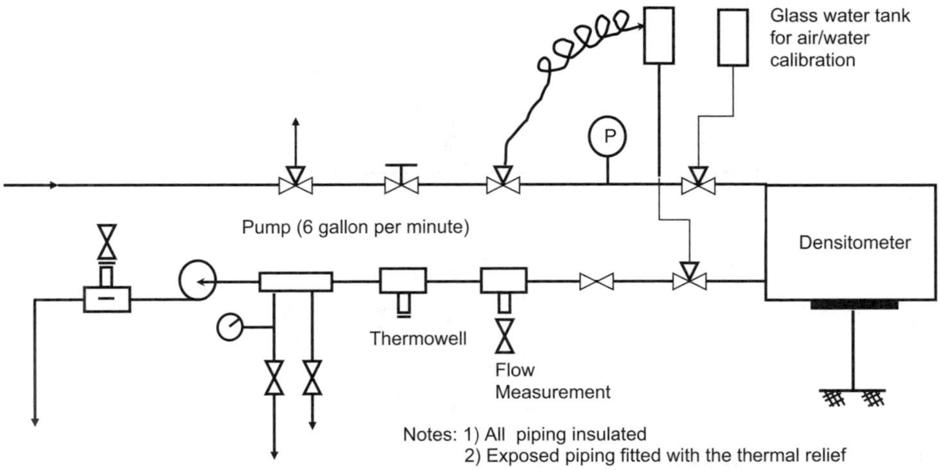

Figure 9-33. Densitometer piping arrangement

9.6 VALVES AND OTHER APPURTENANCES

Block valves for CO_2 applications are typically ball valves with the following material specifications:

Dry CO_2:

Ball	0.001" ENP;
Seat Rings	Carbon Steel;
Seal Areas	Untreated;
Through Bore	Untreated;

Typically Welded Body Valve (Figure 9-34).

Wet CO_2:

Ball	0.003" ENP;
Seat Rings	0.003" ENP or 410SS;
Seal Areas	Weld inlay or Phenolic coating;
Through Bore	Weld inlay or Phenolic coating;

Typically Bolted Body Valve (Figure 9-34).

Valve seats are usually Vitan as CO_2 eats Teflon. Other various elastomers have been

Figure 9-34. Welded vs. bolted body ball valves

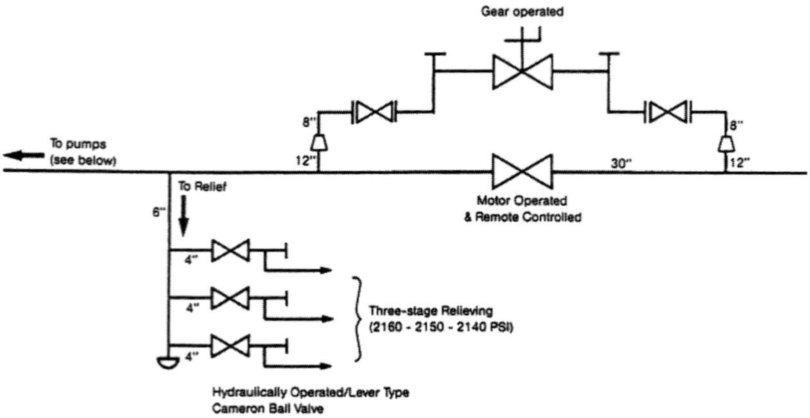

Figure 9-35. Block valve and blowdown piping arrangement, and three-stage relief system

used. EPDM with 85 hardness is the most suitable but cannot withstand hydrocarbon carryovers. It is good in decompression and dynamic works. Attention needs to be paid to the method of manufacturing elastomers. Extruded, machined, moulded, or formed elastomers all respond differently to dynamic conditions. Dense chrome plating is best for hard seats. Anodized aluminum should be used if in contact with CO_2. For valve sealants or lubricants, petroleum-based products are not recommended as they react with CO_2 and decompose. Some sealants like Valtec 80 are not used as they harden in contact with CO_2. Valve gaskets are typically Flexotalic or equivalent.

Relief valves must seal under high CO_2 pressure. Pressure relief arrangement must be installed on all exposed piping. Pump stations typically have three pressure relief systems at three different relief settings (e.g., 14.55 MPa, 14.62 MPa, and 14.69 MPa) due to the spongy effect of CO_2. Typical station piping relief system is schematically shown in Figure 9-35.

Control valve materials are the same as the block valve described above with the exception that the body material is SS 316 L. The trim materials are either HVD1, Ferralium or Titanium.

Blowdown valves are usually double sitting ball valves with Viton seals. These seals vary. All block valves have blowndown assemblies. Blowdowns are sized for blowing down a 20 mile (32 km) section of pipeline in 6 to 8 hours to avoid dry ice formation. Slow opening of all blowdown valves is recommended.

REFERENCES

[1] Kunz, O., Klimeck, R., Wagner, W. and M. Jaeschke, 2007, "The GERG-2004 Wide-Range Equation of State for Natural Gases and Other Mixtures," Groupe Européen de Recherches Gaziéres (GERG), Technical Monograph, GERG TM15.

[2] NIST Standard Reference Database 23, NIST Reference Fluid Thermodynamic and Transport Properties Database (REFPROP): Version 8.0, www.nist.gov/ts/msd/srd/nist23.cfm.

[3] AspenTech, Aspen Plus® V7.2 Simulation Engine, 200 Wheeler Road, Burlington, Massachusetts 01803, [http://www.aspentech.com/]

[4] Turner, W. J. and Simonson, M. J., "Compressor Station Transient Flow Modeled," Oil and Gas Journal 79-83, May 20, 1985.

[5] Gas Processors Suppliers Association, Engineering Data Book, Tulsa Oklahoma, 2006.

[6] Mohitpour, M., Botros, K. K. and Van Hardeveld, T., 2008, "Pipeline Pumping and Compression Systems: A Practical Approach," Chapter 10, ASME Press, 2008.

[7] Mohitpour, M., Golshan, H. and Murray, A., 2003. Pipeline design & construction: a practical approach. 2nd Ed. The American Society of Mechanical Engineers, Three Park Av., New York, NY 10016.

[8] WRI, 2008. CCS guidelines: guidelines for carbon dioxide capture, transport, and storage. Washington, DC. World Resources Institute.

[9] IEA GHG, 2009. CO_2 storage in depleted gas fields. Report Number 2009/01, International Energy Agency R&D Programme.

[10] International Energy Agency (IEA), 2007, "Capturing CO_2," Paris.

[11] Moore, J. J. and Nored, M. G., 2008, "Novel Concepts for the Compression of Large Volumes of Carbon Dioxide," Proceedings of ASME Turbo Expo 2008, Berlin.

[12] Romeo, L. M., Bolea, I., Lara Y., and Escosa J. M., 2008, "Optimization of Intercooling Compression in CO_2 Capture Systems," Appl. Therm. Eng., Available Online.

[13] Rao, A., 2007, "An Assessment of CO_2 Compression Options for Near Zero Emission Power Plants," Panel Presentation, ASME Turbo Expo 2007, Montreal.

[14] Botero, C., et al., "Thermoeconomic Evaluation of CO_2 Compression Strategies For Post-Combustion CO_2 Capture Applications," Proceedings of ASME Turbo Expo 2009: Power for Land, Sea and Air, Paper # GT2009-60217, Orlando, Florida, USA, June 8–12, 2009.

[15] Caroll, J., "Compression and High Pressure Pipeline Transportation of Carbon Dioxide Rich Fluids," 1st Regional Symposium on Carbon Management — Carbon management Challenges and Opportunities for Petroleum Industry, May 22-24, 2006.

[16] Black, H. F. and Jenssen, D. N., "Effects of High Pressure Ring Seals on Pump Rotor Vibration," ASME, 71-WA-FE-36.

[17] 49CFR192, 2008, "Pipeline Safety: Pipeline Integrity Management in High Consequence Areas (Gas Transmission Pipeline)," Final Rule, Department of Transportation, Revision 04/08.

[18] 49CFR195, 2008, "Pipeline Safety: Pipeline Integrity Management in High Consequence Areas (Hazardous Liquid Operators With 500 or More Miles of Pipeline)," Final Rule, Department of Transportation, Revision 06/08.

[19] API 5L, 2007, "Specification for Line Pipe," ANSI/API Specification 5L, 44th Edition.

[20] ASME B31.4, 2006, "Pipeline Transportation Systems for Liquid Hydrocarbon and Other Liquids," ASME B31.4-2006, American Society of Mechanical Engineers, New York, USA.

[21] ASME B31.8, 2007, "Gas Transmission and Distribution Piping Systems," ASME B31.8-2007, American Society of Mechanical Engineers, New York, USA.

[22] Anheden, M., Andersson, A., Bernstone, C., Eriksson, S., Yan, Liljemark, S. and Wall, C., 2005, "CO_2 Quality Requirement for a System with CO_2 Capture, Transport and Storage." *Proceedings of the 7th Greenhouse Gas Technologies Conference GHGT-7*, Vancouver, 2 pp. 2559-2563.

[23] Barrie, J., Brown, K., Hatcher, P. R. and Schellhase, H. U., 2005, "Carbon Dioxide Pipelines: A Preliminary Review of Design and Risks," *Proceedings of the 7th International Conference on Greenhouse Gas Control Technologies*, Vancouver, Canada, 2005.

[24] Bentham, M, 2006, "An Assessment of Carbon Sequestration Potential in the UK — Southern North Sea Case Studies," Tyndall Centre for Climate Change Research Working Paper 85, BS8010-1(2004) "Code of practice for pipelines. Steel pipelines on land," British Standards Institute, 2004.

[25] Decker, L. C., Checak, J. R., Heath, W. H. and Warren, R. N., 1985, "CO_2 Pipeline Design, Construction and Operation" in Facilities, Pipelines, and Measurements: A Workbook for Engineers, presented at 41st Petroleum Mechanical Engineering Workshop and Conference, Kansas City, MO, USA.

[26] Downie, M. J., Seevam, P. N. and Race, J. M., 2007, "Transport of CO_2 for Carbon Capture and Storage in the UK," In: *Proceedings of Offshore Europe* 2007, Aberdeen, UK.

[27] Fradet, A., Saysset, S., Odru, P., Broutin, P., Ruer, J. and Bonnissel, M., 2007, "Technical and Economic Assessment of CO_2 Transportation for CCS Purposes," *Journal of Pipeline Engineering*, 6(3), pp. 173-180.

[28] Hein, M. A., 1985, "Rigorous and Approximate Methods for CO_2 Pipeline Analysis in Facilities, Pipelines and Measurements: A Workbook for Engineers," 41st Petroleum Mechanical Engineering Workshop and Conference, Kansas City, MO, USA.

[29] Hendriks, C., Hagedoorn, S. and Warmenhoven, H., 2007, "Transportation of Carbon Dioxide and Organisational Issues of CCS in the Netherlands," Report prepared by Ecofys for EnergieNed, the Ministry of Economic Affairs and the Ministry of Housing, Spatial Planning and the Environment, March 2007.

[30] IGE/TD/1, 2001, "Recommendations on Transmission and Distribution Practice — Steel Pipelines for High Pressure Gas Transmission" IGE/TD1/1 Edition 4, Institution of Gas Engineers, London.

[31] IPCC, 2005, "IPCC Special Report on Carbon Dioxide Capture and Storage," Working Group III of the Intergovernmental Panel on Climate Change, 2005, Eds. Metz, B., Davidson, O., de IEA GHG. Coninck, H.C., Loos, M. and Meyer, L. A., Cambridge University Press, Cambridge, United Kingdom and New York, NY, USA.

[32] Kantar, K. and T. E. Connor, 1984, "Turkey's Largest Oil Field Poised for CO_2 Immiscible EOR Project," Oil and Gas Journal, 82(48), p. 95.

[33] McCullough, D. E., 1986, "Central Basin Pipeline: A CO_2 System in West Texas," Energy Progress, 6(4), p. 230.

[34] Najera, G., 1986, "Maintenance Techniques Proven on CO_2 Line," *Oil and Gas Journal*, 84(46), p55.

[35] Oosterkamp, T. and Ramsen, J., 2008, "State of the Art Review of CO_2 Pipeline Transportation with Relevance to Offshore Pipelines," Report No. POL-0-2007-138-A, Polytec, Norway.

[36] PD 8010-1, 2004, "Code of Practice for Pipelines, Part 1 — Steel pipelines on Land," *British Standards Institute.*

[37] PSR, 1996, "The Pipelines Safety Regulations 1996 — Statutory Instrument 1996 No. 825," *The Stationery Office Limited.*

[38] Seevam, P. N., Race, J. M. and Downie, M. J., 2008, "Transporting the Next Generation of CO_2 for Carbon, Capture and Storage: The Impact of Impurities on Supercritical CO_2 Pipelines," *Proceedings of IPC2008, 7th International Pipeline Conference*, Calgary, Alberta, Canada: ASME.

[39] Seevam, P. N., Race, J. M. and Downie, M. J., 2009, "Transient flows of CO_2 streams in pipelines," to be published.

[40] Seiersten, M. and Kongshaug, K. O., 2005, "Materials Selection for Capture, Compression, Transport and Injection of CO_2," *Carbon Dioxide Capture for Storage in Deep Geologic.*

[41] Formations. D. C. Thomas and S. M. Benson, Elsevier Ltd., 2, pp 937-953.

[42] Visser, E. and Hendriks, C., 2007, "Towards Hydrogen and Electricity Production with Carbon Capture and Storage- DYNAMIS Quality Recommendation," *DYNAMIS*, Project No.: 019672.

[43] Wallace, C. B., 1985, "Drying Supercritical CO_2 Demands Care," *Oil and Gas Journal*, 83(25), p. 98.

[44] White, V., Allan, R., Miller, E., 2006, "Purification of Oxyfuel Derived CO_2 for Sequestration or EOR," *Proceedings of GHGT-8.*

[45] Aspelund, A. and Jordal, K., 2007, "Gas Conditioning — The Interface Between CO_2 Capture and Transport," *International Journal of Greenhouse Gas Control,* 1(3): 343-354.

[46] Bratfos, H. A., Leinum, B. H., Torbergsen, L. E. and Saugerud, O. T., 2007, "Challenges to the pipeline transportation of dense CO_2," *Journal of Pipeline Engineering*, 6(3) 161-172.

[47] Castello, P., Tzimas, E., Moretto, P. and Peteves, S. D., 2005, "Techno-economic assessment of hydrogen transmission and distribution systems in Europe in the medium and long term," EUR 21586 EN. EC-DG JRC-Institute for Energy, Petten, The Netherlands.

[48] Denbury Resources Inc., 2008, Notes to Consolidated Financial Statements, in: Defining Denbury - Annual Report. http://www.denbury.com/assets/docs/DNR_2008AR.pdf. Last consulted on 7 December 2010.

[49] Det Norske Veritas, 2007, Offshore Standard on Submarine Pipeline Systens DNV-OSF101.

[50] Det Norske Veritas, 2010, Recommended Practice on Design and Operation of CO_2 Pipelines DNV-RP-J202, JIP CO_2PIPETRANS Phase 1.

[51] Ecofys Netherlands BV, 2009, Assessment of technical experience and engineering practice for CO_2 pipelines. Deliverable report from DYNAMIS Project.

[52] Fradet, A., Saysset, S., Odru, P., Broutin, P., Ruer, J. and Bonissel, M., 2007, "Technical and economic assessment of CO_2 transportation for CCS purposes," *Journal of Pipeline Engineering*, 6-3: 173-180.

[53] Gale, J. and Davidson, J., 2004, "Transmission of CO_2 — Safety and Economic Considerations," *Energy*, 29: 1319-1328.
[54] Hendriks, C. A., Wildenborg, A. F. B. and Blok, K., 2000, "Costs of Carbon Dioxide Removal by Underground Storage. In: Fifth International Conference on Greenhouse Gas Control Technologies," Cairns, August 13–16, 2000. p. 967–72.
[55] IPCC, 2005, "Special Report on Carbon Dioxide Capture and Storage. Prepared by Working Group III of the Intergovernmental Panel on Climate Change" [Metz, B., O. Davidson, H.C. de Coninck, M. Loos, and L.A. Meyer (eds.)]. Cambridge University Press, Cambridge, United Kingdom and New York, NY, USA, 442 pp.
[56] ISO 13623, 2009, Petroleum and Natural Gas Industries — Pipeline Transportation Systems.
[57] McCollum, D., Ogden, J., 2006, "Techno-economic Models for Carbon Dioxide Compression, TRAnsport, and Storage & Correlations for Estimating Carbon Dioxide Density and Viscosity," Institute of Transportation Studies. University of California.
[58] McCoy, S. T. and Rubin, E. S., 2007, "An engineering-economic model of pipeline transport of CO_2 with application to carbon capture and storage," *International Journal of Greenhouse Gas Control*: pp. 219-229.
[59] McCoy, S. T., 2008, "The economics of CO_2 transport by pipeline and storage in saline aquifers and oil reservoirs," PhD thesis, Carnegie Mellon University.

Chapter 10

CO_2 PIPELINE OPERATION AND MAINTENANCE

10.1 GENERAL OPERATIONAL ISSUES

Review of numerous large natural deposits of CO_2 that have been existing underground for millions of years are testament to stable long-term possibility of storage of CO_2 [1]. Since early 1970, many of these CO_2 reservoirs have been accessed and used mostly for enhanced oil recovery (EOR) operations. An extensive pipeline network (transporting pure CO_2) has been built that now stretches nearly 8000 km, mostly as previously referred to, in the United States. As a result, the technology, operations and risks associated with the transportation of pure CO_2 are well understood and can be relied as a basis for transporting CO_2 containing impurities.

For most pipelines designed for the transportation of supercritical CO_2, the ability to maintain adequate/minimum pressure is important for successful operations. This is typically achieved by compressing CO_2 at source and recompressing it at various locations along the pipeline length (if needed). Not all pipelines require recompression if adequate high pressure is available at the source to overcome frictional and elevation losses while meeting the delivery pressure at the outlet of the pipeline. For example, the Weyburn pipeline, which transports CO_2 over 345 km from an industrial facility in North Dakota from Basin Electric Cooperative Dakota Gasification Project (Figures 10-1 and 10-2) to an EOR site in Saskatchewan, Canada, operates without a recompression system [2]. The Synfuels Plant has been successfully capturing and transporting a portion of its CO_2 emissions to Canada since 2000. However, it is also designed to ensure that the delivery pressure at the terminal point meets the objective of injection pressure. Thus, the delivery pressure is 14800 kPa (2175 psi).

Transportation of CO_2 by pipelines is widely practiced and accepted as a technological application, not only for EOR but also for industrial and commercial purposes. Moreover, operational risks of pipeline failure are well established, but again mostly for relatively pure CO_2. Gathering or extraction, processing, transportation, and injection of (mostly pure) CO_2 are common business operations and have been accepted sometimes to handle larger-scale geologic storage operations.

Experience of EOR and CO_2 pipeline operation and maintenance to 2011 indicates the following:

- Low-volume geologic storage of CO_2 has successfully occurred in the form of enhanced oil recovery since the early 1970s and also under the practice of acid gas injection since 1989 now well over the range of 1500 to 2000 tons/day.
- Specific knowledge and expertise exists for effective management of CO_2 storage as well as pipeline transportation including CO_2 streams containing some impurities.
- Components that have boiling points and densities similar to CO_2 in a flue gas stream can be transported and injected in a reservoir as long as their presence do not cause safety and operational challenges either for pipeline or storage [3]. Through

Figure 10-1. CO_2 sequestration used from the basin electric cooperative Dakota Gasification Project [4, 5]

research, experience, and public outreach, operators and regulators have successfully managed the risks, benefits, and public apprehension associated with CO_2 pipeline and operational activities.

- Pipeline operations and maintenance must be in compliance with rules or regulations of jurisdiction over the enforcement of such rules or regulations.
- Personnel form a central part of CO_2 operations and must be qualified. Personnel must be continuously trained and updated on safety issues, including safety procedures.

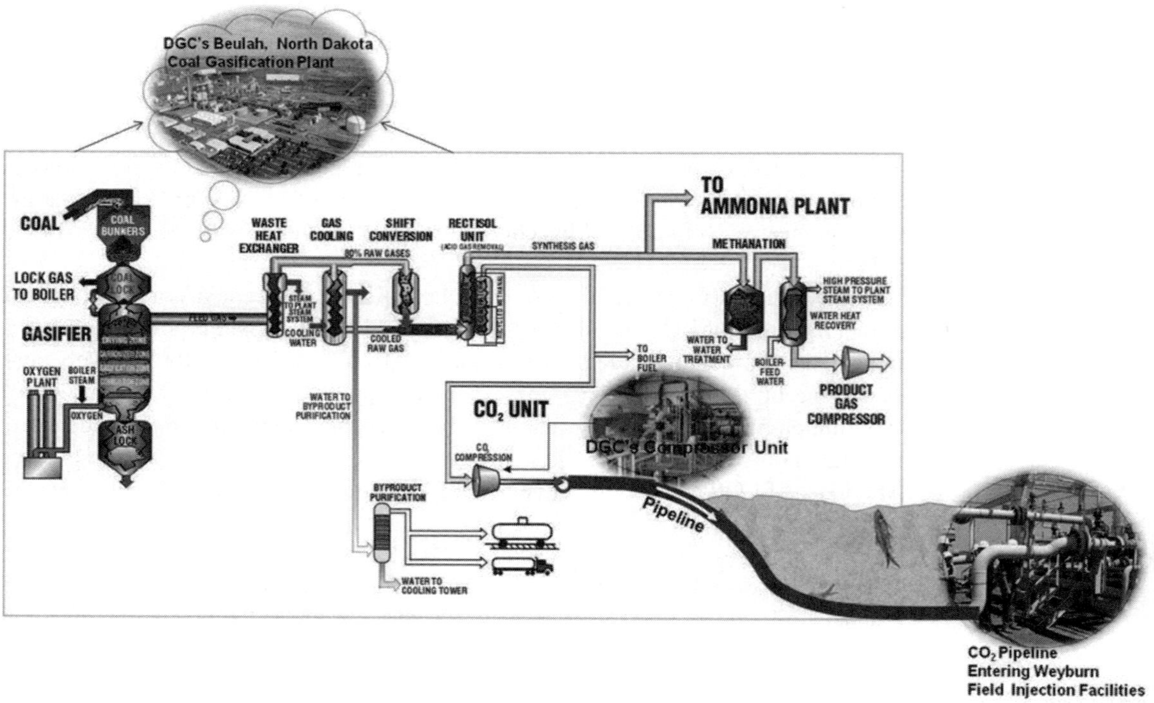

Figure 10-2. The synfuels plant, sources of CO_2 capture for transportation for CCS (reconstructed from Thambimuthu [6])

10.2 CO_2 PIPELINE SYSTEM OPERATIONAL REQUIREMENTS

Operational aspects of pipelines are divided into three areas: daily operations, routine operation, and maintenance, including health, safety, and environment. In general, overall operational considerations include training, inspections, safety integration, signs and pipeline markers, public education/outreach and damage prevention programs, communication, facility security, and leak detection.

Typically, daily pipeline operations are directed from the pipeline control center. In the control center, operators will need to be familiar with the particular properties of CO_2 as well as characteristics of the CO_2 pipeline system to ensure safe and efficient operation free of mishaps.

Routine operation includes:

- CO_2 receipt conditions (pressure/temperature, composition, flow);
- Maintaining steady state operation;
- Maintaining the required minimum operating/delivery pressure;
- Avoiding transient situations as much as possible; and
- Performing routine tasks (hydraulic evaluation and equipment performance assessment);
- Equipment performance reviews.

As with any complex system, CO_2 pipeline operating conditions may not always be normal. A variety of abnormal conditions includes any of the following:

- Operating during a SCADA/communications loss;
- Operating during a scheduled total communication outage;
- Operating without transmitter information;
- Unbalanced line operations (if applicable);
- Responding to transients and diurnal flow fluctuations;
- Operating during maintenance;
- Responding to power outages;
- Responding to unit and station lockouts or failures; and
- Responding to unanticipated valve closures.
- Responding to emergencies (leak, rupture, etc.).

Pipeline system emergencies require an additional level of preparedness and response. An emergency situation is defined as an unforeseen or imminent event which requires:

- Prompt coordination of resources;
- Special communications; and
- Heightened authority for personnel—to protect the health, safety, and welfare of people; or to limit damage to property, the environment or company operations.

For such situations, pipeline companies are required to have an Emergency Response Plan (ERP), which is a document that helps direct and coordinate the actions to be taken in the event of an emergency.

ERP generally must contain the following elements essential for responding to CO_2 pipeline emergencies:

- Critical infrastructure protection, by defining Incident Command System (ICS);
- Preparedness, response, and recovery;
- Communication and information management; and
- Information for the protection of public, personnel, and surrounding infrastructure, including the environment.

CO_2 pipeline emergencies include ruptures and serious leaks, as well as incidents that can arise from blow-downs, where CO_2 release could accumulate in low grounds or in enclosed areas and constitute a hazard if not attended to promptly. Chapter 12 deals with assessment of CO_2 dispersion in such situations and also considers the additional threat posed by H2S in the CO_2 stream. It provides elements that need to be integrated into any ERP in order to provide for effective response to emergency situations involving CO_2 pipelines.

10.2.1 Pipeline Control

As indicated previously for efficient pipeline transportation, supercritical CO_2 will be needed over the entire distance of the pipeline [7]. A typical pipeline control system transporting CO_2 containing impurities for the purpose of injection into a suitable geologic storage is schematically shown in Figure 10-3. It may by noted that the delivery could, on the other hand, for other commercial purposes and controls, include other sophisticated instrumentation for SCADA and safety recognition issues.

To assure a supercritical condition throughout the pipeline system, pipeline operation control features must be capable of directing signals to the compression/cooling and pumping system to adjust the pressures such that the CO_2 condition throughout the pipeline

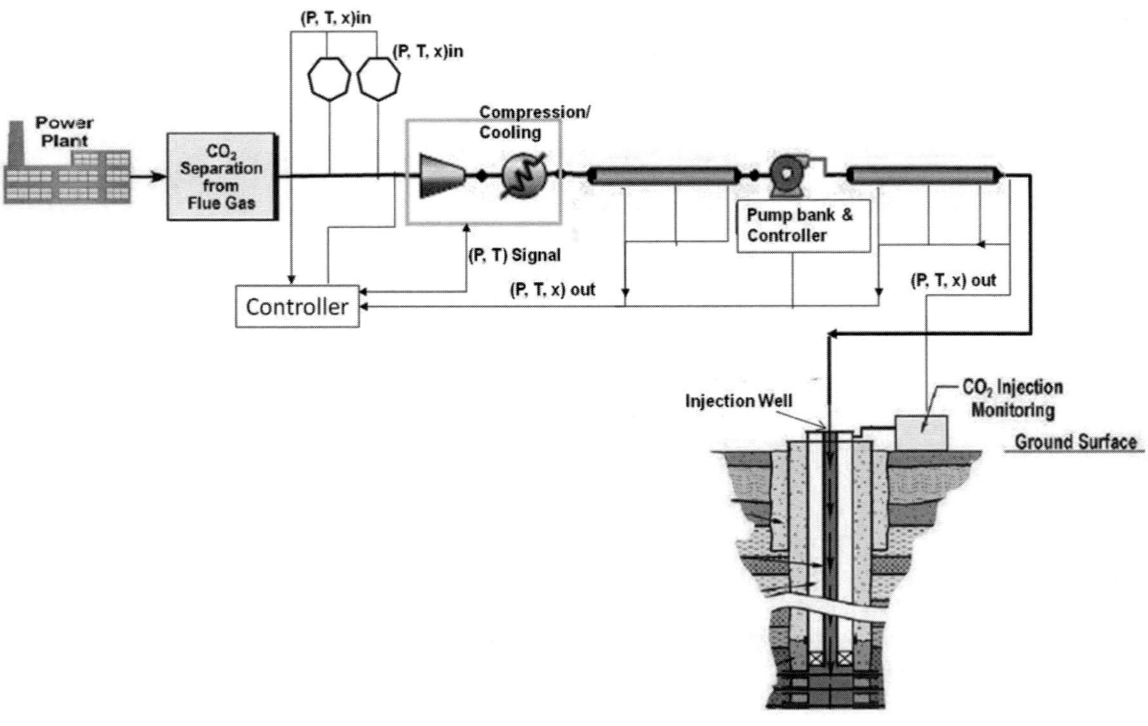

Figure 10-3. Typical CO_2 Pipeline and System Control Schematic

TABLE 10-1. Typical range of CO_2 stream from flue stacks after processing

Component	Example 1 (combustion in 97.5% O_2) oxyfuel	Example 2 (combustion in 95% O_2) oxyfuel	Example 3 (combustion in 92% O_2) oxyfuel	Example 4 Weyburn	Example 5	Example 6	Example 7
	% mole	% mole	% mole	% mole	% mole	% mole	% mole
Carbon dioxide, CO_2	92.9	88.88	89.67	96	90.84	89.67	95
Carbon monoxide, CO	0	0	0	0.1	0	0.14	0
Water, H_2O	0.16	0.15	0.01	0.01	0.01	0.01	0.01
Hydrogen, H_2	0	0	9.93	0	8.07	9.93	0
Nitrogen, N_2	0.58	0.89	0	0.03	0.58	0	0.5
Argon, Ar	4.54	8.35	0	0	0	0	0
Methane, C1	0	0	0	0.66	0.5	0.25	4.49
Ethane +, C_2+*	0	0	0	2.3	0	0	0
Hydrogen sulfide, H_2S	0	0	0	0.9	0	0	0.01
Oxygen, O_2	1.82	1.73	0.4	0.01	0	0	0

* Note: 50% C_2, 25% C_3 and 25% C_4.

and the downstream injection/delivery locations remains above the minimum pressure and temperature conditions that assure a supercritical condition based on generated phase diagram of the particular CO_2 stream [3].

Captured and processed carbon dioxide from flue stacks may contain many impurities/ chemical components. Table 10-1 provides typical examples of effluent gas compositions from power plants with oxyfuel processing capability.

10.2.2 Effect of Impurities on Pipeline Control

CO_2 streams captured from flue stacks are usually processed and have generally components as shown in Table 10-1. CO_2 streams from flue gases (after processing) contain mostly CO_2 with certain percentages of volatiles, nitrogen, oxygen, argon, hydrogen, and H_2S. However, in a CO_2 stream containing impurities, gases like nitrogen (N_2) or argon (Ar) that usually do not account for any safety or operational problems for the pipeline transportation or the final storage, are processed or removed as transportation is both energy and cost intensive [3]. They can affect mechanical design issues.

The level of components present greatly affects the phase behavior of a CO_2 stream containing impurities. However, the effect of impurities (binary and ternary mixtures) are similar. Typical influences of impurities on phase diagrams have been previously described in Chapter 4. To recap, Figure 10-4 below indicates the influence of CO_2 stream impurities for a ternary mixture [8]. The influence of binary CO_2 specification was previously shown in Chapter 4, Figure 4.5. Generally, the presence of impurities (CH_4, CO, N_2, H_2, H_2S, etc.) increases the critical pressure and decreases the critical temperature of the CO_2 mixture, so that high operating pressures may be required to keep the fluid in the dense phase during transportation. On the other hand, high level of NO_x or SO_x, while increasing the critical pressure, also increases the critical temperature.

Depending on the level, some impurities limit efficient transportation of CO_2, as high levels of such impurities can reduce available pipeline capacity, predominantly due to the propensity to reach the two-phase area during operation (because pressure and temperature drop within the pipe). For example, Figure 10-4 would imply a decrease in capacity due to an increase in hydrogen level impurities by as much as 13%. On the other hand, transport capacity might be reduced as impurities occupy space and influence the compressibility of CO_2 negatively [9].

Hydrogen has been found to have the maximum impact in limiting the pipeline capacity. Hydrogen, like argon, nitrogen, and oxygen, is considered a non-condensable gas that

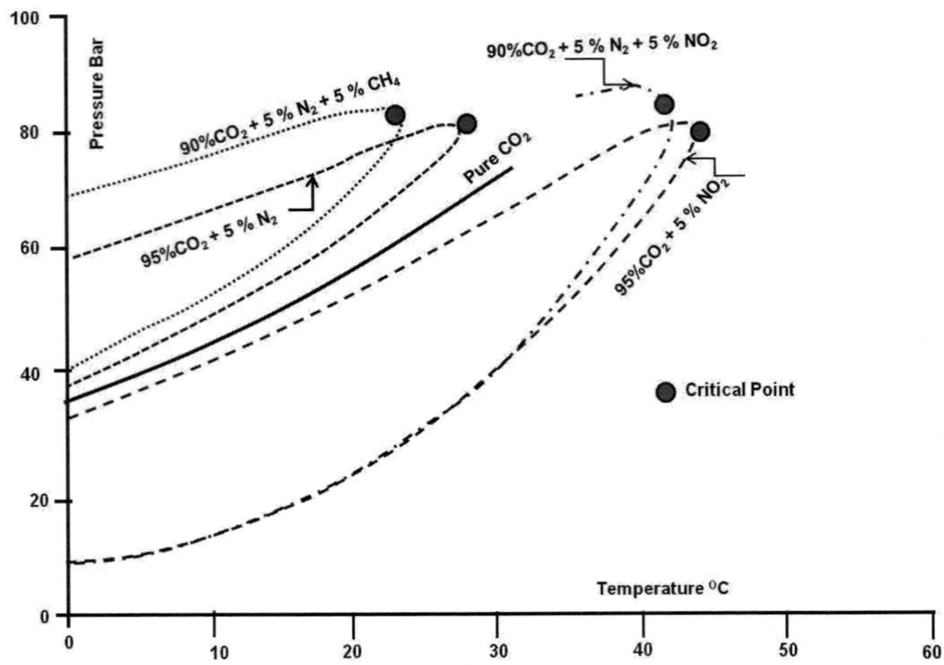

Figure 10-4. Influence of impurities on CO_2 phase diagram–ternary compositions (reproduced from WorleyParsons [8], EOS not reported)

has an elevated solubility with CO_2. Its presence greatly expands the phase envelope of CO_2-rich mixtures, thus increasing the pressure requirements for the transportation of CO_2. For this reason, impurities in general and hydrogen in particular, should be restricted as far as possible in the source streams.

The CO_2 gas composition will influence the required energy for compression and pumping. Compression requirements are affected mainly because higher inlet pressures are needed to avoid reaching the two-phase region during transmission.

It is sometimes more economical to transport CO_2 in the liquid phase to allow the operating pressure to be kept much lower. The presence of impurities can make this kind of transportation difficult to control.

10.2.3 Supercritical State Operation

Knowledge of the phase behavior for assessing the critical region of multi-component mixtures such as CO_2 containing impurities is of value both in pipeline transmission and for the optimum reservoir operation where CO_2 is injected [10].

Therefore, it is essential to determine the phase behavior of a CO_2 stream containing impurities in order to establish the limits of pipeline operation and reservoir injection processes to assure a successful long-term operation and safety.

To be able to control pipeline operation transporting a CO_2 mixture, the limits of phase changes for the mixture must be determined and two-phase region clearly defined. The limits in the case of gas–liquid phase changes are the bubble point and the dew point.

- The bubble point is the point at which the first drop of a liquid mixture begins to vaporize.
- The dew point is the point at which the first drop of a gaseous mixture begins to condense.

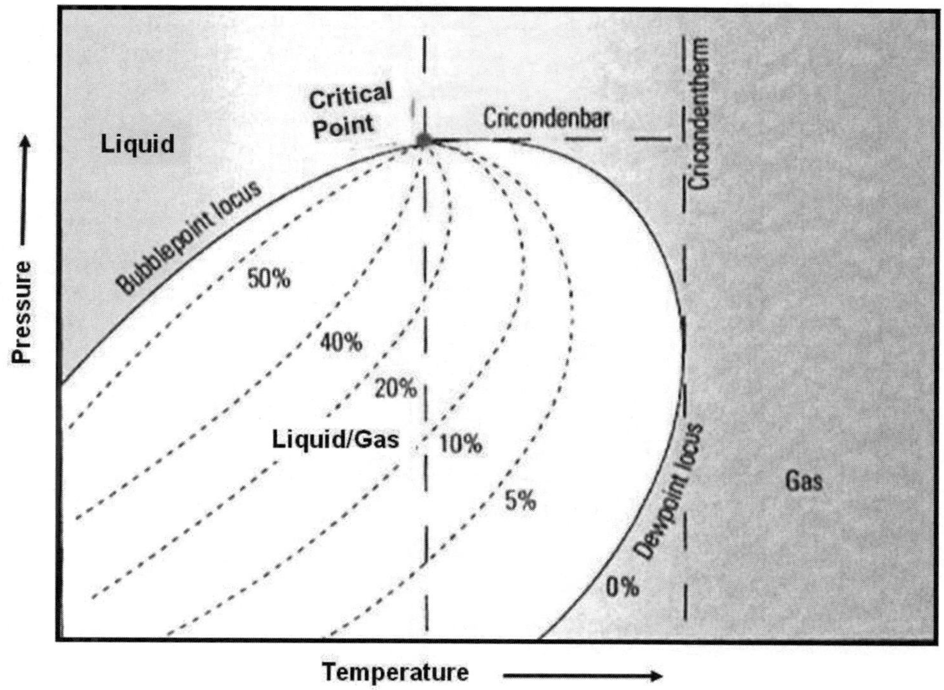

Figure 10-5. Phase diagram and demarcation regions

CO_2 stream two phase region is demarcated by dew point curve at the bottom and a bubble point curve at the top, as indicated in Figure 10-5. The loci of critical points, or the critical loci, lie on a line of higher pressure and lower temperature.

Additionally, the limits where two phases of gas and liquid mixtures can also coexist must be defined. These are *Cricondentherm* and *Cricondenbar:*

- Cricondentherm is the maximum temperature at which two phases (liquid and vapor) can coexist
- Cricondenbar is the maximum pressure at which two phases (liquid and vapor) can coexist

Carbon Dioxide Supercritical State: For a CO_2 stream containing impurities to be efficiently transported and injected into a subterranean reservoir, it is preferable for it to be in a supercritical phase so that the volume of the rock occupied by the CO_2 can be minimized. This state is considered to be most efficient and economical mode of transportation.

A supercritical fluid is any substance above its critical temperature and pressure. In the supercritical phase, the fluid will possess both gas- and liquid-like properties. It will have the density of a liquid and the properties of a gas to allow it to fill the maximum pore space available. For pipeline transportation, supercritical state can be defined as a top right quadrant area where vertical and horizontal lines cross at the critical point, see Figure 10-5.

Figure 10-6 (reproduced from Chapter 2) illustrates an example of critical phase in a pure CO_2 phase diagram and the temperature and pressure in a typical underground formation/reservoir. At a depth of approximately 2200 m, the reservoir pressure is 22 MPa and temperature is 100°C. Under these conditions CO_2 will have a density of 550 kg/m^3 (refer to Figure 5-6, Chapter 5) compared with fresh water with a density of 970 kg/m^3 and

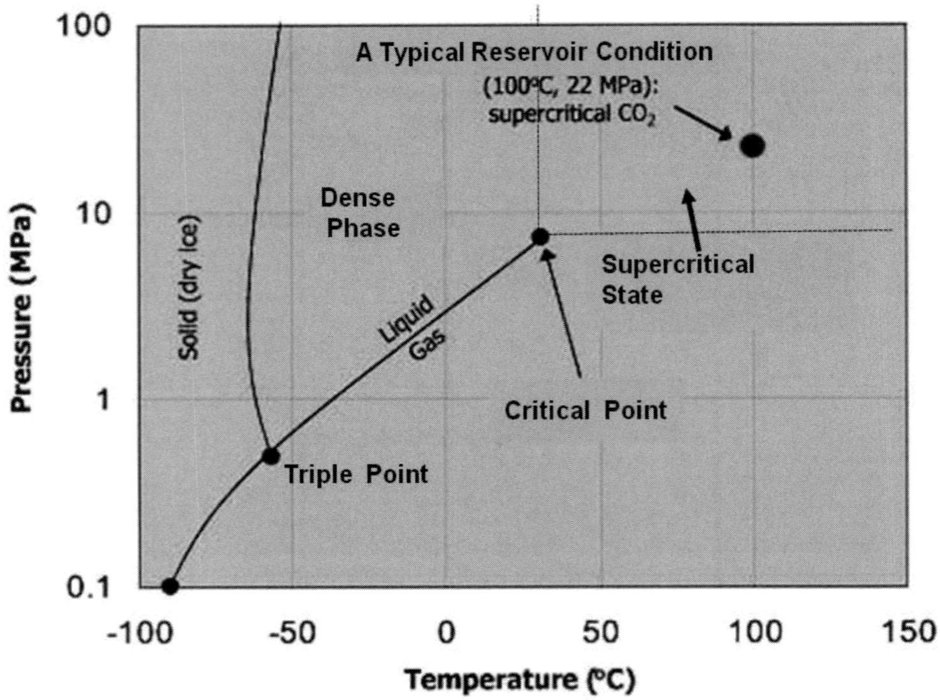

Figure 10-6. Supercritical state operation

normal ocean water with a density of about 1030 kg/m^3. However, it may be noted that as it is less dense than the waters that may already be contained in the subterranean formation, supercritical CO_2 will tend to rise vertically due to buoyancy forces.

10.2.4 Phase Diagrams and Pipeline Control

An essential part of pipeline control is the prediction and use of phase diagram to assure a super critical state (ex; Figure 10-6) during the transportation of CO_2 containing impurities. If the operating conditions are such that the temperature and pressure are higher than critical pressure and temperature, respectively, then CO_2 stream would be in a supercritical state. In the supercritical state and in the liquid state (above the vapor liquid line, Figure 10-6), the pressure drop in the pipeline is low. However, if the CO_2 stream exists in a state within the two phase region/boundary at any point along the pipeline, then pressure drop is considerably higher. This significantly affects the transportation/injection economies and may provide a situation unacceptable for pipeline design and delivery conditions if pipeline operational configuration is not changed.

10.2.4.1 Phase Diagrams Applications

Captured and processed CO_2 from flue stacks may contain many components impurities as previously indicated in Table 10-1. Also, as indicated previously, phase diagram for a CO_2 stream is very sensitive to the level of impurities. This in turn influences the effective boundaries where CO_2 pipelines can be operated without impacting the facilities design as well as delivery conditions. Therefore, it is important to utilize the appropriate phase diagram that predicts correctly the thermodynamic behavior and fluid properties commensurate with the CO_2 stream mixture impurities.

Selection of an appropriate phase diagram correlation would be critical in determining the following and the changes that may occur as consequence of operational issues effectively:

- Pipeline pressure drop conditions;
- Dynamic situations (such as valve closure, compression/pump trips);
- Pressure surge/line break controls;
- New CO_2 stream intake;
- Pipeline leak/failure.

Chapter 4 discusses the phase equilibrium calculations for a CO_2 stream containing impurities. Several equations of states have been referred to and treated with discussions. As indicated in Chapter 4, many existing CO_2 EOS are optimized for pure CO_2 but not for CO_2 stream mixtures (i.e., CO_2 stream containing impurities). An example is the use of BWRS EOS [11].

While the applicability of different EOS is explained in Chapter 4, the use of the following EOS will illustrate their applicability to CO_2 streams containing impurities:

a. Benedict–Web–Rubin–Starling (BWRS) EOS,
b. GERG model,
c. Peng–Robinson (PR),
d. VMG APRD EOS: — (Virtual Materials Group Advanced Peng–Robinson Dynamic EOS),
e. Virtual Model Group (VMG) Peng–Robinson EOS.

BWRS equation of state (EOS) (Eq. 10-1) has been in used over the last three decades for pressures up to 300 bar on a regular basis and up to 540 bar with CO_2+ Hydro carbon (HC) gas mixture in some specific cases and as well in the supercritical region

$$
\begin{aligned}
P = RTD &+ (BoRT - Ao - Co/T_2 + Do/T_3 - + Eo/T_4)D_2 \\
&+ (bRT - a - d/T)D_3 + \alpha(a + d/T)D_6 + c/T_2)(1 + \gamma D2)D3e\text{-}\gamma D2 \qquad (10-1)
\end{aligned}
$$

Experience has shown that at high pressure (above 480 bar) and temperatures, BWRS equation of state (EOS) requires careful verification of literature data and is not suitable for liquid–vapor equilibrium calculations [11]. This is further illustrated when determining the phase envelope for the example 1(Combustion in 97.5 % O_2) as listed in Table 10-1. The corresponding phase diagrams for the three equation of state (EOS) listed above (i.e., d-Virtual Model Group-Advanced Peng–Robinson Dynamic (VMG-APRD)EOS, e.-VGM (Virtual Model Group) Peng–Robinson EOS and a.-BWRS EOS) are compared as shown in Figure 10-7. Further, the results were also produced with GERG and Peng–Robinson EOS (equation b and c in above list), using REFPROP, Version 8.0 [12], Figure 10-8.

From Figures 10-7 and 10-8, it is evident that BWRS equation of state (EOS) would not result in any meaningful phase envelope for CO_2 mixture containing impurities.

From Figure 107, it may be noted that the phase diagrams predicted by the two correlations "VMG-Advanced Peng–Robinson Dynamic (VMG-APRD)" EOS and "VMG Peng–Robinson" EOS provide similar results. However, it is reported that VMG-APRD is very good in calculating liquid densities and also calculations of data close to critical point and the dynamic technique implemented makes it run significantly faster [13]. GERG and Peng–Robinson EOS also produced similar results.

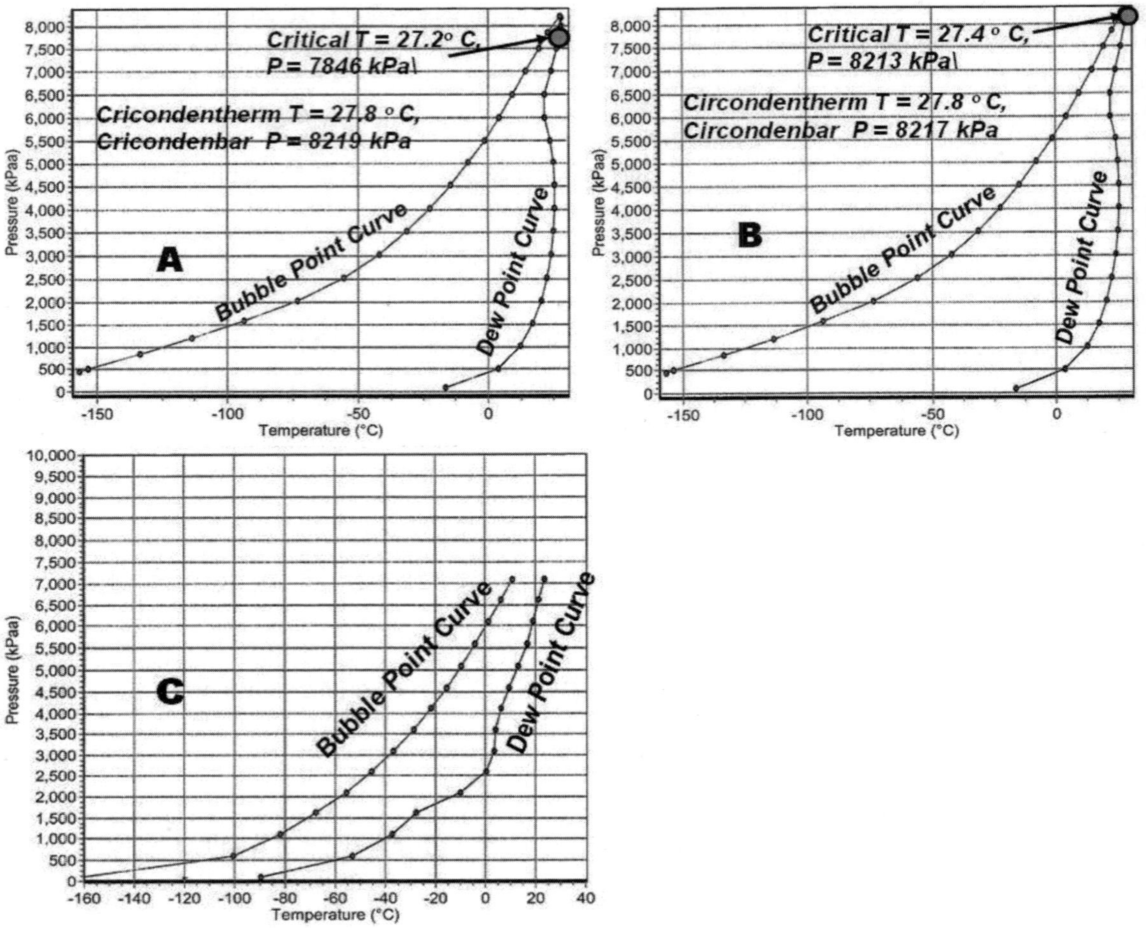

Figure 10-7. Comparison of phase diagrams generated by A: VGM Peng_Robinson Advanced, Detailed, B: VGM –Peng Robinson as C: BWRS: refer to Example 1, Table 11-1 for CO_2 mixture composition (courtesy: Golshan H, & Dueck, A., TransCanada PipeLines Limited, 2011)

10.2.4.2 CO_2 Stream Composition Implications

It may be noted from Figures 10-7 and 10-8 that the applicability of EOS for a particular mixture must be fully investigated to ascertain operational limits/boundaries of a pipeline system with confidence. If CO_2 mixture compositional changes are envisioned, then such boundaries that would affect their implication on design and operational conditions must be investigated. Even with EOS's producing similar phase boundaries/results, there is no guarantee that operational limitations are unaffected unless the EOS results are backed up with experimental data.

Figures 10-9 through 10-14 provide phase diagrams for examples 2 through 7 cited in Table 10-1.

Table 10-2. (based on "VMG-APRD EOS") provides a summary of critical temperatures and pressures for CO_2 stream examples listed in Table 10-1. VGM-APRD EOS correlation provides slightly higher critical point pressures and temperatures which is more pertinent to pipeline design and operation under supercritical state.

Referring to Table 10-1 and Table 10-2, it is evident that operating range of a pipeline under supercritical conditions, very much depends on the impurities contained in the CO_2 stream from flue gases. Therefore, conditions throughout the pipeline must be monitored to assure effective pipeline and product delivery controls.

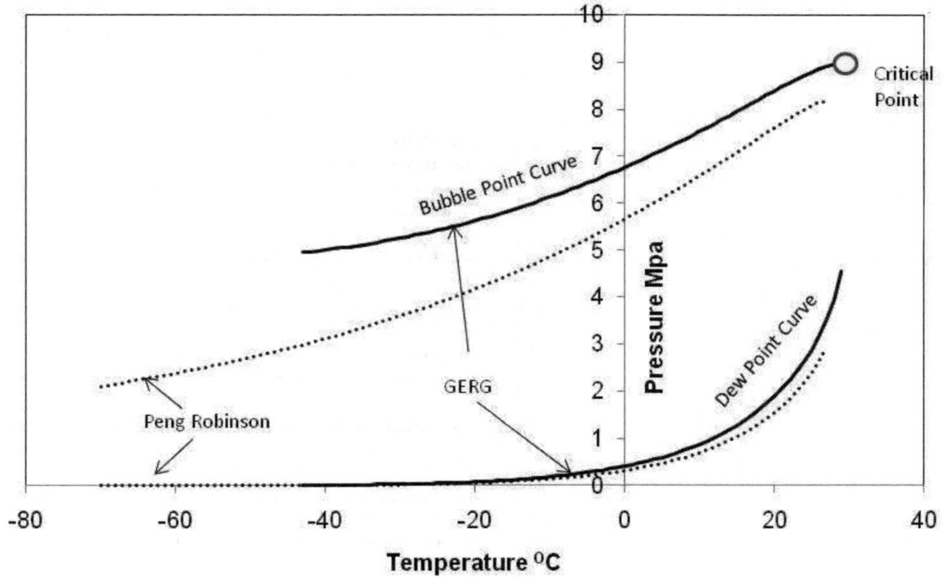

Figure 10-8. Comparison of phase diagrams generated by REFPROP, Version 8.0, GERG and Peng–Robinson EOS (refer to Example 1, Table 11-1 for CO_2 mixture composition)

From Figure 10-14 (representing phase diagrams from 3 equation of state, EOS) and comparing the phase envelopes with Figures 10-7 to 10-13), it is conclusively evident that impurities present in CO_2 stream mixture significantly impact the calculation of thermodynamic properties of the fluid and hence its design and operating conditions of the pipeline.

10.2.4.3 Operational Control Through Pipeline Hydraulics

An effective operational control of a pipeline transporting CO_2 containing impurities is illustrated through steady state and dynamic analyses of a pipeline system from a flue gas supply location (with processing compression facilities), stretched on mountainous terrain conditions, representative of typical CO_2 supply–delivery chain for injection into a reservoir for enhanced oil recovery purpose.

The following data are applicable;

Terrain profile	Refer to Figure 10-15
Pipeline length	370 km
Pipeline diameter	NPS 16 (406.4 mm OD)
Pipeline grade	API 5L X70
Pipe wall thickness	9.6 mm
Pipe roughness	0.01905 mm
Ground temperature	5°C
Inlet pressure	101.325 kPa
Inlet temperature	10°C
Inlet pressure to pipeline after compression	14500 kPa
Design pressure	15306 kPa (Limit of ASME ANSI 16.5, 1500# Flange at 38°C)
Maximum operating pressure	14790 kPa
Inlet temperature	10°C
Maximum operating temperature	50°C (coating limit)
Depth of cover (DOC)	1 m

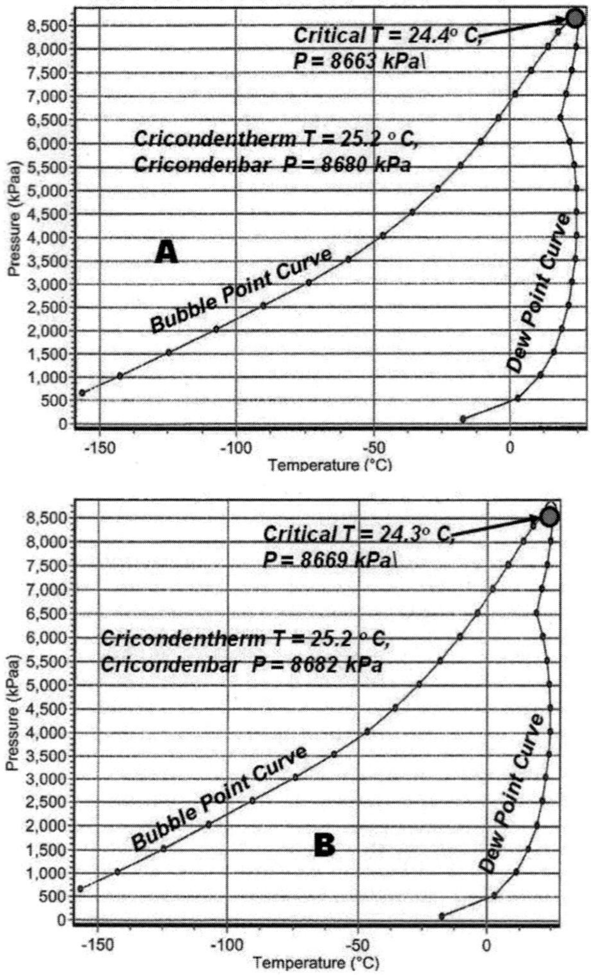

Figure 10-9. Phase diagrams for Example 2, Table 11-1 (A: VGM Peng–Robinson and B: VMG-Advanced Peng–Robinson Dynamic (VMG-APRD)) (courtesy: Golshan and Dueck [14], TransCanada PipeLines Limited, 2011)

Over all heat transfer film coefficient	1.1356 W/m^2 OK
CO_2 stream composition	As per Table 10-1, Example 4, Figure 10-11 (Weyburn), critical temperature and pressure 27.9°C and 7128 kPa
Minimum operating/delivery pressure	8958 kPa (Injection Pressure Requirement)
Standard pressure and temperature	101.325 kPa and 15°C

Salient pipeline configuration/features are shown in Figure 10-16. Gas simulator (TG-NET) of ESI PipelineStudio version 3.2.5.6 was used for the steady state and dynamic simulations [15]. Peng–Rubinson78 was chosen as the equation of state and Colebrook–White for pressure drop calculation [16].

Based on pipeline data, steady state hydraulics and other CO_2 flow condition profiles are provided in Figures 10-17 and 10-18.

From Figures 10-17 and 10-18, average values for density, viscosity, and compressibility were deduced for average line pressure and line temperature conditions and were then

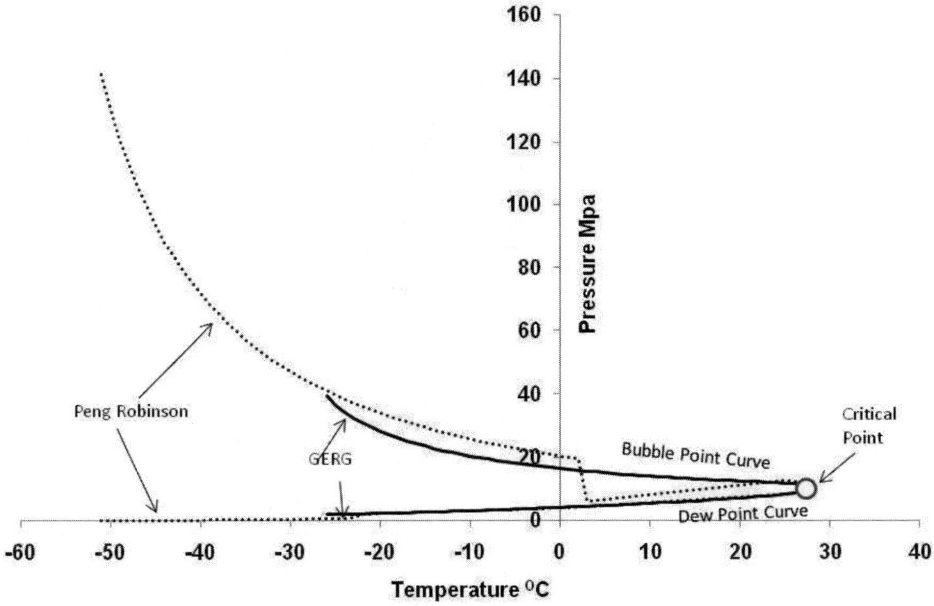

Figure 10-10. Phase diagrams for Example 3, Table 11-1 (generated by REFPROP, Version 8.0 GERG and Peng–Robinson EOS (refer to Example 3, Table 11-1 for CO_2 mixture composition)

compared with values of pure CO_2 at the corresponding pressure and temperature, Table 10-3. For pure CO_2 values, refer to Chapter 5, Figure 5-4, 5-6 and 5-7.

It may be noted that while density and compressibility compare well for pure CO_2 and CO_2 containing impurities, predicted viscosity is higher for CO_2 mixture.

In order to assure that the Gas Simulator, TGNET provided reasonable results, the pipeline was simulated for transportation of pure CO_2, refer to Table 10-3 for comparison of results. Row 2 of Table 10-3 provides values for density, viscosity and compressibility from graphs of Chapter 5, Figures 5-4, 5-6 and 5-7. Row 3 of Table 10-3 provides simulated results from Gas Simulator TGNET for the same pipeline but transporting pure CO_2. Only initial conditions (i.e., the values of density and viscosity at the entry to the pipeline) are included.

The calculated viscosity in PipelineStudio Gas Simulator, uses the empirical correlation of Lee, Gonzalez and Eakin (LGE) [15].

The standard LGE viscosity correlation gives the dynamic viscosity as a function of fluid molecular mass, density, and absolute temperature.

For CO_2 in supercritical state, the viscosity increases with increasing temperature and pressure Figure 10-19 (see also Chapter 1, Figure 1-34). The industry usually uses the correlation devised by Carr, Kobayashi, and Burrows [17], which include corrections for H_2S, CO_2, and N_2. But the Standard Lee, Gonzalez, and Eakin (LGE) formulation does not account for these impurities (H_2S, CO_2, and N_2), therefore, it is likely higher viscosities are predicted here [18].

However, ESI, Gas Simulator TGET, provides two alternate sets of parameters for the LGE viscosity correlation (LGE-3 and LGE-4)both suitable for high pressure gasses.

The correlation that best predicts prediction of viscosity of CO_2 containing impurities is the modified LGE indicated as Eq. 10-2 below and is referred to as LGE-4 in Pipeline Studio Gas simulator

$$\mu = 10^{-4}\left(\frac{(-3.837+0.485M)T^{3/2}}{-1047.0+T+59.08M}\right)e^{K} \qquad (10-2)$$

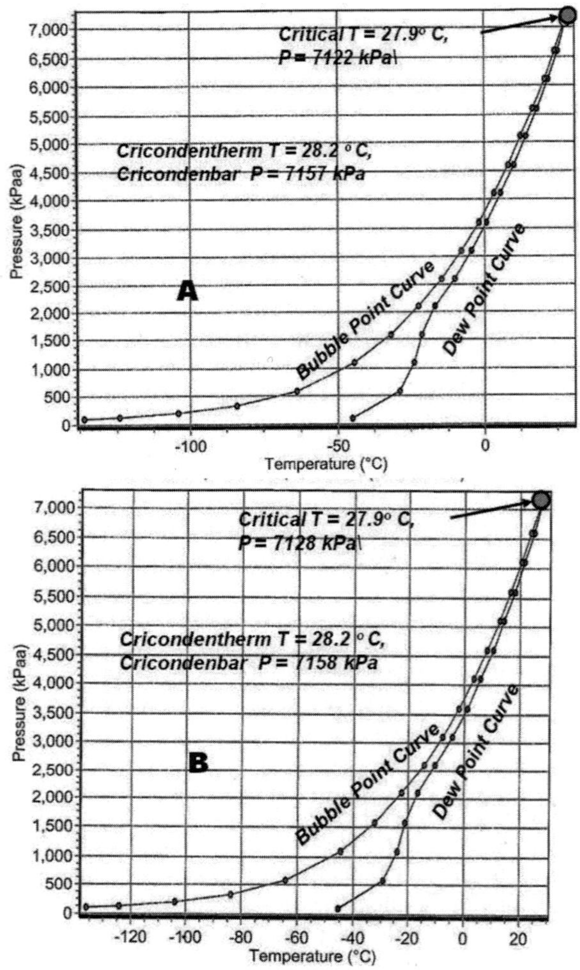

Figure 10-11. Phase diagrams for Example 4, Table 11-1 (A: VGM Peng–Robinson and B: VMG-Advanced Peng–Robinson Dynamic (VMG-APRD)) (courtesy: Golshan and Dueck [14], TransCanada PipeLines Limited, 2011)

Where:

$$K = A\rho^{B}$$

and

$$A = \left(-5.576 + \frac{6467}{T} - 0.05M\right)$$

$$B = 0.7491 + 0.090\left(-5.576 + \frac{6467}{T} - 0.05M\right)$$

Where:

T is the gas temperature in degrees R
M is the gas molecular mass in kg/kmol
ρ is the gas density in g/cm^{-3}
μ is the gas viscosity in centipoises (cP)

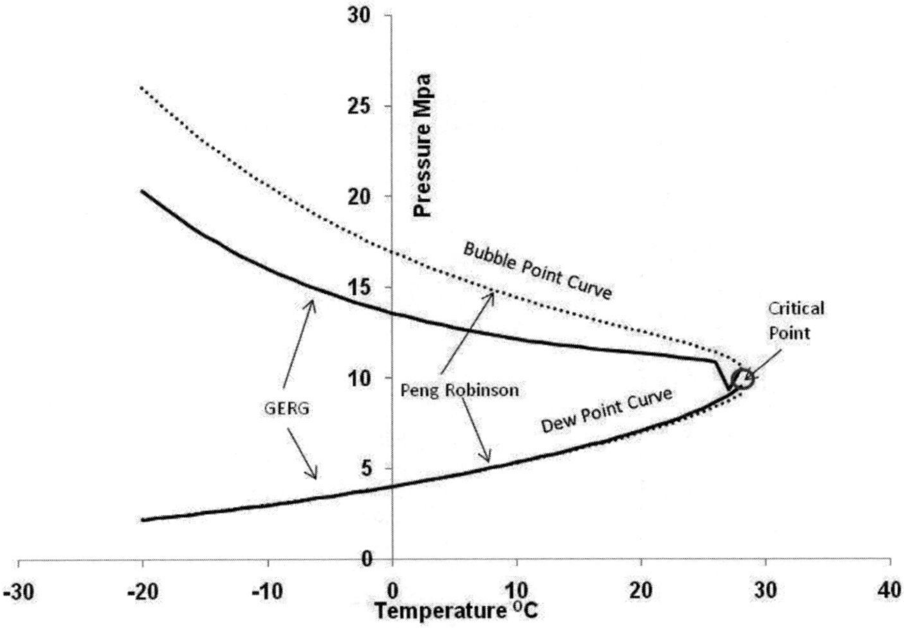

Figure 10-12. Phase diagrams for Example 5, Table 11.1 (generated by REFPROP, Version 8.0, GERG and Peng–Robinson EOS)

As indicated in Figure 10-19, existing data on pure CO_2 viscosity however confirms high values at higher pressures and temperature [19, 20]. Comparison of PipelineStudioGas Simulator output for pure CO_2 indicated LGE4 to be the most relevant for simulation of high-pressure dense phase CO_2 containing impurities.

It may be noted that gas viscosity is a measure of the resistance the gas has to flow. Higher values indicate more resistance to flow as shown in Figures 10-17 and 10-18. For the simulated pipeline, nowhere along the pipeline does the pressure fall below bubble

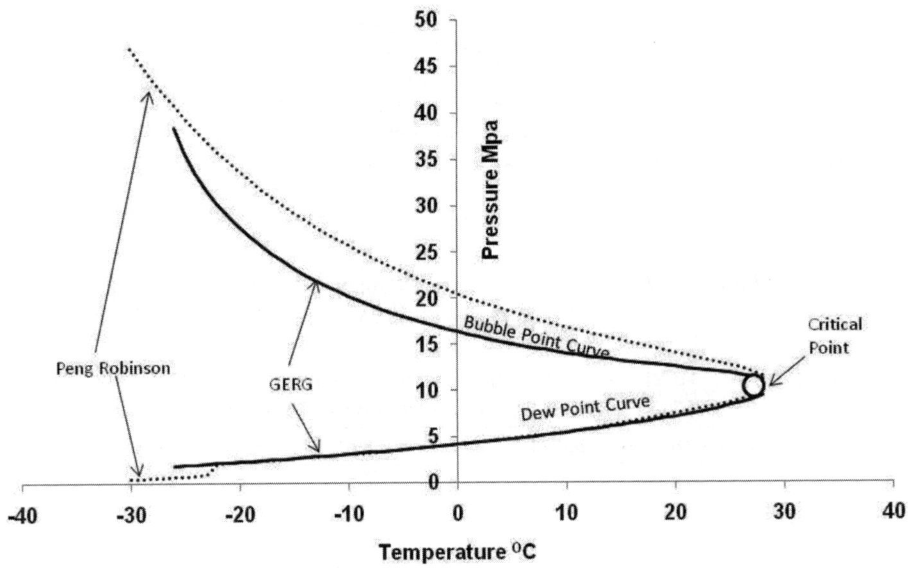

Figure 10-13. Phase diagrams for Example 6, Table 11-1 (VGM Peng–Robinson) (courtesy: Golshan and Dueck [14], TransCanada PipeLines Limited, 2011)

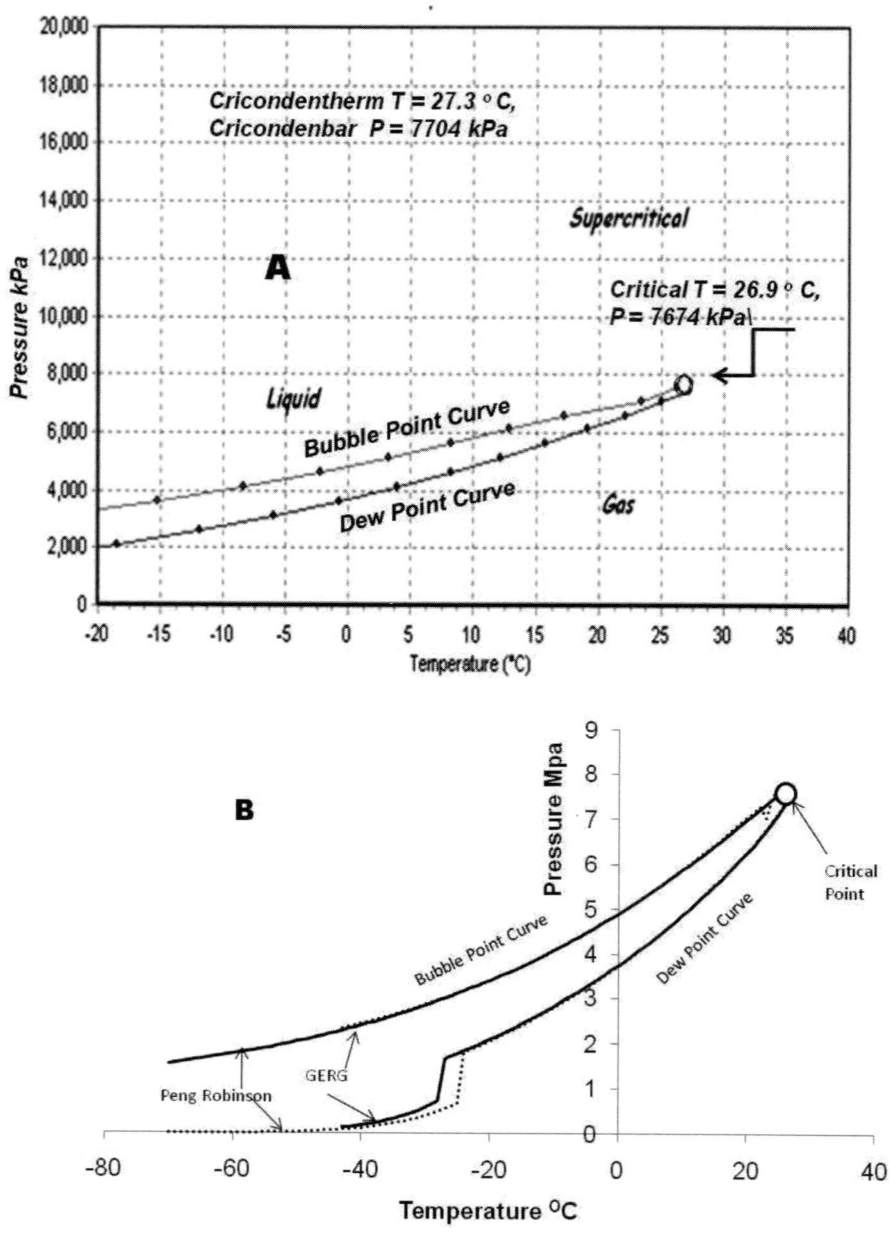

Figure 10-14. Phase diagrams for Example 7, Table 11-1 (A — VGM Peng–Robinson, and B — generated by REFPROP, Version 8.0 GERG and Peng–Robinson EOS)

TABLE 10-2. Variations of critical temperatures and pressures for various CO^2 streams containing impurities (refer to Table 10-1 for impurities)

Example (refer to Table 11-1)	Critical temperature, °C	Critical pressure, kPa
1	27.2	7846
2	24.4	8663
3	25.4	9695
4	27.9	7422
5	20.4	11804
6	20.3	12896
7	26.9	7674

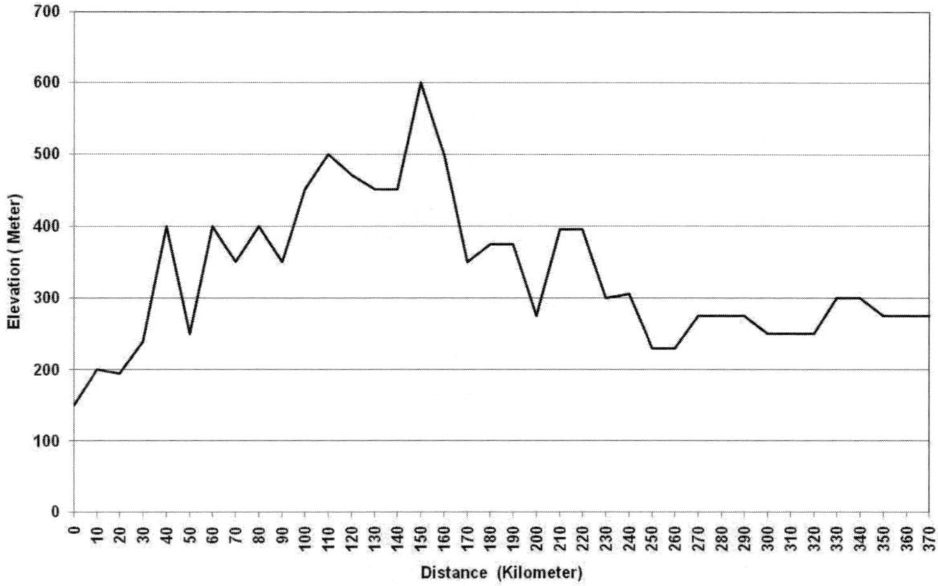

Figure 10-15. Pipeline profile

point. It may may noted that the flow is limited to 5026 km^3/day. Higher volumes or lower delivery pressures would cause the pipeline at some locations to dip into the two phase region particularly at higher elevation points. Therefore, it is significant to control the pipeline through installation of check and control valves at strategic locations as well as pressure and temperature transmitters to monitor and assure supercritical flow along the pipeline. As indicated previously, pipeline control features (Figure 10-3) must be capable of directing pressure and temperatures signals (via RTU through an appropriate SCADA) to the compression/cooling and pumping system (via central control room) to adjust the pressures such that the CO_2 condition throughout the pipeline remains in supercritical condition under normal and emergency situations.

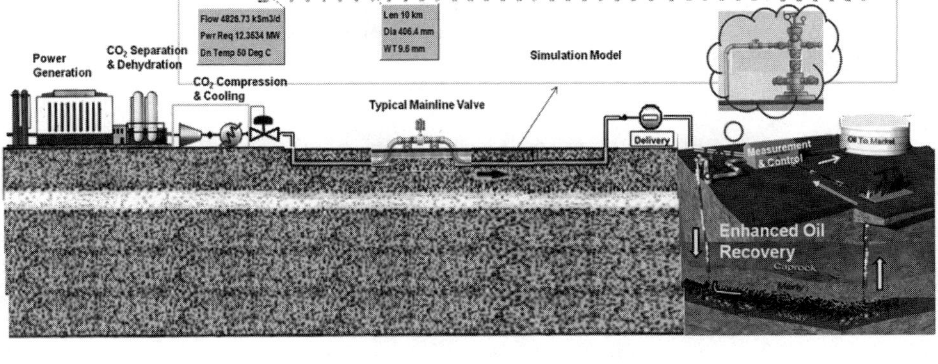

Figure 10-16. Typical CO_2 pipeline for EOR and a TGNET Simulation Model with typical hydraulics/flow results

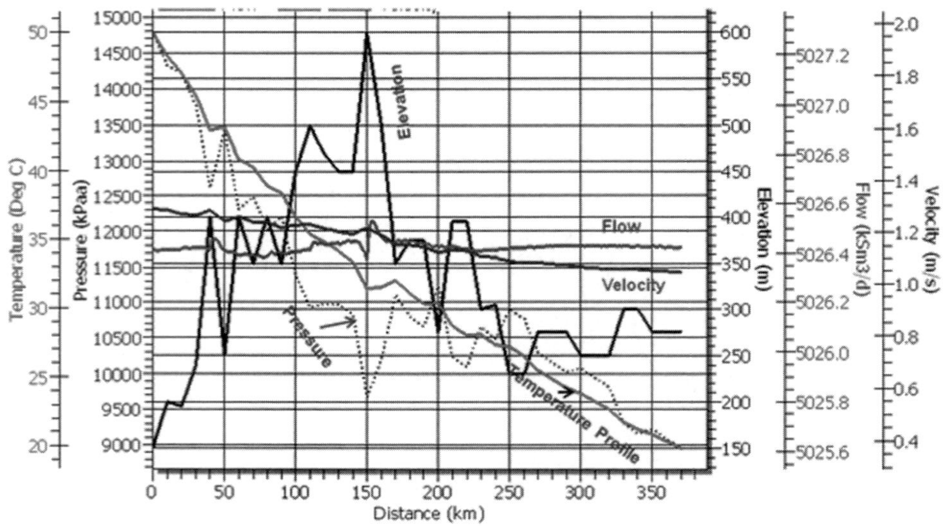

Figure 10-17. Hydraulics profile, CO_2 containing impurities (Example 4, Table 10-1)

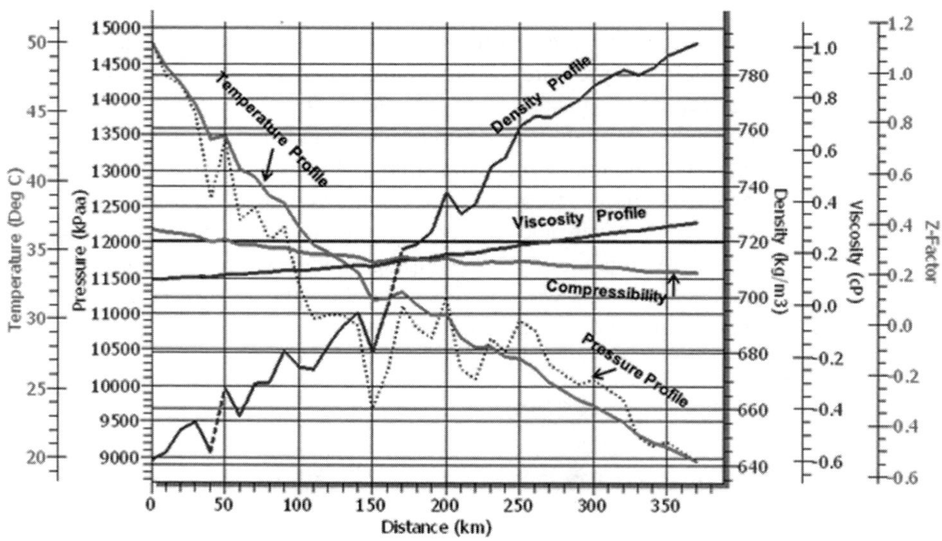

Figure 10-18. Flow condition profiles for CO_2 containing impurities (Example 4, Table 10-1)

TABLE 10-3. Comparison of simulated results and third party data

	Pressure, kPa	Temperature, °C	Density, kg/m^3	Viscosity, cP	Compressibility factor
Example 4 average	10996.13	31.65	716.75	0.189492964	0.265
Pure CO_2	11000	32	800	0.07	0.24
Pure CO_2 simulated initial	14790 (at entry to pipeline)	50	650.377	0.096	0.37249

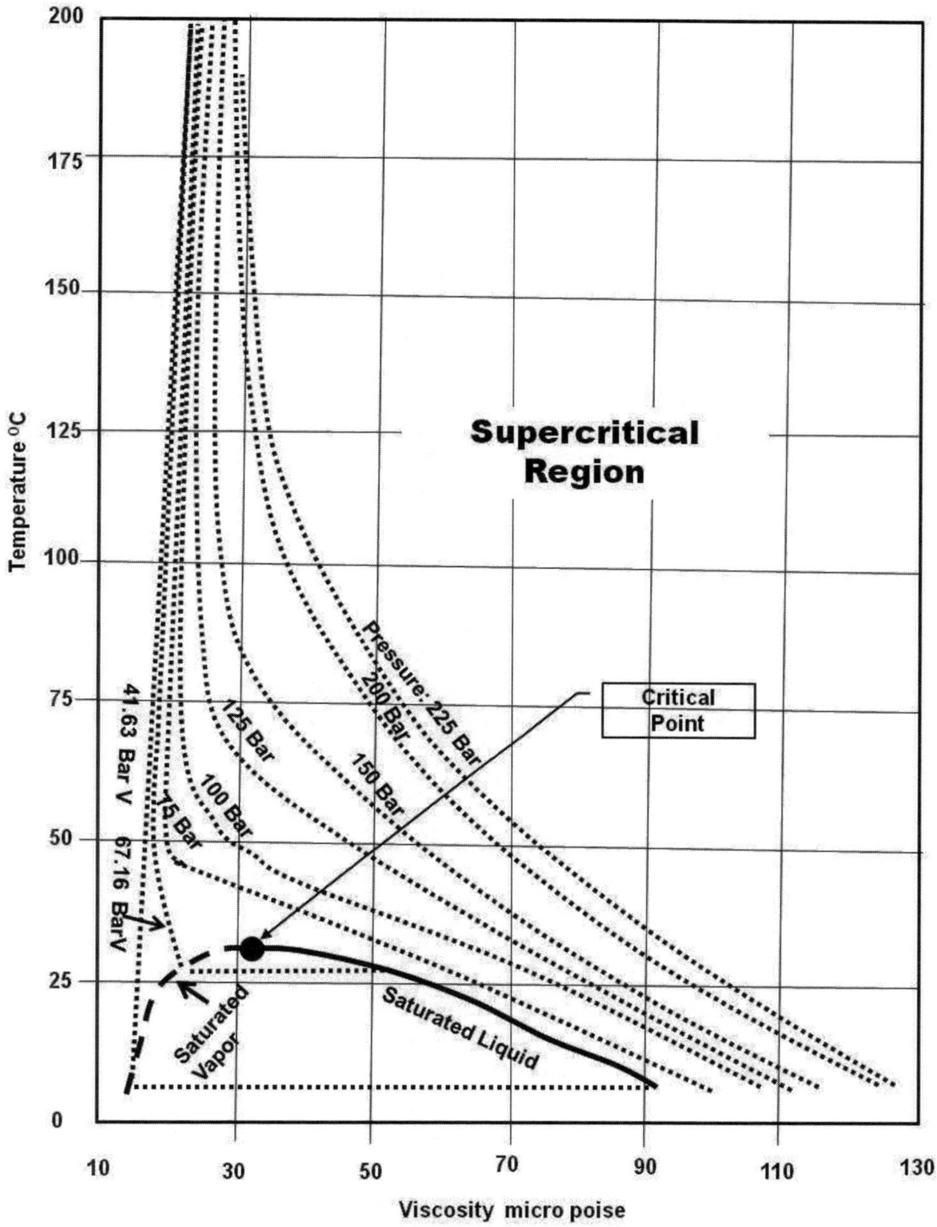

Figure 10-19. Variation of viscosity of pure CO_2 as a function of pressure and temperature for typical operating range of pipelines [19, 21, 22]

10.3 SALIENT OPERATIONAL AND MAINTENANCE REQUIREMENTS

Salient features that assures long term integrity of a CO_2 mixture pipeline transmission system is listed below [16]:

- Keep impurities in the CO_2 stream as low as possible through processing prior to admission into the pipeline. The presence of impurities, such as methane (CH_4), nitrogen (N_2), hydrogen (H_2), hydrogen sulfide (H_2S), Argon (Ar) affect the phase diagram, increases the critical pressure and decreases the critical temperature, so that high-operating pressures may be required to keep the fluid in the super critical dense phase during transportation.

- Keep pipeline pressure at all location through the pipeline length above the supercritical phase. CO_2 is handled as a dense phase (supercritical) liquid in pipeline operation with pressures between 8274 kPa (1200 psi) and 17500 kPa (2500 psi) and temperatures below 30°C (80°F).
- Excessive water content in CO_2 can cause formation of highly corrosive carbonic acid. Keep moisture and water levels of between 18 and 30 lb/MMSCF (288-480 mg/m3).
- Concentrated CO_2 causes deterioration and explosive decompression in petroleum based sealing materials (O-rings, seals, valve seats, etc.) when they are taken from line pressure down to atmospheric pressure.
- Petroleum based greases and many synthetic greases (for use in valves) are deteriorated by CO_2 stream such that they become hardened and ineffective. Use specially selected inorganic sealing materials.
- Blow-downs and pipeline loading must be controlled over significantly longer times than in normal natural gas pipeline procedures, to prevent excessively low temperature gradients.
- Due to the sensitivity of CO_2 properties (such as compressibility and density) to temperature, ambient temperatures can have a significant effect on the operation of a CO_2 pipeline (particularly for above ground facilities).
- Air or nitrogen leak tests do not confirm a system will be leak tight when commissioned with CO_2. All mechanical joints, seals, packings, etc., must be carefully designed, installed, and tightened to prevent leaking of CO_2.
- Pipeline pigging in CO_2 service is difficult without the aid of a precursor lubricant, such as diesel, because dry CO_2 has very poor lubricating characteristics.
- CO_2 pipeline operation may cause transient situations (which can cause commonly referred to as "slinky effect" a variation on the common "water hammer" experienced in liquid pipelines). Therefore, avoid rapid valve closures, rapid pump shut down, and other abnormal operational conditions that can cause general high-pressure surges affecting pipeline system integrity.
- Slow opening of all blow-down valves is recommended.
- For maintenance purposes, pressurized piping should not be worked on. Pipeline segments and valve bodies must be depressurized prior to any work including removal of valves (as required). It is not recommended to shut two valves on a short section/run (especially exposed piping) that have not any thermal relief; as high ambient temperatures can cause over-pressuring of the pipe.
- Existing CO_2 pipeline facilities have been designed to meet current gas and/or oil pipeline systems codes and standards. No specialty fluid transmission code is applicable or available.
- Pipeline transport of CO_2 through populated areas will require special attention to design factors, overpressure protection, and leak detection. However, there is no indication that potential problems for CO_2 pipelines are any more challenging than hydrocarbon pipelines, or that they cannot be resolved [23].
- The two primary safety considerations for CO_2 facilities are:
  (1) to avoid suffocation in areas where CO_2 may be blown down, leaking, and displacing oxygen (especially in enclosed or low areas); and (2) to exercise extreme

caution when operating or maintaining high pressure CO_2 facilities due to the compressibility and potentially violent expansion (150 to 1) of CO_2 as it changes phase (refer to Figure 5-2, Chapter 5).

10.4 CO_2 PIPELINE MAINTENANCE

The most economical means of transporting CO_2 is in the supercritical state because of its low viscosity and high density. However, a typical large diameter pipeline can have a very large inventory (e.g., 100 km of NPS 32 can have an inventory of approximately 9000 tons of CO_2).

In order to assure continued successful business operation, maintenance activities must be performed on the pipeline on a routine basis. There are also activities that concern emergency situations such as the case of pipeline rupture due to unforeseen circumstances of environmental and geotechnical activities, third party damage, etc.

Maintenance activities can therefore be divided in three groups:

- Daily maintenance,
- Routine maintenance, and
- Emergency response activities.

However, it may be noted that there are situations that overlaps those of routine maintenance activities that can lead to an emergency maintenance condition.

10.4.1 Daily Operational Maintenance

Daily operations include daily maintenance, scheduled planning, and communications, policies for inspecting, maintaining and repairing all equipment including the pipe itself. This equipment and support includes valves, compressors, pumps, tanks, rights of way, public signs and line markers as well as periodic pipeline flyovers.

Long-distance pipelines are instrumented at various locations for flow, pressure monitoring purposes. The monitoring points, compressor stations and block valves are tied back to a central pipeline operations control center. Computers control much of the operation, and manual intervention is necessary only in unusual upsets or emergency conditions. CO_2 pipeline systems generally have inbuilt redundancies to prevent loss of operational capability for unforeseen component failures.

10.4.2 Routine Maintenance

Routine maintenance are those planned activities that do not involve an incident, so these are scheduled as per equipment manufacturer's requirement or planned in a timely manner to allow for specific studies to be done to assure hazard control. Routine Maintenance Activities include the following:

- Environmental protection,
- Right of way (ROW) and site maintenance,
- Depth of cover,
- Air patrol and leak detection,
- Erosion control including inspection and review of ROW stabilization system such as diversion berms,

- Cathodic protection maintenance,
- Pipeline integrity assessment,
- Pipeline repair and modifications,
- Pipeline encroachment assessment and requirement for reclassification (if applicable),
- Facilities identification.

There are also maintenance activities that are not usually routine in nature, but are performed as a consequence of undertaking the routine maintenance ,such as excavation to assess external corrosion damage. These include the following:

- Trenching and excavation,
- Locating pipelines,
- Pipeline realignment/relocation,
- Foreign crossing,
- Hot tapping and stub installation,
- Welding including
  - Welders qualification
  - Maintenance and repair welding, (including welding on pressurized pipeline),
  - Cutting
- Safety management Including;
  - Safe working practices
  - Safety gearing
  - Safe work permit
  - Confined space entry
  - Fire protection
  - Material and hazardous material handling and disposal
  - Safe equipment/vehicle/tool operation
  - Personal protective facilities
  - Incidence recording & reporting
- Record keeping
- Training

10.4.3 Emergency Maintenance

These are activities that usually result from a pipeline failure such as a rupture or a leak, due to unforeseen events such as geo-technically induced pipe movement or a third party damage (Figure 10-20) and often involve replacing a section of pipe or installing a repair

Figure 10-20. Some causes of CO_2 pipeline failure (A: Geo-hazard, B:Third party Damage)

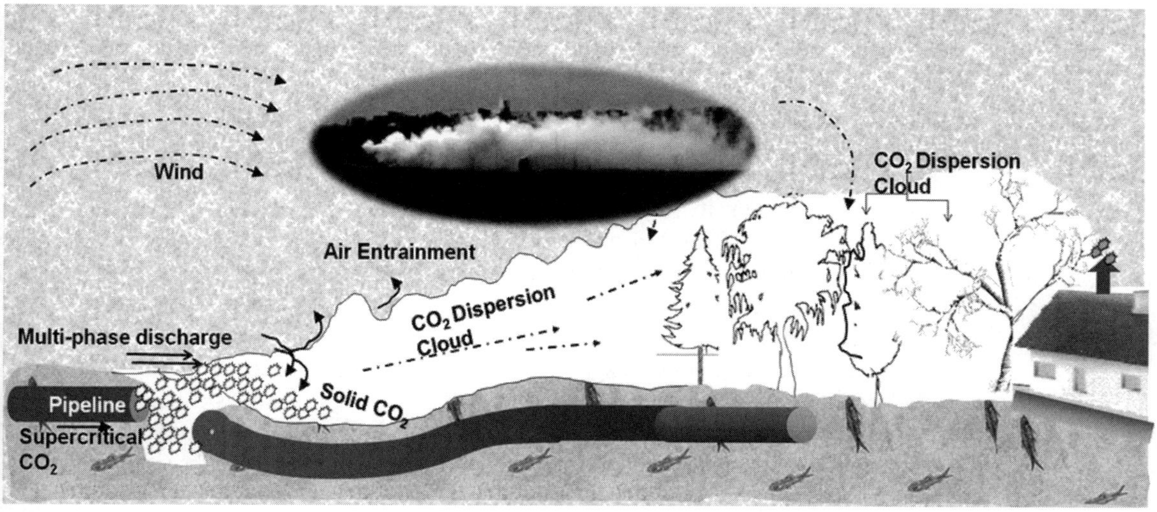

Figure 10-21. Significance of CO_2 release in a pipeline subjected to full bore rupture (Inset: typical Blow-down release)

sleeve. It may be noted that CO_2 pipelines in general have experienced few serious accidents that have resulted in emergency situations. For example, according to US DOT Office of Pipeline Safety (OPS) statistics, there were 12 leaks from CO_2 pipelines reported from 1986 through 2006, none resulting in injuries to people.

However, in the event of a pipeline failure, for example a full bore rupture, based on the hyperbolic discharge behavior synonymous with such failures, a significant proportion of the inventory would be discharged in a very short time (a matter of a few minutes). At a concentration of 10%, an exposed individual would lapse into unconsciousness in 1 minute. Furthermore, if the concentration is 20% or more, the gas is instantaneously fatal, refer to Chapter 13, Table 13-2. The ability of CO_2 to collect in depressions in the land, in basements, and in other low-lying areas, such as valleys near the pipeline route (Figure 10-21) presents a significant hazard if leaks continue undetected. Also, impurities, such as hydrogen sulfide (H_2S) in a CO_2 stream which is potentially toxic (if in high concentration), whose natural dispersion can be extended over a large distance, may cause serious repercussion. Also, impurities in a CO_2 stream, such as hydrogen sulfide (H_2S) which is potentially toxic (if in high concentration), whose natural dispersion can extend over a large distance, may cause serious repercussions. Dispersion of such toxic components might be impeded by the dense CO_2 vapor layer close to the ground, further increasing hazards.

10.4.4 Maintenance Activities and Schedule

Generally, pipeline regulations or acts as well as codes will require the pipeline operators to maintain their pipeline to the prescribed standards. These inspections are recommended in order to reduce potential third-party interference on such pipelines. The codes generalize the requirement for routine maintenance which cover the inspection of all pipeline rights-of-way and facilities, etc. Furthermore, industry's practice may entail augmenting such a requirement to assure continued business operation and indeed public safety and minimum operational impact on the environment.

Maintenance activities for CO_2 pipelines generally fall into two categories, monitoring (including patrols), and preventive/emergency maintenance. An additional activity that can be considered maintenance is locating the pipeline for a third party for excavation purposes on the right-of-way or crossing the pipeline.

Listed below are typical scheduled maintenance tasks that are generally performed on CO_2 pipelines by the industry per minimum code requirement or by experience with their typical frequency:

Maintenance Task	Frequency
• Aerial patrols/leak detection/monitoring survey	26 times/year
• Population density survey,	Once/two years
• Right of way (ROW) inspection,	26 times/calendar year
• Valve maintenance (for repacking, seat inspection, etc.) and inspection,	Twice/calendar year
• Relief valve inspection	<5 years
• Emergency systems check	Once per year
• Rectifier maintenance,	6 times/calendar year
• Cathodic protection survey and coupon inspection,	Once/calendar year
• Internal corrosion monitoring	<6 months (if pigged)
• Internal inspection of the pipeline using an electronic tool,	every five years or more frequently if necessary
• Overpressure protection devices,	Once/calendar year
• Remote control/shutdown devices	Once/calendar year
• Instrumentation inspection and recalibration	As required
• Public awareness and damage prevention program	Once/calendar year

Others include routine maintenance tied up with operation that covers the followings:

- Compression/pump unit selection;
- Station bypass;
- Station bypass for a pig run;
- Welding;
- Tightline operations;
- Diurnal delivery fluctuations;
- Drain-up/blow-down;
- Shutdown; and
- Operating pressures for a flowing pipeline.

10.4.5 Maintenance Planning

Good planning is essential to all pipeline maintenance and repair work. It ensures that the work is done safely, on schedule and with minimum environmental impact and/or safety issues.

Scheduling and preparation are important for timely maintenance and safe work operation. This covers the following:

- Maintenance schedule preparation;
- Availability of tool and equipment in good and safe working condition;
- Availability of necessary resources for the job;
- Previous on the job training of personnel to assure work safety and efficiency;
- Availability of safety equipment (as necessary), all in a good working condition;
- Necessary environmental protection measures;
- Permit to work.

10.5 CO_2 PIPELINE LEAK DETECTION/BLOWDOWN

Monitoring any leak from a CO_2 pipeline (particularly transporting CO_2 containing impurities) is an integral component of pipeline design, operation, and maintenance [6]. Visual inspection is a part of such as a leak detection process through a biweekly survey that considers the following:

- Unusual noise,
- White vapor cloud producing dry ice, and
- Frozen liquid around suspected piping leak location (Figure 10-22).

However, while any unplanned CO_2 release to the surface or near-surface environment from a pipeline or surface facilities or unplanned migration of CO_2 from deep underground storage/fault is considered to result in a low risk to fauna, it can affect human presence and activities as a result of the level or magnitude and direction of release [24]. Generally, the following are also surveyed or monitored:

- Air quality,
- CO_2 emission and dispersion extension and hazard zone assessment,
- Impact of CO_2 dispersion on the adjoining area,
- Evaluation of controlling techniques,
- Strategic planning to reduce environmental risk and maintain compliance.

When CO_2 is evacuated from the pipeline during an intentional blow-down operation, e.g., for maintenance, the decrease in pressure and temperature can be controlled, enabling the operation of the pipeline without hydrate formation, and other hazards. Generally, a fast blow-down is preferred because it would reduce pipeline downtime. However, in an unintended blow-down, e.g., due to a valve malfunction, operational error, or pipeline external damage

Figure 10-22. Typical CO_2 leak and release from a pipeline

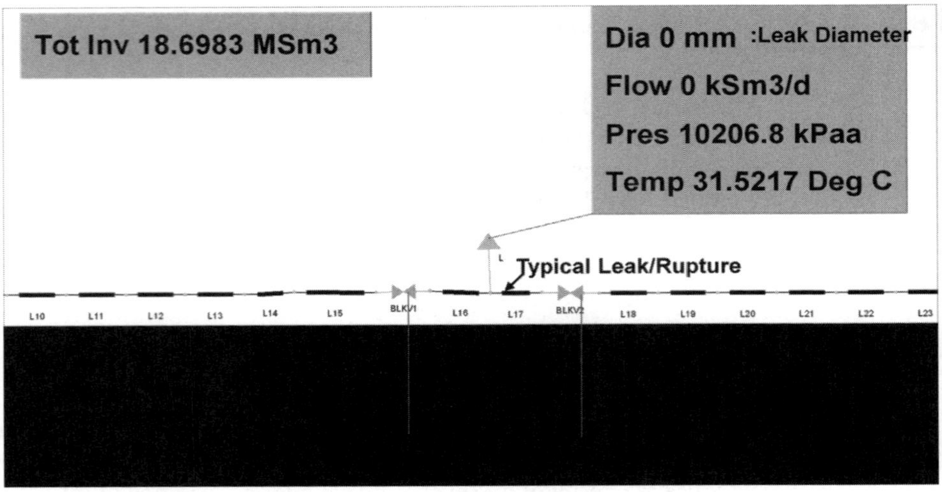

Figure 10-23. Pipeline condition at rupture location prior to rupture

leading to rupture, the pressure decrease and flow is not controlled. Therefore, mitigation/monitoring measures, such as Supervisory Control and Data Acquisition (SCADA) system and a pipeline control as indicated in Figure 10-3, will provide a degree of protection. If there is a leakage or valve malfunction, pressure and temperature could decrease rapidly.

To illustrate the temperature and pressure condition during leak/rupture, the previous pipeline example given in Section 10.2.4.3 (elevation profile as given in Figure 10-15) was dynamically simulated for rupture at location KMP 160, Figure 10-23. This location was selected as it was soon after the highest point in the pipeline with highest consequence of CO_2 mixture discharge after rupture.

The outcome of dynamic simulation with 200 mm diameter is shown in Figure 10-23. Again, as with steady state simulation, ESI PipelineStudio Gas Simulator (TGNET) was utilized for this dynamic simulation.

Considering phase diagrams for Example 4 (Figure 10-11), the following conditions may be deduced from the simulation results:

Temperature, °C	Dew Point Pressure, kPa
−50	0
−40	200
−20	1500
0	3500
15	4500
20	6200
28	7128

By referring to Figure 10-24, it can be concluded that during the rupture, most CO_2 gas mixture will escape from the 200-mm diameter rupture.

10.5.1 CO_2 Measurement and Monitoring Approaches

CO_2 is different from any other gas (particularly dense phase CO_2) when attempts are made to measure it [25].

CO_2 generally arrives at entry to pipelines at 38°C which gives a density of approximately 800 kg/m^3 (95% Pure). Under this condition, the pipeline can operate in the "dense

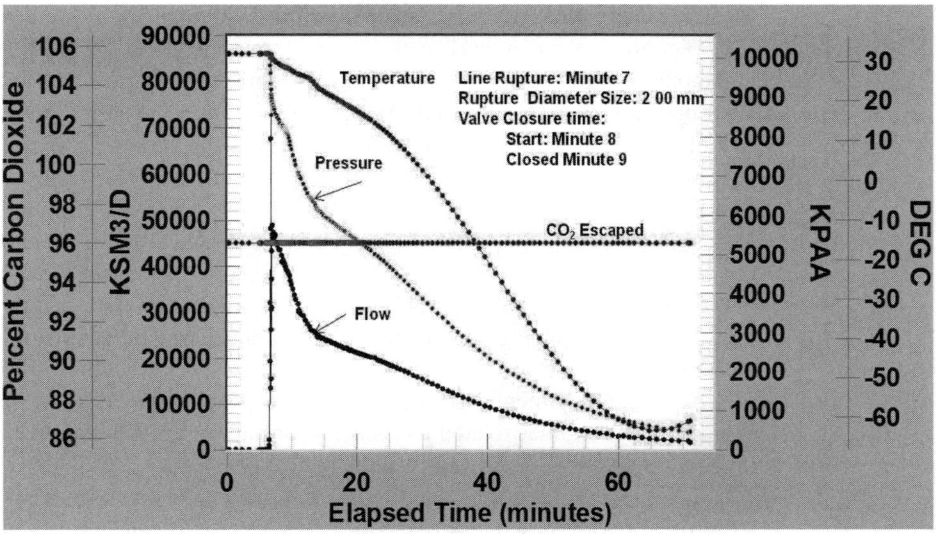

Figure 10-24. Typical pressure, temperature conditions reached due to uncontrolled rupture/blow-down

fluid" Super-critical State. Because of its fairly constant conditions at pipeline inlet, CO_2 can be measured easily using an orifice/corriolis meter and densitometer with a flow computer. Generally, pipeline metering stations are located far away from the compression systems in order to minimize flow-generated pulsation effects. Orifice meters are generally used for both the liquid and gaseous CO_2 measurement with densitometer for liquid and Fpv corrections for the CO_2 gas. The liquid is calculated based on differential pressure and density as the measured variables and yielding mass per hour. The gas is calculated using a Flow Computer using corrections for super compressibility values. While flow is measured in volume, mass flow is calculated as follows:

$$\text{Mass} = \text{Turbine Volume} \times \text{Meter Factor} \times \text{Density}$$

10.5.1.1 General Measurement Method

Table 10-4 provides a summary example of methods for measurement for monitoring purposes for pipeline and injection well piping [26].

TABLE 10-4. Summary of measurement and monitoring methods for CCS

Injected/transportation volume	Flow meter
Injection/transportation pressure or rate	Pressure meter
Composition of CO_2	Chemical analysis of gas stream, gas chromatography
Spatial distribution of CO_2 plume	Time-lapse seismic, satellite radar interferometry (InSAR), and for injection well monitoring equipment, such as micro-seismic, micro-gravity, vertical seismic profiling, crosswell seismic, formation fluid chemistry from monitoring zone
Reservoir pressure and temperature	Down-hole pressure sensor, Bragg fiber-optic grating, thermocouples
Well integrity	Cement casing and imaging logs, vertical seismic profiling, well-head detection devices, mechanical integrity testing
Amount of any measurable leakage	Soil–gas surveys, LIDAR atmospheric eddy co-variance. For Well leakage, groundwater sampling, CO_2 monitoring, artificial tracers/isotopes,

10.5.1.2 Monitoring Methods for Pipeline Leakage

CO_2 pipeline leaks can be detected through an appropriate SCADA system that controls and monitors pipeline operational conditions (such as flow, pressure, and temperature, etc.). As a first line of defense, SCADA systems use flow or pressure deviation calculations or rate of change monitoring on flow and pressure measurements and alarm a sudden change in flow or pressure as a possible leak [27].

Smaller leaks are, however, harder to detect and may not necessarily generate an analog deviation or rate of change alarm. A computational pipeline monitoring (CPM) system can be used in combination with a SCADA system to detect such leaks.

Computational pipeline monitoring (CPM), API 1130 is a widely used technique for leak detection, requiring flow, pressure, temperature, and other data provided by a SCADA system.

API 1130 document describes various detection methods used in the pipeline industry. It also describes other related technical information, operational recommendations, training, documentation, and testing related to CPM systems.

API 1130 provides guidance on the design, implementation, testing and operation of CPM (or software-based leak detection) systems that use an algorithmic approach to detect hydraulic anomalies in pipeline-operating parameters. The monitoring systems would provide an alarm and display and other related data to the pipeline controllers in detecting and aiding decision-making in case of commodity releases that are within the sensitivity of the algorithm and can be divided into four types [28]:

- Flow or pressure change;
- Mass or volume balance;
- Dynamic model based system;
- Pressure point analysis.

API-1130 is always used in conjunction with API-1149 [29]. It may be noted that API 1155 [30], has been included as Annex C of API RP 1130:2007 [31].

API1149 approach can be used to establish theoretical leak detection sensitivity based on measurement instrumentation inaccuracies (Eq. 10-3). For an operating pipeline that is controlled through a SCADA system, real time data are collected for various leak scenarios and modeled off-line using various leak detection systems. API-1130 approach can then be used to evaluate various leak detection systems (generally supplied by different vendors) against each other and calculated against the API-1149 benchmark (Figure 10-25).

$$\frac{Q_{\text{leak rate}}}{Q_{\text{pipeline flowrate}}} \geq \sqrt{(\text{Flow}_{\text{in}}\,\text{Uncertainty}^2 + \text{Flow}_{\text{out}}\,\text{Uncertainty}^2) + \left(\frac{\text{Linefill Uncertainty}}{\Delta\text{Time} \times \text{Flowrate}}\right)^2} \tag{10 – 3}$$

It may be noted that in the real world while Flow Uncertainty is constant over time, Line fill uncertainty diminishes over time. The uncertainties are those due to inaccurate pressure and temperature profiles along the pipeline. Therefore it may be noted that, in the short-term, leak detection sensitivity is more dependent on measurement meter accuracy and the degree one can accurately determine temperature and pressure along the pipeline. However, in the long term, the leak detection (LD) sensitivity converges to meter accuracy [32].

For compositional analysis of the CO_2, gas chromatography can be used. The components that need to be measured depend on the specification for purity of CO_2. From the compositional analysis, it is possible to measure the density of the gas/fluid. Gas detectors may also be used as part of the risk management strategy for a pipeline. A risk-

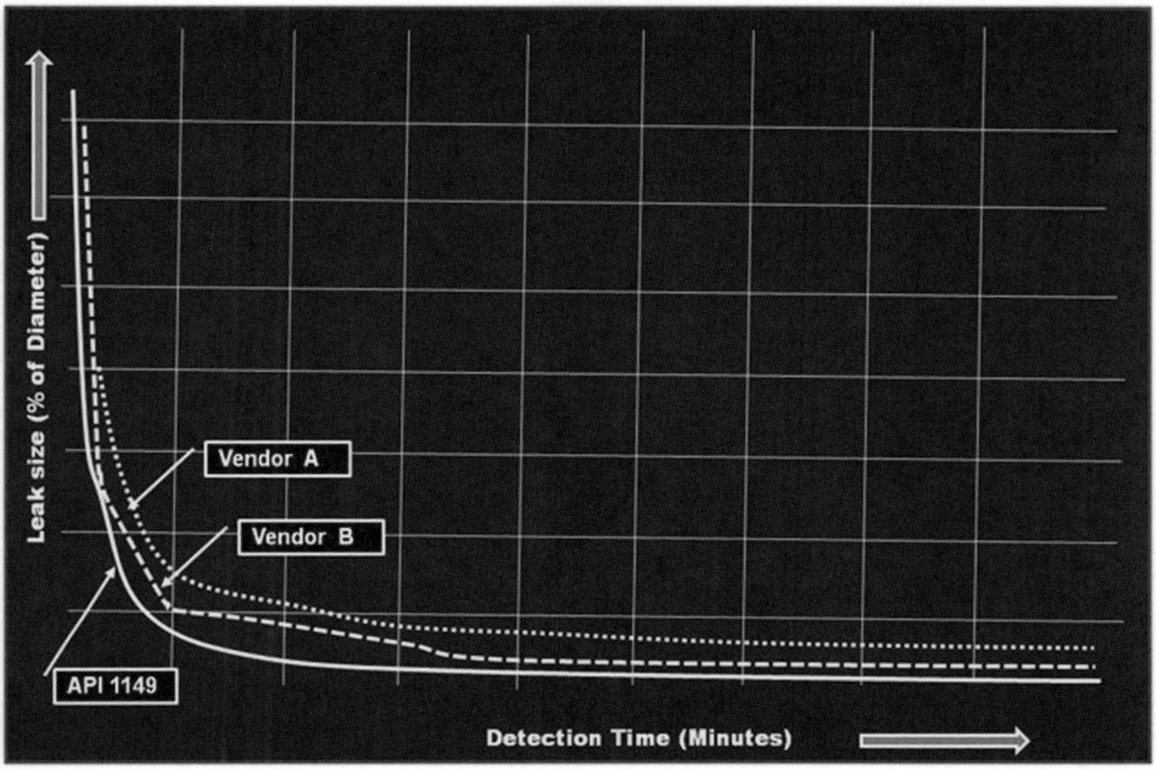

Figure 10-25. API 1149 leak detection sensitivity benchmark

based approach is recommended to determine the need for, and location of, gas (and other) detectors.

Moisture analyzer is generally used to measure and analyze the water content. There are several choices of moisture analyzers on the market that can be used with CO_2, both contact as well as non-contact.

10.6 PIPELINE COMMISSIONING AND STARTUP

Activities that are related to putting a CO_2 pipeline into operation are summarized in Table 10-5.

10.6.1 Pre-Commissioning

Pre-commissioning activities are performed to prepare the pipeline system to be ready to be put into operation. This requires cleaning of the system (from any debris), setting,

TABLE 10-5. Activities summary: (pre-commissioning to on-streaming of pipeline)

Activity	Definition
Settings/calibrations/servicing/operability checks	Pre-commissioning completion
Purge/drying/pressure/equalize/final calibration and operability checks	Commissioning activities and completion
Operation	On-Stream

calibration and functioning verification of all equipment/instrumentation and system testing activities and preparation to fill the pipeline system with the CO_2 mixtures intended for operation.

Pre-commissioning also include activities for preservation of the pipeline system prior to full commissioning if required. The need for preservation of the pipeline between pre-commissioning and commissioning phases should be assessed based on duration before the pipeline is put into operation. Gas such as nitrogen (N_2) and dry air may be used for preservation of the pipeline, but the requirement to the gas quality needs to be assessed.

Preservation should be selected with proper consideration toward the commissioning requirements. This may include requirements for internal pressure.

10.6.2 Commissioning

The operator needs to perform the required risk assessment for the particular system and to establish proper commissioning procedures. The constraints for operation of the pipeline need to be taken into considerations in the commissioning. Assurance that the pipeline is filled with a CO_2 mixture quality in accordance with the requirements at the receiving end concludes the commissioning phase.

10.6.2.1 Planning to Purge and Load a CO_2 Pipeline

a. Planning

In commissioning CO_2 pipelines it is important that proper planning for purging and loading the system is undertaken to avoid incidents of uncontrolled CO_2 gas release that could cause serious damage including personal injuries or death, etc. To plan a "Purge and Load," the following components must be considered:

- Scope,
- Plan and Schedule
- Purge and load team organization,
- Tool requirements,
- Safety,
- Environment and health,
- General purging and loading procedures,
- Specific purging and loading procedures.

These are described below in detail:

***Scope*:** defines the facilities that are affected by the commissioning, for example, pipeline segment, interconnecting piping, measurement facilities, etc.

***Schedule*:** Schedule describes the "Commissioning" pressure, regulation, and sequence of the loading each segment of the pipeline. It would describe the method and location of CO_2 gas measurement, as well as provide an estimate of time required for loading different line segments. Such information will provide a basis for determining work schedule and for estimating the manpower required to initiate and complete the "commissioning" program.

***Organization and Plan*:** Once plan and schedule for "commissioning" is completed, the organization needs to be defined which would handle the program including operation, (any maintenance), emergency response, documentation, reports, and requirements for pre-job training and meetings as well as familiarization (with equipment, facilities, location, communities affected, etc.). This organization describes commissioning team members jobs and duties/responsibilities during the process of "commissioning" program.

The plan includes a pre-job meeting to undertake the following:
- Ensure all necessary tools and safety equipment are available,
- Assign responsibilities to team members,
- Review the Work Plan and to resolve outstanding issues,
- Review all appropriate Safety Procedures,
- Review the Emergency Response Plan.

Tools: Tools for the job need to be identified. Typically, this includes the following:
- CO_2 gas mixture detectors;
- Pressure gauges;
- Other pipeline maintenance equipment, such as tools to operate valves, fittings, etc.;
- First aid equipment;
- Hearing protection equipment;
- Self-contained breathing apparatus (SCBA).

***Safety & Health*:** Safety is a very important when undertaking "commissioning" operations. Safety considerations include:
- Employee training to ensure competency;
- Requirement for safe work Permit;
- Requirement for placement of warning signs;
- Requirement for breathing protection and to assure presence of equipment for oxygen deficient atmosphere and containment exposure;
- Safe use of equipment;
- Periodic air testing around venting areas including testing of air to assure oxygen adequacy and low level of H_2S;
- Testing of equipment and fittings subject to CO_2 gas mixture leaks.

Protection of the environment and health of personnel during the process of "commissioning" is of vital importance to the success of a completed system.

Procedures must be devised and implemented to assure minimal impact on the environment and the health, as well as the safety of personnel and residences in the area where pipeline traverse.

Environment: Avoidance of, CO_2 gas mixture emission as much as possible from affected facilities during commissioning reduces the impact on the atmospheric environment and nearby residence. For pipeline facilities, programs and procedures such as the simulation detailed for Purge & Load of a gas pipeline [16], would help to reduce the impact with respect to noise and disturbance in the communities where this activity is carried out.

***General Procedures*:** General "commissioning" procedures that are prepared usually include pre-requisite procedures that ensure:
- Pipeline is dry and clean;
- Permits to proceed with commissioning are obtained;
- Surrounding residents are informed well in advance;
- CO_2 gas mixture at the required pressure and condition is available to commission the line.

General purge procedures generally provide guidelines including the following:
Preparation for purge air to establish/ensure that
- Communications are in place;
- Time of operation and nitrogen or CO_2, gas volume requirement are known;
- Pressure gauges are installed;
- Means for nitrogen or CO_2 gas stream sampling is arranged.

TABLE 10-6. Typical specific purge procedure [16]

(1) Confirm communication line between the Inlet and Outlet ends and the supply Control Valve. Synchronize watches. (2) Confirm that copies of the "Commissioning" plan are available at the Inlet and the Outlet ends as well as the supplier Control Valve	
At the Inlet End	**At the Outlet End**
• Ensure every one who is on site has hearing protection in place	• Ensure everyone on site has hearing protection in place
• Contact Outlet End—"Prepare to receive purge"	• Ensure Nitrogen Gas Sampler is installed and functioning.
• Contact 1st Control Valve operator-"Prepare to commence purge"	• Contact Inlet End — "Ready to receive nitrogen"
	• Introduce nitrogen. Open #1 Valve, by-pass/ control valve
• Notified by Outlet End-"Ready to receive nitrogen"	
• Open Valve, #	
• Slowly Open Valve—to start air purge-quickly bring pressure up to—kPa. Record time and pressure. Control piping vibration using valve #	
• Record Start air purge time	• Notified from Inlet End-"Nitrogen purge started"
• Contact Outlet End and nitrogen Supplier/1st Control Valve-"nitrogen purge started"	• Closely monitor vent. Record time of first air movement
	• Communicate pressure to Inlet End as pipeline pressure rises.
• Monitor/record pressures and inform Outlet side of pressure reading every min	• Monitor and record nitrogen gas content
• Notified by Outlet end-"First Detection of Gas"	• Record first detection from the vent
	• Contact Inlet End-"First Detection"
	• Continue monitoring and recording % nitrogen gas readings
	• When 100% nitrogen gas detected-record time and continue purging for one more minute. Close Vent.
• Notified by Outlet End-"Detection of 100% nitrogen gas and vent valve closed"	• Contact Inlet End-"Detection of 100% nitrogen gas and Vent valve closed"
• Remove plug from Valve # — open valve and purge valve control piping until 100% gas is detected	• Contact Inlet End-"air purging and nitrogen loading complete"
• Close Valve #	
• Notified by Outlet End-"Purging Complete"	

Commissioning actual operation to ensure that:

- Valves are in correct operating positions, opened and closed, or locked as appropriate;
- Monitoring of discharging of any gas and its condition taking a note of gas color and piping vibrations, etc.;
- Reporting and documentation procedures are in place and performed;
- Permissions obtained for loading;
- Communications are established between pipeline segment ends;
- Leak tests are conducted as applicable;
- Recording of appropriate readings are taken.

Specific "Purge and Load" Procedure: These procedures must address specified requirements for each section of pipeline that will be purged of air and loaded with nitrogen and subsequently with CO_2 gas in a supercritical state. A typical specific purging procedure for removing air from pipeline segment with Nitrogen may include those shown in Table 10-6. Similar procedures can be prepared for loading the line with CO_2 gas.

b. Sequence

b-1 Pipeline Drying: When a CO_2 pipeline construction is completed, it must undergo hydrostatic testing before commissioning and operation. While air or nitrogen can be used as alternative test medium, treated water is generally used for hydrostatic purpose. After testing, water must be completely removed to assure water concentration level between 40-500 ppm [33]. To prevent a corrosive CO_2 mixture and water, the pipeline should be completely dry prior to commissioning. Methods described by Barden et al [34] can be adopted for drying and assuring the residual water content in a CO_2 pipeline prior to commissioning. The most effective methods are

1. Hot air drying,
2. Nitrogen purge drying,
3. Vacuum drying.

As commissioning of a pipeline intended to transport CO_2 mixture will require to be commissioned with nitrogen back pressure (see below), the nitrogen as an inert purge gas is the best way to dry a CO_2 pipeline. Two advantages of using nitrogen gas are:

- Nitrogen is very dry gas and therefore has a high capacity for water adsorption.
- Nitrogen is inert and can be used to prepare and subsequently commission pipelines or process vessels/piping.

The same procedure is used as in hot air drying where the comparison between the feed gas minus the outlet gas dew points are measured to compare the dryness of the pipeline. It

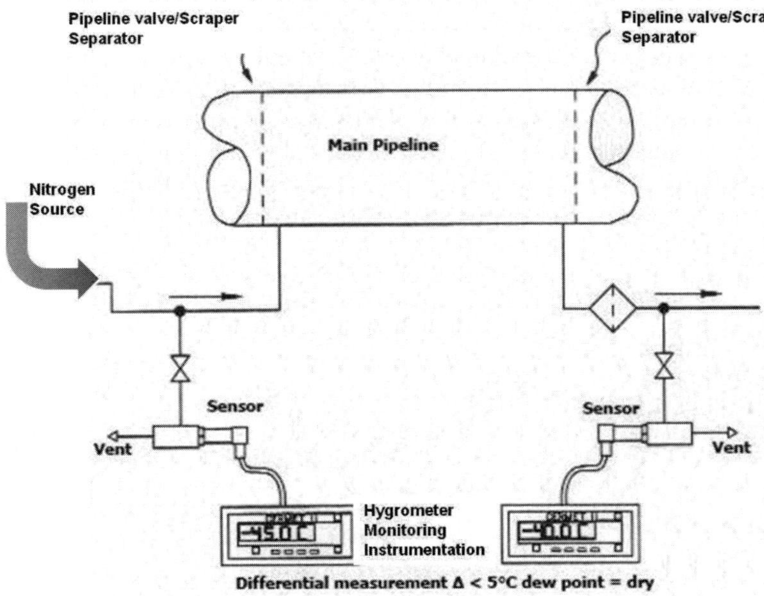

Figure 10-26. An example of nitrogen purge drying (after Kahn Instruments Inc. www.kahn.com)

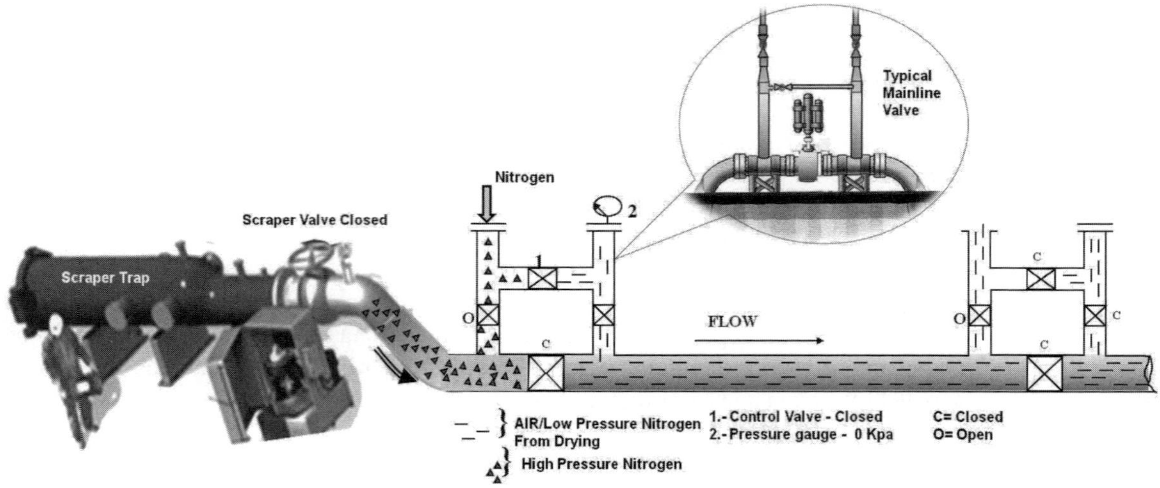

Figure 10-27. CO_2 pipeline air removal and initial nitrogen fill

is common for a differential of <5°C dewpoint to be used as the signal that the pipeline is dry. Typical nitrogen drying process is schematically shown in Figure 10-26.

b2- Initial Filling With Nitrogen

Before initial filling, the pipeline must be in the condition as defined by the pre-commissioning. Mixing of CO_2 and the pre-commissioning substance (e.g., nitrogen) should be considered as part of this activity. Experience shows that dense phase CO_2 and N_2 do not always mix well and, therefore, a pig between the two media is generally required. To commission the CO_2 pipeline, the best option after drying with nitrogen is to pressure the nitrogen and then commission the pipeline with a CO_2 mixture through the scraper trap launching facilities with an appropriate commissioning pig as shown in Figure 10-27.

Removal of air and filling with nitrogen (for drying purposes as indicated in *b* above) can follow the methods described by Mohitpour et al [27, 35] for commissioning of a gas pipeline.

b3- Filling the Line with Supercritical CO_2 Mixture

Once the pipeline is full of pressurized nitrogen, a separation pig can be inserted in the pipeline scraper launcher and launched with CO_2 mixture against the nitrogen back pressure (Figure 10-28).

Low CO_2 temperatures must be considered during first fill operation as the CO_2 stream may have to be throttled downstream at the delivery location or a compressor until a suf-

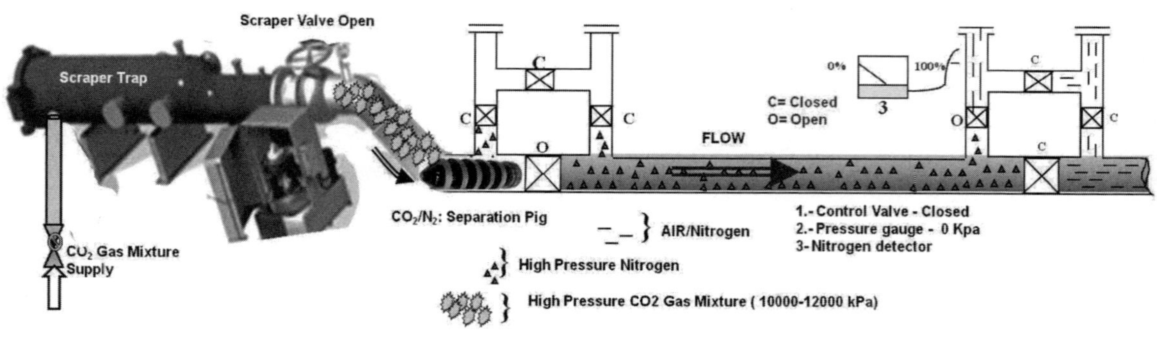

Figure 10-28. Commissioning of CO_2 line with separation pig with nitrogen back pressure

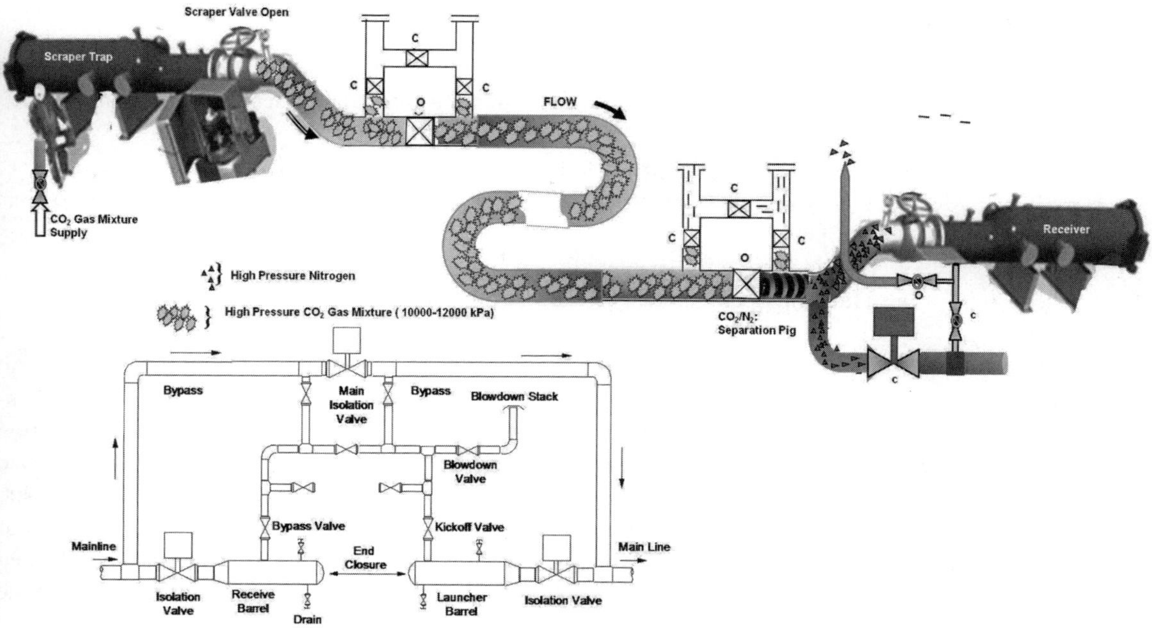

Figure 10-29. Schematic representation of completed commissioning of a supercritical CO_2 mixture pipeline receiving the CO_2/N_2 separation pig at a downstream receiving station (Inset: Typical piping arrangement for launcher/receiving station, [16])

ficiently high pressure is reached. Risk of forming solid CO_2 as a result of the throttling process must therefore be considered.

It may be noted that CO_2 mixture will have a higher pressure (upstream of the separation pig) than nitrogen back pressure acting downstream of the separation pig. CO_2 will then move the separation pig towards the next/downstream pipeline valve.

When nitrogen pressure at the next block valve reaches a preset value, the downstream control valve can be opened to pressurize the downstream pipeline segment with high pressure nitrogen. The pipeline valve can then be opened upon balanced pressure between up/down stream of the valve. This will allow the separation pig to go through and fill the next segment of the pipeline with high pressure supercritical CO_2 mixture. In this way, nitrogen use is maximized and its wastage/emission is reduced. It may be noted that this assumes that the subsequent pipeline segments are already dried and filled with low-pressure nitrogen. Figure 10-29 schematically indicates a typical pipeline segment filled with supercritical CO_2 mixture while receiving the CO_2/N_2 separation pig at the upstream receiving station. The inset diagram shows typical piping arrangement for launcher/receiving station [16].

b4 -Other Alternative

CO_2 mixture in gaseous state, but at a lower pressure can be used during pipeline filling/commission process, however, as the pipeline fills and its pressure is increased, the CO_2 enters a liquid phase, Subsequently, as the pressure continues to increase, the CO_2 returns to gaseous phase and enters a supercritical state [36]. This is the best alternative with CO_2 mixture that have close bubble and dew points demarcation curves, as previously shown in Figure 10-11 (refer to Example 4, Table 10-1 for mixture composition). In this situation, a small increase in pressure will be required to move the CO_2 mixture condition out of a two-phase region and hence into liquid/gaseous state.

10.6.3 Startup-Restart

With reference to startup and restart of a pipeline segment, the particular aspects related to supercritical phase flow assurance. The following particular issues should be considered:

- Effect on flow capacity of fluid temperature and pressure;
- Transportation in supercritical (liquid like) phase (i.e. above critical pressure) or as two-phase;
- Hydrate formation, potentially causing pipeline blockage;
- Transient operation (shutdown, startup and restart).

CO_2 gas mixture must be dehydrated well below its saturation point to ensure that no condensation will occur in the pipeline under unusual or upset conditions. Typically, the industry specifies the water content of supercritical CO_2 gas mixture at field delivery location as no more than 50 ppm (by weight) even though saturation is about 2500 ppm. It is also indicated that even with water content of about 1000 ppm (1%) and H_2S level of 800-900 ppm (0.08-0.09%), the steel pipe corrosion is minimal. However, rapid corrosion can occur if free water content is high (i.e. over 450 mg/m^3) as indicated in Chapter 1 and is collected at low points. Continuous monitoring of water content of CO_2 gas mixture in the pipeline during startup and restart is therefore highly essential.

Prior to any start/re-start, pipeline or any repaired segment must be thoroughly cleaned and dried prior to charging with CO_2 gas mixture. Once a segment is repaired, it can be tested and dried using nitrogen as per commissioning procedure.

As for any newly constructed pipeline, the following procedures must be used for getting a segment of pipeline that has been repaired and needs to be recharged:

1. Hydrotesting (with treated water) or air testing,
2. Cleaning with multiple mechanical scrapers,
3. Nitrogen drying,
4. Loading with nitrogen,
5. Packing with CO_2 gas mixture.

It has been reported that when hydrotesting with water and drying with methanol wash, plugging by a solid corrosion product can be expected from the pipeline soon after startup. This is usually caused by inadequacy of drying by methanol wash and presence of displacement air which can cause the large internal area of the pipeline to act as an "iron sponge" during the prolonged charging period. Therefore, nitrogen drying is the preferred method for drying and after hydrostatic testing and methanol wash.

REFERENCES

[1] Holloway, S., Heederik J. P., Van der Meer, L. G. H., Czernichowski-Lauriol, I., Harrison, R, Lindeberg, E., Summerfield, I. R., Rochelle, C., Schwarzkopf, T., Kaarstad, O., and Berger, B., 1996, "The Underground Disposal of Carbon Dioxide: Summary Report," JOULE II Project No. CT92-0031, British Geological Survey, Keyworth, Nottingham UK, http://www.ipcc.ch/pdf/special-reports/srccs/srccs_chapter1.pdf.

[2] Hattenbach, R. P., Wilson, M., and Brown, K. R., 1999, "Capture of Carbon Dioxide from Coal Combustion and its Utilization for Enhanced Oil Recovery", Greenhouse Gas Control Technologies, Elsevier Science Ltd, pp. 217-221. http://books.google.ca/books?id=B-DS4P_J8KUC&pg=PA217&lpg=PA217&dq=%22capture+of+carbon+dioxide+from+coal+combustion+and+its+utilization&source=bl&ots=JTjEw2tS1T&sig=zhbTKTbQDjQjvOn03EyNE1Tc5HQ&hl=en&ei=XWKLT-

tPeDoGqsQL-2KipBA&sa=X&oi=book_result&ct=result&resnum=5&sqi=2&ved=0CEYQ6AEwBA#v=onepage&q=%22capture%20of%20carbon%20dioxide%20from%20coal%20combustion%20and%20its%20utilization&f=false.

[3] Aspelund, A., and Jordal, K., 2007, "Gas conditioning—The interface between CO_2 capture and transport," International Journal of Greenhouse Gas Control, Elsevier Ltd, www.elsevier.com/locate/ijggc.

[4] IPCC, 2006, "Special Report – Carbon Dioxide Capture & Storage" Cambridge, UK, http://www.ipcc.ch.

[5] Basin Electric (Basin Electric Power Cooperative), 2010, "Gasification," http://www.basinelectric.com/Gasification/index.html.

[6] Thambimuthu, K., 2010, "IEA Weyburn CO_2 Monitoring and Storage Project," http://www.iea.org/work/2004/zets/conference/presentations/thambimuthu.pdf.

[7] Robertson, E. P., 2007, "Analysis of CO_2 Separation from Flue Gas, Pipeline Transportation, and Sequestration in Coal," Idaho National Laboratory (INL), Report # NL/EXT-08-13816, Prepared for the U.S. DOE, Fossil Energy, under Contract DE-AC07-05ID14517, http://www.inl.gov/technical publications/Documents/4010767.pdf.

[8] WorleyParsons, 2009, "Carbon Dioxide Specification Study," Report procuded for Australian Department of Resources, Energy and Tourism (DRET), Aug., http://www2.ret.gov.au/resources/Documents/Programs/cst/WP%20-%20Carbon%20Dioxide%20Spec%20Study.pdf.

[9] DYNAMIS, 2007, "DYNAMIS, D 3.1.3 DYNAMIS CO_2 quality recommendations; Project Title: Towards Hydrogen and Electricity Production with Carbon Dioxide Capture and Storage, June, http://eippcb.jrc.es/reference/.

[10] Grieve, R. B., and Thodos, G., 1964, "The Cricondentherm and Cricondenbar Pressures of Multicomponent Hydrocarbon Mixtures," Society of Petroleum Engineers Journal, pp. 240-246, July, http://www.aimehq.org/search/docs/Volume%20231/231-075.pdf.

[11] Minotti, M., 2009, "CO_2 compression capabilities," GE Presentation, National Institute of Standards and Technology (NIST), http://www.nist.gov/pml/high_megawatt/upload/3-3-Minotti-Approved.pdf.

[12] NIST (National Institute of Standard and Technology), 2010, "Standard Reference Database 23, Fluid Thermodynamic and Transport Properties Database (REFPROP): Version 8.0," www.nist.gov/ts/msd/srd/nist23.cfm.

[13] NEOTEC, 2011, "Personal Communication with TransCanada NEOTEC SPT Group Company" Feb., Calgary, AB, Canada.

[14] Golshan, H., and Dueck, A., 2011,"Personal Communication with NEOTEC SPT Group Company" Feb.

[15] ESI (Energy Solutions International), 2010, "Gas & Liquid Pipeline Simulation" http://www.energy-solutions.com/products/pipeline_studio.php.

[16] Mohitpour, M., Golshan, H., and Murray, M. A., 2007, "Pipeline Design and Construction—A Practical Approach," 3rd Edition, Chapter 12, ASME Press, NY.

[17] Jeje, O., and Mattar, L., 2004, "Comparison of Correlations for Viscosity of Sour Natural Gas," Presented at 5th Canadian Int. Pet. Conference, June 8-10 paper # 2004-214, http://www.fekete.com/resources/papers/comparison_correlations_viscosity_paper.pdf.

[18] Fekete Associates, 2011, "Gas Viscosity (mg)," http://www.fekete.com/software/welltest/media/webhelp/Gas_Viscosity.htm.

[19] Bachu, S., 2003, "Screening and ranking sedimentary basins for Sequestration of CO_2 in Geological Media in response to Climate Change," Environmental Geology, 44, pp. 277-289, http://www.geology.wmich.edu/bachu_Barnes_2003.pdf.

[20] Enick, R. M., 1998, "A Literature Review of Attempts to Increase the Viscosity of Dense Carbon Dioxide," Report for US DOE, FETC, Oct., http://www.netl.doe.gov/publications/others/techrpts/co2thick.pdf.

[21] Hamouda, A. A., 2007, "Study of CO_2 Properties and Transportation for Injection well into Reservoir," MSc Thesis, Stavanger University Faculty of Science and Technology, http://brage.bibsys.no/uis/bitstream/URN:NBN:no-bibsys_brage_5591/1/Sugianto%20ipe%202007.pdf.

[22] Ghazi, N., 2010, "Techno-Economic Modelling and Analysis of CO_2 Pipelines," MSc Pipeline Engineering Dissertation, Newcastle University, UK.

[23] CSLF (Carbon Sequestration Leadership Forum), 2008, "CO_2 Transportation — Is it Safe and Reliable?" http://www.cslforum.org/publications/documents/CSLF_inFocus_CO2Transport.pdf.

[24] Heinrich, J. J., Herzog, H. J., and Reiner, D. M., 2003, "Environmental Assessment of Geologic Storage of CO_2" Laboratory for energy and the environment, Report MIT LFEE 2003-002, http://lfee.mit.edu/publications/.

[25] DeBusk, F., 1984, "Summary of Measuring and Metering CO_2 for Enhanced Oil Recovery Methods," CO_2 Measurement Conf Hobbs New Mexico, USA, Nov. 13, http://www.triad-measurement.com/images/triad/Tech%20Doc's/Measuco2.pdf.

[26] Forbes, S. M., and Ziegler, M. S., 2010, "Carbon Dioxide Capture and Storage and the UNFCCC Recommendations for Addressing Technical Issues," WRI Issue Brief Nov.

[27] Sharoki, S., Maschmeyer, K., and Yoon, M. S., 2004, "Selecting a pipeline leak detection system, a practical approach," http://consipio.com/doc/Selecting%20a%20pipeline%20leak%20detection%20system.pdf.

[28] DNV (Det Norske Veritas), 2010, "Recommended Practices J202: Design and Operation of CO_2 Pipelines," CO2PIPETRANS Project.

[29] API (American Petroleum Institute), 1993, "API RP 1149 Pipeline Variable Uncertainties and Their Effects on Leak Detectability."

[30] API (American Petroleum Institute), 1995, "API-1155 Evaluation Methodology for Software Based Leak Detection Systems (Now Appendix C of API-1130)."

[31] API (American Petroleum Institute), 2007, "API-1130-Computational Pipeline Monitoring for Liquids Pipelines" 2nd Edition.

[32] Nagala, D., and Verret, J., 2002, "Practical Application of API 1149 and 1155" UTSI Conference,http://www.utsi.com/papers/Presentations/april_2002/Nagala_Verret_2002.pdf.

[33] Buitt, L., Ahmad, M., Mallon, W., and Hage, F., 2010, "CO_2 EuroPipe Study of occurance of free water in dense phase CO_2 transport" ScienceDirect, Energy Procedia, paper GHGT-10, Elsevier, www.sciencedirect.com.

[34] Barden, A. J., Powers, M. M., and Karklis, P., 1996, "Evaluation of Pipeline Drying Techniques," Proceedings Pipeline Pigging Conference, Jakarta, May 27.

[35] Mohitpour, M., Kazakoff, J., Jenkins, A., and Montemurro, D., 2000, "Planning for Purging and Loading of Newly Constructed Gas Pipeline System Using a Pipeline Simulator," 3rd ASME International Pipeline Conference (IPC), October, Calgary, Alberta.

[36] IEA, 2011, "CO_2 Capture and Storage" Greenhouse Gas R&D Programme R&D Project Data Base, Weyburn," http://www.co2captureandstorage.info/project_specific.php?project_id=70.

Chapter 11

RISK ASSESSMENT AND INTEGRITY MANAGEMENT

11.1 HAZARD AND RISK

Risk is a compound measure of the probability and consequences of failure (Figure 11-1). If probability (or expected frequency per unit of exposure) is expressed quantitatively, then risk can be evaluated as the product of the probability and consequence values. If either the probability is zero or consequence non-existent (i.e., zero), then risk becomes zero.

Hazard is defined as a characteristic or a group of characteristics that provide or cause the potential for a loss. Toxicity is an example of such characteristics.

Therefore, by definition, risk can be changed without changing the hazard. However, it may be noted that risk assessment and analysis cannot predict future events; it can only provide an estimate of the *chance* of specified events.

There are calculable and perceived risks for pipeline transportation of CO_2 containing impurities. Calculable risks are those risks that can be estimated based on historical information and past events that have led to adverse outcomes, e.g., a pipeline failure that caused a spill which in turn led to environmental and other consequential damages.

Perceived risks are those that may be perceived (generally by the public) to happen (such as security risks) but are technically very unlikely to happen and are not calculable easily. Nevertheless, they need to be dealt with in some way (for example, to assist in managing public perception). Security risk, however, is not considered here.

The relationships between risk analysis, risk assessment, and risk management are defined as follows:

- Risk Analysis = Estimation and combination of frequency and consequences;
- Risk Assessment = Risk Analysis + Risk Evaluation (is the risk significant?);
- Risk Management = Risk Assessment + Risk Control (including establishment and implementation of maintenance strategy, monitoring and communication).

Pipelines systems generally have established a good safety record, in particular when compared with other potential means of transporting hazardous materials over long distances. CO_2 systems are usually compared with the existing systems for long-distance pipeline transportation of other fluids, such as natural gas, which have comparable failure and accident rates [1]. For the existing pipeline transportation system, these incidents seem to be perceived by the broad community as acceptable in spite of occasional energy transportation incidents that cause serious pollutions such as the Exxon Valdez oil spill [2]. Nonetheless, pipelines have been impacted by high-profile incidents such as those in Bellingham, Carlsbad, and San Bruno that have had an immense impact on the way the pipeline industry is perceived and regulated.

Because the potential consequences of CO_2 pipeline incidents are perceived to be significant [3] (Figure 11-2), stricter regulations for CO_2 pipelines than those for natural

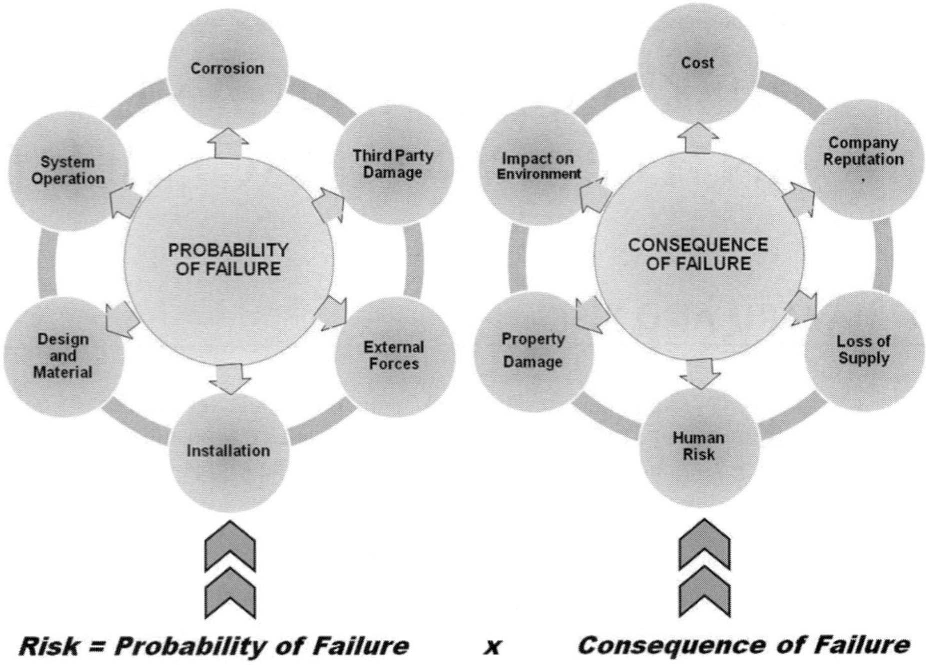

Figure 11-1. Interrelationships of risk value to failure and consequence

gas pipelines are generally in force. However, the incidence of failure is relatively small. For example, the European Gas Pipeline Incident Data Group (2002) indicate that the incidence of failure has markedly decreased from an incident rate of almost 0.0010 /km-year in 1972 to below 0.0002 per km-year^{-1} in 2002 [4]. A similar drop in the incidence of gas pipeline failures has been reported elsewhere [5]. Numbers of reportable incidents due to the operation of CO_2 pipelines have also been in decline [6] (refer to Chapter 13 for details).

Leaks from a pipeline transporting carbon dioxide containing impurities could constitute potential physiological hazard for humans and animals (see Chapter 13 for details of physiological effects). The consequences of CO_2 incidents can be modeled and thus assessed by taking into account local topography, meteorological conditions, population density, and other local conditions.

A property of CO_2 that needs to be considered when selecting a pipeline route is the fact that CO_2 is denser than air and can therefore accumulate to potentially dangerous concentrations in low-lying areas. Any leak or rupture of a pipeline will disperse CO_2 and other associated components (such as H_2S) to the atmosphere.

If substantial quantities of impurities, particularly H_2S, are in the CO_2 stream, these could affect the potential impacts of a pipeline leak or rupture. For example, the exposure threshold at which H_2S is immediately dangerous to life or health, according to the National Institute for Occupational Safety and Health (NIOSH), is 100 ppm, compared to 40,000 ppm for CO_2.

If a CO_2 stream containing impurities is transported over significant distances in densely populated regions, the number of people potentially exposed to risks from CO_2 transportation facilities may be greater than the number exposed to potential risks from CO_2 capture and storage facilities. Consequently, risk assessment must be performed for the entire pipeline system to assure not only adequacy in design but also adequacy in operational control and maintenance and monitoring programs to assure acceptable performance over

Figure 11-2. A CO_2 accident and discharge from a pipeline (collaged from http://www.eutawfire.com/co2-spill.htm)

the entire life of the system. In considering system-wide risks, it is significant that, unlike most energy pipeline systems, major receipt-end processing plants will usually be in or adjacent to heavily populated areas, while the delivery point will usually be in relatively remote areas.

11.2 PIPELINE RISK ASSESSMENT

As previously indicated, risk is defined as a combination of the probability of occurrence of a scenario and the severity of its consequences. The following process is therefore appropriate to define the magnitude of risk in a pipeline system:

- Identify scenarios or events that may occur and have adverse consequences (hazard identification);
- Estimate the likelihood of this event occurring;
- Estimate potential consequences from the event;
- Evaluate the risk.

11.2.1 Sources of Risk to a Pipeline

The risk associated with a pipeline, in terms of vulnerability of people, damage to the environment, and loss of facilities and hence income, depends on the expected frequency and the associated consequence. Such a risk is directly related to the type of fluids transported and the sensitivity of locations of the pipeline. In this context, *pipeline failures* are defined as *loss of containment.*

The risks associated with gaseous emission of CO_2 with impurities must be considered for the pipeline transportation from the sources of supply (e.g., power station flue stack) to the delivery locations (e.g., an injection site) due to incidents that can be potentially caused by:

- System Design
  - o Design deficiencies
  - o Material inadequacy
  - o Inadequate design (e.g., improper/inadequate valving arrangements)
- Operations
  - o Over-pressurization
  - o Equipment malfunction
  - o Incorrect operation
  - o Human error
- Third Party Activities
  - o Damage inflicted by third parties
  - o Vandalism
- Environmental/terrain impacts
  - o Earth/soil movement
  - o Heavy rain/flooding
  - o Corrosion
- Maintenance activities
  - o Blow-down
  - o Repairs
  - o Re-commissioning.

Other facilities that can be considered are [7]:

- Systems initiating station,
- Pipeline laterals,
- Systems Emergency Flow Restricting Devices (Line Valve Stations),
- Terminal/Measurement/Delivery Stations.

Failures of the CO_2 engineered system that must be evaluated for risk should include the following [8]:

- Pipeline rupture;
- Pipeline puncture (i.e., releases through a small hole);
- Accidental releases that could potentially affect the general public in the vicinity of a release.

Also, risks associated with non-failure releases (such as blowdowns) need to be considered at the system design stage.

A pipeline rupture release would occur if the pipeline was completely severed, for example, by heavy equipment during excavation activities. In a rupture scenario, the escaping dense CO_2 gas from the pipeline is usually assumed to escape as a horizontal jet at ground level, which is typically the worst case event for heavier-than-air gases [9]. However, this will only usually occur for a long fracture or where pipe ends are significantly misaligned. Otherwise, as with natural gas, the release is initially momentum-dominated and the CO_2 goes up, potentially slumping to ground later. A rupture could also result from a seam-weld failure or other sources of a critical through-wall defect, refer to Chapter 7. In rupture cases, the entire contents of the pipeline between the two nearest (upstream and downstream) emergency block valves could be discharged from the severed pipeline within minutes. The release time is much shorter than for blow-down or a puncture.

The most common cause of pipeline failure is third party damage. A pipeline puncture release is typically from a roughly 75 mm by 25 mm hole made by an excavator tooth. In such a case, all of the contents in the pipeline between the two nearest control valve stations would discharge into the atmosphere, but the release would occur over an extended period of time, up to several hours, as the opening is small relative to the total volume and the pressure declines as the fluid escapes.

11.3 RISK ASSESSMENT METHODS FOR PIPELINE TRANSMISSION SYSTEMS

11.3.1 Risk Assessment Steps

The general steps in a risk assessment within the overall risk management process are as follows:

1. **Data gathering** — Identifying the location and characteristics of the pipeline system and surrounding terrain, land-use and development.
2. **Hazard identification** — Characterization of the pipeline route in sufficient detail to formulate potential accident scenarios and to permit subsequent evaluation of accident probability, likely release amount, and nature and magnitude of resulting impacts.
3. **Probability analysis** — Determining the likelihood of an event, expressed in relative (typically referred to as likelihood) or quantitative terms (typically referred to as probability or expected frequency).
4. **Consequence analysis** — Examining the potential physical impacts and derivative consequences (e.g., harm to people or the environment) of a pipeline failure and accidental release of product.
5. **Risk Estimation** — Generating a numerical or matrix-based combination of both the probability of an event and its consequences.
6. **Risk Presentation** — Risk results may be expressed in many ways, such as single values (numerical or qualitative), plots of risk contours along the entire pipeline route, individual risk transects at specific points, and societal risk values or risk spectra associated with specific locations and developments.
7. **Risk Evaluation** — This is the process of evaluating the risk analysis results to determine their significance, for example, by comparison with external or internal standards or with results for other pipeline segments or pipelines.
8. **Risk control** — Risk control consists of prevention and mitigation measures, respectively, to reduce the probability that a release of pipeline product will occur and to minimize the impacts of any release that might occur.
9. **Risk monitoring** — Consisting of a method of monitoring the effectiveness of the risk assessment and management.

It may be noted that items 1 to 6 are included in risk analysis, items 1 to 7 constitute risk assessment, and items 1 to 9 make up the complete process of risk management.

There are three main types of risk assessment methods that are typically used in the pipeline industry. These are

- Fully quantitative methods, in which failure frequency and consequences are estimated probabilistically and multiplied to produce a risk value in terms of expected frequency of an adverse effect (e.g., annual probability of a certain level of injury to a hypothetical person permanently present at a certain distance from a pipeline).
- Risk index methods, in which individual factors that influence failure frequency and consequences are assigned numerical index values which are combined, usually using weighting factors, to give an overall risk value;
- Risk matrix methods, in which frequency and consequences are each assigned to one of a number of severity categories, and combined in a two-dimensional matrix of risk categories (ranging from the highest, requiring immediate corrective action, to the lowest, requiring further monitoring only).

11.3.2 Quantitative Methods

Fully quantitative risk assessment imposes considerable demands relative to the availability of numerical data and calculation methods, in order to produce meaningful results. Appropriate methods have been developed for natural gas pipelines (see, for example, IGEM [10]), but significant difficulties can exist in quantitative consequence estimation for cases in which the position of the receptors most at risk may be at some distance from the failure site. Nevertheless, in jurisdictions in which quantitative standards of tolerable safety risk apply (see, for example, HSE [11]), these difficulties will need to be overcome.

Failure frequencies can be estimated quantitatively on the basis of historical data, though it is important to ensure that frequencies are determined on the basis of a population of pipelines that closely matches the characteristics of the pipeline being analyzed. Assuming that new CO_2 pipeline systems are properly designed and managed for the prevention of internal corrosion, then failure statistics related to modern pipelines carrying non-corrosive fluids would be appropriate. Alternatively, failure frequencies can be calculated using analytical models for each potential failure cause in a reliability analysis (see Nessim and Zhou [12]), which takes into account all of the properties of the pipeline, the contained fluid, and the external environment to provide an estimate of the probability of failure. For this application, the annual failure frequency is taken as equal to the probability that failure occurs in any given year of service life. A further technique that can be useful in determining failure frequency in pipeline systems is fault tree analysis [13]. In line with the failure definition, the top event for the fault tree can be defined as "a loss of service in the pipeline system." The analysis then concentrates on identifying and developing the various contributing events and causes which could lead to such a failure; the analysis proceeds downwards through the tree until the basic or initiating events are reached. The analysis generally considers all modes of failure which could lead to a service outage or loss of containment. The frequencies of the various events or conditions are combined using Boolean logic, through "and," "or" and conditional gates, to reach the frequency of the top event. All possible paths are quantified based on the probability of each branch of the tree [14].

For CO_2 pipelines, quantitative safety consequence analysis poses some problems. The basic tools are available. Models to determine the outflow profile over the course of a release exist, as do dispersion models that can be used to determine the relationship between the concentration of CO_2 (and any entrained toxics) and time at specific distances from a release (see Chapter 12). A Probit equation has been proposed that provides a dose–response function (in terms of lethality probability) for CO_2, and is consistent with the SLOT and SLOD criteria referenced in Chapter 13 [15]. Given these tools, it is possible to determine the usual measures of individual and societal risk (the latter in terms of expected casualties or F-N curves) for specific locations, relative to the pipeline alignment. Where difficulty is likely to arise is in dealing with situations in which topography has a significant potential effect, possibly leading to the accumulation of high concentrations in low-lying areas at greater distances than would be anticipated on the basis of conventional atmospheric dispersion models. A secondary model that accounts for potential gravity draining to, and concentration in, low-lying areas is needed for this purpose.

For the purposes of integrity maintenance planning and prioritization, index methods and matrix methods are widely used for natural gas and hydrocarbon liquid pipelines, and their adaptation to CO_2 pipelines does not pose such major problems. These methods are considered in some detail in the following sections.

11.3.3 Index Methods

11.3.3.1 Background

The index method is a semi-quantitative approach to risk assessment, which allows comparison of risks associated with a section of pipe to the other sections or other pipelines. For each pipe segment, risk is calculated as the product of the aggregate indices representing probability and consequences of failure.

The probability of failure is calculated as a weighted algebraic sum of the threats, and the consequence of failure is calculated as the weighted algebraic sum of the consequences. The relative weights of each variable in the algebraic sums represent the relative importance of each in contributing to total risk. This is algebraically defined in the following equations:

$$R = P \times C \tag{11 – 1}$$

$$P = \Sigma_{\mathrm{i=1}}^{\mathrm{n}} aiAi \tag{11 – 2}$$

and

$$C = \Sigma_{i=1}^{n} biBi \tag{11 – 3}$$

where;

R = Risk (or Risk Value)
P = Probability of failure
C = Consequence of failure
Ai = Probability of failure due to threat i
ai = Relative weight of threat i
Bi = Consequence of failure due to threat i
bi = Relative weight of consequence i
n = Number of threats or consequences.

The index n in turn depends on a number of conditions, including the variables that influence the threat or consequence. For example, the threat corrosion is dependent on soil type, the type of pipe coating, the efficiency of cathodic protection, etc. The model is completed by assigning values to each of these variables depending on the properties of each segment

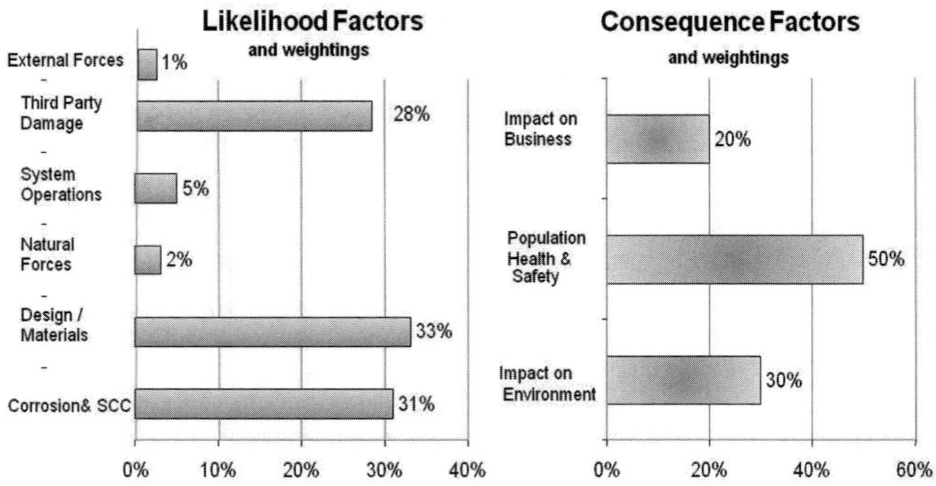

Figure 11-3. General scheme of threats and consequences, with examples affecting total risk (*Risk = Likelihood × Consequence*)

of the pipe. For example, Figure 11-3 shows the 6 threats (among many threats as identified under this Figure) that can contribute to the probability of failure and the 3 effects contributing to the consequence of failure.

Some of these likelihoods and consequences are explained below:

A — Likelihood Categories (partial)

Corrosion: This accounts for defect severity based on inline inspection and digs, historical internal and external corrosion and coating condition evaluation as well as CP assessments along the pipeline right of way. In the event that in-line inspection results are not available, then qualitative risk assessment factors such as coating type, soil conditions, etc. may be included in the risk assessment. For those pipeline sections where in-line inspection data exist, the more quantitative (through engineering assessment) in-line inspection data such as defect severity, density, and growth may be used in the risk assessment.

Design and Materials: This includes information regarding the design and quality of the pipe materials used in the construction of the pipeline system and includes data such as seam type, toughness, age, maximum operating pressure versus pipe strength, measure of pressure cycling, and outer diameter to wall thickness ratio.

Stress Corrosion Cracking: Stress corrosion cracking includes information which identifies areas which may be susceptible to stress corrosion cracking and includes pressure cycling, coating susceptibility to breakdown, shielding, soil pH, soil moisture content, and effectiveness of cathodic protection.

System Operations: The System Operations category covers operational errors. This is based on the assumption that there is always some degree of human error in any operational environment. The information can be used to gauge incorrect operational issues which include inadequate employee training, leak history, near miss data, and pressure safety systems, etc.

External Forces: External forces considers the possibility of a pipeline rupture from a naturally occurring event such as those associated with river scour, flood plain areas, seismic areas and/or fault lines, and ground movement (slumping, landslides, settlement).

Third Party: Third party includes those conditions which indicate the degree of third party activity in the area of the pipeline right of way and considers construction and farm activity, third party leak history, right-of-way condition, line marking, depth of cover, and additional pipeline protection. It may be noted that land use/location class are usually the most significant factors.

B – Consequence Categories

Impact on Population: Impact on population identifies areas where a pipeline rupture would have a direct consequence to population and population sensitive areas and includes information such as population density, population sensitive areas (schools, hospitals, sports facilities, etc.), pipe diameter, and type of product (toxicity, flammability, etc.).

Environmental Impact: Impact on environment identifies areas where a pipeline rupture would have a direct consequence to the environment and includes information, such as environmentally sensitive areas, adjacent land use, pipe diameter, and response effectiveness.

Business Impact: Impact on business identifies the impact that a rupture would have to both customers of the pipeline and to the pipeline company itself. Information that is considered in this analysis includes redundancy of supply for customers of the pipeline, adjacent land use, and cost to repair the pipeline.

Some of the probability/likelihood factors can be attributed to:

- Design Considerations (Pressure, Temperature, Throughput);
- Pipeline specification and operating environment;
- Pipe Grade and wall thickness, material;
- Construction/installation techniques and date;
- Route condition and hazard potentials.
    - o Erosion
    - o Flood
    - o Landslide, settlement
    - o Earthquake
    - o Volcanic
- Pipe span;
- Buoyancy;
- Leak History;
- CP Conditions;
- Inspection and hydrostatic tests.

With the following consequence factors included, then it is possible to estimate the pipeline risk segment by segment and then assess the entire pipeline system risk:

- Rupture,
- Volume loss,
- Dispersion,
- Population impact,
- Environmental impact,
- Business operation impact (including loss of property and income).

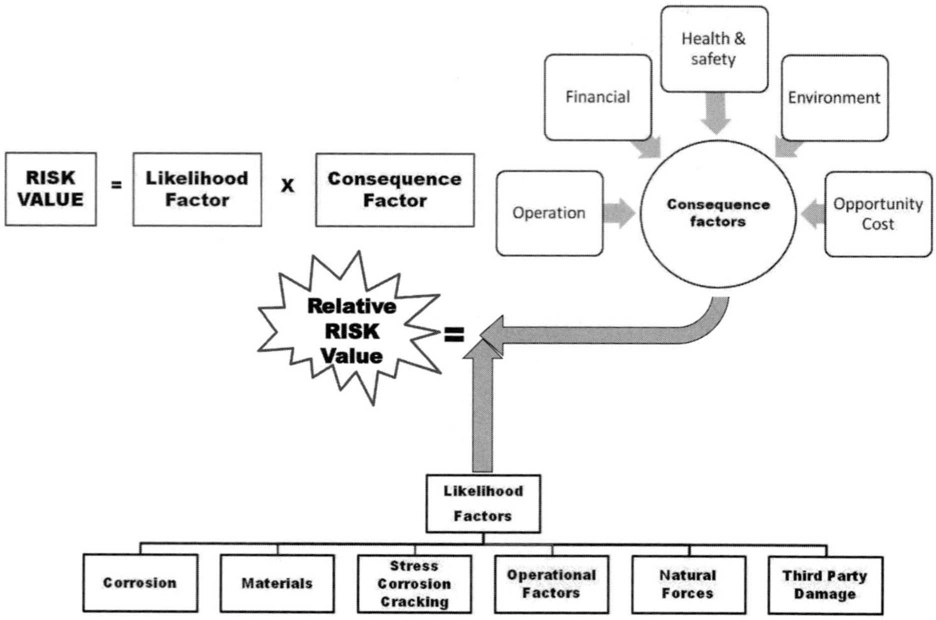

Figure 11-4. Typical risk assessment model

11.3.3.2 Numerical Index-based Risk Assessment Example

The following explains an index-based method of calculation of risks. Figure 11-4 depicts an example of consequence categories that can be used in calculating a risk value.

Typical "Likelihood and Consequence" factor values for a typical pipeline are shown in Figures 11.5 and 11.6. The values are based on industry experience for a typical large-scale liquid pipeline. The percentage factors indicated will vary according to the age of facilities, the type of fluid transported, and the extent of the network. However, these charts are included to illustrate how data can be arrived at and used in pipeline risk calculations.

11.3.3.3 Numerical Risk Evaluation Calculation

As an example for a given section of pipeline the **Likelihood Score*** can be determined as follows:

- **Third Party Damage (Category Weighting = 28%)**

Construction Activity (Variable Weighting = 17% or 0.17)

Activity is "Medium" — Scores as 7(See APPENDIX A to this chapter), therefore $7 \times 0.17 \times 0.28 = 0.3332$

- System Operations =(5%)
- Natural Forces =(3%)
- Design / Materials =(32%)
- External Corrosion =(20%)
- Internal Corrosion =(10%)
- Stress Corrosion Cracking (SCC) =(2%)

Sum = Likelihood Score

The **Consequence Score*** is determined in the same manner using variables for Impact on Business (20%), Health and Safety (50%), and Environmental (30%).

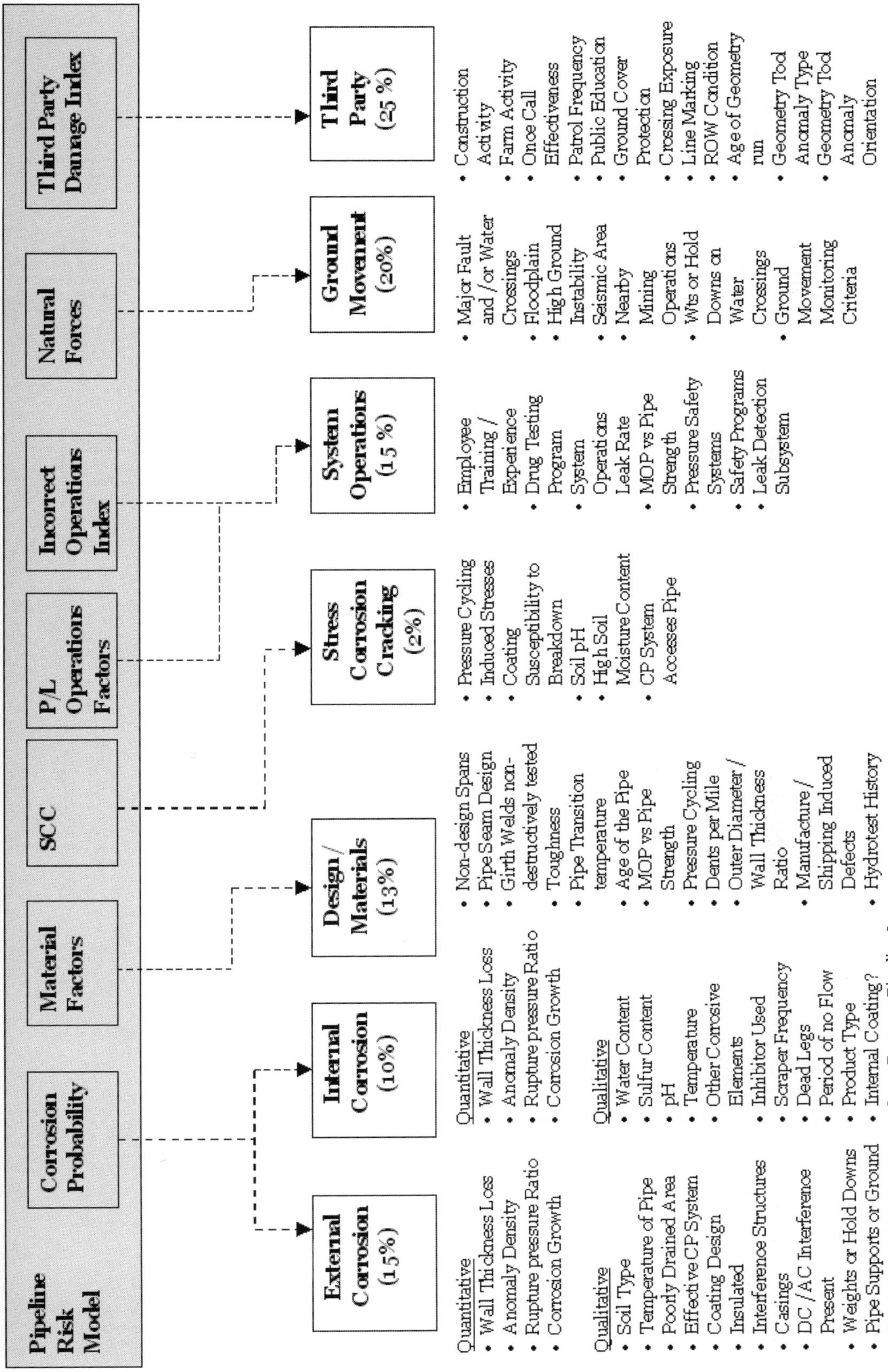

Figure 11-5. Typical likelihood threat factors

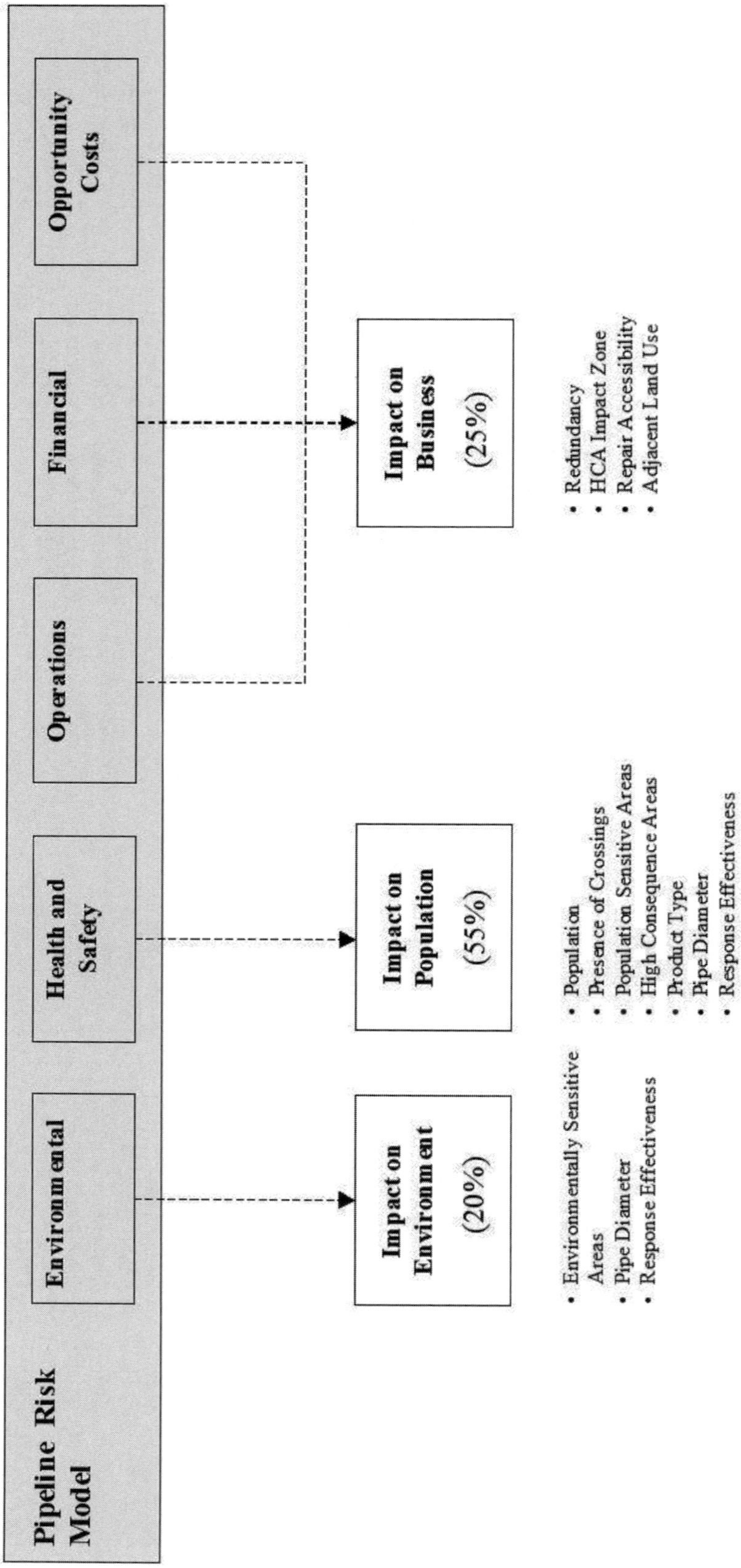

Figure 11-6. Typical consequence factors

*It may be noted that the score must be based on the "worst case" within a given risk assessment section.

11.3.3.4 Benefits and Applications

The above rigorous pipeline risk assessment method:

- Ensures focus on high risk areas;
- Serves as a mechanism for identifying new hazards;
- Adds structure, transparency, and process to risk management activities;
- Addresses "Whole Risk" picture;
- Facilitates demonstration of regulatory compliance;
- Can be used as a decision tool for various activities, such as:
  - o Prioritization/optimization of expenditures for maintenance
  - o Identification of risk mitigation and/or prevention activities
  - o Analysis of the placement of valves and flow restrictive devices(check valves, control valves, surge mitigation, etc.)
  - o Dig prioritization
  - o Vulnerable area assessment and prioritization (e.g., due to potential earthquake, land slide, flooding, etc.).

The strength of this risk assessment technique is thus its ability to facilitate the identification of suitable risk mitigation and prevention activities. The results facilitate focus of expenditures on areas identified as higher risk.

The risk assessment results can be beneficial in two ways:

- o Passively — obtain a risk score from the risk profile for a given pipeline system.
- o Actively — use the risk assessment to identify systems that will reduce any increase in risk.

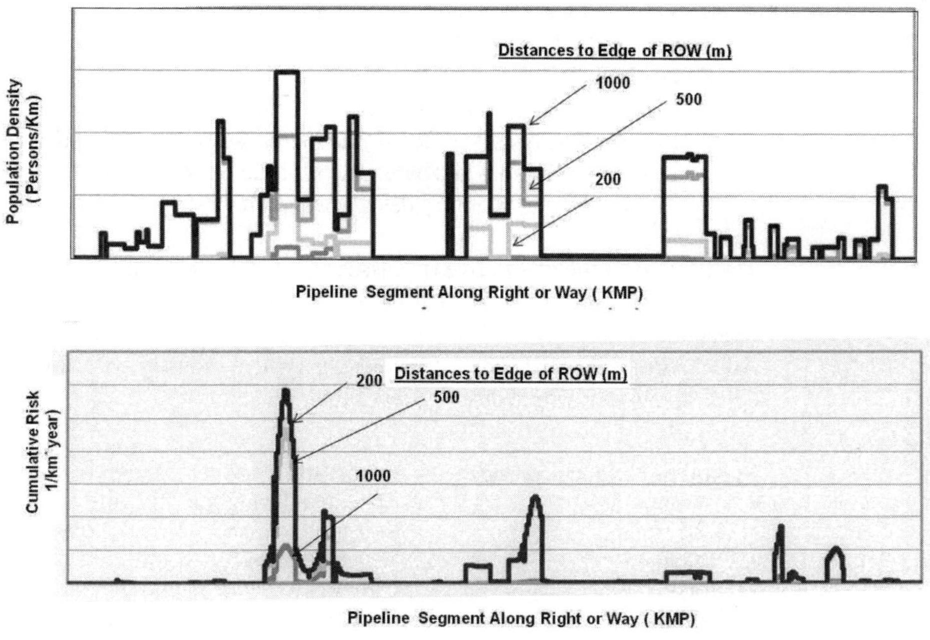

Figure 11-7. An example of calculated societal risk profile along a pipeline ROW [7]

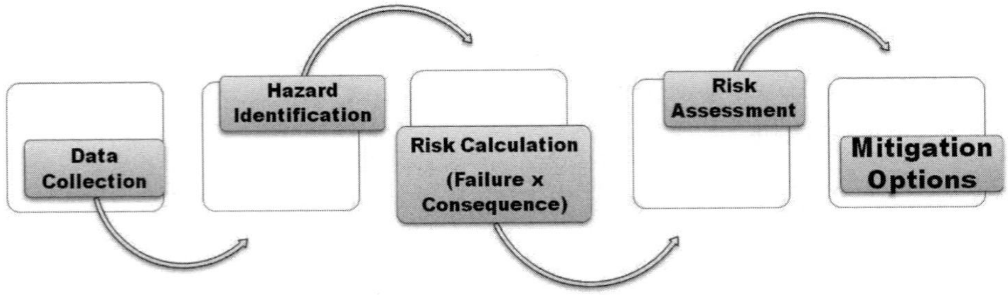

Figure 11-8. Qualitative risk assessment methodology

An example of a typical societal risk profile/plot based on population density and hazard severity along a pipeline route is illustrated in Figure 11.7.

11.3.4 Decision Matrix Risk Assessment

The decision matrix model provides an evaluation of a qualitative (rather than quantitative) approach to risk assessment. It relates each pipeline risk according to the likelihood and the potential consequences of an event by a simple scale, such as high, medium, or low. This method focuses on the most crucial threats [16] [17]. The qualitative approach for combining failure probabilities and consequences often uses numerical scoring methods or grading technique (H (High), L (Low), I (Intermediate), . . . etc.) to generate a relative risk ranking for various pipeline segments and lengths, along a pipeline route.

In undertaking the risk assessment with this technique, the following parameters are generally considered:

- **Route Location Analysis**—to provide the basis for the identification of critical areas.
- **Threat List**—A list of physical threats and hazard events to the pipeline at each location along the route. Some threats are location specific, while others are more uniformly distributed.
- **Options for External Protections**—to provide controls for threats (through design, physical protection, and monitoring measures).
- **Failure Analysis**—to determine the causes of past damage to the pipeline due to an identified threat and assessment of the consequence. This is also for threats that cannot be controlled (by design or monitoring).

The process of the qualitative risk assessment is depicted in Figure 11-8.

This risk assessment methodology combines an estimate of the frequency of occurrence of each hazardous event with an estimate of severity of the hazardous event to produce a risk class/metrics [18]. The following lists typical industry hazardous event frequency categories, together with a description of their meaning.

Frequency of Occurrence	**Description**
• Continual	Expected throughout the operation
• Frequent	Expected to occur (typically once/year or more)
• Occasional	Expected to occur several times in the life of pipeline
• Unlikely	Unlikely to occur within the expected the life of pipeline
• Remote	Very unlikely to occur within the life of the pipeline
• Improbable	Historically but not anticipated for this line or segment
• Hypothetical	Possible, but never occurred on a similar pipeline

Table 11-1 below indicates generally industry accepted categories for severity of consequences for use in a pipeline risk matrix.

TABLE 11-1. Typical consequence severity class for pipelines for use in risk matrix

Severity class/consequence	Description
Catastrophic	Applicable to pipeline routes/segments where the number of humans within the range of influence of the pipeline would result in many fatalities.
Major	Event causing few fatalities or loss of continuity of supply or major environmental damage.
Severe	Event causing hospitalization injuries or restrictions of supply.
Minor	Event causing no injuries and no loss or restriction of supply.

The "qualitative risk assessment" then combines an estimate of the frequency of occurrence of each hazardous event with the estimated severity of the hazardous event in a risk matrix format, Table 11-2 which can then be used to produce a risk class and action plan as shown in Table 11-3.

TABLE 11-2. Risk matrix (Legend: H = High risk, I = Intermediate risk, L = Low risk, N = Negligible)

	Risk Class			
Frequency of occurrence	Severity Class			
	Catastrophic	**Major**	**Severe**	**Minor**
Frequent	H	H	H	I
Occasional	H	H	I	L
Unlikely	H	H	L	L
Rare/remote	H	I	L	L
Improbable	H	I	L	N
Hypothetical	I	L	N	N

TABLE 11-3. Risk class and management actions

Risk Class	Action Required
High	Modify the hazardous event, the frequency or the consequence to ensure the risk class is reduced to intermediate or lower.
Intermediate	Repeat the risk identification and risk evaluation processes (and accuracy) to verify and, quantify, the risk estimation For confirmed intermediate risk class, modify the hazardous event, the frequency or the consequence to ensure that the risk class is reduced to low or negligible.
Low	Determine the management plan/process(for the hazardous even)t to prevent occurrence. Monitor changes which could affect the classification.
Negligible	Review at the next review interval.

In relation to Tables 11-2 and 11-3, it may be noted some standards (e.g., AS 2885.1 [19]) include an "Extreme" class for the three worst combinations and as well include the required timing of remedial action, e.g., extreme and high risks require mitigation "immediately" and "ASAP" (interpreted to mean within no more than a few weeks), respectively.

In terms of a pipeline transporting CO_2 containing impurities, the US DOE potential health effects for workers and the public who may be exposed to releases of CO_2 during the transportation process can also be considered in arriving at severity classes, such as those described below [20]:

- **Possible:** Accidents estimated to occur one or more times in 100 years of facility operations (typical frequency $\geq 1 \times 10^{-2}$ per year).
- **Unlikely:** Accidents estimated to occur between once in 100 years and once in 10,000 years of facility operations (frequency from 1×10^{-2} to 1×10^{-4} per year).
- **Extremely Unlikely:** Accidents estimated to occur between once in 10,000 years and once in 1 million years of facility operations (frequency from 1×10^{-4} to 1×10^{-6} per year).
- **Incredible:** Accidents estimated to occur less than one time in 1 million years of facility operations (frequency $< 1 \times 10^{-6}$ per year).

There are variations to the definition of hazardous events criteria [21] which may assign numerical values to frequency and consequence categories (Tables 11-4 and 11-5) and combine them by multiplication to produce a numerical overall risk index, which determines the required disposition (Table 11-6). They can also be displayed on a matrix, as shown in Table 11-7.

TABLE 11-4. Frequency index

Range	Frequency	Description
7	**Continual**	Expected to present itself during every point of operation
6	**Very Frequent**	Once a month; Can be expected to occur in most operational circumstances
5	**Frequent**	Once in three month; May occur in most operational circumstances
4	**Infrequent**	Once a year; May occur at some time
3	**Possible**	Once in 5 years; Could occur at some time
2	**Rare**	Once in 50 years; May only occur in exceptional circumstances
1	**Extremely Rare**	Once in 100 years; May only occur in exceptional circumstances

TABLE 11-5. Consequence severity range/index [21]

Severity range	Consequences	Health impacts	Critical services interruption	Organizational outcomes/objectives	Non-compliance
7	Extreme (Catastrophic)	Multiple deaths and serious prolonged health impacts	Facility closure and cessation of all activities	Complete performance failure	Serious, willful breach; criminal negligence or act
6	Large (More than Severe)	Multiple severe health crises/injury or death	Indeterminate prolonged suspension of work; nonperformance	Non achievement of objective/outcome; performance failure	Serious, willful breach; criminal negligence or act
5	Medium (Severe)	Severe health crisis	Prolonged suspension of work—additional resources required; performance affected	Significant delays; performance significantly under target	Deliberate breach or gross negligence; formal investigation
4	Intermediate (Small /Moderate)	Routine medical attention required	Short term temporary suspension—backlog cleared < 1 day	Minimal impact on organizational objectives	Breach; objection/complaint lodged; minor harm with investigation
3	Low (Minute)	First aid or equivalent only	No material disruption	Minor impact on organizational objectives	Negligible procedural breach; no evidence of poor action, good mitigation measures; minor impact to no impact on sensitive receptors (schools, hospitals, etc.)
2	Limited (Negligible)	May require first aid or equivalent only	No material disruption	Insignificant impact on organizational objectives	Negligible procedural breach; evidence of good faith and proper mitigation measures; no impact
1	None (No Consequence)	None	None	None	None

TABLE 11-6. Risk index [21]

Range	Consequences	Description
1 to 7	Acceptable	Acceptable risk without requiring any changes
8 to 20	Acceptable with controls	Acceptable risk after recommended changes and modifications are made to reduce risk
21 to 34	Undesirable	Undesirable risk requires design changes or safety evaluation
35 and above	Unacceptable	Unacceptable risk major modifications and design changes required

TABLE 11-7. Typical risk matrix—Combination of the indices of Tables 11-4 and 11-5. Shading depicts the required disposition shown in Table 11-6 [21]

		Frequency						
		Extremely Rare	Rare	Possible	Infrequent	Frequent	Very Frequent	Continual
Consequence		1	2	3	4	5	6	7
Extreme (Catastrophic)	7	7	14	21	28	35	42	49
Large (More than Severe)	6	6	12	18	24	30	36	42
Medium (Severe)	5	5	10	15	20	25	30	35
Small (Moderate)	4	4	8	12	16	20	24	28
Low (Minute)	3	3	6	9	12	15	18	21
Limited (Negligible)	2	2	4	6	8	10	12	14
None (No Consequence)	1	1	2	3	4	5	6	7

11.4 CO_2 PIPELINE FAILURE FREQUENCY CALCULATION BASED ON STATISTICAL DATA

The largest concentration of CO_2 pipelines is in North America (mostly in the USA); the oldest one was constructed in 1963 [21]. Indicative pipeline failure frequency calculations for similar pipelines can be based on statistical data from the US Office of Pipeline Safety (Figure 11-9).

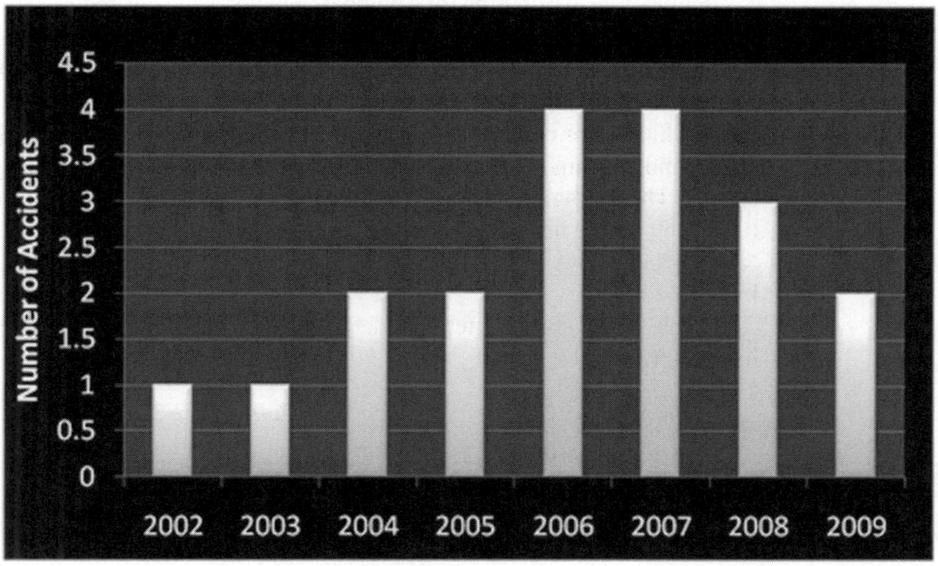

Figure 11-9. 2002–2009 Distribution of reportable accidents for CO_2 (and other non-flammable, non-toxic fluid at ambient conditions (loss of 5 $\geq$ barrels))

TABLE 11-8. CO_2 pipelines historical failure rate distribution (1986 to 2008) [21]

Failure Mode	Total Number of Accident	Percentage	Historical Failure Rate per km (Mile) of Carbon Dioxide Pipeline Per Year
Equipment failure	6	46	4.82E-05 (7.77E-05)
Corrosion	2	15.5	1.67E-05 (2.70E-05)
Operation error	2	15.5	1.67E-05 (2.70E-05)
Unknown	3	23	2.41E-05 (3.89E-05)
Total	13	100	1.05E-04 (1.69E-04)

TABLE 11-9. Typical causes leading to a release from CO_2 pipelines [21]

Deviation	Cause	Consequence	Example Safeguards
Pipe and valve failure	• Corrosion • Maintenance errors • External impacts (including third-party damage)	• Release of carbon dioxide	• Pipeline controls; Pressure and temperature indicators • Automated block valves for isolation • Inspections and preventative maintenance • Cathodic protection • Pipeline will be pressure rated to greater than the maximum discharge pressure of carbon dioxide centrifugal compressor
Operator Error	• Maintenance errors • External impacts (including third-party damage)	• Release of carbon dioxide	• Supervision of all personnel; including inspection of maintenance and operation activities • Appropriate training and experience requirements for all personnel

Failure rate distribution for a number of major accidents reported that involved a major release (i.e., ≥4000 barrels) of CO_2 is indicated in Table 11-8.

The above Table 11-8 (based on a total exposure of 3500 miles of CO_2 pipelines for 22 years) shows that 46% of the accidents were caused by equipment failure, the majority of which were caused by failure of a subcomponent (such as a valve or gasket). The second most common cause was "Unknown," accounting for approximately 23% of all accidents. The average failure rate for this period of time was 0.000105 failure per kilometer (0.000169 failure per mile) of carbon dioxide pipeline per year [20].

Typical causes of CO_2 releases of from pipelines and associated equipment are shown in Table 11-9.

Table 11-10 shows safety incidents between 1988 and 2008 related to CO_2 pipelines in the U.S. [8].

TABLE 11-10. US CO_2 pipelines safety records,1988 to 2008 [8]

• Approximate Total Length km (miles)	5,581 (3,468)
• Number of Incidents	26
• Number of Fatalities	0
• Number of Injuries	1
• Incidents/1000 km per year	$0.23 = 26/5.581 \times 20$

As shown in Table 11-10, incidents involving CO_2 pipelines have not resulted in any fatalities and the annual incident frequency is 0.23 per 1000 km. The major cause of pipeline failure is damage (*rupture* or *puncture*) during excavation for repairs or replacement or third-parties working on new pipelines [22].

The overall accident data from 1988 to 2008 indicates the following categories of release

- Rupture Release: Involving 4 accidents with the largest carbon dioxide releases (including 4000 barrels and 7408 barrels).
- Puncture Release: Involving next 7 highest releases (300 barrels to 3600 barrels).
- Other 15 incidents: Involving minor occurrences, (<100 barrels).

From statistical data including the amount of release, OPS estimated CO_2 pipeline failure rates and the probabilities of pipeline release incidents. Using the total length of CO_2 pipelines involved of 5581 km (3,468 miles), the annual *rupture* failure frequency was calculated to be 3.41×10^{-5} (kilometer-year)$^{-1}$, i.e., 0.0000341 per kilometer-year (or $1/5581 \times (4/21)$ i.e., for 4 ruptures over 21 years 1988–2008 inclusive).

The annual failure frequency from puncture (for 21 year period) was calculated to be 5.97×10^{-5} (kilometer-year)$^{-1}$, i.e., 0.0000597 per kilometer-year (or $1/5581 \times (7/21)$, i.e., for 7 punctures over 21 years).

It may be noted that the annual pipeline failure frequencies and the probability of at least one failure over a 20-year lifetime of the pipelines is calculated assuming the probability of failure to be uniformly distributed with the failure rate equal to the product of the failure frequency and the pipeline length.

Incident data between 1988 and 2008 from above data can be used to calculate the frequency and probability of pipeline ruptures and punctures.

11.4.1 Sample Calculations

Assuming 15 km 12" outside diameter pipeline,

- Annual rupture frequency = $15 \times (3.41 \times 10^{-5}) = 0.00051$ per year;
- Annual puncture frequency = $15 \times (5.97 \times 10^{-5}) = 0.0009$ per year.

In reality, the annual pipeline failure frequencies for a particular assessment must be calculated based on the site-specific pipeline lengths and likely conditions.

Based on the above available and recorded data, failure frequencies can be calculated for different ranges of likely volume of CO_2 released from a rupture or puncture and accidental or even planned release.

11.5 USE OF RISK ASSESSMENT IN PIPELINE MAINTENANCE PLANNING

The key to any successful risk management program is the pro-active selection of pipeline maintenance activities based the integrated measure of harm-risk. Therefore, an effective Risk Assessment process needs to identify, quantify, and analyze areas of high concern. The appropriate steps are as follows:

Step 1 — Locate Highest Risk Pipeline in the System
Step 2 — Locate Highest Risk Portion of the Line

Step 3 — Determine Risk Increasing Variables
Step 4 — Design Risk Reduction Methods/Systems
Step 5 — Evaluate Risk Reduction Techniques
Step 6 — Confirm Results
Step 7 — Export Risk Results

11.6 PIPELINE RISK MANAGEMENT

The consequences associated with accidents and unforeseen incidents that may occur in a pipeline transmission system can be most serious and alarming because in general they can have influence on personnel, public, and the environment and company operation. It is therefore important to have in place an effective risk management plan, from assessment to mitigation which will require a reliable data collection, risk analysis, and prediction tool as well as response. There are publications that include a description of the essential elements of a risk management system for a pipeline system (e.g., Canadian Standard Association (CSA) Z662 Annex A) that can be useful in devising a proper plan.

For the pipeline industry, risk management is the set of actions taken from risk assessment to risk control, including continuous monitoring and communication. It implies the process for assessing this risk through execution of an action plan to control and reduce future risks.

The pipeline industry, while it considers the enterprise wide risk management, generally focuses on pipeline operational risks (Figure 11-10) to assure continued safe and reliable operation benefitting the industry, specifically the owning company, its personnel, and the public with least impact on the environment [23].

Figure 11-10. Interrelationship of enterprise wide risk management

Risk assessment is one component of risk management. It is a decision support tool providing a structured approach which identifies and documents possible cost-effective actions to improve facilities integrity and thus public safety and environmental protection. It is a decision framework that captures information from the existing pipeline system, conditions and operation to aid allocation of funds to improve the system safety and reliability. It is a potential alternative to prescriptive regulations and thus is an "evolving" or "living" process for the reasons identified below:

- To address public safety and environmental concerns while managing expectations of greater efficiencies and cost control
- Rapid pace of change — past does not always predict the future
- Increasingly complex web of inter-related factors
- Increasingly high stakes with incidents
- Change from being reactive to proactive — concentrating on incident prevention and reduction in incident frequency
- To increase consistency and level of objectiveness in overall pipeline risk assessments including decision making. To adjust and focus operation, maintenance, and rehabilitation activities on the most important risks inherent to pipeline operations.
- To effectively communicate and formally demonstrate that risks are systematically identified, analyzed, and managed.

11.6.1 Reducing Risk in Operating Pipelines Transporting CO_2 Containing Impurities

Steps can be taken to minimize risks from CO_2 pipelines (as for any hazardous material pipelines) [24]. Steps include:

- Site pipelines away from densely populated areas when possible;
- Review site conditions (e.g., hilly terrain, valleys, depressions) influencing heavy gas dispersion;
- Avoid routing pipelines near homes in sheltered, populated valleys where leaking CO_2 could accumulate to dangerous levels;
- Optimize pipeline systems based on realistic CO_2 stream properties;
- Design against running facture. Implement fracture propagation control efficiency with respect to wall thickness, steel grade, and steel toughness. Select appropriate materials properties respecting occurrence of low temperatures under normal operating conditions and abnormal operation, including the case of line depressurization;
- Perform route related risk assessment/ minimization studies;
- Review pipeline operation for unexpected and transient situations;
- Review and implement the requirement for flow restrictive devises (i.e., sectionalizing line valves and check valves) based on minimizing containment losses and dispersion in event of a unforeseen failure. Increase the number of valves for leak volume reduction. The additional risks for leakages introduced by the line valves themselves should also be considered. Optimize valving location arrangements considering elevation profile and locations of populated areas;
- Consider a highly efficient and reliable leak detection and localization systems including the installation of a pipeline-related Geographical Information System (GIS);
- Monitor pipelines regularly for corrosion and other abnormalities;
- Monitor constantly for leaks;

- Consider adding odorant to CO_2, as is done with natural gas, to allow people to notice small leaks;
- Develop and implement appropriate operating, maintenance and emergency procedure. Consider and implement a rigorous training program for operators to take necessary actions to avoid/minimize potential releases of CO_2 and impurities;
- Implement a systematic, comprehensive and integrated integrity management program to investigate/maintain/improve the safety of the CO_2 pipeline system (see, for example, the applicable sections of ASME B31.8S-2004);
- Develop Emergency Response Plans including implementing appropriate measures to alert the public on existence and routing/location of below ground pipelines; use of pipe markers and warning tape. Introduce measures including a one call system to prevent accidental pipeline damage by third parties carrying out digging and construction activities nearby. Establish and maintain liaison with police and other relevant organizations to quickly respond to a CO_2 release. Establish and perform a continuing training program for emergency situations.

11.7 INTEGRITY MANAGEMENT OF CO_2 PIPELINES

Pipeline Integrity Management is a process for identifying, assessing, and evaluating (associated risk) for mitigating threats to the integrity of a pipeline system for the intended function over the expected design life. It is a dynamic and cyclical process. It is a comprehensive, systematic and integrated program providing the means to improve safety of pipeline systems with the least impact on the public and the environment [25]. Most importantly, it is a requirement of most regulations and standards.

11.7.1 Risk and Integrity Relationship

Risk assessment is a tool that is used in the integrity management process. From operational points of view risk and integrity are interrelated tools that help to identify weaknesses in a pipeline system, leading to corrective action or monitoring means for assuring continued safe pipeline operation, see for example, CSA Z662-11 Annex N [26]. They are a management tool that can help in making informed decisions for timely expenditures based on optimization of maintenance requirements versus upgrades, replacements, and monitoring systems investments.

Integrity assessment is a combination of the following:

- Pipeline and right of way inspection and data collection,
- Defect/damage assessment,
- Repairs/Upgrades,
- Monitoring.

The integration of pipeline integrity and risk assessments is implemented through an active and continuous integrity management program that brings the following critical programs together [27]:

1. Maintenance programs detailing maintenance activities and schedules
2. Defect Assessment Methodology: methods operators used to assess defects and make repair decisions, e.g., ANSI/ASME B31.G-, RSTRENG,- etc,
3. In-Line Inspection (ILI) policy: decisions with respect to types of inspection tools selection (e.g.: high resolution MFL and UT inspection, crack detection or other technologies) and their frequency of use

4. Corrosion prevention management program: the methods used to prevent both internal and external corrosion and for system measurement (e.g., CP — off potential and external coating surveys) as well as protection monitoring, etc.
5. Repair techniques: types of repair methods used (composite wrap, sleeves/clamps, cut-out, weld deposition/grind out, etc.)
6. Spill detection system: methods for detection of system leaks and ruptures
7. Emergency planning and facilities: emergency plans and facilities in place and location
8. Third party damage prevention: procedures in place to prevent 3rd party interference
9. Failure data and history base
10. Risk assessment techniques: method(s) used to assess risk and analysis to identify high risk areas and design plans to mitigate through integrity assessment (items 1 to 9 above)

In reality, risk and integrity assessments are a part of an integrity management program for all pipelines. Risk assessment is a tool that is used, wherever there is the potential for non-trivial failure consequences, as a part of the integrity management program. In such a program, the principal use of risk assessment is to establish pipeline vulnerabilities and areas and to prioritize schedules of physical integrity assessments and repairs. The physical assessments are:

- In-service pressure testing
- Inline inspection with intelligent inspection pigs (ILI),
- Direct assessment (DA) comprising:
  - o Pre-assessment (risk assessment)
  - o Indirect examinations
  - o Direct examination
  - o Analysis and information integration,
- Site investigation and monitoring programs for environmentally related outside forces including geotechnical loadings.

11.7.2 Code Requirements for Integrity Assessment

Internationally recognized standards are commonly used for the design, construction, operation and maintenance of pipeline system facilities. In relation to pipeline structural integrity assurance, there are a number of standards and documents that set out / outline the assessment requirements. Some of the applicable documents (as related to pipeline transportation of CO_2 containing impurities) are listed below [28], [30]:

- ASME B31G — *Manual for Determining the Remaining Strength of Corroded Pipelines: A Supplement to ASME B31 Code for Pressure Piping*
- API Std 1160 — *Managing System Integrity for Hazardous Liquids Pipelines*
- API RP 1162 — *Public Awareness Programs for Pipeline Operators*
- API Std 1163 — *In-Line Inspection Systems Qualification Standard*
- API Std 5T1 — *Imperfection Terminology*
- API RP 1110 — *Pressure Testing Liquid Pipelines*
- API 1156 — *Effects of Smooth and Rock Dents on Liquid Petroleum Pipelines*
- API RP 579 — *Fitness for Service*
- CSA Z662 Annex N — *Guidelines for Pipeline Integrity Management Programs*
- PHMSA 2001 — *Liquid Pipeline Integrity Management Rule*
- NACE 35100 — *In-Line Nondestructive Inspection of Pipelines*

- NACE RP0102 — *In-Line Inspection of Pipelines*
- NACE RP 0169 — *Control of External Corrosion on Metallic Piping Systems*
- NACE SP0502 — *Pipeline External Corrosion Direct Assessment Methodology*
- NACE SP0204 — *Stress Corrosion Cracking Direct Assessment Methodology*
- NACE SP0206 — *Internal Corrosion Direct Assessment Methodology for Pipelines Carrying Normally Dry Natural Gas (DG-ICDA)*
- *NPR-CEN/TS 15173:2006 Gas supply systems — Frame of reference regarding Pipeline Integrity Management System (PIMS)*
- US Department of Transportation (DOT) Title 49 Code of Federal Regulations (CFR) Part 195, *Integrity Management for Hazardous Liquid Pipeline Operators (specifically Appendix C)*
- PHMSA Pipeline Library http://primis.phmsa.dot.gov/comm/PipelineLibrary.htm

API Standard 1160 well defines the framework for a hazardous liquid pipeline structural integrity management program as indicated in Figure 11-11. This framework can be utilized in establishing an integrity management program for pipeline transportation of CO_2 containing impurities. Integrity assessment techniques and comparisons of various global codes addressing pipeline integrity issues are further addressed elsewhere [29].

11.7.3 Integrity Management Planning

The interrelationship between risk/integrity assessments and integrity management planning are illustrated in Figure 11-11. As noted previously, the risk assessment described in this chapter is an integral part of pipeline system integrity management planning in which the integrity of the system is determined through the use of physical integrity assessments.

The basic purpose of risk assessment is to guide other actions to be taken to assure the integrity of the pipeline system. This guidance will principally be in determining the risk-prioritized order for physical integrity assessments and mitigative measures. The basic integrity management process is summarized in Figure 11-12 [27, 29].

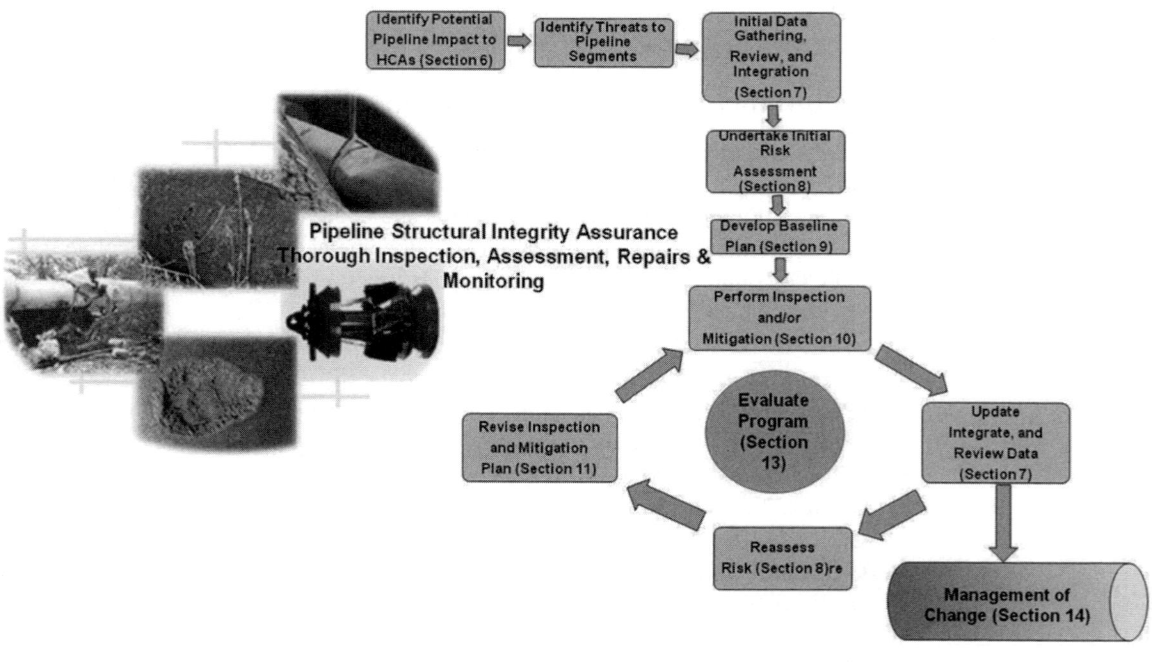

Figure 11-11. Typical pipeline integrity assurance planning program/process — API 1160 [25, 29]

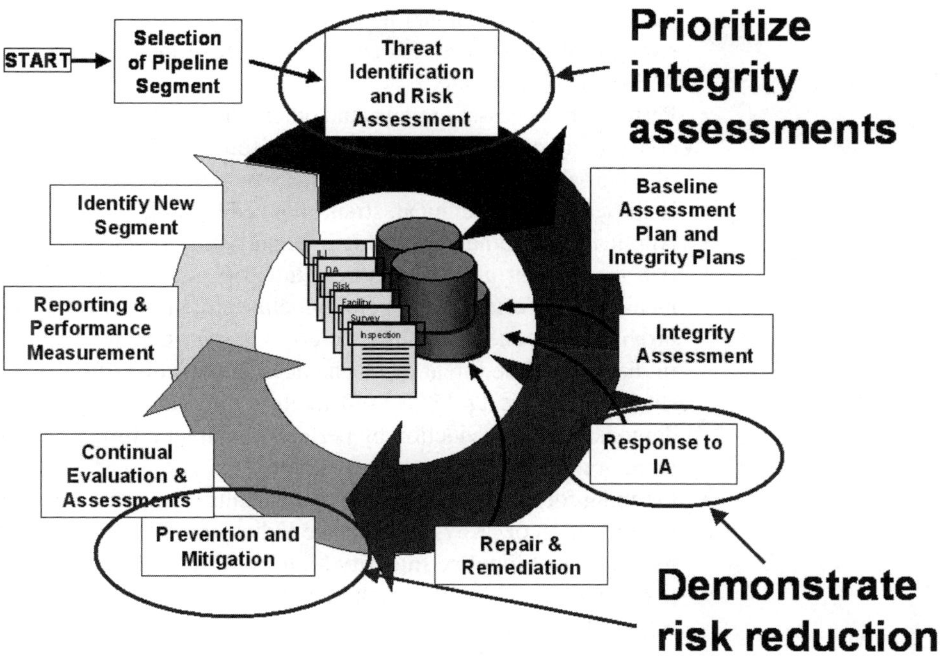

Figure 11-12. Pipeline risk and integrity assessments and interrelationship processes [27, 29]

The entire integrity assessment and management process is described by Mohitpour et al [29] and is summarized in the following main steps:

- Identification of pipeline segments in high consequence areas (HCAs) or other risk receptors.
- Identification of the applicable threats to pipeline structural integrity
- Performing a risk assessment to prioritize physical/structural integrity assessments. The threats to structural integrity would generally include the following:
  - o Mechanical damage (third party interference)
  - o External corrosion
  - o Internal corrosion
  - o Stress corrosion cracking
  - o Manufacturing defects (especially those associated with seam-welded pipe)
  - o Construction defects (especially those associated with pipe joining methods)
  - o Weather-related and outside force. This category of threat includes loads applied to buried pipelines due to earth movement. Note that such loads may be especially damaging in the presence of construction defects associated with pipe joining methods such as girth weld defects.
  - o Incorrect operations (principally those that could lead to over-pressure of the pipeline)
  - o Equipment failure
- Assemble a plan to perform physical integrity assessments on the pipeline system. This plan is often referred to as the Baseline Assessment Plan (BAP) and typically will cover a rolling 5-10-year period.
- Perform the pipeline structural integrity assessments
- Respond to the results of the structural integrity assessments and respond to the anomalies found during the assessments. For example, in the case of metal loss

defects due to corrosion, this step will usually include an estimate of the remaining strength of the defective area according to some pre-determined criteria such as ASME B31G [30], etc.

- Perform the necessary repairs and other remedial measures to the structural integrity concerns associated with the anomalies found. Such measures might include the use of repair sleeves in the case of metal loss defects and slope stabilization measures in the case of geotechnical, strain-induced loadings.
- Application of other risk reduction and risk mitigation measures indicated by the risk assessment and the physical integrity assessments. These typically will be measures that affect the ongoing pipeline structural integrity protection such as the provision of concrete slabs over crossings or increased right-of-way surveillance in the case of the threat of mechanical damage due to third party interference. As shown in Figure 11-11, it most likely will be necessary at this stage to explicitly demonstrate risk reduction by performing either a further system-wide risk assessment or a site-specific risk assessment.
- Planning for continual integrity assessments.
- Identification of "Key performance indicators (KPIs)" to measure and track the effectiveness of Pipeline Integrity Management Plan and determine if the plan needs to be revised. Such KPIs might include:
    - Number of leaks/ruptures
    - Lengths inspected versus lengths planned for inspection
    - Delivery interruptions
    - Corrective and predictive maintenance spending trends
- New segments that could affect HCAs or other risk receptors identified during the (annual) application cycle of the pipeline integrity assurance management plan are added to the integrity management database and included in the next system-wide risk assessment as the integrity management process cycle continues.

Again, the central position of the database in Figure 11-11 should be noted (and emphasized). Risk assessment in common with all other activities in the Pipeline Integrity Management Plan process relies heavily on high quality, currently updated data. The execution of each step of the process also continually produces data and integrity-related information. A well-structured data management system is essential to an effective pipeline integrity management planning.

REFERENCES

[1] Serpa, J., Morbee, J., and Tzimas, E., 2011. "Technical and Economic Characteristics of a CO_2 Transmission Pipeline Infrastructure" European Commission, Joint Research Centre, Institute for Energy, http://publications.jrc.ec.europa.eu/repository/bitstream/111111111/16038/1/reqno_jrc62502_aspub lished.pdf

[2] Bluemink E., 2010. "Size of Exxon spill remains disputed." Anchorage Daily News. June 10th http://www.adn.com/2010/06/05/1309722/size-of-exxon-spill-remains-disputed.html.

[3] HSE (UK Health and Safety Executive), 2010. "Analysis of responses to the consultation on proposed amendments to the pipelines Safety Regulations 1996 and the Health And Safety (Fees) Regulations."

[4] Guijt, W., 2004. Analyses of incident data show US, European pipelines becoming safer. Oil and Gas Journal, January 26, pp. 68–73.

[5] Tuft. P., 2005. "Review of pipeline incident data to 2004." The Australian Pipeliner — January., http://pipeliner.com.au/news/review_of_pipeline_incident_data_to_2004/043339/.

[6] Smith R., 2010. "Safety Perspective and Technical Challenges for CO_2 Pipelines Panel on CO_2 Pipelines. International Pipeline Conference, Calgary Alberta, Canada, Sept 27th.

[7] Kaufmann, K.D., 2008. " Pipeline Design -Key Issues in Pipeline Design to Avoid Pipeline Failure," Workshop Energy Institute and CCSA, London, UK, March 11.

[8] US DOT OPS (Office of Pipeline Safety), 2009. "Pipeline Safety Incident Online Database and Supporting Information. Statistics and Incident Data for Natural Gas and CO_2 pipelines from 1988-2008." Pipeline and Hazardous Materials Safety Administration (PHMSA), April http://ops.dot.gov/stats/IA98.htm.

[9] Hanna, S.R. and Drivas, P.J., 1988. "Guidelines for Use of Vapor Cloud Dispersion Model," American Institute of Chemical Engineers Center for Chemical Process Safety. New York.

[10] IGEM (Institution of Gas Engineers and Managers), 2008. IGEM/TD/2 "Application of pipeline risk assessment to proposed development in the vicinity of Natural Gas PIpelines."

[11] HSE (UK Health and Safety Executive), 2001, "Reducing Risk, Protecting People," HSE Books, Sudbury, Suffolk, UK.

[12] Nessim, M. and Zhou, W., 2005. "Guidelines for Reliability-Based Design and Assessment of Onshore Natural Gas Pipelines." Gas Research Institute Report 04/0230. Des Plaines, IL.

[13] Mohitpour, M., Szabo, J., and Van Hardeveld, 2010. " Pipeline Operation & Maintenance — A Practical Approach" 21st Edition, ASME Press, New York pp. 623–666.

[14] D'Atri, M.F., Rodriguez, D., and García-Martínez, R., 2008. "IMPROVING PIPELINE RISK MODELS BY USING DATA MINING TECHNIQUES" www.igu.org/html/wgc2009/papers/docs/wgcFinal00663.pdf.

[15] McGillivray, A. and Wilday, J., 2009. "Comparison of risks from carbon dioxide and natural gas pipelines," HSE Report RR749, HSE Books, Sudbury, Suffolk.

[16] Muhlbauer, W.K., 2004. "Pipeline Risk Management Manual Ideas, Techniques and Resources," Elsevier Inc. UK.

[17] Nyborg, M., Arvidsson, K., Johansson, J., Liljemark, S., and Olsson, L., 2011. "Risk analysis methodology for CO_2 transport including quantified risk calculation' Energy Procedia Volume 4, Pages 2816–2823.

[18] ERS (Environmental Risk Solutions Pty Ltd), 2004. "CHEVRONTEXACO GORGON DEVELOPMENT- Public Risk Assessment Technical Report, Appendix:E3," Document # J9765 PRA_2, Rev. A4.Nov. http://www.chevronaustralia.com/Libraries/Chevron_Documents/E3_Public_Risk_Assessment.pdf.sflb.ashx.

[19] Standards Australia, 2007. "Pipelines — Gas and liquid petroleum. Part 1: Design and Construction," Sydney, Australia.

[20] AEP (American Electric Power), 2011. "Human Health and Safety Impacts from Potential CO_2 Releases" Appendix G report Mountaineer Commercial Scale Carbon Capture and Storage Project Mason County, West Virginia, USA," Contract # 326849x215 February http://www.netl.doe.gov/technologies/coalpower/cctc/EIS/mountaineer_pdf/Appendix_G_Volume2.pdf.

[21] HECA (Hydrogen Energy California), 2009. "Carbon Dioxide Pipeline Risk Analysis- Appendix E," May 19., http://www.energy.ca.gov/sitingcases/hydrogen_energy/documents/applicant/revised_afc/Volume_II/Appendix%20E.pdf.

[22] PHMSA OPS Personal Communication, 2011

[23] Meier, R.L, 2000. "Integrating Enterprise-Wide Risk Management Concepts into Industrial Technology Curricula" Journal of Industrial Technology • Volume 16, Number 4 • August, www.nait.org.

[24] Price, P.N., McKone, T.E., and Sohn, M.D., 2007. "Carbon Sequestration Risks and Risk Management," Environmental Energy Technologies Division ,Lawrence Berkeley National Laboratory, Berkeley, California 94720, http://eetd.lbl.gov/ie/pdf/LBNL-513E.pdf.

[25] Reynolds, L., 2010, "Pipeline Integrity Standards & Regulations" AUCSC Integrity Management Workshop — Period1, http://www.aucsc.com/_aucsc%20speaker%20files/PM1%20-%20Pipeline%20Integrity%20Standards%20and%20Regulations.pdf.

[26] Canadian Standards Association, 2011. Standard Z662, "Oil and gas pipeline systems", Canadian Standards Association, Mississauga, Ontario.

[27] Colquhoun I.R., Calvi C., and MacPherson, H. 2006, "Management System Approach to Pipeline Integrity." Proceedings of IPC2006. International Pipeline Conference, Calgary, Alberta. September 25-29, 2006. Paper IPC 2006-10531.

[28] Mora, R.G., 2007. "Pipeline Integrity Management Program (IMP) — Key Documentation" http://www.tecnologiatotal.net/pipeline_integrity.pdf.
[29] Mohitpour, M., Murray, A., McManus M., and Colquhoun, I., 2010. " Pipeline Integrity Assurance- A Practical Approach," ASME Press, New York.
[30] ASME (American Society of Mechanical Engineers), 2009. "ASME B31.G Manual for Determining the Remaining Strength of Corroded Pipelines." A Revision of ASME B31.G 1991.

APPENDIX TO CHAPTER 11 AN EXAMPLE OF INDUSTRY INDEXING NUMBERS (BASED ON LARGE PIPELINE NETWORK)

1. Design

	Maximum Index Points	Assigned Index	Justifying Feature	KMP
1.1 Operating Stress Level			46% SMYS	
72–80% SMYS	20			
60–72% SMYS	10			
40–60% SMYS	5	5		
less than 40%	0			
1.2 Material Properties			Old Pipe	
Low Impact Toughness	10	10		
Acceptable Impact Toughness	0			
1.3 Hydrostatic Test Pressure				
Test Pressure less than 1.25 ´ MAOP	10	10		
Test Pressure greater than 1.25 ´ MAOP	0			
1.4 River Weights				
Buoyancy Present	5			
No Buoyancy	0	0		
1.5 Road and Rail Crossings				
Cased Crossing	10			
Uncased Crossing	5			
No Crossing	0	0		
1.6 Weld Design- Materials and Procedure				
Does Not Meet Code Criteria	20	20		
Does Meet Code Criteria	0			
1.7 Above Ground Facilities			Station Discharge	
High Potential for stress/fatigue	10	10		
Well Designed	0			
1.8 Thermal Stresses				
Potential for Thermal Stresses	5	5		
No Thermal Stresses	0			
1.9 System Age				
greater than 30 years	10	**10**		
10 to 30 years	5			
less than 10 years	0			
TOTAL INDEX	100	70		
FINAL INDEX (Total Weighting %)	30%	21*	Note*= $0.3 \times 70 = 21$	

2. Third Party Damage and Encroachment

	Maximum Index Points	Assigned Index	Justifying Feature	KMP
2.1 Activity Level				
High Construction Activity	45			
Normal Residential	15			
Agricultural	5	5		
Pasture	0			
2.2 Depth of Cover				
Insufficient depth of cover	10			
Minimum depth of cover	5	5		
Excess depth of cover	0			
2.3 Above Ground Facilities			Shell Jumping Pound	0.00
Above Ground (A/G)Facility within 50 m of vehicular traffic	10	10		
A/G Facility fenced and signed	5			
A/G Facility with barrier	3			
No A/G Facility	0			
2.4 Public Education				
Mail outs/internet	10			
And Meetings with First Responders	5			
And Meetings with Contractors	3			
And Landowners	0	0		
2.5 Dent/Gouge ILI				
No ILI Information	10	10		
ILI within 5 years	5			
ILI within 2 years	0			
2.6 Right-of-way condition				
Not signed, unclear	5			
Signed, clear	0	0		
2.7 Patrol Frequency				
Never	10			
Monthly	5	5		
Weekly	0			
TOTAL INDEX	100	35		
FINAL INDEX(Total Weighting %)	**20%**	7	$=0.2 \times 35$	

3. Coating System Condition

	Maximum Index Points	Assigned Index	Justifying Feature	KMP
3.1 Coating Type/Brand			Bare Section Of Pipe	0.000 to 0.570
Bare Pipe	50	50		
Tape	30			
Coal Tar	10			
Extruded Tap	0			
FBE	0			
3.2 Coating Installation			N/A	
Over the Ditch Applied Winter	10			
Over the Ditch Applied Summer	5			
Shop Applied	0	0		
3.3 Coating Condition			N/A	
Poor Condition — Wrinkled/ Disbonded	20			
Fair Condition	10			
Good Condition	0	0		
3.4 System Operating Temperature			Has Operating Coolers	0.00
Greater than 65 °C	10			
35–65 °C	5			
Less than 35 °C	0	0		
3.5 Repair			N/A	
Tape (Repairs, Tie-ins, Girths, etc.)	10			
No Tape	0	0		
TOTAL INDEX	100	50		
FINAL INDEX (Total Weighting %)	20%	10	= 0.2 × 50	

4. Surface Hydrology

	Maximum Index Points	Assigned Index	Justifying Feature	KMP
4.1 Surface Drainage			Creek	15.235 to 15.250
Rivers	35			
Creeks	30	30		
Intermittent Creeks	15			
None	0			
4.2 Flood Plane				
Evidence Of Flood Plane	40	40		15.215 to 15.275
No Evidence	0			
4.3 Cross Slope Drainage				
Drainage Along Pipeline	25			
Drainage Perpendicular To Pipeline	15	15		
No Evidence Of Drainage Pattern	0			
TOTAL INDEX	100	85		
FINAL INDEX (Total Weighting %)	5%	4.25	= 0.05 × 85	

5. Soils and Terrain Geology

	Maximum Index Points	Assigned Index	Justifying Feature	KMP
5.1 Composition				
Clays	30	30		
Loamy Clays	20			
Loam	10			
Sands and Gravels	0			
5.2 Subsidence				
Rivers or Creeks	30			
Rolling Hills	20			
Flat Terrain	10			
No Subsidence	0	0		
5.3 Drainage				
Very Poorly Drained	15			
Poorly Drained	10			
Moderately Well Drained	5	5		
Well Drained	0			
5.4 Organics				
High	25			
Medium	15			
Low	0	0		
TOTAL INDEX	100	35		
FINAL INDEX (Total Weighting %)	5%	1.75	$= 0.05 \times 35$	

6. Stress Corrosion Cracking

	Maximum Index Points	Assigned Index	Justifying Feature	KMP
6.1 SCC Susceptibility				
Class 1 Susceptibility	40			
Class 2 Susceptibility	20			
Class 3 Susceptibility	10			
Class 4 Susceptibility	5			
No Class Susceptibility	0	0		
6.2 In Line Inspection Results				
No Crack Tool ILI	20	20		
Crack Tool ILI Run Within 5 Years	10			
Crack Tool ILI Run Within 2 Years	0			
6.3 Investigative Excavation Results				
No Investigative Excavations	20	20		
1–5 Investigative Excavations	10			
More Than 5 Investigative Excavations	0			
6.4 SCC Hydro Strength Test				
No Hydrotest Within 10 Years	20	20		
Hydrotest Within 10 Years	0			
TOTAL INDEX	100	60		
FINAL INDEX (Total Weighting %)	10%	6		

7. Corrosion

	Maximum Index Points	Assigned Index	Justifying Feature	KMP
7.1 Cathodic Protection Potentials			T.P. 1 950-1200 mV	0.000 to 0.570
less than 850 mV	40			
greater than 1200 mV	20			
850–950 mV	10			
950–1200 mV	0	0		
7.2 Rectifier Outages				
More than 12 months outage	5			
Less than 12 months outage	0	0		
7.3 Rock Shielding				
Evidence of Bedrock	5			
No Bedrock	0	0		
7.4 Cathodic Protection Interference				
Evidence of Interference	10			
No Evidence of Interference	0	0		
7.5 In Line Inspection Frequency				
No ILI	10	10		
ILI run within 5 years	5			
ILI run within 2 years	0			
7.6 Investigative Excavation				
No Investigations	20	20		
1–5 Investigations	10			
6 or more Investigations	0			
7.7 Shorted Casings				
Possible Shorted Casing	10			
No Shorted Casings	0	0		
TOTAL INDEX	100	30		
FINAL INDEX (Total Weighting %)	10%	3	$=.1 \times 30$	

Chapter 12

ATMOSPHERIC DISPERSION OF CO_2 CONTAINING IMPURITIES

12.1 CO_2 EMISSIONS RISK

Dispersion modeling of CO_2 release scenarios needs to be included as part of a quantitative risk analysis when designing or operating CO_2 pipelines, as discussed in Chapter 11. Analysis of release conditions, and subsequent dispersion, can help evaluate factors that increase the risk of a loss due to physical or toxicological impacts (e.g., emergency shutdown valve spacing). Simulation of a CO_2 release presents new challenges when compared to simulation of natural gas releases. One of these challenges is the dispersion characteristics of a CO_2 pipeline release to atmosphere. Although a CO_2 release does not pose an explosion risk as with a natural gas pipeline, the released CO_2 disperses as a nonflammable dense (i.e., heavier than air) cloud that can potentially create hazardous conditions for some distance from the release site. Dispersion modeling techniques need to address the dense-gas specific behavior of a CO_2 release, for example, gravity slumping that leads to increased dispersion and air entrainment in the cross-wind direction. As CO_2 pipelines are a relatively young technology, compared to hydrocarbon pipelines, the dispersion modeling techniques are still being refined to account for CO_2-specific phenomena such as solid particle (dry ice) formation in the cold depressurized jet.

Risk analysis of CO_2 dispersion must address the potential toxic or asphyxiant hazards posed by a CO_2 cloud, as well as the potential toxic hazards posed by any impurities (e.g., H_2S, NH_3, CO) that are mixed within the CO_2 cloud. The CO_2 release scenarios that should be simulated are dependent on the risk analysis application; a new-construction risk analysis would need to account for accidental rupture, puncture of the pipeline, or equipment failure, whereas a maintenance risk analysis could require that dispersion modeling be performed for a planned pipeline depressurization (blowdown).

The following chapter provides a discussion on the release hazards specific to CO_2 pipelines, a brief background on atmospheric dispersion and pipeline release dynamics, and an introduction to dispersion modeling techniques. Atmospheric dispersion results from simulating controlled blowdowns of a CO_2 pipeline with impurities as well as results from modeling a full pipeline rupture are also presented.

12.1.1 CO_2 Dispersion Hazards

A high-pressure CO_2 pipeline release to atmosphere can present a toxicological hazard as well as an asphyxiation hazard to humans and livestock exposed to the dispersing plume. CO_2 concentrations in excess of 40,000 ppm_v are considered to be immediately dangerous to life and health by the United States (U.S.) National Institute of Safety and Health (NIOSH). The toxicological hazards posed by impurities within the CO_2 mixture must also be evaluated in dispersion modeling even if these impurities are minor components of the dispersing plume, as these impurities may determine the hazard distances of concern.

Although CO_2 does not present a combustion hazard upon release, the potential for overpressure near the jet must also be considered as a release risk, particularly in the case of an unforeseen pipeline rupture. Other hazards that could be present in the vicinity of the high-momentum release include extremely cold temperatures due to the depressurization, friction heating of the surrounding air potentially creating an ignition source for hydrocarbons, and solids formation within the high-velocity jet.

Supercritical CO_2 is a solvent frequently employed in the manufacturing and food preparation industries. However, CO_2 is not a solvent in its gaseous state, and therefore any impurities that have been absorbed by the supercritical CO_2 could "fall-out" upon depressurization [1], therefore increasing the concentration of impurities in the vicinity of the release. Assuming that any impurities released within the CO_2 plume remain fully mixed in the dispersing cloud, the dense-gas dispersion behavior of CO_2 could lead to accumulation of toxic impurities in low-lying areas, and increased cross-wind impact distance due to gravity-driven spreading as compared to a less-dense release (e.g., methane). A discussion on CO_2 pipeline hazards is presented in Chapter 13, Section 13.3.

The U.S. Occupational Safety and Health Administration (OSHA) and the NIOSH exposure limits for CO_2 and H_2S are listed in Table 12-1 below. The occupational exposure limits for H_2S set by Canadian provinces are summarized in a Canadian Association of Petroleum Producers guide [2].

12.1.2 Concentration Averaging Times

The concentration values presented for the simulations in this chapter are calculated by ERCBSLAB [5] with a 3-minute averaging time; exposure limits are often given in terms of longer averaging times (e.g., 30,000 ppm_v CO_2 averaged over 15 minutes). Generally, Probit functions that describe the relationship between concentration, exposure duration, and degree of lethality can be applied to scale exposure limits to varying concentration averaging times; however, this relationship is complex for CO_2, and the applicable concentration and exposure time ranges for the Probit analysis should be well understood before the analysis is used for quantitative risk analysis as determined by the Netherlands National Institute for Public Health and the Environment [6]. The United Kingdom Health and Safety Executive (HSE) has proposed Specified Level of Toxicity due to Dangerous Toxic Load (SLOT DTL) and Significant Likelihood of Death due to Dangerous Toxic Load (SLOD DTL) levels in HSE [7] that can be used to derive a Probit function:

$$Pr = A + B \ln (C^n t) \tag{12 – 1}$$

TABLE 12-1. CO_2 and H_2S exposure levels and human health effects [3, 4]

	Exposure limit and duration	
Exposure limit	**CO_2**	**H_2S**
NIOSH REL TWA[a]	5,000 ppm_v[b] 8 hours/day	–
OSHA PEL[a]	5,000 ppm_v 8 hours/day	20 ppm_v ceiling 50 ppm_v maximum peak 10 minutes
NIOSH REL STEL[a]	30,000 ppm_v 15 minutes	10 ppm_v ceiling 10 minutes
NIOSH IDLH[a]	40,000 ppm_v Immediate	100 ppm_v Immediate

[a] REL = recommended exposure limit, PEL = permissible exposure limit, STEL = short-term exposure limit and IDLH = immediate danger to life or health concentration value.
[b] ppm_v = parts per million on a volume basis.

Where:

Pr	=	Probit value, used to determine the probability of mortality (a *Pr* of 5 implies a 50% probability of mortality)
A, B	=	Probit constants
C	=	Concentration of CO_2 in ppm_v
n	=	Toxic load exponent (substance-dependent)
t	=	Exposure time

The SLOT DTL and SLOD DTL parameters are discussed in Chapter 13.

The CO_2 cloud dissipation time should be considered when evaluating the appropriate exposure and exposure duration limits. If the plume duration is short, a comparably short averaging time (i.e., averaging time less than or equal to the source duration) exposure limit would represent the actual exposure of an individual as compared to a long averaging time (i.e., averaging time greater than the source duration) exposure limit [8].

12.2 ATMOSPHERIC DISPERSION

12.2.1 Meteorological Effects on Dispersion

Meteorological conditions play a key role in determining the downwind behavior of emissions. The plume behavior is governed through two primary mechanisms:

- advection, where the plume is "stretched" in the wind direction;
- and mechanical or buoyant turbulence, where mixing in the atmosphere results in plume vertical and cross-wind dispersion and air entrainment.

The advective conditions are primarily governed by the mean horizontal wind speed. Short-term dispersion modeling generally assumes that the bulk air is travelling at a constant speed and in a constant direction and neglects the vertical component of the wind vector. More complex dispersion models can account for transient meteorological conditions and even the change in wind velocity and direction in the vertical direction [9].

Atmospheric turbulence is generated by eddies of various magnitudes and directions that result in the plume being mixed with ambient air. Turbulence is generated through flow over roughness elements or wind shear—termed *mechanical turbulence*—or through changes in the buoyancy of the air through heating and cooling—termed *buoyant turbulence*.

A categorization of the degree of mechanical and buoyant turbulence in the atmosphere has been proposed by Pasquill [10] and later modified by Turner [11]. This categorization is used to apply empirical cross-wind and vertical dispersion coefficients as functions of

TABLE 12-2. Pasquill atmospheric stability characterization

	Daytime			Nighttime	
	Incoming Solar Radiation			Cloudiness	
Surface wind speed at 10 mAGL[a] (m/s)	**Strong**	**Moderate**	**Slight**	**≥ 4/8**	**≤ 3/8**
< 2	A	A–B	B	–	–
2–3	A–B	B	C	E	F
3–5	B	B–C	C	D	E
5–6	C	C–D	D	D	D
> 6	C	D	D	D	D

[a] mAGL = meters above ground level.

Where: A is extremely unstable, B is moderately unstable, C is slightly unstable, D is neutral, E is slightly unstable, and F is moderately stable.

downwind distance for use in Gaussian plume modeling. The categorization assigns a stability class to the atmospheric conditions through consideration of the surface wind speed (as faster winds generate more mechanical turbulence) and the heat flux to and from the surface. The surface heat flux is an indicator of buoyant turbulence because as the surface is heated through solar radiation, the lower-level air will be heated by the surface and the warmed-air packet will rise. As the air packet rises, it will cool if a temperature inversion is not present, and return to ground level. This cycle—termed a *convective cell*—generates buoyant turbulence in the atmosphere. On a clear night, the Earth will radiate heat to the upper air layer, thereby resulting in a cooler ground surface that then cools the air near ground level, and a convective cell is not formed. This condition is considered to be stable as little buoyancy-driven turbulence is generated.

The Energy Resources Conservation Board in Alberta (ERCB) has noted that stable atmospheric conditions—class F—coupled with low wind speed is generally the worst condition with regards to plume dispersion [5]. It is important to screen multiple meteorological conditions when evaluating dispersion hazards. A stable atmospheric condition with low wind speed may produce high concentrations due to the plume dispersion suppression; however, an unstable atmospheric condition may increase the cross-wind dispersion, and a fast wind speed could increase the range of territory affected by the release. The latter case is of consequence when impurities with relatively low exposure limits are present in the plume.

12.2.2 Terrain Effects

Special consideration needs to be accorded to dense gas releases in low-lying areas, for example valleys, streams, and ditches. The dense CO_2 release may settle in these areas and dispersion will be dependent primarily on mechanical turbulence. The low-turbulence wake formed behind physical structures can also result in dense-gas accumulation. Norway's Det Norske Veritas (DNV) recommends that numerical dispersion simulation tools should be employed for assessing dispersion in areas where the effects of topographical characteristics, physical structures, and the release properties of CO_2 could combine to present a significant hazard to people and/or livestock [12].

The terrain surface roughness impacts the generation of mechanical turbulence in the atmosphere. The surface roughness represents the height of the surface boundary layer and is caused by the presence of physical objects such as plants or buildings. The horizontal wind velocity increases with height within the surface boundary layer and approaches the free stream velocity at the surface roughness height above ground level. The surface roughness ranges from 0.01 m over prairie grassland to greater than 1 m in urban environments [5]. The surface roughness parameter can be used to generalize the vertical wind profile through a power law relation [9].

Dispersion over water bodies differs from dispersion over land due to the negligible surface heat flux from the water surface, and alternative dispersion models need to be employed when the release is offshore. Unless specifically accounted for, it is assumed that the plume is reflected off of any impinged terrain with no deposition of plume mass.

12.3 RELEASE DYNAMICS

12.3.1 Source Flow Rate

Upon pipeline depressurization initialization (for example, through opening a blowdown valve, accidentally puncturing a pipeline, or upon full guillotine rupture), the flow rate from the opening will decay exponentially with time.

Kruse et al. [13] approximated the time-dependent outflow rate through the following:

$$\dot{m}e = A_e\sqrt{P(t)\rho(t)c} \qquad (12-2)$$

Where:

$\dot{m}_e$	=	Mass flow rate from one end of pipeline
A_e	=	Throat area at the vena contracta
$P(t)$	=	Transient pipeline pressure
$\rho(t)$	=	Transient pipeline density
t	=	Time
c	=	Material constant

In this equation, the mass of CO_2 released in time step dt would be evaluated and then the physical properties of the fluid in the pipe at the new lower pressure would be determined. From this approximation, a 30 km, gaseous, 3.5 MPa and 0.65 m diameter pipe section would be 90% empty within 163 seconds of full rupture [25].

The volume of fluid vented will be dependent on the pipeline segment size, the initial fluid conditions, and the release mitigation measures. If automated emergency shut-down (ESD) valves are employed on the pipeline, the decompression wave will be transmitted down the length of the pipeline until the ESD valves are triggered through either a pressure rate-of-change signal or through surpassing a minimum pressure setting. Once the ESD valves are triggered, the valves will close and the release will be limited. The total volume released is then dependent on the decompression wave speed, the length of the segment, the ESD trigger settings, and the ESD closure time. An "effective" pipeline segment length accounts for these factors.

Wilson has developed a double-exponential approximation to the transient mass flow from one end of an isothermal pipeline [5]:

$$\dot{m}_e = \left(\frac{1}{1+\alpha}\right)\left[\dot{m}_{ei}\exp\left(-\frac{t}{\alpha^2\beta}\right)+\frac{M_i}{\beta}\exp\left(-\frac{t}{\beta}\right)\right] \qquad (12-3)$$

$$\dot{m}_{ei} = K_{inertia}\rho_{ei}A_e c_{ei} = K_{inertia}\left(\frac{Z_{ei}}{Z_{oi}}\left(\frac{2}{k+1}\right)\right)^{\frac{k+1}{2(k-1)}}\left(\frac{k}{Z_{oi}RT_{oi}}\right)^{\frac{1}{2}}P_{oi}A_e \qquad (12-4)$$

$$\alpha = M_i/\beta\dot{m}_{ei} \qquad (12-5)$$

$$\beta = \frac{2}{3}\frac{\rho_{oi}A_p}{3B_A\dot{m}_{ei}}\left[(1+B_A L_{eff})^{\frac{3}{2}}-1\right] \qquad (12-6)$$

$$B_A = \left(\frac{2}{k+1}\right)^{\frac{k+1}{k-1}}\left(\frac{K_{inertia}A_e}{A_p}\right)^2\left(\frac{Z_e}{Z_{oi}}\right)\frac{kf}{D} \qquad (12-7)$$

Where:

α	=	Initial time constant adjustment factor
$\dot{m}_{ei}$	=	Inertia-adjusted initial mass release rate
β	=	Final isothermal inertialess pipeline release time constant
M_i	=	Initial mass in the effective pipeline segment length
ρ_{ei}	=	Density of the moving gas at sonic velocity at the throat

c_{ei}	=	Initial speed of sound at adiabatic exit temperature
Z_{ei}	=	Initial gas compressibility at the exit plane
R	=	Ideal gas constant
T_{oi}	=	Stagnation temperature at initial conditions
P_{oi}	=	Stagnation pressure at initial conditions
ρ_{oi}	=	Initial pipeline density at stagnation conditions
A_p	=	Pipeline cross-sectional area
L_{eff}	=	Effective pipeline length (accounting for ESD closure)
k	=	Ratio of specific heats (assumed constant)
K_{inertia}	=	Inertial delay factor (constant, user-determined)
Z_e	=	Compressibility factor at throat conditions
Z_{oi}	=	Compressibility factor at initial conditions

Assuming a finite effective segment length, the pressure will decay within the pipeline from an initial mass flow rate. The initial mass flow rate can be reduced through application of the inertial delay constant to approximate the effects of inertia within the pipeline. The equivalent *steady* release time, t_{eq}, for use in dispersion modeling is a sum of the time constants:

$$t_{eq} = \beta + \alpha^2 \beta \qquad (12-8)$$

12.3.2 Jet Dynamics

A high momentum jet resulting from depressurization of a supercritical CO_2 pipeline will undergo the following phases as depicted in Figure 12-1 for a vertical release and Figure 12-2 for a horizontal release:

- choked flow at the release throat creating an under-expanded jet,
- near-adiabatic, non-reversible expansion to atmospheric pressure and a decrease in temperature (thereby possibly increasing the solid and/or liquid mass fraction in the jet),

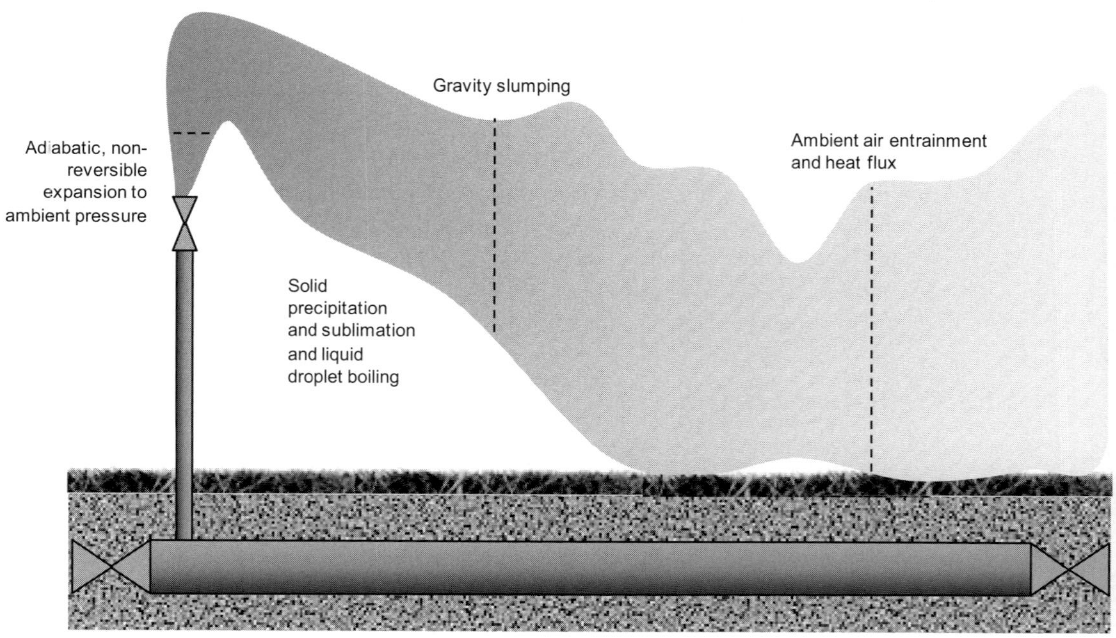

Figure 12-1. Vertical jet release behavior for a denser-than-air plume (pipeline blowdown)

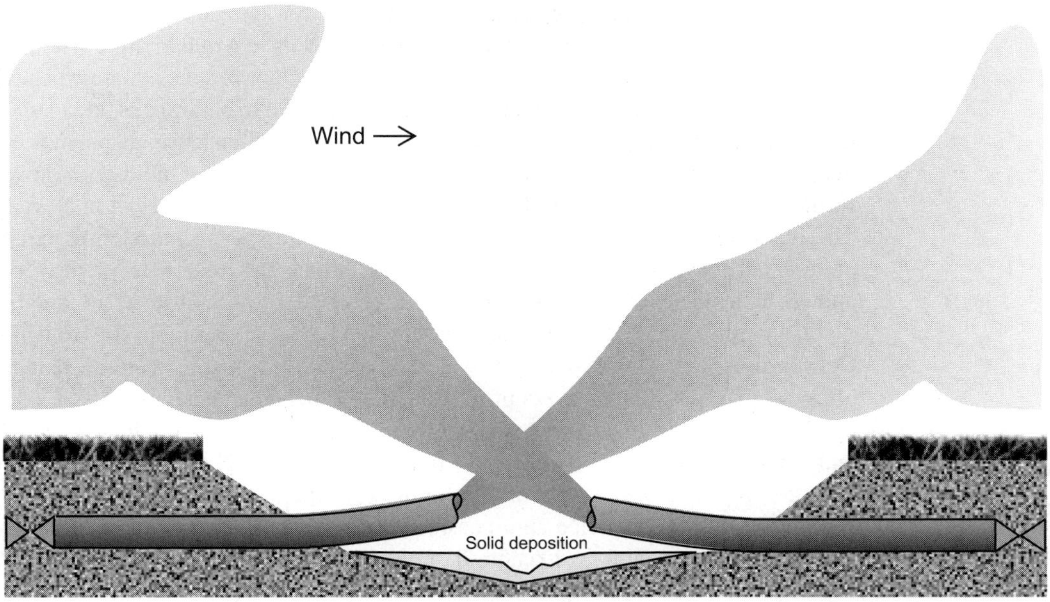

Figure 12-2. Horizontal jet release behavior for a denser-than-air plume (full pipeline rupture). Although one of the pipeline ruptured ends may direct the jet against the dominant wind direction, as the jet velocity decreases the wind advection will dominate and potentially reverse the jet direction

- potential impingement on surrounding structures leading to a decrease in jet momentum and possible change in jet orientation,
- solid precipitation and sublimation and liquid droplet boiling, and if the jet release is denser than ambient air, gravity slumping,
- and dispersion through advection and turbulent air entrainment with warming of the plume through surface and solar heat flux.

In the case of a buried CO_2 pipeline, rapidly expanding gas would blow out the surrounding ground cover and create a crater. Crater formation is a concern in the natural gas transportation industry since ignition of the released combustible material could impact neighboring exposed pipelines as described in Leis et al. [14]. The crater size and shape formed by a pipeline rupture will be dependent on the pipeline release dynamics [15]. It is noted in an Energy Institute guidance [16] that the crater formation impacts the subsequent plume dispersion since some of the jet's momentum may be lost when the surrounding soil is ejected and as the jet impinges on the crater walls.

12.3.3 Release Experiments

The thermodynamic properties of pure CO_2 are difficult to represent with a traditional cubic equation of state due to the high temperature and pressure of the fluid's triple point as compared to methane. CO_2 solids may form during depressurization due to the significant cooling effect from expansion. Another simulation challenge exists if there are impurities, such as H_2S, CO, NH_3, or N_2, present as impurities—even in very low concentrations—can significantly change the thermodynamic behavior of the CO_2 mixture as compared to pure CO_2. (A discussion on equations of state for CO_2 with impurities can be found in Chapter 4 of this book.)

Therefore, the validation of dispersion models for application to CO_2 mixture releases is important when applying the results of these models to a quantitative risk analysis. CO_2 release experiments have been conducted in the past; however, the dynamic behavior of CO_2 in the atmosphere is not as well understood as compared to natural gas due to the aforementioned thermodynamic complexity as well as the fluid's dense gas dispersion behavior, as exemplified in Figure 12-3. CO_2 release experiments ("Kit Fox" field-tests) have been performed as part of a Petroleum Environmental Research Forum project in Nevada [17, 18]. The intent of these experiments was to improve the dispersion modeling parameters (including vertical entrainment into the dense gas plume and the effects of surface roughness and obstacles) used in dense gas models such as HEGADAS, DEGADIS, and SLAB.

Cameron-Cole LLC performed a controlled release experiment for Howell Petroleum Corporation [19]. This experiment was used to field-test results from air dispersion modeling of a CO_2 wellhead blowout in moderately unstable (Pasquill class C) atmospheric conditions with a 3 m/s wind speed. CO_2 monitoring devices were located at ground level and 1.5 mAGL between 16 m and 137 m from the source. The CO_2 source release rate was 4.9 kg/s from a 7 MPa (1000 psi), 50 mm (2") pipeline vent. The conclusions from this study were that the modeled and measured concentrations were within an order of magnitude of each other, with the exception of one location where the disparity was attributed to shifting winds and topographical effects.

Several CO_2 release experiments are in development with the recent interest in investment in CO_2 pipelines due to carbon capture and sequestration. These projects have been summarized in [20] and include:

- DNV's CO_2PIPETRANS joint industry project with an objective to develop recommended practices with regards to offshore and onshore CO_2 pipelines. A knowledge gap has been identified in the area of CO_2 release modeling validation data and this will be an area of research and development work in the current phase of work [21].

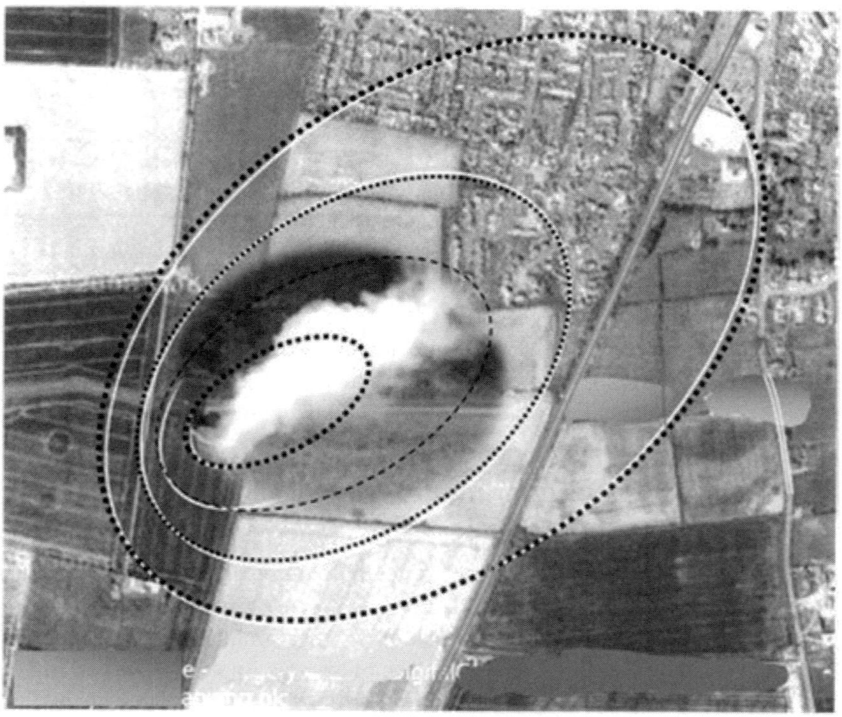

Figure 12-3. A high-pressure CO_2 release from a pipeline (composition from internet sources)

- The European Commission Seventh Framework Programme (FP7) funded CO2PIPEHAZ project, *Quantitative failure consequence hazard assessment for next generation CO_2 pipelines* that is focusing on discharge and dispersion modeling of CO_2 releases [22].

12.4 MODELING TOOLS

The primary air dispersion modeling techniques can be classified as:

- Gaussian,
- numerical (computational fluid dynamics),
- or dense gas.

The selection of dispersion models for simulating CO_2-pipeline releases should consider the following:

- the model's equation of state that is used to determine the pressure, temperature and phase of the released CO_2 and impurities,
- the model's limitations on pipeline dimensions and failure modes,
- the model's treatment of plume rise and subsequent dispersion (including meteorological conditions),
- and the model's incorporation of automated release mitigation.

The selection of a dispersion model also needs to address the level of complexity that is appropriate for the risk analysis application by considering:

- the effect of topography on dispersion
- and the degree of risk associated with a release (e.g., population density).

In Gaussian dispersion modeling, the downwind concentration profile is considered to be directly proportional to the emission source mass flow rate, inversely proportional to the mean wind speed, and directly proportional to the cross-wind and the vertical dispersion factors [9]:

$$\bar{c}(x, y, z, H) = \frac{Q}{\bar{u}} F_y F_z \qquad (12-9)$$

$$F_y = \frac{1}{\sqrt{2\pi}\sigma_y} \exp\left[\frac{-y^2}{2\sigma_y^2}\right] \qquad (12-10)$$

$$F_z = \frac{1}{\sqrt{2\pi}\sigma_z} \left\{ \exp\left[\frac{-(z-H)^2}{2\sigma_z^2}\right] + \exp\left[\frac{-(z+H)^2}{2\sigma_z^2}\right] \right\} \qquad (12-11)$$

Where:

$\bar{c}$ = Mean mass concentration at a point x, y, z and release height H
H = Effective source release height
Q = Source mass emission rate
$\bar{u}$ = Mean transport velocity across the plume

F_y = Dispersion factor in the cross-wind direction
F_z = Dispersion factor in the vertical direction
σ_y = Gaussian plume dispersion parameter in the cross-wind direction
σ_z = Gaussian plume dispersion parameter in the vertical direction

Mass conservation is maintained in Gaussian modeling through Equation (12–12):

$$\int_0^{\infty}\int_{-\infty}^{\infty} \overline{cu}\,dydz = Q \qquad (12-12)$$

The primary assumptions made in the Gaussian plume model include:

- The release duration is long, the terrain is not complex, and meteorological conditions are constant, therefore the plume dispersion can be considered to be steady-state.
- Normal distribution of concentration within the plume.
- Conservation of species within the plume (i.e., no deposition).
- Advection is the dominant dispersion mechanism and turbulent diffusion is negligible in the downwind (x) direction.

A complete treatment of the Gaussian model formulation and assumptions inherent within can be found in Arya [9]. The Pasquill categorization of atmospheric stability provides Gaussian plume dispersion parameters for the cross-wind and vertical directions as a function of downwind distance and stability class.

Numerical solutions to the conservation of mass, momentum, and energy equations through computational fluid dynamics (CFD) can also be employed to solve for the dispersion behavior of fluids in the atmosphere. As mentioned in Section 12.2.3, numerical dispersion simulation tools should be employed when the release risk is elevated due to dispersion over complex terrain, near physical structures, or where the risk of exposure to the public is high [12]. Numerical methods are generally used when atmospheric reactions are of importance, as in the case of ozone creation through photochemical reactions of NO_x and hydrocarbons in the atmosphere. CFD is computationally expensive, and while the tools can provide a more complete solution to a specific dispersion scenario, Gaussian models or dense gas models can screen dispersion scenarios to determine if a more refined solution is required.

The behavior of dense gas plumes requires special treatment as compared to passive, neutrally-buoyant plumes as discussed in Section 12.2. Dense plumes may rise due to the momentum of the source; however, the plumes will sink to ground level due to the effects of negative buoyancy. The SLAB dense gas model solves the cross-wind averaged conservation equations of mass, momentum, and energy (and species) through numerical integration with the Runge–Kutta method [8]. The SLAB model assumes a constant concentration gradient across the box half-width, B and box height, h as shown in grey in Figure 12-4.

An actual dense gas plume will have decaying concentrations profile from the centerline along the cross-wind and vertical directions. The box model dimensions are chosen such that the ground-level concentration along the plume centerline is the uniform concentration across the dimensions of the box through a simple mass balance (including air entrainment from the sides of the box as shown in Figure 12-4). If the plume is neutrally buoyant, the dense gas box model approximates a Gaussian plume model. However, if density effects are non-negligible, the dimension h will decrease as the plume slumps, thereby increasing the cross-wind dimension B and increasing the lateral plume spread as compared to a passive plume. The centerline concentration of a dense gas plume is not necessarily higher than for a passive plume due to air entrainment into the top and side of the plume as described in Appendix B of the ERCB model technical guidance document [5].

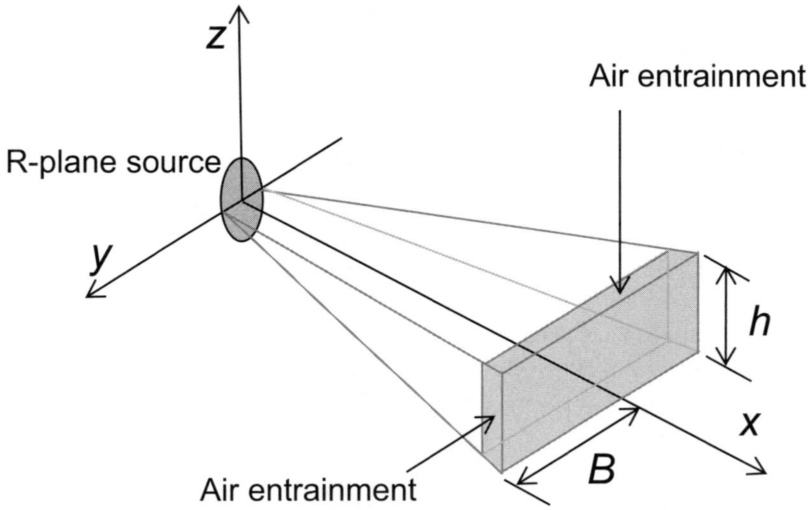

Figure 12-4. SLAB dense gas box model parameters. The plume concentration is constant over the *y–z* plane

The steady dispersion model concentrations can be transformed to account for the decaying mass flow rate from the source for a dense release [5]:

$$X_{o,transient} = \left(\frac{\dot{m}_e}{\dot{m}_E}\right)\left(\frac{F_N(\phi)}{F_N(\phi_E)}\right)^2 X_{o,steady} \tag{12 – 13}$$

where:

$X_{o,transient}$	=	Transformed transient release concentration of contaminant
$X_{o,steady}$	=	Steady release concentration of contaminant
$\dot{m}_e$	=	Transient mass release rate of source gas
$\dot{m}_E$	=	Modeled steady mass release rate of source gas
$F_N(\phi)$	=	Effect of density on the dense gas box dimensions ($2Bh$) normalized to the neutrally-buoyant case
ϕ	=	Dimensionless variable for a continuous steady release of dense gas, dependent on relative density difference between the source and ambient conditions, the source volumetric release rate, and the reference wind speed

Numerous dispersion models are available for simulating dense gas releases, including:

- the publically available Areal Locations of Hazardous Atmospheres (ALOHA) model, provided by the U.S. National Oceanic and Atmospheric Administration (NOAA) and the U.S. Environmental Protection Agency (EPA);
- the publically available Dense Gas Dispersion Model (DEGADIS), provided by the U.S. EPA for *zero-momentum*, gravity-driven dense gas releases;
- and the commercial consequence modeling package PHAST developed by DNV Software.

The PHAST package has been enhanced to allow for simulation of solid formation in CO_2 releases as described in Witlox et al. [23]. The Energy Institute used PHAST software to simulate dispersion of CO_2 resulting from pipeline ruptures [16].

Two publically available dense gas dispersion models are described in more detail below: the ALOHA model and Alberta's Energy Resources Conservation Board (ERCB) $ERCBH_2S$ model.

12.4.1 ALOHA Dispersion Model

The ALOHA model can calculate the rupture release rate of a single chemical component from a pipeline and the subsequent atmospheric dispersion at user-defined meteorological conditions. ALOHA assumes that one end of the pipeline segment is either capped, thus providing a finite reservoir, or provides an infinite reservoir, with the other end opened to atmosphere. The model can output threat zone distances (i.e., the maximum distance from the source to where a given concentration is found) as well as concentration as a function of time at a chosen downwind location. More details regarding the ALOHA model can be found in [24].

12.4.2 $ERCBH_2S$ Dispersion Model

The $ERCBH_2S$ software package was developed by the ERCB in Alberta to address the need for calculating emergency planning zones for sour (i.e., with H_2S) gas/liquid well and pipeline operation. The $ERCBH_2S$ package couples a source release rate model and a thermodynamic properties database with a modified version of SLAB (called ERCBSLAB), a dense gas atmospheric dispersion model provided by the EPA. SLAB models the dispersion of a given chemical mixture. $ERCBH_2S$ outputs the emergency planning and awareness zones for H_2S releases as well as the ERCBSLAB chemical mixture dispersion results. The concentration and cloud duration predictions are written to the ERCBSLAB output file. $ERCBH_2S$' source release rate model includes the effects of ESD valve closure time and runs through multiple meteorological conditions to determine the worst-case dispersion scenario (as determined by threat zone distances). Technical documentation of the $ERCBH_2S$ package is provided in [5].

The $ERCBH_2S$ model does not account for solid formation at the low temperatures local to the release source. Instead, the release is modeled as a two-phase flow with liquid droplet evaporation downstream of the source as heat is transferred to the plume and the plume's kinetic energy decreases. In the U.S. Department of Energy's FutureGen Project Environmental Impact Statement it is estimated that 26% of the volume of a 95% CO_2/100 ppm_v H_2S pipeline fluid would solidify. The FutureGen Project also employed SLAB as a dispersion model. In order to account for the decreased volume of gas that would actually disperse downwind, the project input 74% of the volume released by a pipeline rupture to the SLAB model [25]. In the $ERCBH_2S$ simulations described in the following sections, it was assumed that the entire release would remain in the vapor or liquid phases with no solid formation or precipitation. Modeling in Koornneef et al. [26] indicates that solid precipitation and sublimation results in higher maximum concentrations near the source as compared to a low solid formation release of the same total mass.

The $ERCBH_2S$ model assumes that the pipeline depressurization is isothermal (i.e., heat transfer into the pipeline balances the temperature decrease due to expansion). In the technical guidance document [5], it is stated that this assumption of maintaining isothermal gas temperature in the pipeline is generally valid except for the last few 100 diameters of distance to the blowdown location (presumably this assumes that the initial pipeline fluid is in a steady state with the surrounding environment temperature). The fluid properties during depressurization and discharge are computed by an internal solver (ERCBFLASH) for the Peng–Robinson equation of state.

The double-exponential blowdown model discussed in Section 12.3.1 is the exact analytical solution to the mass release rate from a quasi-steady, inertialess flow if the inertial delay factor is unity. This assumption means that upon depressurization initiation, the pipeline fluid is instantly accelerated to the mass flow rate at the exit point. This assumption allows for an analytical solution of the pipeline blowdown instead of numerically integrating the real equations of motion as in Botros et al. [27]. The application of an inertial delay factor to the initial mass release rate in the double-exponential blowdown model can be used to more closely approximate the real inertia conditions within the pipeline [5].

The ERCBH$_2$S discharge model assumes that the pipeline fluid isentropically expands to critical conditions at the pipeline release plane. The associated drop in temperature at the throat is approximately 25°C for the conditions modeled in this chapter. Downstream of the release plane, the simulated CO_2 jet adiabatically and non-reversibly expands to atmospheric pressure at the equalization plane to possibly supersonic velocities. The temperature is expected to drop significantly—to approximately −90°C—at the equalization plane for the CO_2 pipelines simulated herein. ERCBH$_2$S predicts this low temperature to occur only once the CO_2 has exited from the blowdown valve and has undergone expansion in the open atmosphere. A complete discussion on blowdown pressure and temperature transients within the pipeline is presented in Chapter 6.

The ERCBSLAB model treats finite duration releases as a steady-state Eulerian plume while the source is active; when the source is turned off, the model treats the emitted fluid as a Lagrangian puff. The model can simulate two-phase (vapor/liquid) flow, however, it does not account for liquid droplet deposition. The SLAB model has been enhanced to produce the ERCBSLAB model; these enhancements include a vertical or horizontal transient jet source type, revised buoyancy flux and dense plume rise models, and a transformation from steady release concentrations to transient concentrations.

12.4.3 Comparison of ALOHA and ERCBH$_2$S Dispersion Models

A comparison of the two aforementioned models was carried out by simulating the rupture scenarios outlined in Table 12-3.

TABLE 12-3. Simulation parameters for comparison of ALOHA and ERCBH$_2$S models

Parameter	ALOHA	ERCBH$_2$S (1)	ERCBH$_2$S (2)
Rupture location	End of segment	Middle of segment	Middle of segment
Segment length	7.5 km	15 km	15 km
Pipe outer diameter	406 mm (16 in.)	406 mm (16 in.)	406 mm (16 in.)
Pipe wall thickness	n/a	12.7 mm (0.5 in.)	12.7 mm (0.5 in.)
ESD valve closure time	Instantaneous	1.5 min	6 min
Composition	100% CO_2	96% CO_2, 0.9% H_2S	96% CO_2, 0.9% H_2S
Pressure	30.60 MPa (4440 psig)	15.30 MPa (2220 psig)	15.30 MPa (2220 psig)
Temperature (pipeline)	50°C	50°C	50°C
Wind speed	2 m/s (7.2 km/hr)	2 m/s (7.2 km/hr)	2 m/s (7.2 km/hr)
Atmospheric stability class	F (stable)	F (stable)	F (stable)
Temperature (ambient)	5°C	5°C	5°C

ALOHA calculates the source release rate by assuming that at the time of rupture initiation the non-ruptured segment end is capped (i.e., instantaneous ESD valve closure); in order to match this condition in ERCBH$_2$S, the model's minimum ESD valve closure time, 1.5 minutes, was selected. For purposes of comparison, this rupture scenario was also computed with ERCBH$_2$S for a 6-minute ESD valve closure time. The pipeline pressure in ALOHA was set to twice the expected operating pressure with half the actual segment length to simulate a rupture occurring half-way between two ESD valves separated by 15 km (note that the maximum segment length allowed in ALOHA is 10 km).

The model results were evaluated by comparing the predicted CO_2 threat zone distance for 5,000 ppm$_v$, 40,000 ppm$_v$ and 100,000 ppm$_v$ concentrations of CO_2. Tables 12-4 and 12-5

TABLE 12-4. Threat zone maximum distances for ALOHA/ERCBH_2S comparison

CO_2 Concentration Llevel	ALOHA	ERCBH_2S (1)	ERCBH_2S (2)
100,000 ppmv	321 m	194 m	194 m
40,000 ppmv	727 m	683 m	683 m
5,000 ppmv	2,700 m	3,790 m	3,810 m

TABLE 12-5. Times when maximum concentration occurs at threat zone maximum distances for ALOHA/ERCBH_2S comparison

CO_2 Concentration Level	ALOHA	ERCBH_2S (1)	ERCBH_2S (2)
100,000 ppm_v	5 min	6 min	7 min
40,000 ppm_v	6.5 min	7.5 min	8.5 min
5,000 ppm_v	19 min	23 min	26 min

show the maximum threat zone distances and the times of maximum concentration, respectively, at these threat zone outer edges for the two models.

With reference to Tables 12-4 and 12-5, the ALOHA and ERCBH_2S(1) CO_2 dispersion results were found to agree within 40% for threat zone size (referenced to ALOHA results) and within 21% for time of maximum concentration (referenced to ALOHA results). The ERCBH_2S(2) scenario predicted longer times of maximum concentration for all threat zone levels, as expected due to the delayed valve closure time calculated in ERCBH_2S(2). The decrease in valve closure time modeled in the ERCBH_2S(1) results, compared with the ERCBH_2S(2) results, appears to result in a longer cloud duration (on the order of 1–3 minutes for the additional 4.5-minute delay in closure time) and, when considering far-field concentrations (i.e., 5000 ppm_v threat zone), longer threat zone distances. Based on the results (in Tables 12-4 and 12-5) for the cases outlined in Table 12-3, ERCBH_2S was chosen to perform modeling of the release scenarios presented in this chapter for the following reasons:

- ERCBH_2S has a longer total simulation time (ALOHA is limited to a 1 hour release whereas ERCBH_2S' maximum outdoor exposure time is 3 hours).
- ERCBFLASH, the model's thermodynamics package, can calculate properties of gas mixtures and ERCBSLAB can model a chemical mixture, whereas ALOHA can only model a gas composed of one chemical component.
- ERCBH_2S will adjust the source release rate based on the specified ESD valve closure time, whereas ALOHA assumes the volume of gas to be released is isolated instantaneously at the time of rupture resulting in no valve closure time, hence, a less realistic emissions source model.
- Unlike ERCBH_2S, ALOHA cannot model the total line volume for a rupture midway between two ESD valves as it can only model the emission from the end of a pipe section. To model the actual volume of such a rupture location, the half section volume of the pipe length is modeled with a pressure that is artificially increased to allow the same initial mass of gas to be present at the start of the simulation.

12.5 CO_2 PIPELINE BLOWDOWN MODELING

12.5.1 Blowdown Model Assumptions

It is important to include dispersion modeling as part of any operations and emergency response plan for blowing down a CO_2 pipeline. Although high-pressure natural gas pipeline blowdown jets are also initially dense due to the low temperatures following depressurization, the high vertical momentum flux allows for the plume to warm, entrain air and

TABLE 12-6. CO_2 pipeline properties and meteorological conditions used in blowdown and rupture simulations with $ERCBH_2S$

Model Parameter	Units	
Pipeline segment length	km	20 km
Pipeline outer diameter	mm	406 mm (NPS 16)
Pipeline wall thickness	mm	9.6 mm
Fluid composition	–	96 mol% CO_2, 0.9 mol% H_2S[a]
Initial pressure (pipeline)	kPa	10.2 MPa
Initial temperature (pipeline)	°C	31.5°C
Ambient wind speed	m/s	2 m/s (4.5 mph)[b]
Ambient stability class	–	F (stable)[b]
Ambient temperature	°C	5°C
Model averaging time	min	3 min

[a] as per composition of Example 4 in Chapter 10.
[b] with the exception of Case 5, where a class D stability class with 5 m/s wind speed was simulated.

become neutrally buoyant. CO_2 releases are denser than the surrounding air even at ambient temperatures; therefore, what goes up must come down. Dispersion modeling can provide insight on the safety distances to maintain from a blowdown area and on the atmospheric conditions that ensure that the plume is quickly mixed with ambient air.

The $ERCBH_2S$ model was used to simulate a controlled CO_2-pipeline blowdown. The simulated CO_2 pipeline properties and the meteorological conditions applied are listed in Table 12-6. The initial pipeline composition is similar to the Example 4 composition in Table 10-1 in Chapter 10 (Weyburn pipeline), with an H_2S component of 0.9 mol%. In the dispersion model ERCBSLAB, it is assumed that the pipeline composition is composed of one pseudo-component that remains perfectly mixed with a single boiling point.

The blowdown scenario modeling parameters are documented in Table 12-7. As the blowdowns are controlled, the segment to be depressurized is considered to be isolated and ESD valves are closed. The drag plane coefficient was set to zero, indicating that the blowdown fluid is not expected to lose momentum by impinging on any surrounding structures. The blowdown valve is assumed to be on the top of a 2 m tall stack, where the internal diameter of the stack is the same as the main pipeline internal diameter, therefore the blowdown valve is where the fluid chokes. The blowdowns simulated here are from vertical stacks; it should be noted that DNV [12] has recommended that CO_2 pipelines be blown down from stacks angled at 45° away from densely populated areas. The blowdown valve coefficient of discharge was assumed to be 0.9.

Quantitative risk assessments should consider a matrix of meteorological conditions; however for brevity in the simulations presented here, the assumed worst-case weather conditions as specified in [5], with atmospheric stability class F (stable conditions) and

TABLE 12-7. Blowdown and rupture scenario modeling parameters

	Units	Case 1	Case 2	Case 3	Case 4	Case 5
Blowdown valve size	in.	4	6	4	16	16
Number of blowdown valves	–	2	2	1	n/a	n/a
Throat Diameter	mm	102.26	154.05	102.26	(rupture)	(rupture)
Discharge Coefficient, Cd	–	0.9	0.9	0.9	(rupture)	(rupture)
Hole Area, Ae (with Cd)	mm^2	7392	16,775	7392	(rupture)	(rupture)
Gas Pipeline Release Hole Size Fraction, Ae/Ap	–	0.0628	0.1425	0.0628	2.0000	2.0000
Inertial Blowdown Delay Factor, k	–	1	1	1	0.5	0.5
D Plane Drag Coefficient	–	0	0	0	0.5	0.5
Release Height	mAGL	2	2	2	0	0
Jet Direction	–	Vertical	Vertical	Vertical	Horizontal	Horizontal

mean horizontal wind speed of 2 m/s have been used. Results for simulations of multiple blowdown configurations are presented:

- Case 1: one 4" valve on either end of the pipeline segment;
- Case 2: one 6" valve on either end of the pipeline segment;
- Case 3: one 4" valve on one end of the pipeline segment.

In the case of the two-stack blowdown the emission from each stack was assumed not to affect the emissions and resulting downwind concentrations from the other stack due to the long (20 km) segment length.

12.5.2 Case Study 1 and 2: Two 4″ Valves or Two 6″ Valves

Pipeline blowdowns can be performed by isolating both ends of the pipeline and then relieving the pressure through a blowdown valve on top of a stack; in case studies 1 and 2, it has been assumed that the pipeline is being blown down from both ends of the segment. It has been assumed that since the pipeline segment is long (20 km), the emissions from the blowdown valves on either end of the segment are independent.

Valve sizing influences the depressurization time as well as the dispersion behavior of the released fluid. The maximum CO_2 and H_2S concentration contours are shown in Figures 12-5 and 12-6 for a blowdown performed with two 4" valves, and the contours resulting from using two 6" valves are shown in Figs. 12-8 and 12-9. Figures 12-7 and 12-10 illustrate the maximum concentration along the plume centerline and the plume centerline height for the 4" and 6" valves, respectively. The blowdown jet is predicted to rise to 70 mAGL to 100 mAGL before falling to ground level at 300 m to 400 m downwind.

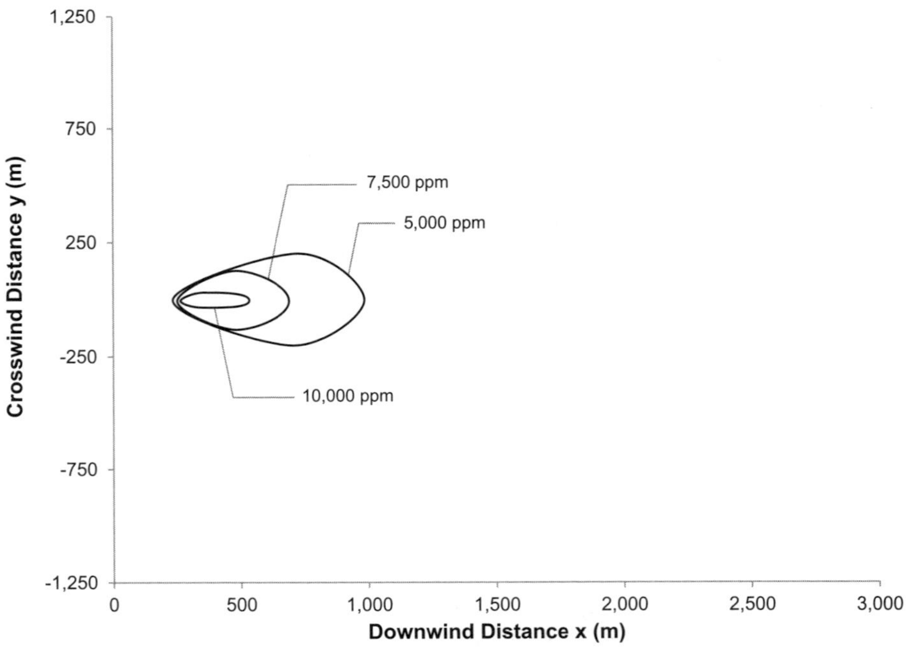

Figure 12-5. Maximum CO_2 concentration contours from the simulation of a pipeline blowdown with two 4" valves in stable meteorological conditions (Case 1)

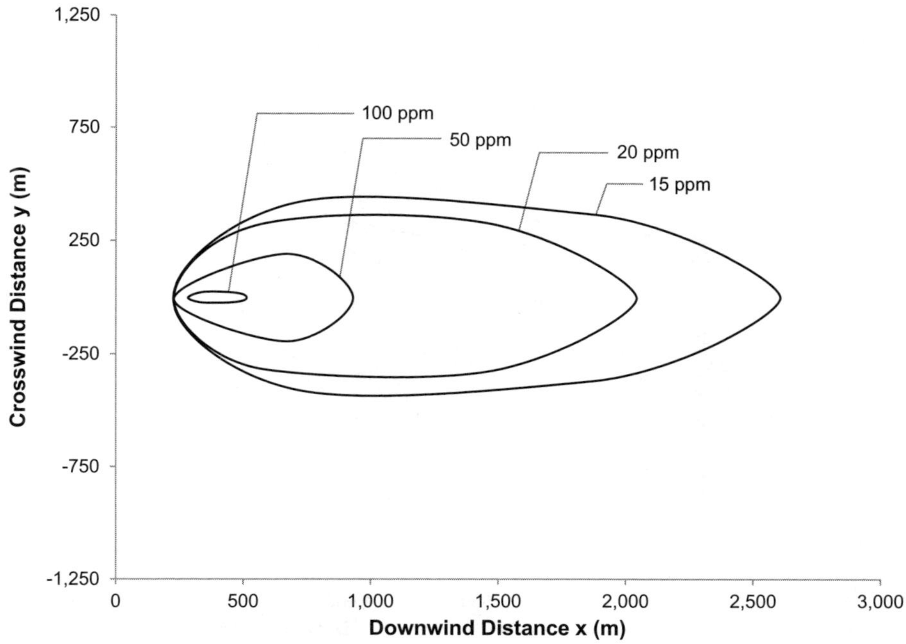

Figure 12-6. Maximum H_2S concentration contours from the simulation of a pipeline blowdown with two 4" valves in stable meteorological conditions (Case 1)

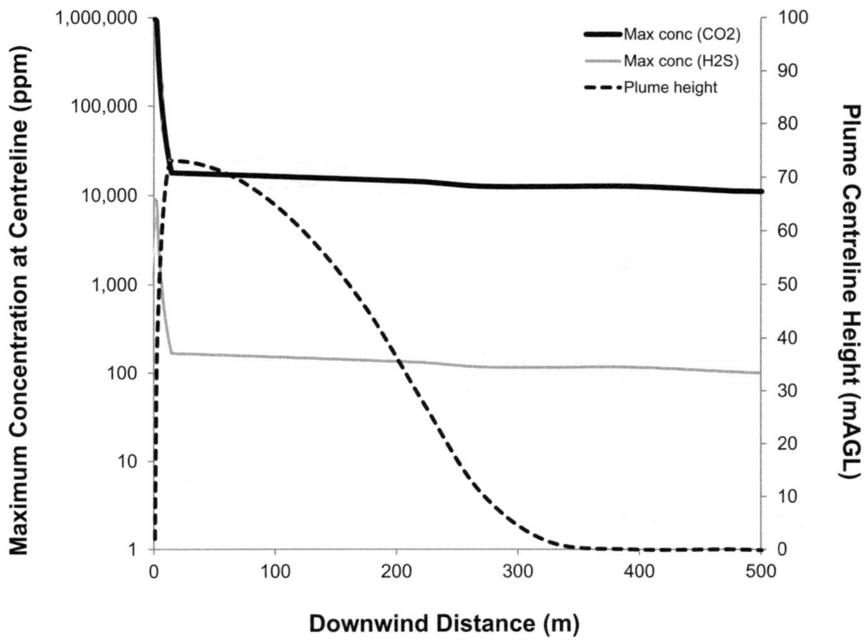

Figure 12-7. Maximum CO_2 and H_2S centerline concentrations and the plume centerline height AGL from the simulation of a pipeline blowdown with two 4" valves in stable meteorological conditions (Case 1)

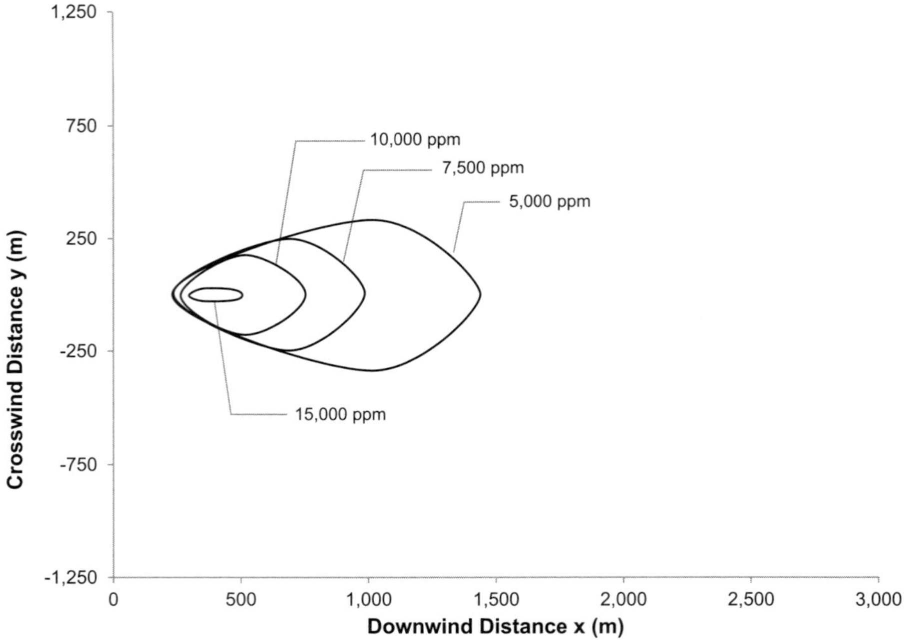

Figure 12-8. Maximum CO_2 concentration contours from the simulation of a pipeline blowdown with two 6” valves in stable meteorological conditions (Case 2)

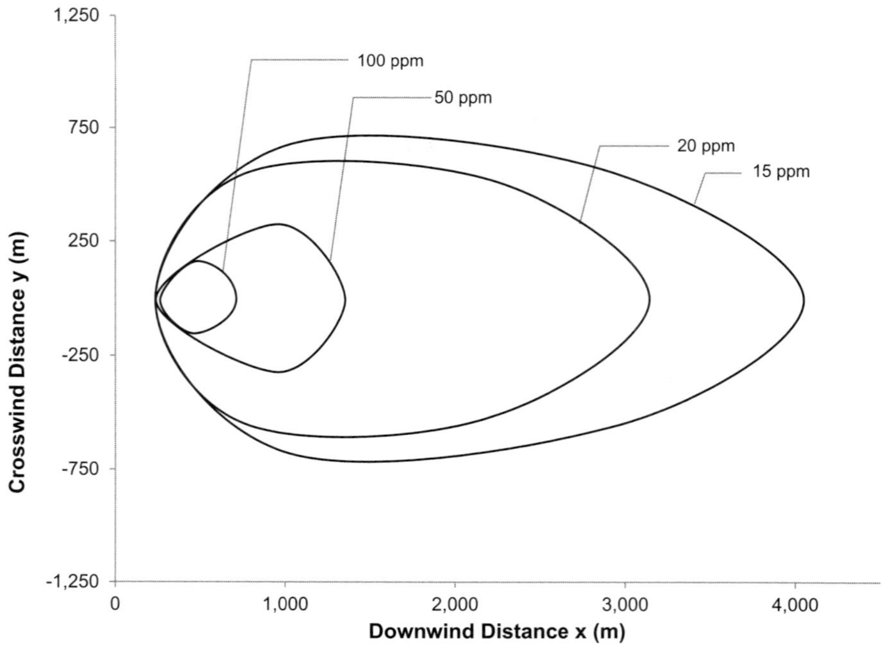

Figure 12-9. Maximum H_2S concentration contours from the simulation of a pipeline blowdown with two 6” valves in stable meteorological conditions (Case 2)

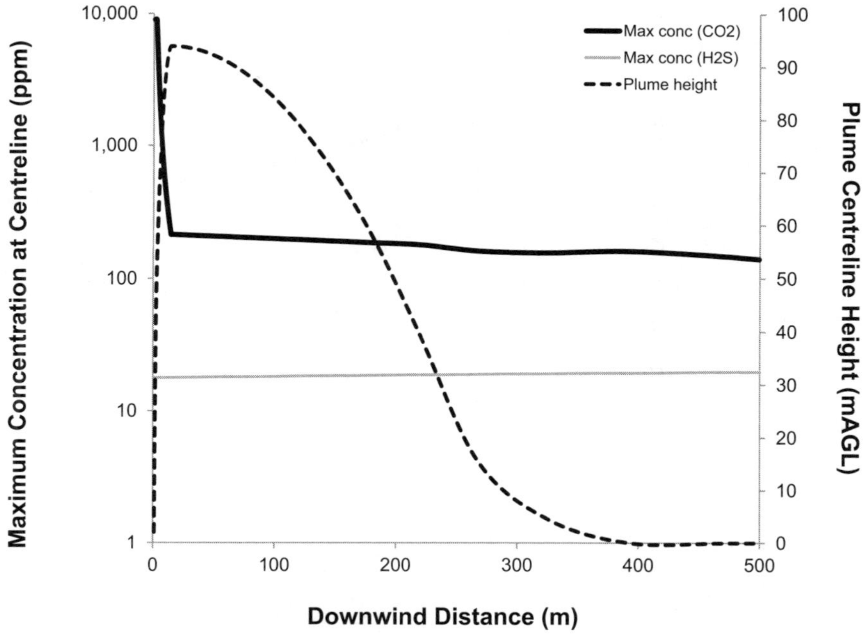

Figure 12-10. Maximum CO_2 and H_2S centerline concentrations and the plume centerline height AGL from the simulation of a pipeline blowdown with two 4" valves in stable meteorological conditions (Case 2)

Although the concentration profile as a function of time is not readily available from the simulation results, the time of maximum concentration (i.e., duration that the maximum concentration exists at a given location) can be determined from Figs. 12-11 and 12-12 below. The maximum ground level CO_2 and H_2S concentrations increase at approximately 200 m when the plume begins to touch down after the initial momentum-driven plume rise.

Table 12-8 summarizes some of the thermodynamic and source properties of the expanding jet for the two-valve blowdown cases, where Case 1 is with the 4" valves and Case 2 is with the 6" valves. Before the release is initiated, the pipeline fluid density is 650 kg/m^3 and the fluid is supercritical. Upon initial release to the exit plane, the jet has not fully expanded to atmospheric pressure and the density and jet temperature decrease. The jet is then flashed to atmospheric pressure, resulting in a vapor–liquid jet at –90°C.

The expanded jet then enters the "drag zone" where the jet could lose momentum due to potential impacts with surrounding structures. As the vertical blowdown jets are unrestricted, the drag coefficient has been set to zero. The source-plane results are input to the ERCBSLAB module for dispersion modeling. The initial mass source rate is nearly doubled for Case 2 as compared to Case 1, as is the initial expanded jet source area. The source duration is consequently halved for Case 2 as compared to Case 1.

12.5.3 Case Study 3: One 4″ Valve

The results from using one valve at one end of the pipeline segment instead of one valve at each end are shown in the following figures. The modeling parameters were kept constant for Case 3 as compared to Case 1, including the use of a 4" valve at a 2-m elevation, with the exception of total segment length being blown down through the valve. Figures 12-13 and 12-14 depict concentration contours that are very similar to Figures 12-5 and 12-6

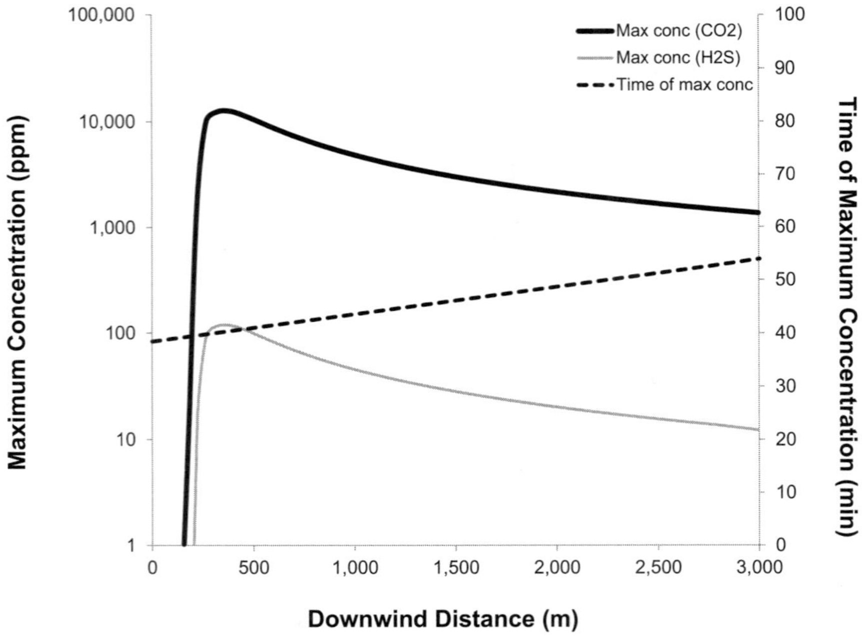

Figure 12-11. Maximum CO_2 and H_2S ground-level concentrations and the time of maximum concentration at ground level from the simulation of a pipeline blowdown with two 4" valves in stable meteorological conditions (Case 1)

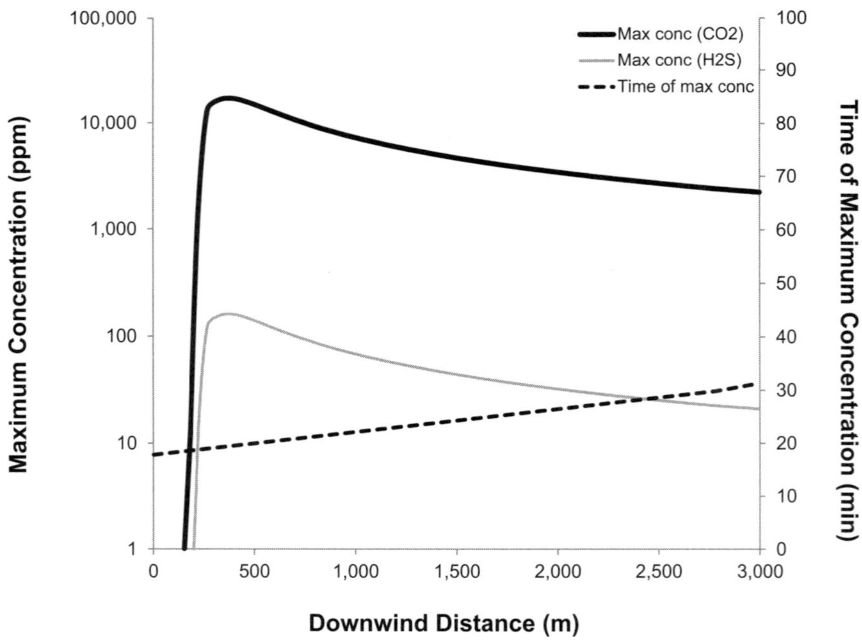

Figure 12-12. Maximum CO_2 and H_2S ground-level concentrations and the time of maximum concentration at ground level from the simulation of a pipeline blowdown with two 6" valves in stable meteorological conditions (Case 2)

TABLE 12-8. Results summary from ERCBFLASH for Cases 1 to 4

ERCBFLASH Outputs	Units	Case 1	Case 2	Case 3	Case 4
Initial density	kg/m³	650	650	650	650
Initial liquid fraction	–	1	1	1	1
Average release plane pressure	kPa	4239	4239	4239	3628
Average release plane temperature	K	279	279	279	273
Average release plane density	kg/m³	291	291	291	231
Isenthalpic flash temperature	K	183	183	183	183
Isenthalpic flash density	kg/m³	3.7	3.7	3.7	3.7
Isenthalpic flash liquid mass fraction	–	0.3	0.3	0.3	0.3
Source plane velocity	m/s	244	244	244	178
Source plane temperature	K	183	183	183	183
Source plane density	kg/m³	4.1	4.1	4.1	3.9
Source plane liquid mass fraction	–	0.38	0.38	0.38	0.34
Initial mass release rate	kg/s	334	757	334	5314
Steady mass source rate	kg/s	166	358	165	1360
Source area	m²	0.17	0.36	0.16	1.96
Source duration	min	77	36	155	20

from the Case 1 simulation. The exception lies in Figure 12-16, where the time of maximum concentration for each threat zone is almost doubled when compared to the time in Figure 12-11. The thermodynamic release properties for Case 3 are listed in Table 12-8. It should be noted that, as expected, the source duration for the one-valve case is doubled when compared to the two-valve case. As the increased blowdown time does not reduce the ground-level concentrations, it would likely be of benefit to use more than one valve to blowdown a CO_2 pipeline.

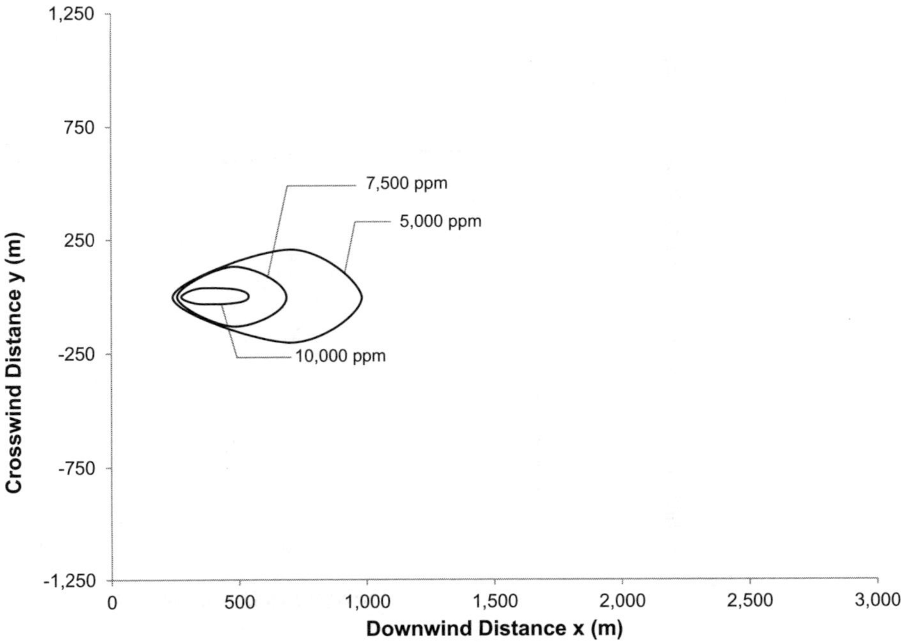

Figure 12-13. Maximum CO_2 concentration contours from the simulation of a pipeline blowdown with one 4” valve in stable meteorological conditions (Case 3)

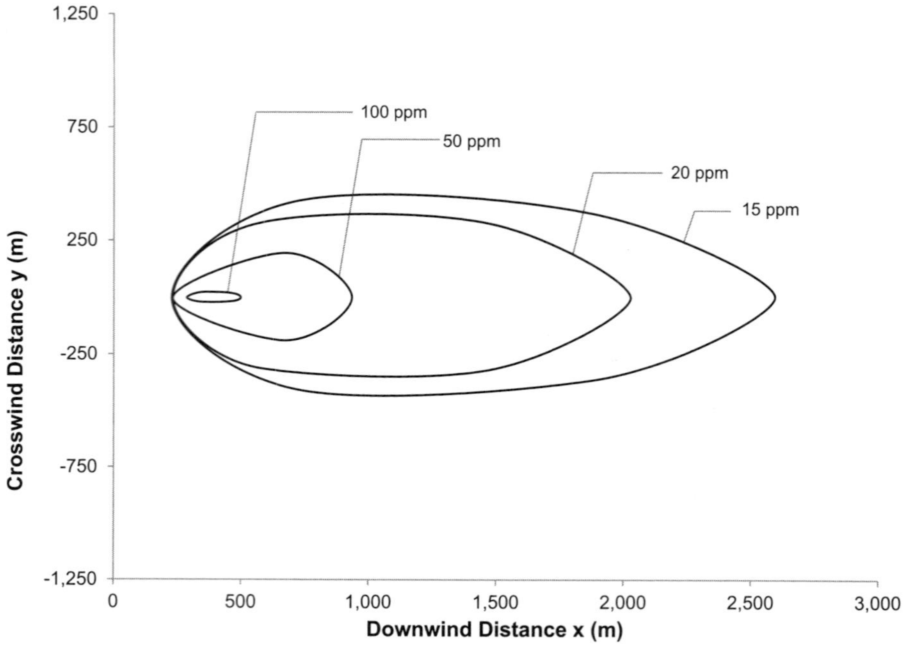

Figure 12-14. Maximum H_2S concentration contours from the simulation of a pipeline blowdown with one 4" valve in stable meteorological conditions (Case 3)

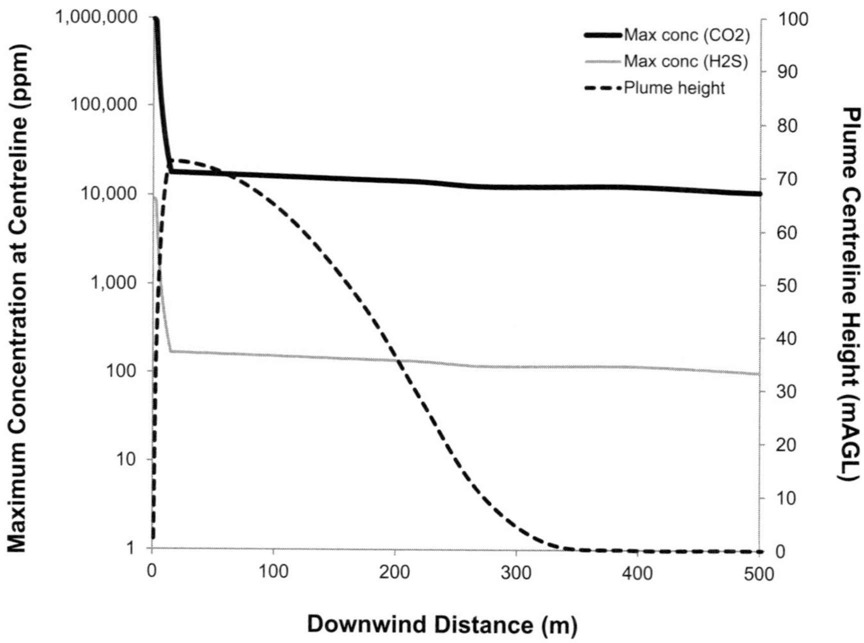

Figure 12-15. Maximum CO_2 and H_2S centerline concentrations and the plume centerline height AGL from the simulation of a pipeline blowdown with one 4" valve in stable meteorological conditions (Case 3)

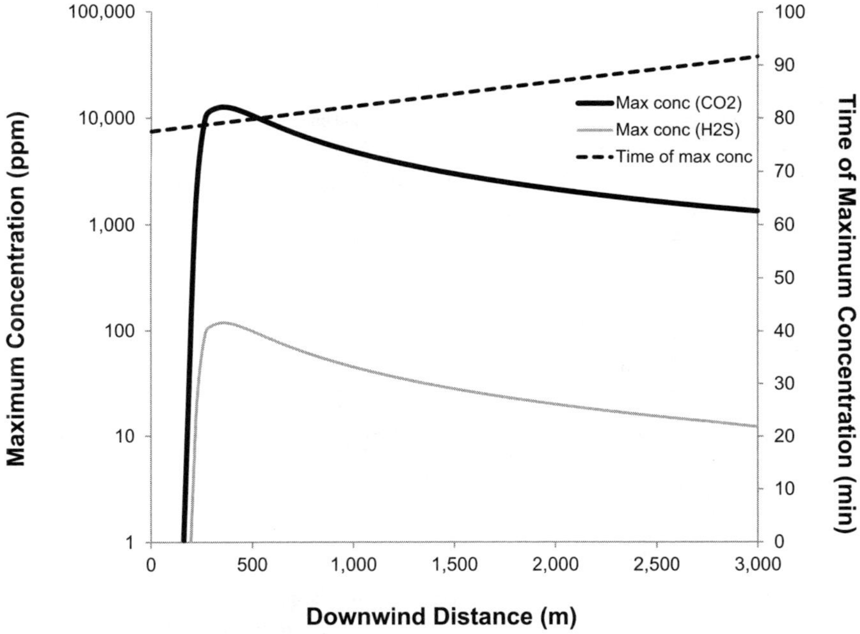

Figure 12-16. Maximum CO_2 and H_2S ground-level concentrations and the time of maximum concentration at ground level from the simulation of a pipeline blowdown with one 4" valve in stable meteorological conditions (Case 3)

The maximum threat zone distances and the times of maximum concentration at these distances for Cases 1–3 (blowdowns) are summarized in Table 12-9 below.

12.6 CO_2 PIPELINE RUPTURE MODELING

12.6.1 Rupture Model Assumptions

A pipeline rupture is a safety risk that needs to be understood and planned for when routing and designing any type of pipeline. CO_2 pipelines present dispersion challenges when compared to natural gas, as addressed in Section 12.1, in particular when toxic impurities are present. One of these challenges is the required automatic or manual mitigation of the large volume of pipeline fluid released by a full rupture to mitigate the release hazard and the loss of material. A full pipeline rupture would most likely also lack the vertical momentum of a planned blowdown, resulting in higher ground-level concentrations for a larger distance and longer time.

The ERCBH_2S release model has been used in the following sections for simulation of a full rupture scenario. As with the blowdown simulations (Cases 1, 2, and 3) presented in

TABLE 12-9. Threat zone maximum distances and times of maximum concentration in brackets for blowdown of a CO_2 pipeline

	Concentration Level (3-min averaging time)	Case 1	Case 2	Case 3
CO_2	10,000 ppm_v	534 m (41 min)	756 m (21 min)	531 m (80 min)
	5,000 ppm_v	983 m (43 min)	1,439 m (24 min)	979 m (82 min)
H_2S	100 ppm_v	498 m (41 min)	706 m (21 min)	496 m (80 min)
	15 ppm_v	2,607 m (52 min)	4,050 m (35 min)	2,594 m (90 min)

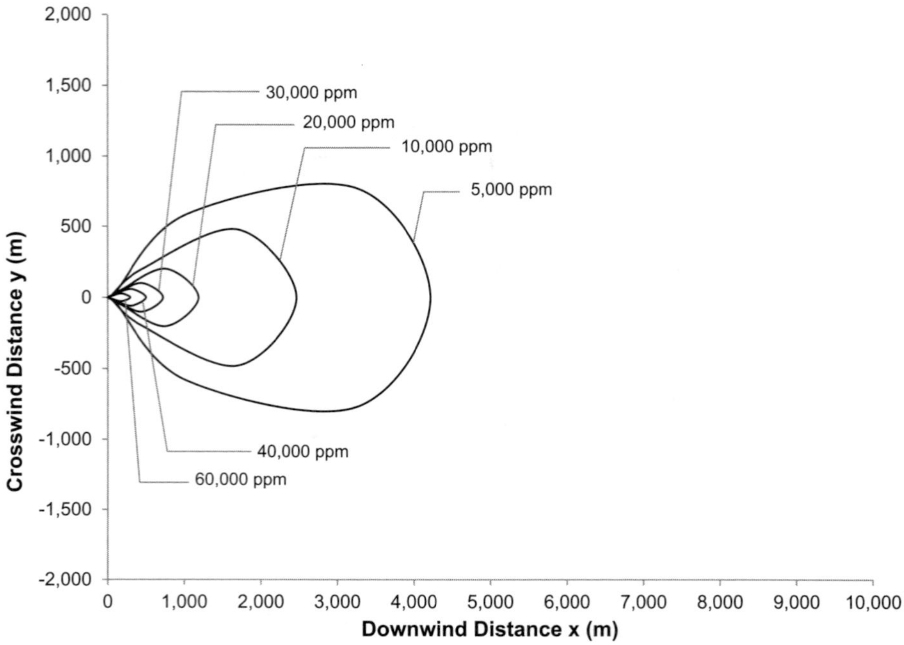

Figure 12-17. Maximum CO_2 concentration contours from the simulation of a full pipeline rupture in stable meteorological conditions (Case 4)

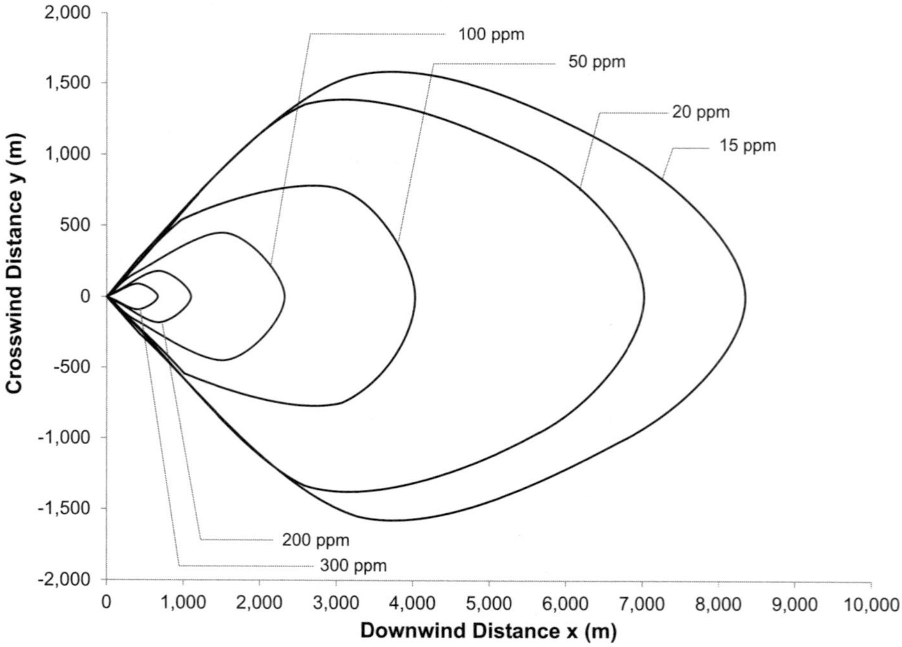

Figure 12-18. Maximum H_2S concentration contours from the simulation of a full pipeline rupture in stable meteorological conditions (Case 4)

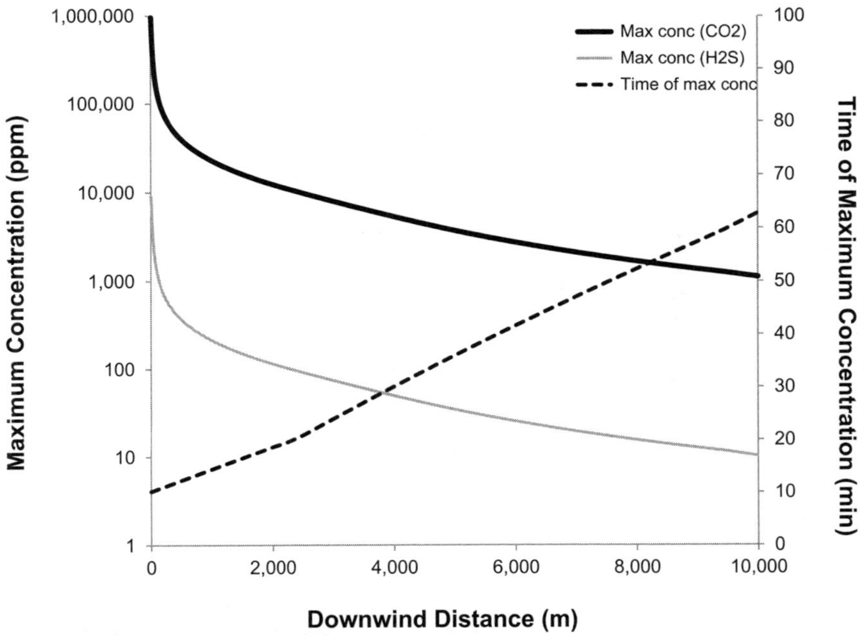

Figure 12-19. Maximum CO_2 and H_2S ground-level concentrations and the time of maximum concentration at ground level from the simulation of a full pipeline rupture in stable meteorological conditions (Case 4)

Section 12.5, the Example 4 composition described in Table 10-1 in Chapter 10 has been used as an example. This pipeline fluid is comprised of 96% CO_2 and 0.9 mol% H_2S, with the balance described in Table 10-1. If the H_2S concentration in the simulated pipeline was doubled, the downwind concentrations could be approximated by doubling the initial case concentrations. This approximation assumes that the thermodynamic behavior of the CO_2 mixture is not significantly impacted by a small change in the H_2S content.

The initial pipeline conditions are described in Table 12-6. The rupture release conditions are described in Table 12-7. These conditions are similar to the blowdown conditions, with the major exceptions of:

- Release height = 0 m (ground level),
- Ratio of exit-to-pipeline areas = 2 (guillotine rupture),
- Jet direction = horizontal,
- Drag plane coefficient = 0.5,
- Inertial delay factor = 0.

The source release angle was horizontal (i.e., the jet was released parallel to the ground), with a drag plane coefficient of 0.5 indicating that some of the jet's momentum would be lost to the impacted structures/crater as described in [5]. As the rupture depressurization case will be very fast when compared to controlled blowdowns, the depressurization time where the flow is dominated by inertia effects will be non-negligible. An inertial delay factor has been applied to account for the acceleration of the pipeline fluid.

The ESD valve low pressure trigger was set to 6 MPa with a 60-second closure time once triggered. With the given pipeline conditions, the ESD valves were closed 9 minutes after initial rupture. Given these ESD valve settings, the effective pipeline length is 21.6 km instead of 20.0 km due to the extra mass released through the valves before they close.

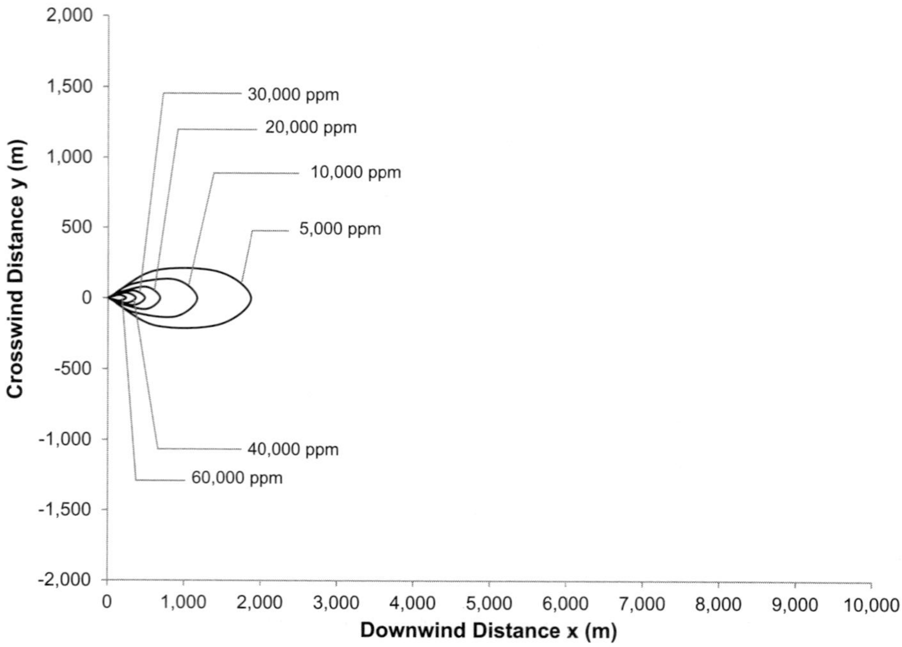

Figure 12-20. Maximum CO_2 concentration contours from the simulation of a full pipeline rupture in neutral meteorological conditions (Case 5)

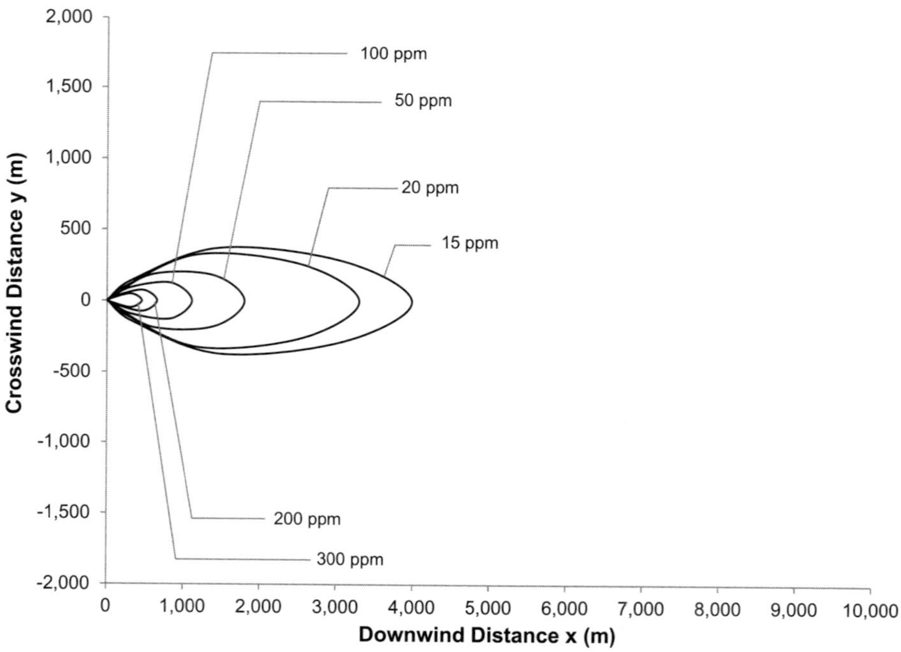

Figure 12-21. Maximum H_2S concentration contours from the simulation of a full pipeline rupture in neutral meteorological conditions (Case 5)

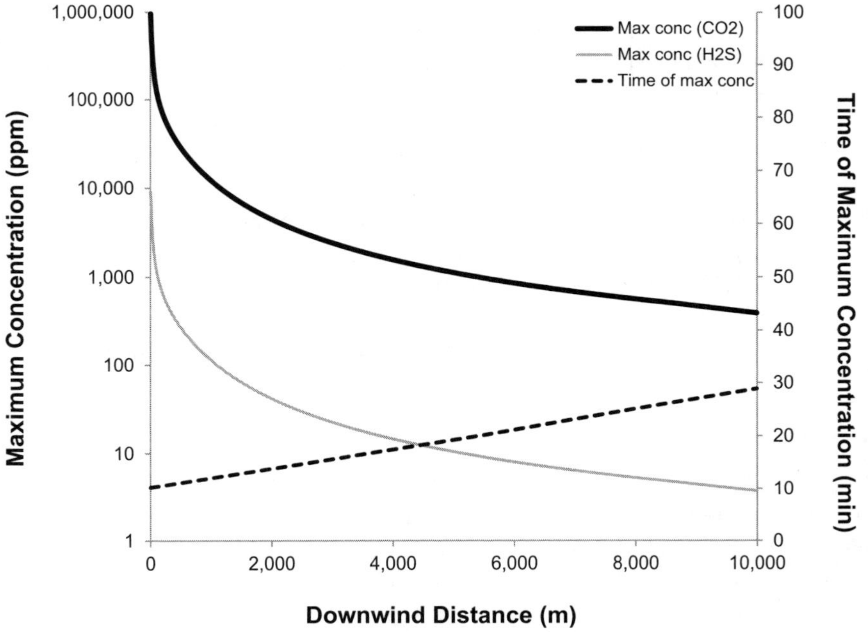

Figure 12-22. Maximum CO_2 and H_2S ground-level concentrations and the time of maximum concentration at ground level from the simulation of a full pipeline rupture in neutral meteorological conditions (Case 5)

The rupture is modeled as two halves of the pipeline segment, each depressurizing in a similar fashion. To model the flow from both sections, the mass release rate from one half of the pipeline segment was doubled to simulate two halves depressurizing.

12.6.2 Case Study 4: One 16″ Full Pipeline Rupture, Stable Conditions

Figures 12-17, 12-18 and 12-19 depict the maximum ground-level concentrations of CO_2 and H_2S resulting from a full pipeline rupture in stable atmospheric conditions as described in Section 12.6.2 (Case 4).

12.6.2 Case Study 5: One 16″ Full Pipeline Rupture, Neutral Conditions

Atmospheric stability is a key parameter in downwind dispersion of the released fluid. If the plume air entrainment is increased through advection (faster wind speed) and turbulence (atmospheric instability), the downwind exposure distances could decrease, as could the

TABLE 12-10. Threat zone maximum distances and times of maximum concentration in brackets for a CO_2 pipeline rupture

	Concentration Level (3-min averaging time)	Case 4	Case 5
CO_2	100,000 ppm_v	149 m (11 min)	132 m (10 min)
	40,000 ppm_v	502 m (12 min)	365 m (11 min)
	5000 ppm_v	4218 m (31 min)	1869 m (14 min)
H_2S	200 ppm_v	1099 m (15 min)	652 m (11 min)
	100 ppm_v	2324 m (20 min)	1112 m (12 min)
	15 ppm_v	8356 m (54 min)	3985 m (17 min)

time required for the plume to dissipate. The results from simulating a rupture in neutrally stable atmospheric conditions with 5 m/s wind speed are shown in Figures 12-20, 12-21 and 12-22 (Case 5).

In Figure 12-20, the 5,000 ppm_v CO_2 threat zone size is decreased from 4 km to 2 km in neutral conditions vs. stable conditions. The 15 ppm_v H_2S maximum downwind distance is similarly halved in neutral conditions in Figure 12-21. The plume begins to dissipate from 5000 ppm_v CO_2 at 4 km in approximately 40 minutes in stable atmospheric conditions, whereas the 5,000 ppm_v CO_2 level at 2 km in the neutral conditions begins to decay after only 15 minutes (Figure 12-22).

The maximum threat zone distances and the times of maximum concentration at these distances for Cases 4 and 5 (ruptures) are summarized in Table 12-10:

REFERENCES

[1] HSE, 2011, "General hazards of carbon dioxide," Health and Safety Executive, United Kingdom, retrieved from http://www.hse.gov.uk/carboncapture/carbondioxide.htm.

[2] CAPP, 2003, "Guide: Occupational Health and Safety of Hydrogen Sulphide (H_2S)," Canadian Association of Petroleum Producers, Report No. 2003-0004, retrieved from http://membernet.capp.ca/raw.asp?x=1&dt=PDF&dn=61443.

[3] NIOSH, 1996, "Documentation for Immediately Dangerous to Life or Health Concentrations (IDLHs)-124389," National Institute of Safety and Health, retrieved from http://www.cdc.gov/niosh/idlh/124389.html.

[4] NIOSH, 1996, "Documentation for Immediately Dangerous to Life or Health Concentrations (IDLHs)-7783064," National Institute of Safety and Health, retrieved from http://www.cdc.gov/niosh/idlh/7783064.html.

[5] ERCB, 2008, "ERCBH_2S: A Model for Calculating Emergency Response and Planning Zones for Sour Gas Facilities, Volume 1: Technical Reference Document," Alberta Energy Resources Conservation Board, retrieved from http://www.ercb.ca/docs/public/sourgas/eubmodelsdraft/Volume1_ERCB TechnicalReference_200807.pdf.

[6] RIVM, 2009, "Evaluation of the acute toxicity of CO_2," National Institute for Public Health and the Environment, Netherlands, retrieved from http://www.rivm.nl/milieuportaal/images/20091002_Evaluation_toxicity_CO2.pdf.

[7] HSE, 2011, "Assessment of the Dangerous Toxic Load (DTL) for Specified Level of Toxicity (SLOT) and Significant Likelihood of Death (SLOD)," Health and Safety Executive, United Kingdom, retrieved from http://www.hse.gov.uk/hid/haztox.htm.

[8] Ermak, D.L., 1990, "Users Manual for SLAB: An Atmospheric Dispersion Model for Denser-than-air Releases," University of California, Lawrence Livermore National Laboratory, Livermore, California, U.S., retrieved from http://www.epa.gov/scram001/models/nonepa/SLAB.PDF.

[9] Arya, S.P., 1999, "Air Pollution Meteorology and Dispersion," Oxford University Press, New York, NY.

[10] Pasquill, F., 1961, "Estimation of the dispersion of windborne material," Meteorol. Mag., 90, pp. 33–49.

[11] Turner, D.B., 1970, "Workbook of Atmospheric Dispersion Estimates," Office of Air Programs Pub. No. AP-26, U.S. Environmental Protection Agency, Research Triangle Park, NC.

[12] DNV, 2010, "Design and Operation of CO_2 Pipelines," Det Norske Veritas Recommended Practice DNV-RP-J202, Høvik, Norway, retrieved from http://exchange.dnv.com/publishing/Codes/ToC_edition.asp#Recommended%20Practices.

[13] Kruse, H. and Tekiela, M., 1996, "Calculating the Consequences of a CO_2-Pipeline Rupture," Energy Convers. Mgmt, 37(6-8), pp. 1013–1018.

[14] Leis, B.N., Pimputkar, S.M., and Ghadiali, N.D., 2002, "Line Rupture and the Spacing of Parallel Lines," Battelle Memorial Institute, Pipeline Research Council International, Project No. PR-003-9604.

[15] McGillivray, A. and Wilday, J., 2009, "Comparison of risks from carbon dioxide and natural gas pipelines," Health and Safety Laboratory Research Report RR749, Buxton, UK, retrieved from http://www.hse.gov.uk/research/rrpdf/rr749.pdf.

[16] EI, 2010, "Technical guidance on hazard analysis for onshore carbon capture installations and onshore pipelines" Energy Institute, London, UK http://www.energyinst.org/information-centre/ei-publications/newpubs/hazard-analysis-onshore-carbon-capture.

[17] Hanna, S.R., and Chang, J.C., 2001, "Overview of Petroleum Environmental Research Forum (PERF) dense gas dispersion modeling project," Atmos. Environ., 35, pp. 2223–29.

[18] Hanna, S.R., and Chang, J.C., 2001, "Use of the Kit Fox field data to analyze dense gas dispersion modeling issues," Atmos. Environ., 35, pp. 2231–42.

[19] Cameron-Cole, 2005, "Air Dispersion Modeling of Well Blowout and Pipeline Rupture Scenarios, Salt Creek Field," prepared for Howell Petroleum Corporation, The Woodlands, TX, retrieved from http://www.blm.gov/pgdata/etc/medialib/blm/wy/information/NEPA/cfodocs/howell.Par.9631.File.dat/23apxA.pdf.

[20] COCATE, 2011, "State of the art on safety of CO_2 transport," Seventh Framework Program (FP7), COCATE, retrieved from http://projet.ifpen.fr/Projet/jcms/c_7898/wp3-safety-of-global-capture-to-storage-transport-network.

[21] Leirgulen, S.I., 2009, "Inviting the industry to continue developing CO_2 pipeline guideline," Det Norske Veritas (DNV) press release, retrieved from http://www.dnv.com/press_area/press_releases/2009/invitingtheindustrytocontinuedevelopingco2pipelineguideline.asp.

[22] CORDIS, 2009, "Quantitative failure consequence hazard assessment for next generation CO_2 pipelines," European Commission Community Research & Development Information Service, Seventh Framework Program (FP7), retrieved from http://cordis.europa.eu/fetch?CALLER=FP7_PROJ_EN&ACTION=D&DOC=1&CAT=PROJ&RCN=92935.

[23] Witlox, H., Harper, M., and Oke, A., 2009, "Modelling of discharge and atmospheric dispersion for carbon dioxide releases," J Loss Prevent Proc, 22(2009), pp. 795–802.

[24] U.S. Environmental Protection Agency and the National Oceanic and Atmospheric Administration, 2007. "ALOHA User's Manual," retrieved from http://www.epa.gov/OEM/docs/cameo/ALOHAManual.pdf.

[25] NETL, 2007, "FutureGen Project: Final Environmental Impact Statement (DOE/EIS-0394)," National Energy Technology Laboratory, U.S. Department of Energy, retrieved from http://www.netl.doe.gov/technologies/coalpower/futuregen/EIS/Volume%201%20-%20FINAL.pdf.

[26] Koornneef, J., Spruijt, M., Molag, M., Ramirez, A., Faaij, A., and Turkenburg, W., 2009, "Uncertainties in risk assessment of CO_2 pipelines," Energy Procedia, 1, pp. 1587–94.

[27] Botros, K.K., Jungowski, W.M., and Weiss, M.H., 1989, "Models and Methods of Simulating Gas Pipeline Blowdown," Can. J. Chem. Eng., 67, pp. 529–539.

Chapter 13

HEALTH AND SAFETY AND EMERGENCY RESPONSE

13.1 HEALTH AND SAFETY

13.1.1 CO_2 Safety Issues

CO_2 exists in various concentrations in the atmosphere. The current concentration level is approximately 370 ppm (0.037%) (refer to Chapter 1), although somewhat higher concentrations to 0.06 % by volume may occur in occupied buildings [1]. It is toxic in high concentrations only. Toxicity occurs at about 10% level.

CO_2 is essential for life, being a critical component in photosynthesis. Greenhouses purposely elevate CO_2 levels in order to "fertilize" the plant.

Air in the lungs contains approximately 5.5% (55,000 ppm) of CO_2. At low concentrations (1% by volume), CO_2 causes no ill effects on humans, fauna, or flora. Although it is non-toxic at low concentrations, air containing 10% to 20% CO_2 concentrations by volume are immediately hazardous to life by affecting control of nervous system (dizziness, confusion, drowsiness, headache, and unconsciousness), failure of respiratory muscles (causing shortness of breath), visual impairment, skin sweating, and a change in the pH of the blood stream, etc. [2].

CO_2 at high level of concentration displaces oxygen in air. Oxygen (O_2) level in air is normally about 21%. Life ceases quickly without enough oxygen. When levels of oxygen are reduced below 19.5% (minimal acceptable level), serious health problems begin to occur very quickly. The effect of oxygen depletion is indicated in Table 13-1. If CO_2 level in air is increased to 10%, then oxygen level will be reduced to about 19%. This 19% of oxygen level is regarded as the lowest safe level for working condition. However, it is said that toxic effects of CO_2 dominate below 10% concentration level.

It may be noted that intrinsically a moderate increase in CO_2 concentration level (5% to 10%) in air can lead to increased acidity of the blood. At concentrations of about 6% by volume, CO_2 can cause nausea, vomiting, diarrhea, and irritation to mucous membranes, skin lesions, and sweating. At very high CO_2 concentration level (10% by volume), rapid asphyxiation can occur (Table 13-2) [3].

13.1.2 Hydrogen Sulfide—H_2S

Hydrogen sulfide is more toxic than carbon monoxide. This colorless gas has a characteristic rotten-egg odor at first smell; however, after a short time, the gas renders the olfactory nerve (which controls the sense of smell) ineffective. *A worker may be lulled into a false sense of security because he/she no longer smells the substance*, and yet it causes serious bodily harm in higher concentrations. H_2S exposure effects are indicated in Table 13-3.

TABLE 13-1. Effect of Oxygen Depletion Level

Oxygen level (%)	Effect
20.9	Percent oxygen found in normal air (breathing level).
19.5	Minimum permissible oxygen level (i.e., acceptable breathing level).
15-19	Decreased ability to work strenuously. May impair coordination and may induce early symptoms in persons with coronary, pulmonary or circulatory problems.
12-15	Respiration and pulse increase; impaired coordination, perception, and judgment.
10-12	Respiration further increases in rate and depth, poor judgment, and bluish lips.
8-10	Mental failure, fainting, unconsciousness, ashen-face, blueness of lips, nausea, and vomiting.
6-8	8 minutes—100% fatalities; 6 minutes—50% fatalities; 4—5 minutes—recovery with treatment.
4-6	Coma in 40 seconds, convulsions, respiration ceases — death.

TABLE 13-2. CO_2 safety profile National Institute for Occupational Safety and Health [3]

CO_2 Concentaion			
ppm (vol/vol)	% (vol/vol)	Exposure Duration	Effect
500	0.5	8 hours/day	TWA (Time Weighted Average) Occupational Exposure Standard*
20,000	2	Several Hours	Headache, dyspnoea (shortness of breath) upon exertion
30,00	3	4 × 15 minutes/day	STEL (Short Term Exposure Limit) Occupational Exposure Standard
	3	1 hour	Mild headache, sweating and dyspnoea at rest
40,000	4	30 minutes	No IDLH-Immediately Dangerous to Life and Health, No impairment to escape or long term health effect
40,000-50,000	4-5	Within a few minutes	Headache, dizziness, increased blood pressure, uncomfortable dyspnoea
60,000	6	1-2 minutes	Hearing & visual disturbances
		<16 minutes	Headache, dyspnoea
		Several Hours	Tremors
70,000-100,00	7-10	Few Minutes	Unconsciousness to near unconsciousness
		1.5 minutes to 1 hour	Headache, increased heart rate, shortness of breath, dizziness, sweating, rapid breathing
100,00-150,00	>10-15	1 minute to several minutes	Dizziness, drowsiness, severe muscle twitching, unconsciousness
170,000-300,00	17-30	Within 1 minute	Loss on controlled and purposeful activity, unconsciousness, convulsions, coma, death

*National Institute for Occupational Safety and Health.

TABLE 13-3. H_2S Safety Profile

Level ppm	Effects
0-10	Permissible Exposure Limit over 8 hours (0.01–0.03 odor threshold)
10-50	Headaches, dizziness, nausea/vomiting, coughing/breathing difficulty
50-100	Mild eye and respiratory irritation
200-300	Marked increase in eye, throat/lung irritation, shock, convulsion, coma
500-700	Unconsciousness/death after 30-minute exposure
≥1000	Rapid collapse and death within minutes

It may be noted that prolonged exposures to H_2S gas at lower levels generally can lead to bronchitis, pneumonia, migraine headaches, pulmonary edema, and loss of motor coordination.

13.2 PIPELINE SAFETY ISSUES

External safety is a key aspect of CO_2 pipeline transportation which requires assessment prior to and during the operational phase of a CO_2 transportation project.

Safety issues relate to accidental release as well as pre-planned blowdown of a segment of CO_2 pipeline transmission system. Both involve initial dispersion of CO_2 and subsequent potential for accumulation in low-lying areas (Figure 13-1).

CO_2 pipeline can impact a number of resources and safety issues. Therefore, from a design perspective and operational considerations, the following must be considered [4]:

1. Atmospheric environmental condition.
2. Nearby groundwater resources (including the location and depths of known groundwater resources buried channels and useable aquifers with a total dissolved solids content of 4000 mg/l or less, the locations and depths of all known operating and abandoned water test holes and an inventory of groundwater users within the pipeline ROW, the locations and depths of known structure test holes, operating and abandoned oil and gas wells).
3. Hydrological and aquatic environment.
4. Terrestrial environment (soil and terrain, vegetation and wetlands as well as wild life and habitat).
5. Archaeological, paleontological and heritage resources (i.e., similar to other pipelines, the effect of the pipeline on any structure, site or thing that is of historical, archaeological, paleontological, or architectural significance).
6. Land and resources cultural use, including for traditional use by Aboriginal/native persons (this includes potential environmental effects of pipeline and facilities operation and maintenance activities on lands and resources used for traditional purposes).
7. Land use (including mitigation measures to address adverse environmental effects on crown and agricultural lands, unique sites, road and rail traffic and navigability on waterways).
8. Public health and safety.
9. Socio-economic conditions.

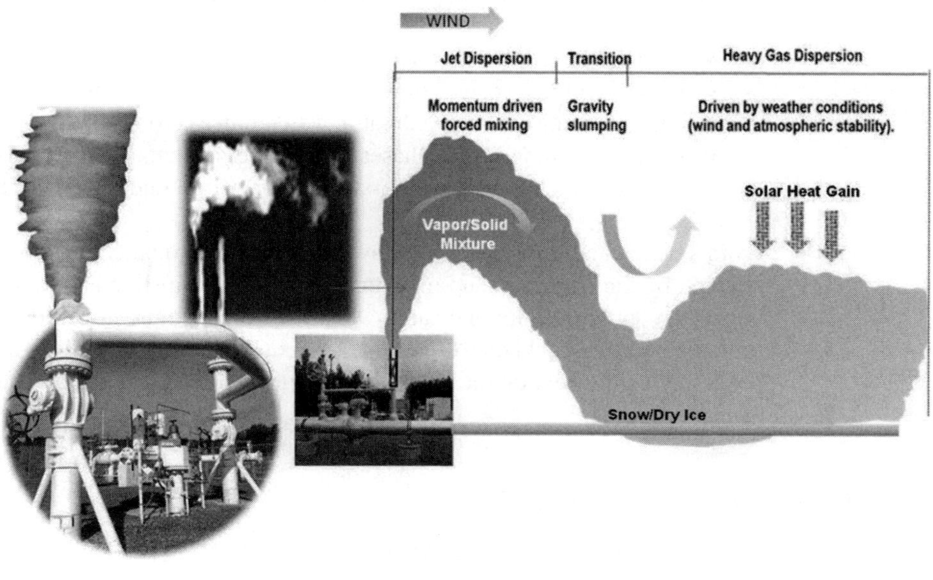

Figure 13-1. CO_2 release at high pressures and possible subsequent dispersion pattern

13.2.1 Safety and Environmental Impact Considerations

Major impacts from some of the issues listed above are highlighted below:

Air, hydro-geologic, and aquatic: While CO_2 pipeline routing and design impact of the pipeline operation and maintenance, consideration must be given to operational and maintenance issues that can result in adverse environmental effects to various components of the atmospheric environment. The potential key environmental effects in the vicinity of the pipeline are:

- Air Quality
- Noise generation and sound quality

CO_2 pipelines also can affect the geologic and hydro-geologic setting in and around the pipeline and facilities. Thus any assessment of environmental impacts must consider the following and provide relevant measures or monitoring techniques to assure long term operational safety;

- Cause of the environmental effect
- Likelihood of the environmental effect occurring
- Nature of the environmental effect
- Magnitude and duration of the environmental effects and their impacts

Any monitoring measure among others must appropriately collect data and consider means that can be put in place to effectively:

- monitor environmental effects to groundwater resources and users,
- mitigate/prevent environmental effects detected in the monitoring program and reported by groundwater users,
- share information with stakeholders/third parties (public and appropriate levels of government),
- to address concerns from groundwater users that may report environmental effects to their groundwater source.

The assessment of the hydrological and aquatic environment must consider:

- drainage areas of watercourses crossed by the project,
- presence of fish species in watercourses crossed by or proximal to the pipeline,
- characterization of fish habitat in watercourses crossed by the pipeline,
- proposed timing for all phases of the pipeline that are of relevance to fish and fish habitat (e.g., the timing of in-stream work or ground disturbance).

From a health and safety point of view, environmental quality and noise generation that is affected by the operation and maintenance of a CO_2 pipeline and facilities have the most impact on "public health and safety" and "socio-economic conditions." Potential impac can only be assessed when CO_2 release composition is known and the impact of each compositional component on the environment, surroundings, and people's health and safety is estimated.

13.2.2 Health and Safety Hazards

***a—Toxicity of CO_2*:** Toxic potency is a quantitative expression that relates concentratior and exposure time to a particular degree of adverse physiological effects (for example death) on exposure of humans or animals [5].

UK HSE in a paper by McGillivray and Wilday [6] indicate that there are a number of criteria against which the level of harm (such as CO_2 toxicity effects) experienced by a population in close proximity to a CO_2 failure can be measured. One of the criteria is the Dangerous Toxic Load (DTL). A DTL describes the exposure conditions, in terms of airborne concentration and duration of exposure, which would produce a particular level of toxicity.

Two levels of harms are defined:

1. Specified Level of Toxicity due to Dangerous Toxic Load (SLOT DTL).
2. Significant Likelihood of Death due to Dangerous Toxic Load (SLOD DTL).

Specified Level of Toxicity Dangerous Toxic Load (SLOT DTL) is defined [7, 8] as a level of toxicity that can cause:

- Severe distress to almost everyone in the area;
- A substantial fraction of the exposed population to require medical attention;
- Serious injury that requires prolonged treatment for some people; and
- Death for highly susceptible people.

The DTL is expressed in ppm^n.min, where ppm is the CO_2 content in the air and min is the exposure time in minutes. In terms of likelihood of death, SLOT DTL corresponds to 1% mortality and Significant Likelihood of Death (SLOD) DTL corresponds to 50% mortality.

The toxic properties for CO_2 are indicated Table 13-4 and Figure 13-2 as SLOT DTL and SLOD DTL. The values of SLOT DTL and SLOD DTL for a number of toxic substances are available on the HSE website [2, 7].

CO_2 begins to affect the respiratory system after several hours' exposure to a concentration of 3%, causing headaches and restricted breathing. Unconsciousness can result within a few minutes of exposure to 7% CO_2 level, while coma and death is possible within a few minutes of exposure to 17% CO_2 [9]. As a result, it is clear that CO_2 has a significant capacity to cause harm, even though it is not flammable.

As can be seen from Figure 13-2, the SLOT DTL corresponds to exposure to a CO_2 concentration of 7.5% for 15 minutes. The SLOD DTL (50% mortality) corresponds to a concentration of 17% for about half a minute.

b—Release of other compounds**:** assessments must be made as to the potential health implications of the compounds that will be released (in a carbon dioxide gas stream containing impurities) to the environment from pipeline and facilities operational and maintenance activities in relation to the exposure limits established to prevent acute and chronic adverse effects on human health.

The potential environmental effects could include/impact:

- Changes in ozone level,
- Greenhouse gas (GHG) emissions including H_2S, CO, Ar, NO_x, and SO_x.
- Potential acid input (PAI),
- Nitrogen deposition,

TABLE 13-4. CO_2 toxic properties [6]

Parameter	Value
SLOT DTL	1.5×10^{40} ppm^8·min
SLOD DTL	1.5×10^{41} ppm^8·min
Toxic exponent "n" value	8

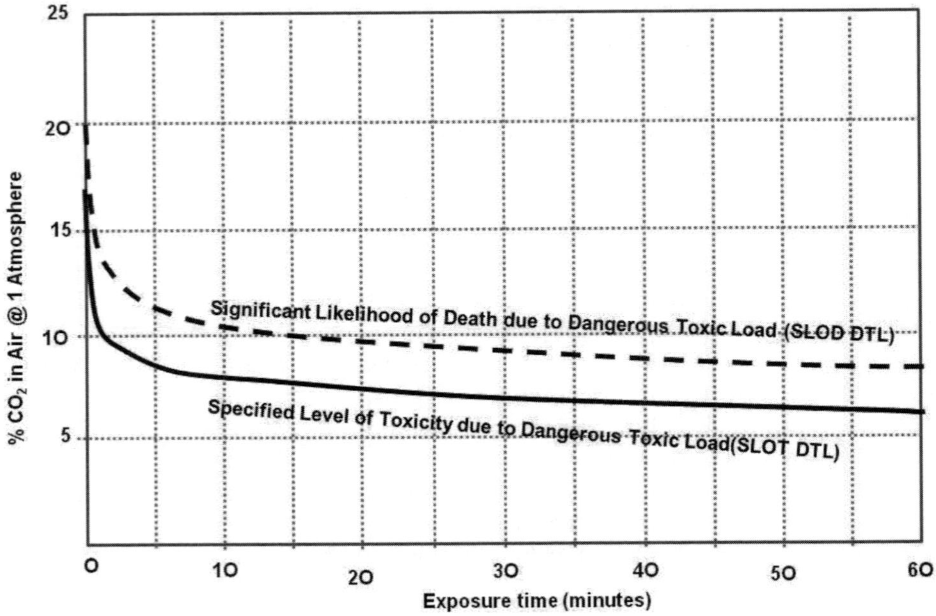

Figure 13-2. CO_2 toxicology level in air (SLOT and SLOD) against exposure time [2, 7]

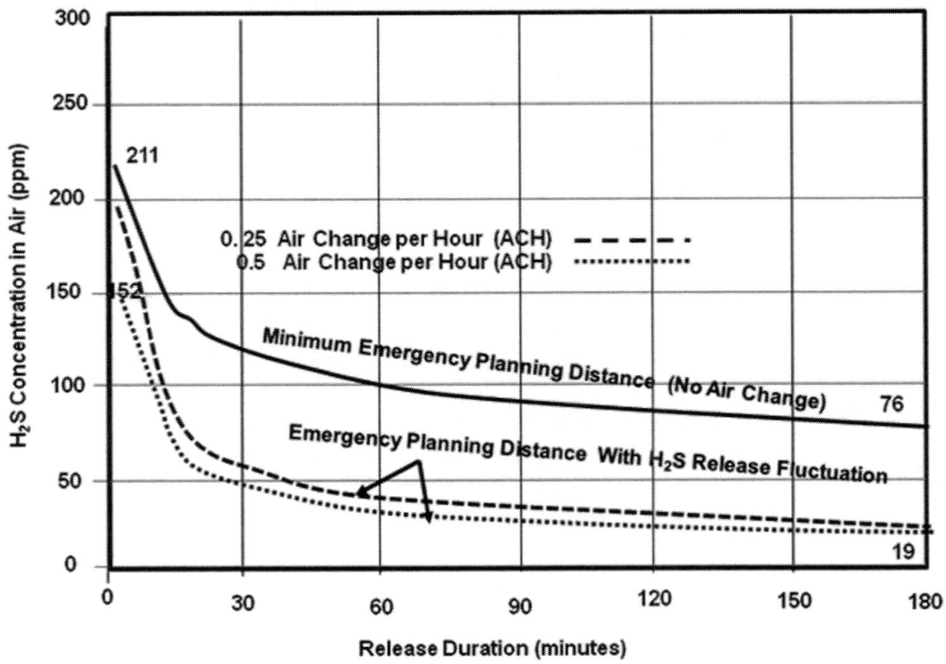

Figure 13-3. H_2S receptor point concentration versus release time from source of release (dat reformatted from Wilson [10])

- Fine particulate matter (PM2.5) and
- Visibility.

While some of the above are the by-product of combustion emitted from power stations or other industrial flue gases, the pipeline transportation of CO_2 that contain above impurities must be subject to scrutiny during design, operation, and maintenance to control or eliminate any negative impact due to their release.

One of the major components of impurities that results in high toxicity is hydrogen sulfide (H_2S). As indicated previously, it is more toxic than any other impurities indicated above. Similar to CO_2, toxicity can be also defined as Dangerous Toxic Load for H_2S. Regulatory bodies generally define set backs for hazardous and toxic fluids (including H_2S) transported through pipelines. An example is the The Energy Resources Conservation Board (ERCB), a regulatory body in Alberta, Canada that defines and regulates set backs for H_2S pipeline and facilities. ERCB defines End Points with and without concentration fluctuations in H_2S releases [10]. Graphs of H_2S exposure concentration level versus release time are shown in Figure 13-3 for the following three cases:

1- no fluctuation in H_2S concentration during release,
2- with fluctuation in H_2S concentration level and 0.25 air change per hour (ACH) at the receptor end, and
3- same as case 2 but with 0.5 ACH.

Cases 2 and 3 consider the effect of sheltering indoors on the time for H_2S to reach the receptor end point.

Generally, regulatory threshold indicates that if the concentration of hydrogen sulfide in a given facility or operation is less than 100 ppm, no further action is required [11]. However, if the concentration of hydrogen sulfide in a given facility or operation is determined to be ≥100 ppm, the radius of exposure must be determined.

The radius of exposure is that radius constructed from the point of H_2S release from a pipeline rupture or leak as the source and its length calculated using an appropriate dispersion formulation. Examples of such formulation are the following Pasquill–Gifford derived equations:

$$\text{Radius of exposure (ROE) for} \leq 100\text{ppm} = [(1.589)(H_2S\ \text{concentration})(\text{release rate})]^{(0.6258)} \qquad (13-1)$$

$$\text{Radius of exposure (ROE) for 500 ppm} = [(0.4546)(H_2S\ \text{concentration})(\text{release rate})]^{(0.6258)} \qquad (13-2)$$

H_2S concentration is the mole or volume fraction of hydrogen sulfide in the gaseous mixture and release rate = SCF/day (at normal atmospheric conditions 1 atmosphere and 15.5°C), ROE is expressed in feet.

The above equations along with the correlation for the dispersion coefficients are known as the "*Pasquill–Gifford*" equation [12]. The equations assumes constant rate of flow from a source.

13.2.3 Noise

Noise generation during pipeline maintenance (e.g., blowdown or unforeseen rupture/leak) can potentially impact maintenance personnel as well as human dwelling receptors (resulting from the associated activities). An appropriate environmental effects analysis must be performed to assess potential impacts of noise on human-dwelling receptors, particularly if the noise is sustained for a long period of time.

Figure 13-4. A typical CO_2 pipeline/piping leak

Since no continuous noise sources are expected from normal CO_2 pipeline and facilities operation, the assessment must focus on abnormal pipeline and associated pumping and compression operation and maintenance activities. ISO standard 9613 "Acoustics—Attenuation of sound during propagation" can generally be followed for such an assessment. ISO 9613 has two parts and these are Part 1: Calculation of the absorption of sound by the atmosphere, and Part 2: General method of calculation.

13.2.4 Socio-Economic Impacts

Conditions must be assessed as to potential impact of population changes along the pipeline ROW that may affect the operation and maintenance of CO_2 facilities [13].

13.2.5 Cumulative/Compounded Environmental Effects

In all above considerations cumulative environmental effects (over time) and effects that are likely to result in an adverse environmental impact from CO_2 pipeline and facilities operation and maintenance activities must be appropriately covered to establish any compounded effects.

Such an evaluation needs also to consider the general environmental effects to and from other similar nearby facilities, provided they have the potential to act in combination with the environmental effects of the CO_2 pipeline and facilities.

13.3 SPECIAL HAZARD CONSIDERATION FOR PIPELINES TRANSPORTING SUPERCRITICAL CO_2 CONTAINING IMPURITIES

13.3.1 Low Vapor Pressure versus High Vapor Pressure and CO_2 Pipelines

In a rupture of a pipeline carrying low vapor pressure (LVP) fluid (such as oil, diesel, kerosene), the liquid spill flows from the pipeline and, depending on topography, drains off until the fluid within the pipeline allowing for the change in the elevation profile is evacuated. However, in a high vapor pressure (HVP) pipeline (such as an LPG or condensate pipeline), the entire inventory of fluid contained in the pipeline section will vaporize over time, irrespective of elevation changes.

In case of the CO_2 pipeline, some of the CO_2 may remain in the pipeline as a liquid depending on the phase diagram and the pressure and temperature equilibrium reached throughout the pipeline.

13.3.2 CO_2 Pipeline Specific Hazards

CO_2 releases from leaks are generally initially directional, whereas rupture releases are momentum dominated and are jetted upwards and then dispersed (like non-ignited releases of natural gas).

The release in a ruptured pipeline transporting supercritical heavy CO_2 gas with impurities will cause the emitting fluid to form a vapor cloud particularly at low spots extending beyond the release location. Pooling of some of the CO_2 in a liquid may also occur close to the release area usually after depressurization.

Most hazards associated with dense phase or supercritical CO_2 arise when the pressure suddenly falls or is lost completely as a consequence of pipeline rupture or containment failure. These events can cause the following [14]:

- Thermal cooling envelope creation,
- Toxic contamination effects,
- Dry ice "grit blasting" effects.

Depressurization of a supercritical CO_2 pipeline can, if not carefully controlled, result in a proportion of the original pipeline inventory to be changed into solid CO_2 at low points within the pipeline, depending on pressure and temperature conditions. It may be noted that pure CO_2 freezes at −78°C (−108°F) at atmospheric pressure. Conversely, if the supercritical CO_2 is reintroduced at a warmer temperature, then there is a likelihood of pipeline over-pressurization due to the rapid increase in volume as the solid sublimes into the vapor phase [15]. For startup/restart, supply of CO_2 must be at constant rate and at controlled pressure and temperature to avoid low temperature as well as over-pressurization of the pipeline [16].

Depressurizing a pipeline in a manner that prevents liquid and solid formation without cooling affecting the pipeline structural integrity can be achieved during normal operations, however, should an uncontrolled depressurization occur, for example, due to a leak (Figure 13-4), the liquid, solids and cooling issues will occur and have to be considered in the design and maintenance processes.

In the event of a pipeline rupture (causing a major pressure loss and hence loss of containment), the depressurization results in CO_2 gas expansion (to atmosphere) by a very large factor of its original volume depending on its containment pressure and temperature at the time of release.

It may be noted that pure CO_2 liquid density (at −20°C, −4°F) and 19.7 bar is 1032 kg/m^3. Its liquid/gas equivalent (1.013 bar and 15°C (per kg of solid)) is 845 vol/vol [17].

Also CO_2 gas density at 1.013 bar and sublimation point (−78°C) = 2.814 kg/m^3 and at 1.013 bar and 15°C (59°F) is 1.87 kg/m^3.

At ambient ground-surface temperature (25°C) and pressure (0.1 MPa), CO_2 is a gas with a density of 1.8 kg/m^3 (the density of air under these conditions is 1.2 kg/m^3). Compressed supercritical CO_2 has a density of 260 to 800 kg/m^3, depending on pressure and temperature [13].

For pipeline pressures between 15 and 20 MPa, expansion can be anywhere between 500 and 600 times its original compressed volume depending on pipeline initial pressure and environmental temperature (air, terrain) equilibrium. Table 13-5 provides an indication of volume expansions that can occur due to blowdown or sudden release of CO_2 containing impurities from a pipeline at pressures 14.7 to 19.7 MPa.

TABLE 13-5. CO_2 volume expansion upon pressure release

Property	Case 1	Case 2	Case 3
Pipeline pressure: P (initial), kPa	14790	19700	14790
Pipeline temperature: T (initial) °C	10	10	10
Density: ρ (initial), kg/m^3	938.8441	976.7163	938.8441
P (flash), kPa	101.325	101.325	101.325
T (flash), °C	−80.08	−80.08	−80.08
Density: ρ (flash, vapor), kg/m^3	2.7871	2.7869	2.7871
vap frac (flash)	0.62	0.61	0.62
T (amb), °C	15	15	5
vap frac (amb)	1	1	1
Density: ρ (amb,vapor), kg/m^3	1.8467	1.8467	1.9144
Density: ρ (initial)/ρ (flash), ρkg/m^3	337	350	337
Density: ρ (initial), kg/m^3/density: ρ (amb), kg/m^3	508	529	490

However, the depressurization causes the phase change as a proportion of the high vapor pressure CO_2 essentially "boils" and becomes a gas while the remainder forms liquids and some solid particles. This rapid, violent expansion causes the temperature of escaping CO_2 to fall very rapidly, frequently below −80°C (Refer to Chapter 10 for an example of blowdown). When particles of solid CO_2 are formed, this can result in projectiles expelled at very high velocities usually at the conclusion of depressurization. This in turn may cause serious hazards to personnel and safety-maintenance equipment. Additionally, pipeline and appurtenances (if any), at the rupture location, can experience structural embrittlement unless appropriately designed and protected.

Figure 13-5. Typical release from a ruptured pipeline transporting CO_2

However, it may be noted that within (the undamaged section of) the ruptured pipe, some of the CO_2 will be in liquid state depending on pressure and temperature equilibrium as well as the terrain profile [16].

Due to depressurization, supercritical CO_2 and its contained impurities (such as H_2S) are released. If rapid mixing with air does not occur (and it is the case for a period of time and distance from the release location, Chapter 12), then the environment within relatively close range of the rupture location will become extremely toxic. Any toxic substance held in such a pressurized pipeline will "precipitate" out on loss of pressure or containment and is likely to result in harmful human exposure or environmental damage due to the contamination of the area of deposition unless appropriate measures are taken.

Dry ice "grit blasting effect" can only be expected when the CO_2 released is directly discharged on the terrain/substrate, thus causing abrasion/grit blasting of the surrounding soil. However, this only would be a problem in a CCS situation when a pipeline carrying the captured CO_2 is ruptured and subsequently retains some solid particles (e.g., sands/soil from the pipeline backfills and other solid debris). In this case, loss of containment may result in these combining with the dry ice formed to produce particles of a much greater abrasive capability than dry ice alone. This would enhance the erosion effects on the pipeline and any pipe works at a leak location which could lead to damage and risk to people. However, CO_2 from a "clean" source should not have this problem [14].

The effects of a typical rupture due to excavation on a pipeline transporting CO_2 with impurities are shown in Figure 13-5.

13.3.3 Safety and Hazard/Risk Assessment

A—Safety Zone

Calculations (based on release and dispersion modeling) can be performed for determining safe zones where related emergency responses equipment can be placed and preparations can be performed. The time calculation for complete pipeline evacuation would be generally longer than such cases as HVP liquids (such as LPG, LNG) where typically appropriate "Emergency Flow Restricting Devices" (i.e., valves) are closed, and the vapor cloud is ignited at safe distance. In addition, the following are applicable to HVP liquids which upon release to atmosphere generate immediate vapors and primarily stay close to the ground and accumulate in low points:

- Liquid spills can be designed to be confined—so the very cold liquid can cool the substrate it covers over time. However, how much the substrate cools depends on its properties and the HVP liquid vapor rate.
- Typical confinement/low point may temporarily confine the LNG/LPG and similar vapors.
- As vapor is formed, it mixes with air in the confined/low point space until gas/air outflows from the confinement/low point.

In case of CO_2, similar experience is expected, but release cannot be confined and/or ignited.

B—Release Hazard/Risk

A review of the hazard and risks associated with pipeline transportation of CO_2 containing impurities will reveal the identification of important information with respect to hazard and the associated risk due to operation and maintenance. This will be only possible through dispersion behavior modeling of CO_2 containing impurities released into the atmosphere providing information with respect to CO_2 and impurities threshold values and (possible) effects of leakages (of CO_2 and impurities) at different distances from the pipeline rupture/leak or blowdown location(s)

To consider the full effect of dispersion and appertaining safety issues due to a release from a pipeline transporting CO_2 containing impurities, any dispersion calculation must take into account the following and the associated implications:

- **Release Form and Direction:** To consider whether an environmental effect on a population or resource is considered to be adverse or neutral.
- **Magnitude:** To consider the implication of the intensity or severity of an environmental effect. Magnitude to indicate whether it is negligible, low, moderate or high.
- **Geographical Extent:** The spatial area affected by a release activity. To be identified as site specific, local or regional (i.e., area affected).
- **Duration:** The length of time over which a release and its effect occurs and whether it is short term, moderate term, long term or permanent.
- **Frequency:** The number of times a release occurs over a period of time, sporadic, regular, or continuous. Typically for a CO_2 pipeline this is sporadic.
- **Reversibility:** To indentify whether an environmental effect is permanent or temporary. Generally it is only irreversible if the resource element (such as flora and fauna, habitats, etc.) cannot be returned to the conditions that existed before the release.
- **Ecological or Socio-Economic Context:** Any impact that may occur as a consequence of a CO_2 release and the general characteristics of the affected area in which the pipeline is located.
- **Threshold standard:** For defining the standards or thresholds for determining the significance of environmental impacts/effects due to the release.

Methods and examples of dispersion modeling are provided in Chapter 12; however, the following example provides a technique for an initial assessment of a hazard due to a sudden release (leak/rupture/blowdown) of CO_2 stream (containing impurities). This example relates to H_2S release based on Eqs. 12-1 and 12-2, but a similar technique can be utilized for assessment of other impurities and the CO_2 as the main component of the gas stream.

C—Example

Considering the example given in Chapter 10 (Sections 10.2.4.3 and 10.5), the safe radius as given by Eqs. 13-1 and 13-2 were calculated for a release due to rupture of a NPS 16 (i.e., 16 inch OD) pipeline transporting CO_2 containing impurities, specifically containing 0.9 % H_2S (as indicated in Chapter 10, Table 10-1, Example 4 for the case simulated), rupture diameter 200 mm.

Time plots of gas flow and corresponding percentages of CO_2 and H_2S (in the gas) released after the initiation of rupture are shown in Figure 13-6. The plots of Radius of Exposure (ROE) using the dispersion Eqs. 13-1 and 13-2 for hydrogen sulfide (H_2S) are shown in Figure 13-7.

As indicated previously, Eqs. 13-1 and 13-2 provide H_2S radius of exposure (ROE) from a constant source of flow. However, in real rupture situations (as indicated in Chapter 11), adjacent upstream and downstream valves are closed (by SCADA system) soon after the occurrence of rupture, leading to the fact that the CO_2 release will be time dependent. Consequently, in a real situation, while H_2S radii are consistent in exposure levels (i.e.,100 ppm H_2S concentration/exposure is experienced at a further distance than 500 ppm H_2S concentration level from the source at the rupture location), the exposure levels are time dependent, and therefore, as the time for CO_2 release goes on, it would be expected that radius of exposure will be reduced, thus allowing emergency equipment and personnel to arrive closer to the rupture scene for investigation and repairs, etc.

For 100 ppm ROE, the above distance–time relationship is indicated in Figure 13-8.

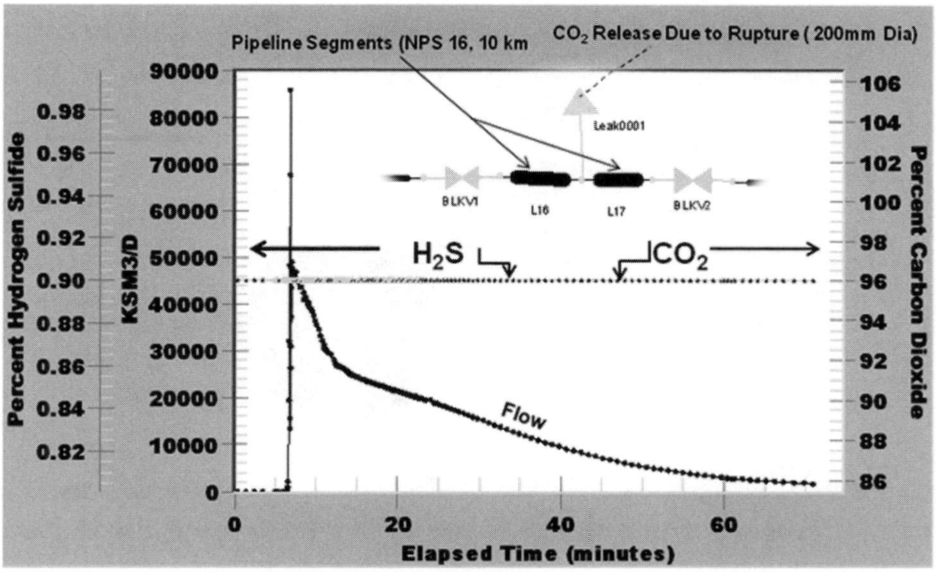

Figure 13-6. Pipeline transporting CO_2 containing impurities (see Chapter 10 for details)

13.4 CO_2 PIPELINE EMERGENCIES

13.4.1 Definitions

Emergency is defined as an unforeseen combination of circumstances or the resulting state that calls for immediate action to prevent, attend to or resolve injury, illness, of damage to facilities, equipment or the environment [18].

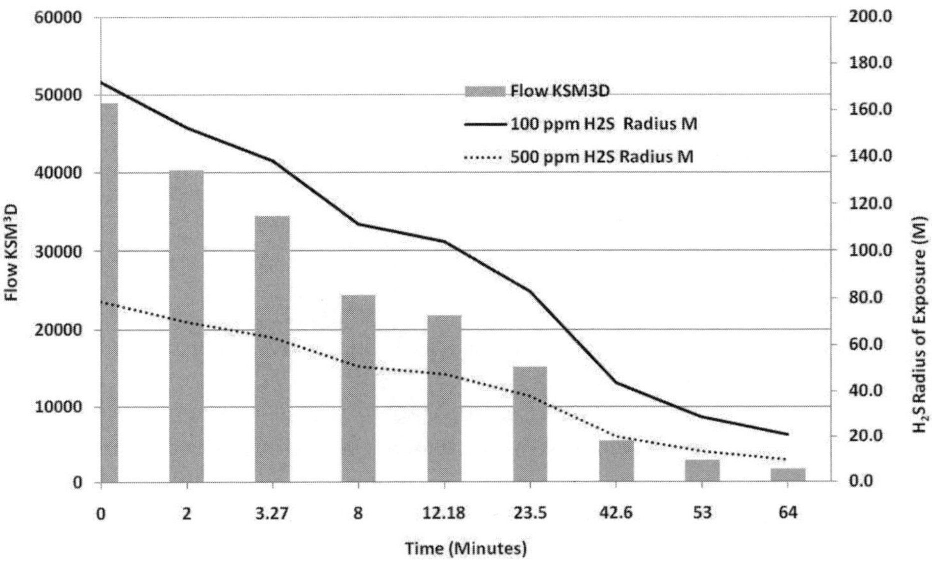

Figure 13-7. H_2S radius of exposure during a sudden release of CO_2 and impurities from a rupture (200 mm equivalent diameter)

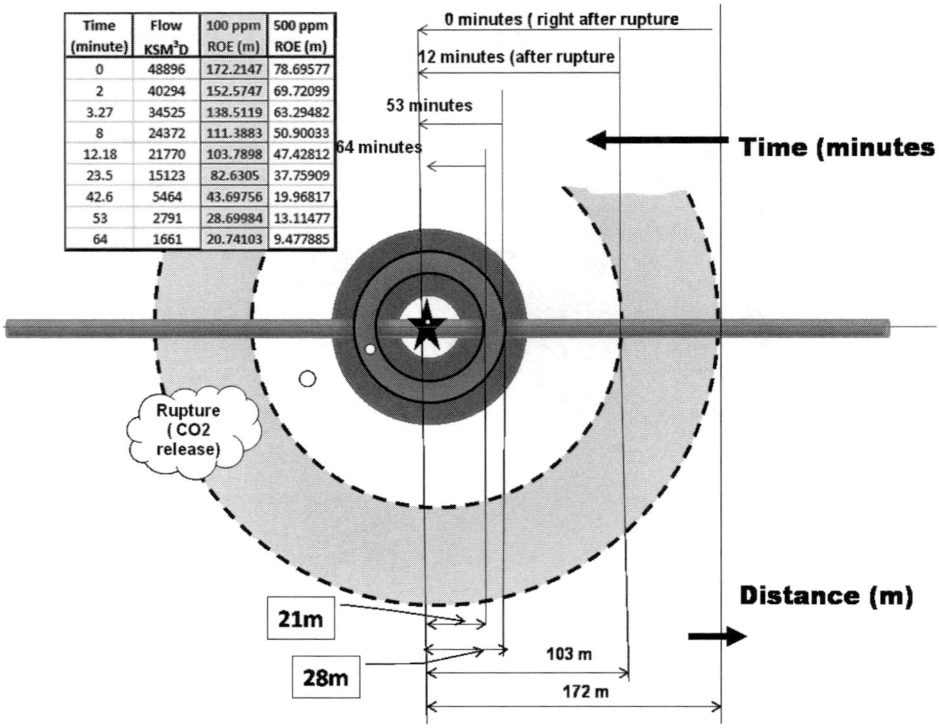

Time (minute)	Flow KSM^3D	100 ppm ROE (m)	500 ppm ROE (m)
0	48896	172.2147	78.69577
2	40294	152.5747	69.72099
3.27	34525	138.5119	63.29482
8	24372	111.3883	50.90033
12.18	21770	103.7898	47.42812
23.5	15123	82.6305	37.75909
42.6	5464	43.69756	19.96817
53	2791	28.69984	13.11477
64	1661	20.74103	9.477885

Figure 13-8. Distance–time relationship for 100 ppm H_2S rate of exposure

Canadian Standard Association (CSA) Z662 defines a pipeline emergency as "an event involving a pipeline system," such as an uncontrolled release of service fluid, that endangers one or more of the following:

(a) life;
(b) the well-being and health of people;
(c) property; and
(d) the environment.

Pipeline industry defines an incident as an event affecting company operations that may be an emergency or crisis. US Department of Transport, Code of Federal regulation DOT 49 CFR 191.3(1) defines emergency as "an event that involves a release of gas from a pipeline," and

(i) A death, or personal injury necessitating in-patient hospitalization; or
(ii) Estimated property damage, including cost of gas lost, of the operator or others, or both, of $50,000 or more

An accident is defined as undesired circumstance(s) which gives rise to ill health, injury, damage, containment/transportation losses or increased liabilities.

13.4.2 Level of Alert

Pipeline industry responds to an emergency depending on the "Level of Alert" and generally it is the pipeline company, either the control center or field offices, which decides whether or not the incident is an emergency. If an emergency condition exists, it will be

TABLE 13.6 Emergency Level of Alert [18]

	Criteria	Example of responses
LEVEL 1	An emergency that has the potential to escalate, but does not meet any Level Two or Three Alert Criteria. Meets all of the following conditions: ✓ No immediate threat to the public ✓ No serious threat to health and safety of workers (personal protective equipment may be required) ✓ Minimal environmental impact ✓ Impact confined to company jobsite ✓ Creates little or no media interest ✓ Handled entirely by company personnel	• First Responder will assess and confirm the situation. • The First Responder confirms level of alert with the Crisis Manager on call. • Additional company personnel may be placed on standby. • External notifications are made as required, complying with appropriate regulatory reporting. • Emergency responders prepare for Level Two or Three Alert responses.
LEVEL 2	An emergency that does not meet any Level Three Alert Criteria, but meets any of the following conditions: ✓ Definite risk to the public, workers or the environment. ✓ Confirmed incident. Expected to be under control quickly. ✓ Requires involvement of external emergency services, state and federal agencies. ✓ Public safety actions are likely required. ✓ Moderate environmental impact that extends or has the potential to extend beyond company right-of-way. ✓ Creates local/regional media interest.	• Level One alert duties are continued as appropriate. • Additional internal and external resources are activated. • Lead regulatory agencies are notified. • Additional notifications are made as required. • Sheltering or evacuation may be required. • Planned ignition or the product may be required. • The Crisis Management Plan is activated. • Emergency responders prepare for Level Three Alert response.
LEVEL 3	An emergency that meets any of the following conditions: ✓ Causes serious threat to the public, workers and /or the environment. ✓ Requires extensive involvement of external emergency services, federal and/or state agencies. ✓ Significant and ongoing environmental impact with extent beyond company right-of-way. ✓ Creates national media interest.	• Level One and Two Alert responses are continued as appropriate. • The company emergency response is at maximum. • The Crisis Management Plan is fully activated including executive involvement. • External government agencies are extensively involved. • The company continues to be a liaison with key government agencies and sends representative(s) to a government command center if established.

determined whether the "Level of Alert" is Level 1, Level 2, or Level 3 with a Level 3 the highest level of alert (Table 13-6) and notifications will begin accordingly.

13.4.3 Causes of Pipeline Emergency

A CO_2 pipeline emergency could thus be a result of the following (Figure 13-9, [18]):

- A damaged pipeline including third part damage or intrusion, encroachment, (Figure 13-10)
- A leaking pipeline caused by an improperly maintained pipeline or abnormal pipeline operation;
- Rupture or blowout near or directly involving a CO_2 pipeline or pipeline facility;
- A natural disaster affecting the pipeline, such as an earthquake, volcanic eruption, flood, or soil erosion.

Abnormal operation can also cause emergency situations, however, such emergencies, while they require repairs and immediate reporting, do not generally require the level of alarm that are caused by hazards. The following events are considered abnormal operations.

- Unintended closure of valves or shutdowns;
- Increase or decrease in pressure or flow rate outside of normal operating limits;

Figure 13-9. Causes of pipeline failure and loss of containment

Figure 13-10. High pressure CO_2 release from a pipeline (Jackson Dome–8" Tinsley Pipeline CO, USA) due to an excavation damage [19]

- Loss of communications;
- Operation of any safety device.

However, if any of the above abnormal conditions is likely to lead to a hazard to persons or property, they would be categorized as an emergency and would require the appropriate response as indicated in Table 13-6.

Emergency response for CO_2 pipelines will require careful planning to account for additional hazards that may be brought about by transportation large volume, high pressure CO_2 containing impurities [20].

13.4.4 CO_2 Hazardous Incident Records and Data

Unlike natural gas and other light and heavy hydrocarbon fluids, hazards associated with an emergency situation (identified in the above list), exclude fire and explosion. CO_2 pipelines could, however, burst due to high pressure under external stress or impact.

A—Incident Records

Regulations such as 49 CFR 195 Part B generally require operators to keep records of accidents and safety-related conditions, and to report them if certain criteria are met.

Figure 13-11. Toxic gas emissions from Dieng volcano [21]

Some past hazardous incidents related to CO_2 release are listed below:

1935	CO_2 seepage along ground fault resulting in CO_2 charged ground water, Paradox Basin, Colorado, USA.
1979	Asphyxiation of 39 residents due to large CO_2 emission from volcanic eruption (Dieng Volcano), Indonesia (Figure 13-11).
1984	Death of 37 people due to CO_2 release from volcanic eruption, Lake Monoun, Cameroon, Africa (Figure 13-12).
1986	Asphyxiation, 1700 people, due to 1.7 MT of CO_2 release, Lake Nyos, Cameroon [22] (Figure 13-12).
1992	Asphyxiation of an individual and his dog due to venting of abandoned coalmine, Northumberland, UK.
1992	CO_2 gas emissions killed one person in a river valley 200 m W of Sikidang Crater, Dieng, 18th March.
1994	CO_2 asphyxiation, dispensing CO_2 trailer tractor.
1994	Fatal injury, CO_2/Methanol vessel operation (at 2000 psi).
1998	Creation of cloud due to high pressure CO_2/H_2S (approximately 82% CO_2) leading to evacuation of 2500 nearby residents, Nagylengyel, Zala Basin (SW Hungary) (Figure 13-13). High-pressure CO_2 eruption from well drilling to 1500 m depth, Cerro Fortunos field, Mendoza Spain, 30 days to drill relief well.
1998	Fatality of an individual and exposure of 15 workers due to accidental release of CO_2 from US DOE reactor maintenance, Idaho, USA.
2011	Toxic gas emissions from Dieng volcano (central Java) forced the evacuation of 1200 residents in June 2011.

B—Pipeline Risk Profile and Hazard Data

Incidents due to CO_2 pipeline and facilities operation are rare. Industrial experience in the USA from 1990 to 2001 indicates a total of 10 incidents with no fatality and no injuries. Property damage recorded amounts to about US$ 15200/1000 km/year and the incident rate per 1000 km/year is 0.32. These statistics represent about 300 km of CO_2 pipeline transporting about 45 Mt/year [8].

Figure 13-12. Seepage of toxic CO_2 from, A—Lake Monoun and B—Lake Nyos, Cameroon, Africa [22, 23]

The risk profile for CO_2 pipelines is somewhat different than for traditional pipelines such as natural gas transmission and hazardous hydrocarbon liquids.

Statistically, since the first construction of a CO_2 pipeline in the USA in early 1974, there has not been a major accident due to the design, operation, and maintenance of CO_2 pipelines.

Since 2002, the number of reportable incidents due to the pipeline transportation of CO_2 in the USA is indicated in Figure 13-14. For these 19 incidents, no fatalities or any major injuries have been reported [25]. Apparent cause and distribution of accidents for spilled CO_2 and other non-flammable, non-toxic fluids (which are gas at ambient conditions) are indicated in Figure 13-15 [26].

Figure 13-13. Outbreak of CO_2 escape—Nagylengyel (11.30 pm on 13 November 1998 [24])

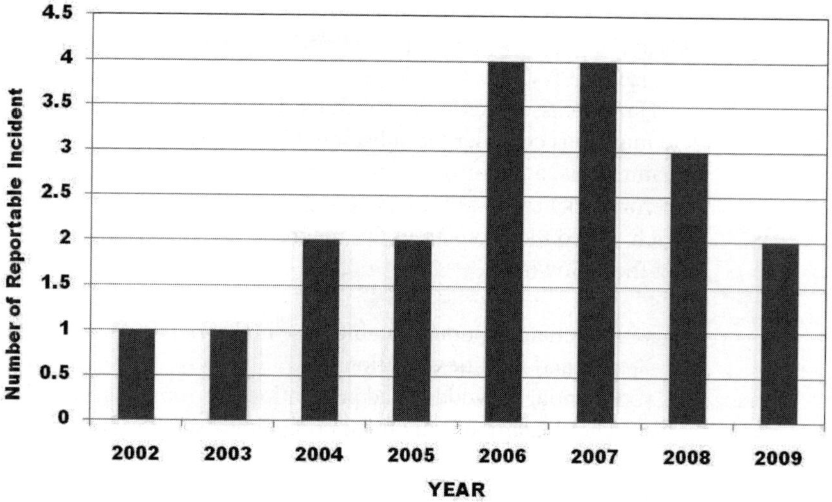

Figure 13-14. Number of reportable incidents due to operation of CO_2 pipelines in the USA [25]

C—Risk Reduction Opportunities

With expanding opportunities for carbon capture and storage, pipeline transportation of CO_2 containing impurities is likely to have inventories of fluid in the order of tens if not hundreds of thousands of metric tons. Such inventories are also transported in the supercritical or dense phase under very high pressure.

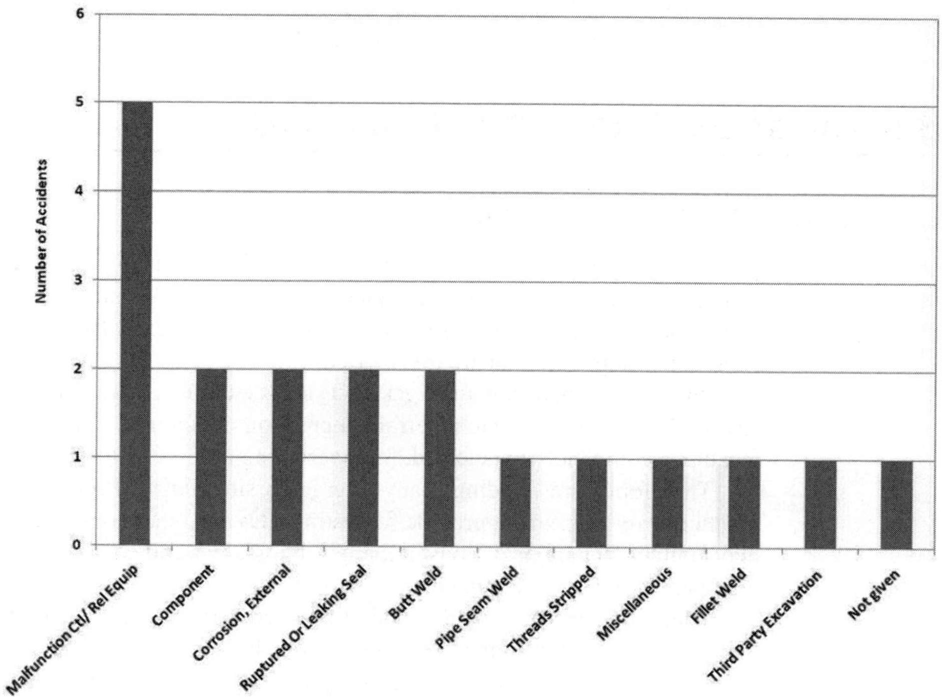

Figure 13-15. Apparent cause and distribution of accidents for CO_2 and non-flammable, non-toxic fluid pipelines in the USA [26]

It is indicated that distances to a given level of risk are roughly comparable between CO_2 and natural gas at similar pressure and temperature range [8]. Increasing the pressure increases the distance to a given risk level for both CO_2 and natural gas.

Hazards associated with the transportation of supercritical CO_2 (that contains impurities) must thus consider the substantial increase in transportation pressure of CO_2 over that of natural gas, a factor of about two to three times higher pressure. The hazard ranges and therefore risks are expected to be substantially larger for CO_2 releases at higher pressure (which would therefore be in the supercritical phase). Any risk assessment must also consider the following:

- Potential for population increase in the vicinity of or along the CO_2 pipeline right of way and facilities location.
- Potential for widespread population exposure to elevated concentrations of CO_2 and other hazardous contained compounds such as H_2S in case of an emergency situation.
- Increased attention must be given to
    - o Toxic contamination spread
    - o Thermal cooling at the worksite and
    - o Approach to CO_2 release site due to potential dry ice "grit blasting" effects.

Factors that could reduce the risk would be through a comprehensive design and predictive risk assessment for effective maintenance procedures that would consider the following:

- Possible failure mechanisms,
- Risk of failures,
- Management systems in place to reduce risk due to operation and maintenance,

Annex N to Canadian Standard Association (CSA) Z662 and/or other similar standards/documents provide options for hazard control and risk reduction.

13.5 EMERGENCY RESPONSE PLANNING

13.5.1 Facilities and Equipment

Similar to other hazardous pipeline systems, CO_2 pipeline and facilities will need careful emergency response planning and the provision of appropriate equipment to account for additional hazards brought about by high pressure and generally large volumes of CO_2 gas containing impurities that are transported.

In comparison with natural gas, CO_2 leaks (small or large) cannot be dispersed easily and will tend to accumulate in terrain depression or confined spaces. CO_2 can stay in these locations, completely undetected, for a very long time.

Therefore, in responding to any emergency situations or even routine maintenance, any practice or response must provide for testing CO_2 and other constituent gas (such as H_2S) it may contain that are hazardous in the low-lying areas and confined spaces before attempting to enter and conduct repair.

The emergency response planning must consider the potential impairment of human response in CO_2 environment because of its asphyxiant and physiological properties. Emergency responders will thus require the appropriate breathing apparatus and any procedure must ensure safety of responders and the general public.

Release issues from a containment failure that relate to pipeline transportation of CO_2 containing impurities are generally illustrated in Figure 13-16.

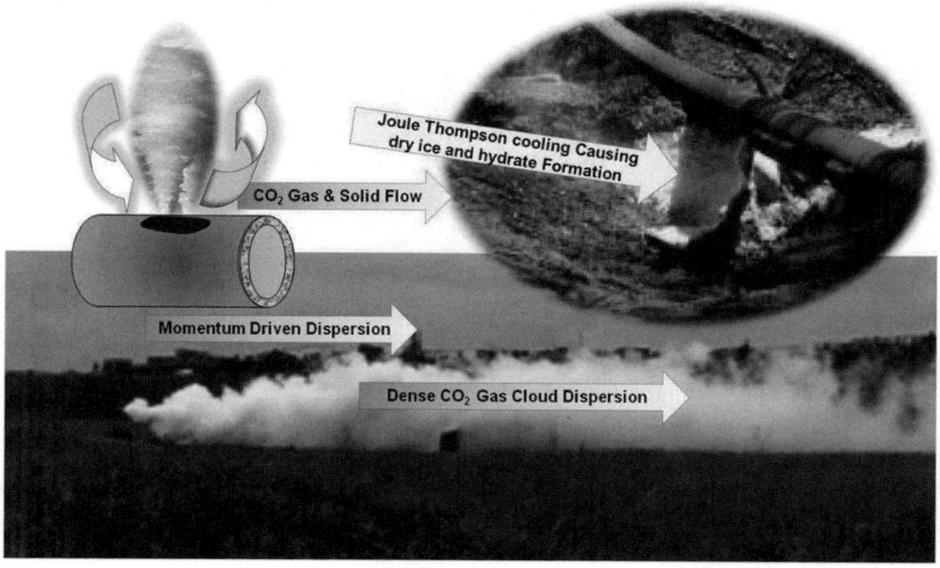

Figure 13-16. CO_2 pipeline containment release/failure

Cause and effect of loss of containment in a high-pressure pipeline transporting CO_2 depends on leak/rupture/blowdown size affecting the rate (thus momentum) and duration of release. This in turn affects the phase behavior, sublimation, dispersion, and cooling [2].

Thus, as can be seen from Figure 13-16, the two primary operational safety considerations for CO_2 facilities are as follows:

1 To avoid suffocation in areas where CO_2 may be blowing down, leaking and displacing oxygen (especially in enclosed or low areas); and
2 To exercise extreme caution when operating or maintaining high pressure CO_2 facilities due to the compressibility and potentially violent expansion (150 to 1) of CO_2 as it changes phase.

Because of safety issues, pipelines that transport CO_2 containing must consider emergency safety facilities

- **Equipment for detection of CO_2 and other associated gases,** particularly air and CO_2 monitors.
- **Blowdown pre-heaters:** CO_2 blowdown can cause auto-refrigeration (i.e., dry ice and hydrate formations at low pressures generally experienced at end of blowdown cycle). Flare pre-heaters and low temperature materials at blowdown locations are thus appropriate.
- **Facility for natural gas injection at blowdowns** to be able to flare the mixture during blowdown.

The following are generally considered by the industry for an effective emergency response procedure and appertaining equipment particularly for CO_2 pipelines that could potentially have high level of H_2S (up to and sometimes upward of 2% H_2S content):

1. Personal Protective Equipment (PPE)—use of supplied air and self contained breathing apparatus type respiratory protective equipment.

2. Portable Gas Detection—to address situations where there may be excessive concentrations of H_2S and CO_2. This would include monitoring requirements and characteristics of the gas (i.e., heavier than air).
3. Building Entry Procedure—to ensure that CO_2 and other hazardous gases are identified as a potential contaminant and that building atmospheres are rendered safe before entry is attempted.
4. Contractor, Visitor, Employee Orientation program.
5. Operating procedures.
6. Inclusion of MSDS (in Emergency Response Manual and lower-level field work practices) for all gas components in the CO_2 gas stream.

Also, the operation of a CO_2 pipeline with significant quantities of H_2S requires the following adjustments to the equipment typically used for pipelines transporting sweet natural gas or LVP liquids:

- Self Contained Breathing Apparatus style RPE would be required to be supplied for those operating the system.
- Escape packs (RPE) would need to be considered.
- Permanent monitoring facilities would need to be considered for facilities (buildings) associated with the CO_2 pipeline.
- A review of the current portable gas monitors would be required to determine if upgrading was required.
- A review of the gas monitor calibration stations would need to be undertaken to ensure gas monitors were calibrated/bump tested at appropriate concentrations.
- Personal gas monitors with alarms would likely be required.
- Respiratory fitness for purpose testing equipment and protocols would require review.

13.5.2 Management System

An effective CO_2 emergency plan must have management systems in place that effectively reduce risk due to operation and maintenance activities. Such a system must include an emergency response plan that encompasses:

A. Emergency procedures;
B. Activation Levels;
C. Maps and drawings of all facilities and pipeline right of way (ROW) terrain profile and access routes;
D. Training and drill details;
E. Organization and coordination with regulatory, government (including state/municipal,) police, and external emergency services, to include contacts and plans;
F. Details and characteristics of CO_2 and its constituents/impurities including a phase diagram;
G. List of all necessary equipment for dealing with emergencies (other than maintenance). These will include signage and markers and the appropriate operation manuals (with date of release and their maintenance/calibration records);
H. Recommended practice, codes and regulatory notices, e.g., API RP-55 "Recommended Practice for Oil and Gas Production Processing Plant Operation Involving Hydrogen Sulfide."

The activation of an emergency response plan must be well defined subsequent to which the level of alarm can be decided and an appropriate emergency response team and equipment dispatched accordingly. The following is a typical example of industry/regulatory rules for activating an emergency response plan for facilities that contain H_2S, [11]:

"At a minimum, the emergency response plan must be activated whenever a release may create a concentration of hydrogen sulfide of more than 100 ppm in any public area, 500 ppm at any public road or 100 ppm, about 3000 ft (900 meters) away from the site of release." New Mexico Oil Conservation Division Rule 118.

13.5.3 Procedure

A typical CO_2 pipeline and facilities must have a plan thoroughly devised for responding to situations arising out of operation, maintenance, third party intrusion/encroachment or unforeseen geohazards and other events. Such an emergency response plan must incorporate as a minimum the following:

- Cover Page Title Pipeline "Emergency Response Plan—For Transportation of CO_2 Containing Impurities"
- Purpose
- Emergency "OUT CALL" System (for notification of nearby residents)
- Definitions
  - o Level of Alert (see previous Table 13-6)
  - o Classifying emergency incidents
  - o Safety-related conditions
- Affected regulation and applicable codes and standards
- Code requirements
- Affected authorities, municipalities etc
- Emergency response organization and notification chart
  - o Line of responsibility
  - o Notifications and means to receive, identify, and classify notice of emergency events
  - o Reporting structure (preliminary and final)
  - o Means to communicate with necessary emergency services
- Emergency Response Procedure covering
  - o Initial notification of a pipeline emergency
  - o Reporting structure
  - o Reports to management/supervisory level
  - o Actions directed toward protecting people first, and property second
  - o Emergency shutdown and pressure reduction procedures, to minimize hazards to life or property;
  - o Making safe any actual or potential hazards to life or property
  - o Establishing Incident Command System, roles and responsibilities
  - o Provisions of emergency equipment, tools, materials, and personnel
  - o Dispatching emergency crew to site
  - o First emergency crew arrival notification and reporting at site
  - o First responders arrive at site
  - o Dispatching additional emergency crews to site
  - o Termination of emergency
  - o Procedures to return the pipeline to safe conditions including testing and certification
  - o Liaison required for establishing and maintaining contact with fire, police, emergency, and other public officials
  - o Required reports and notification of state and municipality and government agencies (as required)

  - o Statements to the news media
- Pipeline operations management contact list
  - o Pipeline operations list (as an attachment)
  - o Facilities management list (as an attachment)
- First responder and emergency services
  - o Contact list
  - o Description of services
- Maps, illustrations, diagrams to cover:
  - o Pipeline route
  - o Valve site locations
  - o Pump and compressor station locations
  - o Access
  - o Receptors locations and contact details
  - o Schematics as necessary
- Guides:
  - o Guide to assess the extent and coverage of CO_2 vapor cloud to determine hazardous area.
  - o Selecting protection techniques for High Consequence Areas
  - o Guide to select clean up techniques for sensitive areas
  - o Safe restoration of outages;
  - o Investigation of failures and post hazard evaluation and effects
  - o Guide to damage prevention
- References (giving list of documents referenced in the development of emergency response procedure):
  - o Operations and Maintenance Plan and related O&M procedures
  - o Management and staffing plans
  - o Maintenance and repairs/ construction and safety plans
  - o Communication plan
  - o Quality assurance/quality compliance plans
  - o Training and operator/responder qualification and procedures

13.5.4 Training Needs

13.5.4.1 Personnel Training

Personnel training and qualifications for responding to emergencies are most important to assure public and personnel safety and environmental protections.

Regulations (such as) 49 CFR Part 195, Section 195.403 are about emergency response training and requires that a continuing educational program be established to enable pipeline users, the public, appropriate government organizations, and persons engaged in excavation-related activities to recognize hazardous pipeline emergencies for the purpose of reporting it to the operator or appropriate public officials.

Additionally, regulations generally require that procedures be established to analyze accidents and failures in order to determine the cause(s) of the failure and to minimize the chance of recurrence.

13.5.4.2 Specific Training Needs and Emergency Responders

Specific training (HAZWOPER—Hazardous Waste Operations and Emergency Response) is required for both pipeline operator personnel and emergency responders. By inference, 49 CFR Part 194 (although applicable to oil pipelines, and its stated purpose is to reduce the impact of oil spills) requires CO_2 pipeline operators to have release or spill response plans and to hold release/spill drills in coordination with local authorities (police, fire departments, regulating authorities, etc.), and to implement public awareness plans addressing the threats coming from hazardous fluid systems under pressure.

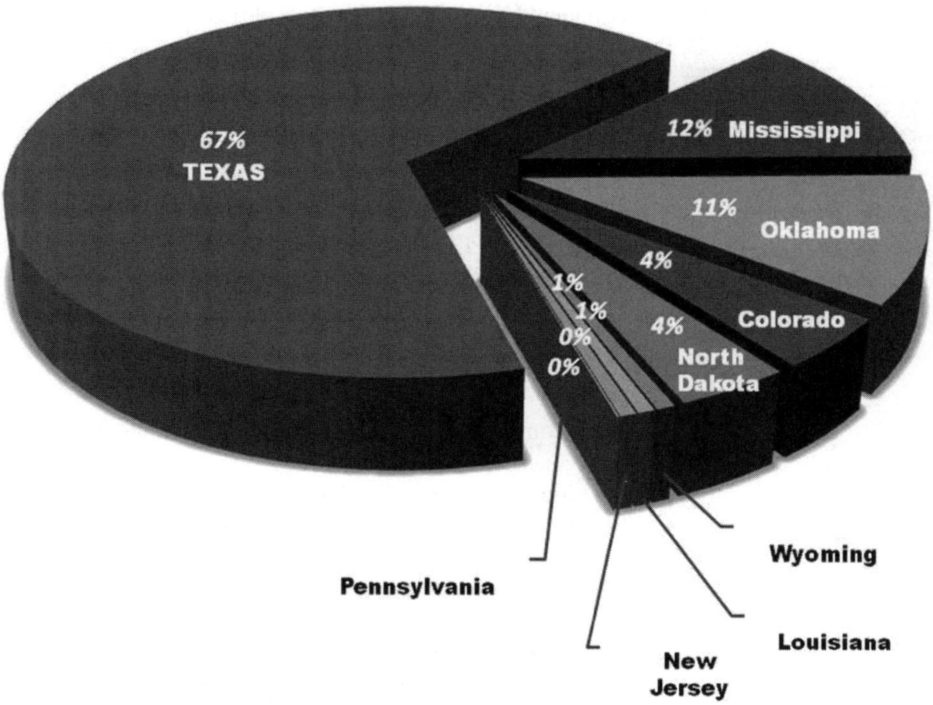

Figure 13-17. Geographical distributions of CO_2 pipelines in the USA

Emergency responders are generally trained to respond safely to pipeline incidents including those arising from CO_2 pipeline incidents. There is emphasis toward safety aspects of responding to emergencies and training needs for use of relevant equipment including the following:

1. Portable gas detection and building entry;
2. Respiratory protective equipment—use, fitting, and maintenance;
3. Gas monitoring equipment—calibration and use;
4. H_2S;
5. CO_2;
6. Emergency levels.

In the USA, the DOE, Pipeline and Hazardous Material Safety Administration (PHMSA)-Office of Pipeline Safety and the National Association of State Fire Marshals (US NASFM) cover training materials for the risks associated with transportation of all pipeline commodities including CO_2 [27].

In the USA, national emergency responder experience in the handling CO_2 pipeline releases may vary geographically due to the concentration of CO_2 pipelines and their irregular placement in specific geographical locations (Figure 13-17).

13.5.5 Effective Emergency Response

Effective emergency response must contain the following steps:

A. Site Management Control and Containment Problem Identification
B. Hazard and Risk Evaluation

C. Selection and Use of Proper Personal Protective Clothing and Equipment
D. Information Management and Resource Coordination
E. Response Actions
F. Decontamination and Clean-Up Operations
G. Incident Termination

Pipeline emergencies are handled safely and effectively when responders are trained on a predetermined and approved system for managing the event.

The above seven-step process can ensure that the basic tactical functions to be evaluated and implemented are in place. It is a flexible guideline and the process can serve as a general framework for systematically resolving a pipeline emergency.

Critical success factors in responding to an emergency are:

- Avoiding commitment or positioning personnel and units in a hazardous position. Consider an escape route.
- Establishing command and an Incident Command Post.
- Establishing a Staging Area.
- Establishing an isolation perimeter.
- Defining a hazardous zone.
- Avoiding entry into the emergency affected area without necessary personal protective equipment appropriate for the incident.
- Isolating all personnel until emergency decontamination can be established.
- Initiating public protective actions (PPAs) as necessary.
- Training.

REFERENCES

[1] Mohitpour, M., Jenkins, A., and Nahas, G., 2008 "A generalized overview of requirements for the design, construction and operation of new pipelines for CO_2 sequestration," Journal of Pipeline Engineering, 7 #4, Dec pp 237–251.

[2] Bilio, M., and Savidis, Y., 2010 "Pipelines for Carbon Capture and Storage," Proceedings of International Forum on the Transportation of CO_2 for CCS," July1-2, Newcastle, UK, Clarion Technical Conf & Tiratsoo Technical.

[3] Turner, R., Hardy, N., and Hooper., B., 2006 "Risks associated with a CO_2 pipeline—Methodology and case study" Green House Gas Control Technology (GHGT) – 8 Conf., Trondheim, Norway,19th-22nd June.

[4] Shell Canada Limited 2010 "Quest Carbon Capture and Storage Project, TOR" http://environment.alberta.ca/documents/Shell-Quest-Carbon-Capt-and-Storage-Proj-PToR.pdf.

[5] Iqbal, N., and Salley, M.H., 2003 "Fire Dynamics Tools (FDTs): Quantitative Fire Hazard Analysis Methods for the U.S. Nuclear Regulatory Commission Fire Protection Inspection Program" Report # NUREG-1805, Vol. 2, Appendices, http://www.nrc.gov/reading-rm/doc-collections/nuregs/staff/sr1805/v2/sr1805v2a.pdf.

[6] McGillivray, A., and Wilday, J., 2009 "Comparison of risks from carbon dioxide and natural gas pipelines" Research Report RR 749 UK Health and Safety Executive, http://www.hse.gov.uk/research/rrpdf/rr749.pdf.

[7] HSE (Health and safety Executive) 2008 "Assessment of the dangerous toxic load (DTL) for specified level of toxicity (SLOT) and significant likelihood of death," http://www.hse.gov.uk/hid/haztox.htm.

[8] HSE (Health and safety Executive) 2009, "Safety in Carbon Dioxide Capture, Transport and Storage" Technical Study Report # 2009/6, IEA, Greenhouse Gas Programme, http://co2storage.org/Reports/2009-6.pdf.

[9] DNV, 2008 "Mapping of potential HSE issues related to large-scale capture, transport and storage of CO_2," DNV report number: 2008-1993, revision 02.

[10] Wilson, D., 2004 "EUBMODELS H_2S Toxic Load Endpoints for Emergency Planning Zones" comments for EUB Board Meeting, Nov 26, http://www.ercb.ca/docs/public/sourgas/Wilson-Presentation.pdf.

[11] Martin, E., 2005 "Hydrogen Sulfide" Issues and Answers Workshop San Juan College, Farmington, New Mexico, USA. http://octane.nmt.edu/sw-pttc/proceedings/H2S_05/HydrogenSulfide_OCD.pdf.

[12] Cameron-Cole 2005 "Air Dispersion Modeling of Well Blowout and pipeline Rupture Scenarios Salt Creek Field" Report Prepared for Howell Petroleum Corporation, Sept 22, http://www.blm.gov/pg-data/etc/medialib/blm/wy/information/NEPA/cfodocs/howell.Par.9631.File.dat/23apxA.pdf.

[13] Jaeger, C.C., Edenhofer, O., and von Flachsland, C., 2005 "Social Conditions of Technological Change—The Case of Carbon Capture and Storage", UNIVERSITÄT POTSDAM, Magisterarbeit.

[14] HSE (Health and safety Executive) 2011, "General hazards of carbon dioxide" (HSE/Guidance/Carbon capture and storage/General hazards of carbon dioxide) http://www.hse.gov.uk/carboncapture/carbon-dioxide.htm.

[15] Eldevik, F., 2008 "Safe Pipeline Transmission of CO_2 "Pipeline & Gas Journal" November Vol. 235 No. 11.

[16] Havelsrud, M., 2010 "Simulation of CO_2 Transport in Pipe Systems" Proceedings of International Forum on the Transportation of CO_2 for CCS, July1-2, Newcastle, UK,Clarion Technical Conf & Tiratsoo Technical.

[17] Air Liquide, 1976 "Gas Encyclopedia," Elsevier Scientific Publishing Co. Holland, pp 333-368, http://encyclopedia.airliquide.com/Encyclopedia.asp?GasID=26.

[18] Mohitpour, M., Murray, A., McManus, M., and Colquhoun, I., 2010 "Pipeline Integrity Assurance—A Practical Approach," ASME Press New York.

[19] Duncan I.J., 2008, " Risk and Monitoring Risk for CO_2 Sequestration in Deep Brine Reservoirs " Gulf Coast Carbon Center (GCCC) Digital Publication Series #08-03l, http://www.beg.utexas.edu/gccc/forum/codexdownloadpdf.php?ID=89.

[20] IEA (International Energy Agency), 2009 "Safety in Carbon Dioxide Capture, Transport and storage" Greenhouse Gas R&D Program.

[21] Seach, J., 2011, Volcano Live, http://www.volcanolive.com/dieng.html.

[22] Krajick, K., 2003 "Defusing Africa's Killer Lakes," Smithsonian, 34(6), pp. 46–55.

[23] BBC 2003 "Cameroon pipes out deadly lake gas," Feb 15, http://news.bbc.co.uk/2/hi/africa/2766487.stm.

[24] EI (Energy Institute), 2010 "Technical guidance on hazard analysis for onshore carbon capture installations and onshore pipelines," 1st edition Sept ISBN 978 0 85293 541 5 http://www.energyinst.org.

[25] Smith, R., 2010 "Safety Perspective and Technical Challenges for CO_2 Pipelines" Panel on CO_2 Pipelines. International Pipeline Conference, Calgary Alberta, Canada, Sept 27th.

[26] PHMSA (US Pipeline and Hazardous Material Safety Administration), 2010 "CO_2 Mileage and Accidents 2002-2009" Office of Pipeline Safety.

[27] NASFM (US National Association of State Fire Marshals) 2011, " Pipeline Emergencies" http://pipeline.mindgrabmedia.com/main.aspx.

APPENDIX TO CHAPTER 13 ABBREVIATIONS AND GLOSSARY OF TERMS/DEFINITIONS

The following provides acronyms and glossaries for a limited terms. For full definition of all terms related to Emergency Response, Safety and Risk refer to Mohitpour et al [18]

ANSI	American National Standards Institute
API	American Petroleum Institute
CO	Carbon monoxide
CO_2	Carbon dioxide
CO_2 plume	The extent, in three dimensions, of the carbon dioxide.
CCS	CO_2 Capture and Storage
CPM	Computational pipeline monitoring
EAP	Employee Assistance Program
EGR	Enhanced Gas Recovery
EIA	U.S. Energy Information Administration. An independent statistical agency within the U.S. Department of Energy
EOR	Enhanced Oil Recovery
EPA	Environmental Protection Agency (U.S.A.)
EPC	Engineering, Procurement and Construction
ETS	Emissions Trading Scheme
EUA	Emission unit allowance
Facility	Any component of a pipeline or pipeline systems through which carbon dioxide moves. This includes pipe, valves, and other appurtenances attached to the pipe, compressor units, pumps, metering stations, regulator stations, delivery stations, holders, and fabricated assemblies and other components
FERC	Federal Energy Regulatory Commission
FTW	Fitness to work.
GF	Gas Forming
GY	Gas Yielding
HAZID	Hazard Identification
HAZOP	Hazard and Operability Study (one method for identifying hazards)
IEA	International Energy Agency
IOIP	Initial Oil in Place
IPCC	Intergovernmental Panel on Climate Change
ISCGT	In-Situ CO_2 Generation Technique
LNG	Liquefied natural gas
LTD	Long Term Disability
MRV	Monitoring, reporting and verification
MSDS	Material Safety Data Sheet(s)
NGO	Non-governmental organization
NIOSH	National Institute for Occupational Safety and Health
NO_x	Nitrogen oxide
OECD	Organization for Economic Cooperation and Development
Operator	A individual or owner responsible for control over the conveyance of fluid in the pipeline
OSHA	Occupational Safety and Health Administration
PHMSA	Pipeline and Hazardous Materials Safety Administration. An agency within the U.S. Department of Transportation
PM	Particulate matter

PPAs	Public Protective Actions
PPE	Personal Protective Equipment
R&D	Research and development
RD&D	Research, development and deployment
SCBA	Self Contained Breathing Apparatus
SDD	Strategic Deployment Document
SO_x	Sulphur oxide
SET	Strategic Energy Technology Plan
SRA	Strategic Research Agenda
TLV©	Threshold Limit Values
TSCA	Toxic Substances Control Act (EPA)
WC	Workers' (or Workmen's) Compensation
WHMIS	Workplace Hazardous Material Information System—Canadian legislation dealing with the preparation and distribution of chemical hazard information

INDEX

Page numbers followed by f and t indicate figures and tables, respectively.